D1201892

The Finite Element Method

Fifth edition

Volume 2: Solid Mechanics

Professor O.C. Zienkiewicz, CBE, FRS, FREng is Professor Emeritus and Director of the Institute for Numerical Methods in Engineering at the University of Wales, Swansea, UK. He holds the UNESCO Chair of Numerical Methods in Engineering at the Technical University of Catalunya, Barcelona, Spain. He was the head of the Civil Engineering Department at the University of Wales Swansea between 1961 and 1989. He established that department as one of the primary centres of finite element research. In 1968 he became the Founder Editor of the *International Journal for Numerical Methods in Engineering* which still remains today the major journal in this field. The recipient of 24 honorary degrees and many medals, Professor Zienkiewicz is also a member of five academies – an honour he has received for his many contributions to the fundamental developments of the finite element method. In 1978, he became a Fellow of the Royal Society and the Royal Academy of Engineering. This was followed by his election as a foreign member to the U.S. Academy of Engineering (1981), the Polish Academy of Science (1985), the Chinese Academy of Sciences (1998), and the National Academy of Science, Italy (Academia dei Lincei) (1999). He published the first edition of this book in 1967 and it remained the only book on the subject until 1971.

Professor R.L. Taylor has more than 35 years' experience in the modelling and simulation of structures and solid continua including two years in industry. In 1991 he was elected to membership in the U.S. National Academy of Engineering in recognition of his educational and research contributions to the field of computational mechanics. He was appointed as the T.Y. and Margaret Lin Professor of Engineering in 1992 and, in 1994, received the Berkeley Citation, the highest honour awarded by the University of California, Berkeley. In 1997, Professor Taylor was made a Fellow in the U.S. Association for Computational Mechanics and recently he was elected Fellow in the International Association of Computational Mechanics, and was awarded the USACM John von Neumann Medal. Professor Taylor has written several computer programs for finite element analysis of structural and non-structural systems, one of which, FEAP, is used world-wide in education and research environments. FEAP is now incorporated more fully into the book to address non-linear and finite deformation problems.

Front cover image: A Finite Element Model of the world land speed record (765.035 mph) car THRUST SSC. The analysis was done using the finite element method by K. Morgan, O. Hassan and N.P. Weatherill at the Institute for Numerical Methods in Engineering, University of Wales Swansea, UK. (see K. Morgan, O. Hassan and N.P. Weatherill, 'Why didn't the supersonic car fly?', *Mathematics Today, Bulletin of the Institute of Mathematics and Its Applications,* Vol. 35, No. 4, 110–114, Aug. 1999).

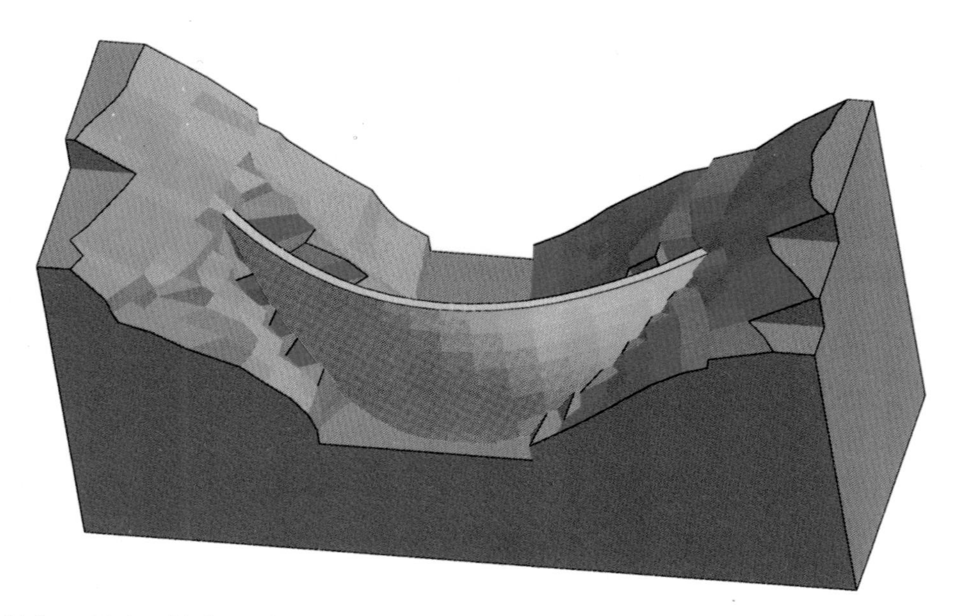

(a) Geometrical model of an arch dam

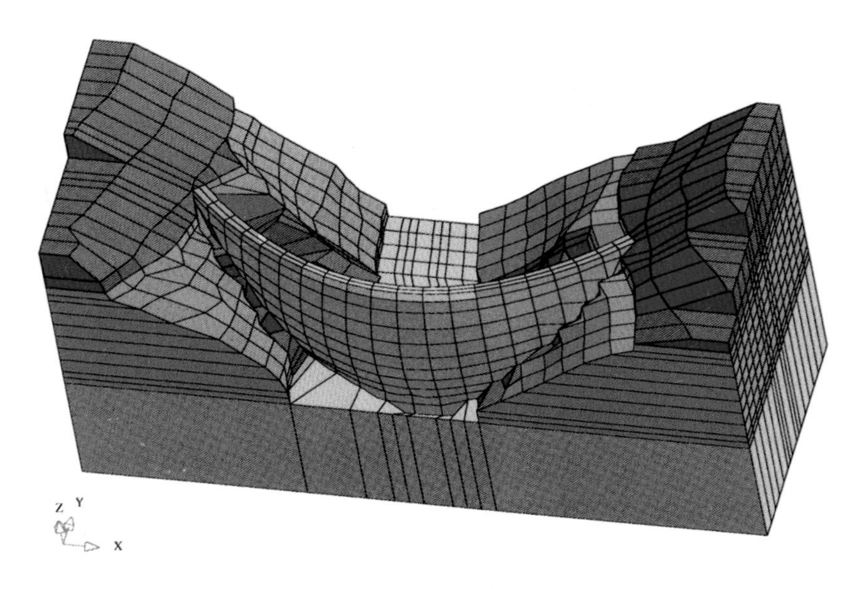

(b) Finite element discretization of an arch dam including the foundation

Plate 1 Three dimensional non-linear analysis of a dam
Courtesy of Prof. Miguel Cervera, CIMNE, Barcelona. Source: B. Suarez, M. Cervera and J. Miguel Canet, 'Safety assessment of the Suarna arch dam using non-linear damage model', *Proc. Int. Sym. New Trends and Guidelines on Dam Safety*, Barcelona, Spain, 1998. L. Berga (ed.) Balakema, Rotterdam.

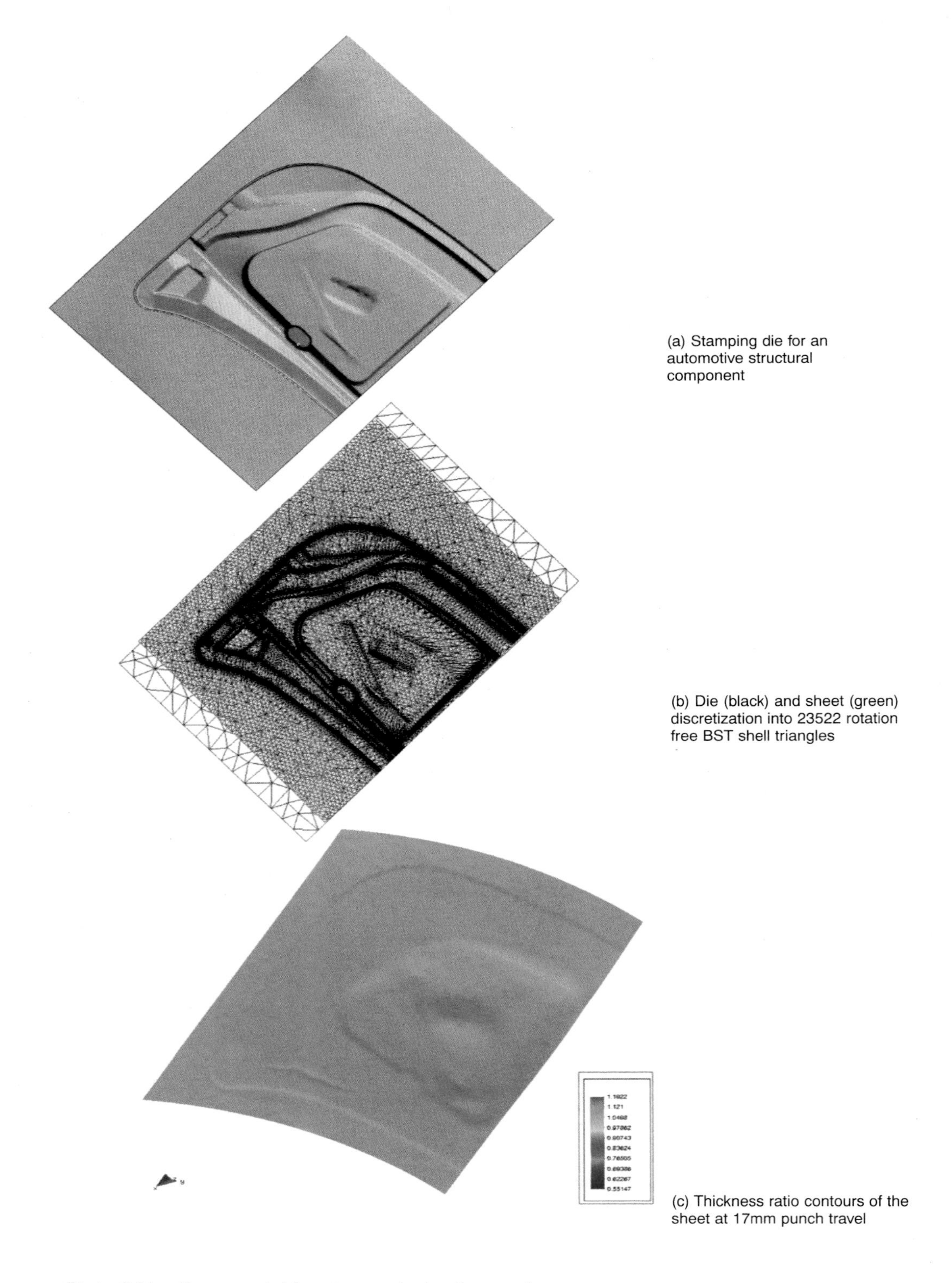

(a) Stamping die for an automotive structural component

(b) Die (black) and sheet (green) discretization into 23522 rotation free BST shell triangles

(c) Thickness ratio contours of the sheet at 17mm punch travel

Plate 2 Non-linear metal forming analysis of a car door

Courtesy of Prof. E. Oñate, CIMNE and DECAD S.A Barcelona. Source: E. Oñate, F. Zarate, J. Rojek, G. Duffet, L. Neamtui, 'Adventures in rotation free elements for sheet stamping analysis'. 4th Int. Conf. Workshop on Numerical Simulation of 3D Sheet Forming Processess (NUMISHEET' 99, Besancon, France, Sept 13-17, 1999)

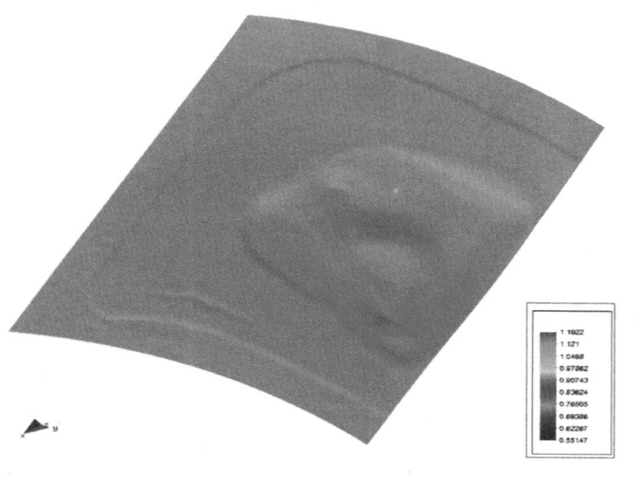

(d) Thickness ratio contours at 35mm punch travel

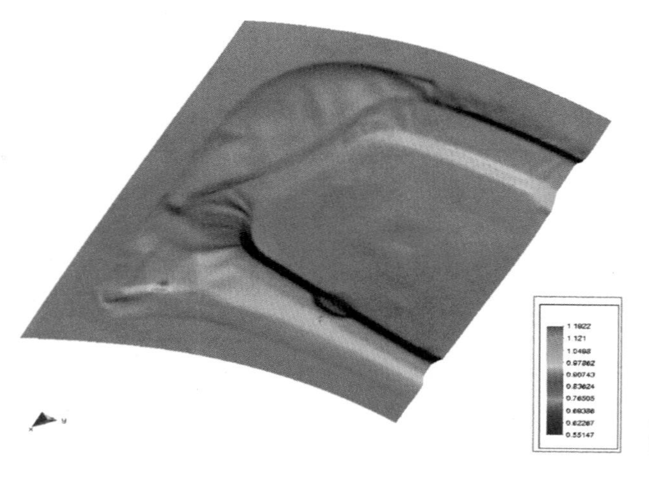

(e) Thickness ratio contours at 54mm punch travel

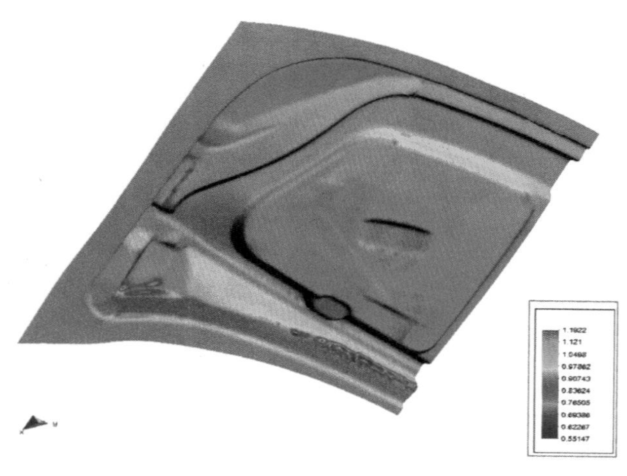

(f) Thickness ratio contours at 72mm punch travel

Plate 2 *continued* Non-linear metal forming analysis of a car door

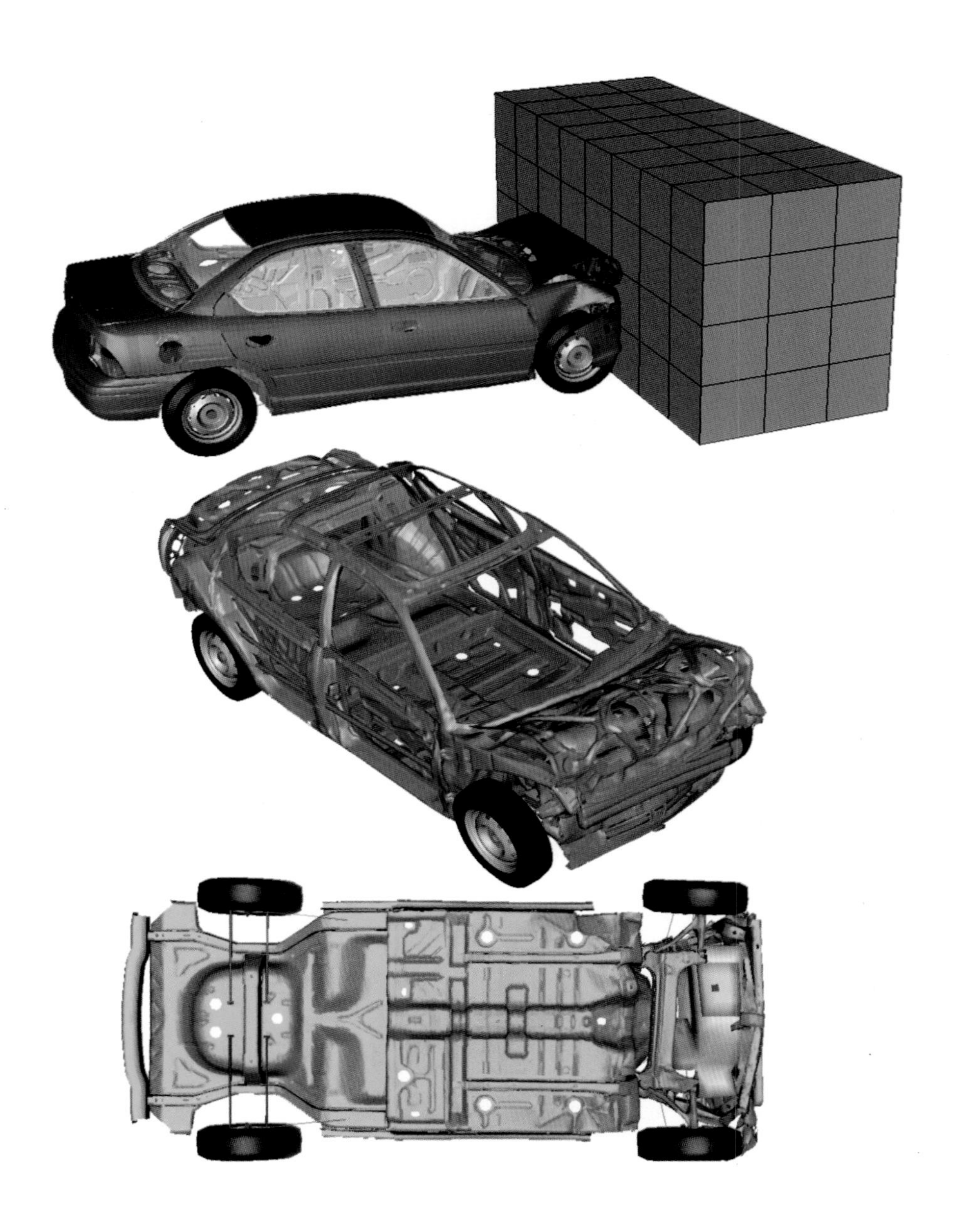

Plate 3 Car crash analysis
Frontal crash of a Neon Car performed using LS-DYNA. Courtesy of Livermore Software Technology Corporation. Model developed by FHWA/NHTSA National Crash Analysis Center of the George Washington University

The Finite Element Method

Fifth edition

Volume 2: Solid Mechanics

O.C. Zienkiewicz, CBE, FRS, FREng
UNESCO Professor of Numerical Methods in Engineering
International Centre for Numerical Methods in Engineering, Barcelona
Emeritus Professor of Civil Engineering and Director of the Institute for Numerical Methods in Engineering, University of Wales, Swansea

R.L. Taylor
Professor in the Graduate School
Department of Civil and Environmental Engineering
University of California at Berkeley
Berkeley, California

OXFORD AUCKLAND BOSTON JOHANNESBURG MELBOURNE NEW DELHI

Butterworth-Heinemann
Linacre House, Jordan Hill, Oxford OX2 8DP
225 Wildwood Avenue, Woburn, MA 01801-2041
A division of Reed Educational and Professional Publishing Ltd

A member of the Reed Elsevier plc group

First published in 1967 by McGraw-Hill
Fifth edition published by Butterworth-Heinemann 2000

British Library Cataloguing in Publication Data
A catalogue record for this book is available from the British Library

Library of Congress Cataloguing in Publication Data
A catalogue record for this book is available from the Library of Congress

ISBN 0 7506 5055 9

Published with the cooperation of CIMNE, the International Centre for Numerical Methods in Engineering, Barcelona, Spain (www.cimne.upc.es)

Typeset by Academic & Technical Typesetting, Bristol
Printed and bound by MPG Books Ltd

Dedication

This book is dedicated to our wives Helen and Mary Lou and our families for their support and patience during the preparation of this book, and also to all of our students and colleagues who over the years have contributed to our knowledge of the finite element method. In particular we would like to mention Professor Eugenio Oñate and his group at CIMNE for their help, encouragement and support during the preparation process.

Contents

Volume 1: The basis

Volume 3: Fluid dynamics

Preface to Volume 2

The first volume of this edition covered basic aspects of finite element approximation in the context of linear problems. Typical examples of two- and three-dimensional elasticity, heat conduction and electromagnetic problems in a steady state and transient state were dealt with and a finite element computer program structure was introduced. However, many aspects of formulation had to be relegated to the second and third volumes in which we hope the reader will find the answer to more advanced problems, most of which are of continuing practical and research interest.

In this volume we consider more advanced problems in solid mechanics while in Volume 3 we consider applications in fluid dynamics. It is our intent that Volume 2 can be used by investigators familiar with the finite element method in general terms and will introduce them here to the subject of specialized topics in solid mechanics. This volume can thus in many ways stand alone. Many of the general finite element procedures available in Volume 1 may not be familiar to a reader introduced to the finite element method through different texts. We therefore recommend that the present volume be used in conjunction with Volume 1 to which we make frequent reference.

Two main subject areas in solid mechanics are covered here:

1. *Non-linear problems* (Chapters 1–3 and 10–12) In these the special problems of solving non-linear equation systems are addressed. In the first part we restrict our attention to non-linear behaviour of materials while retaining the assumptions on small strain used in Volume 1 to study the linear elasticity problem. This serves as a bridge to more advanced studies later in which geometric effects from large displacements and deformations are presented. Indeed, non-linear applications are today of great importance and practical interest in most areas of engineering and physics. By starting our study first using a small strain approach we believe the reader can more easily comprehend the various aspects which need to be understood to master the subject matter. We cover in some detail problems in viscoelasticity, plasticity, and viscoplasticity which should serve as a basis for applications to other material models. In our study of finite deformation problems we present a series of approaches which may be used to solve problems including extensions for treatment of constraints (e.g. near incompressibility and rigid body motions) as well as those for buckling and large rotations.

2. *Plates and shells* (Chapters 4–9) This section is of course of most interest to those engaged in 'structural mechanics' and deals with a specific class of problems in which one dimension of the structure is small compared to the other two. This application is one of the first to which finite elements were directed and which still is a subject of continuing research. Those with interests in other areas of solid mechanics may well omit this part on first reading, though by analogy the methods exposed have quite wide applications outside structural mechanics.

Volume 2 concludes with a chapter on Computer Procedures, in which we describe application of the basic program presented in Volume 1 to solve non-linear problems. Clearly the variety of problems presented in the text does not permit a detailed treatment of all subjects discussed, but the 'skeletal' format presented and additional information available from the publisher's web site[1] will allow readers to make their own extensions.

We would like at this stage to thank once again our collaborators and friends for many helpful comments and suggestions. In this volume our particular gratitude goes to Professor Eric Kasper who made numerous constructive comments as well as contributing the section on the mixed–enhanced method in Chapter 10. We would also like to take this opportunity to thank our friends at CIMNE for providing a stimulating environment in which much of Volume 2 was conceived.

OCZ and RLT

[1] Complete source code for all programs in the three volumes may be obtained at no cost from the publisher's web page: http://www.bh.com/companions/fem

1

General problems in solid mechanics and non-linearity

1.1 Introduction

In the first volume we discussed quite generally linear problems of elasticity and of field equations. In many practical applications the limitation of linear elasticity or more generally of linear behaviour precludes obtaining an accurate assessment of the solution because of the presence of non-linear effects and/or because of the geometry having a 'thin' dimension in one or more directions. In this volume we describe extensions to the formulations previously introduced which permit solutions to both classes of problems.

Non-linear behaviour of solids takes two forms: material non-linearity and geometric non-linearity. The simplest form of a non-linear material behaviour is that of elasticity for which the stress is not linearly proportional to the strain. More general situations are those in which the loading and unloading response of the material is different. Typical here is the case of classical elasto-plastic behaviour.

When the deformation of a solid reaches a state for which the undeformed and deformed shapes are substantially different a state of *finite deformation* occurs. In this case it is no longer possible to write linear strain-displacement or equilibrium equations on the undeformed geometry. Even before finite deformation exists it is possible to observe *buckling* or *load bifurcations* in some solids and non-linear equilibrium effects need to be considered. The classical Euler column where the equilibrium equation for buckling includes the effect of axial loading is an example of this class of problem.

Structures in which one dimension is very small compared with the other two define plate and shell problems. A *plate* is a flat structure with one thin direction which is called the thickness, and a *shell* is a curved structure in space with one such small thickness direction. Structures with two small dimensions are called *beams*, *frames*, or *rods*. Generally the accurate solution of linear elastic problems with one (or more) small dimension(s) cannot be achieved efficiently by using the three-dimensional finite element formulations described in Chapter 6 of Volume 1[1] and conventionally in the past separate theories have been introduced. A primary reason is the numerical ill-conditioning which results in the algebraic equations making their accurate solution difficult to achieve. In this book we depart from past tradition and build a much stronger link to the full three-dimensional theory.

This volume will consider each of the above types of problems and formulations which make practical finite element solutions feasible. We establish in the present chapter the general formulation for both static and transient problems of a non-linear kind. Here we show how the linear problems of steady state behaviour and transient behaviour discussed in Volume 1 become non-linear. Some general discussion of transient non-linearity will be given here, and in the remainder of this volume we shall primarily confine our remarks to quasi-static (i.e. no inertia effects) and static problems only.

In Chapter 2 we describe various possible methods for solving non-linear algebraic equations. This is followed in Chapter 3 by consideration of material non-linear behaviour and the development of a general formulation from which a finite element computation can proceed.

We then describe the solution of plate problems, considering first the problem of thin plates (Chapter 4) in which only bending deformations are included and, second, the problem in which both bending and shearing deformations are present (Chapter 5).

The problem of shell behaviour adds in-plane membrane deformations and curved surface modelling. Here we split the problem into three separate parts. The first, combines simple flat elements which include bending and membrane behaviour to form a faceted approximation to the curved shell surface (Chapter 6). Next we involve the addition of shearing deformation and use of curved elements to solve axisymmetric shell problems (Chapter 7). We conclude the presentation of shells with a general form using curved isoparametric element shapes which include the effects of bending, shearing, and membrane deformations (Chapter 8). Here a very close link with the full three-dimensional analysis of Volume 1 will be readily recognized.

In Chapter 9 we address a class of problems in which the solution in one coordinate direction is expressed as a series, for example a Fourier series. Here, for linear material behavior, very efficient solutions can be achieved for many problems. Some extensions to non-linear behaviour are also presented.

In the last part of this volume we address the general problem of finite deformation as well as specializations which permit large displacements but have small strains. In Chapter 10 we present a summary for the finite deformation of solids. Basic relations for defining deformation are presented and used to write variational forms related to the undeformed configuration of the body and also to the deformed configuration. It is shown that by relating the formulation to the deformed body a result is obtain which is nearly identical to that for the small deformation problem we considered in Volume 1 and which we expand upon in the early chapters of this volume. Essential differences arise only in the constitutive equations (stress–strain laws) and the addition of a new stiffness term commonly called the *geometric* or *initial stress* stiffness. For constitutive modelling we summarize alternative forms for elastic and inelastic materials. In this chapter contact problems are also discussed.

In Chapter 11 we specialize the geometric behaviour to that which results in large displacements but small strains. This class of problems permits use of all the constitutive equations discussed for small deformation problems and can address classical problems of instability. It also permits the construction of non-linear extensions to plate and shell problems discussed in Chapters 4–8 of this volume.

In Chapter 12 we discuss specialization of the finite deformation problem to address situations in which a large number of small bodies interact (multiparticle or granular bodies) or individual parts of the problem are treated as rigid bodies.

In the final chapter we discuss extensions to the computer program described in Chapter 20 of Volume 1 necessary to address the non-linear material, the plate and shell, and the finite deformation problems presented in this volume. Here the discussion is directed primarily to the manner in which non-linear problems are solved. We also briefly discuss the manner in which elements are developed to permit analysis of either quasi-static (no inertia effects) or transient applications.

1.2 Small deformation non-linear solid mechanics problems

1.2.1 Introduction and notation

In this general section we shall discuss how the various equations which we have derived for linear problems in Volume 1 can become non-linear under certain circumstances. In particular this will occur for structural problems when non-linear stress–strain relationships are used. But the chapter in essence recalls here the notation and the methodology which we shall adopt throughout this volume. This repeats matters which we have already dealt with in some detail. The reader will note how simply the transition between linear and non-linear problems occurs.

The field equations for solid mechanics are given by equilibrium (balance of momentum), strain-displacement relations, constitutive equations, boundary conditions, and initial conditions.[2–7]

In the treatment given here we will use two notational forms. The first is a cartesian tensor indicial form (e.g. see Appendix B, Volume 1) and the second is a matrix form as used extensively in Volume 1.[1] In general, we shall find that both are useful to describe particular parts of formulations. For example, when we describe large strain problems the development of the so-called 'geometric' or 'initial stress' stiffness is most easily described by using an indicial form. However, in much of the remainder, we shall find that it is convenient to use the matrix form. In order to make steps clear we shall here review the equations for small strain in both the indicial and the matrix forms. The requirements for transformations between the two will also be again indicated.

For the small strain applications and fixed cartesian systems we denote coordinates as x, y, z or in index form as x_1, x_2, x_3. Similarly, the displacements will be denoted as u, v, w or u_1, u_2, u_3. Where possible the coordinates and displacements will be denoted as x_i and u_i, respectively, where the range of the index i is $1, 2, 3$ for three-dimensional applications (or $1, 2$ for two-dimensional problems). In matrix form we write the coordinates as

$$\mathbf{x} = \begin{Bmatrix} x \\ y \\ z \end{Bmatrix} = \begin{Bmatrix} x_1 \\ x_2 \\ x_3 \end{Bmatrix} \tag{1.1}$$

and displacements as

$$\mathbf{u} = \begin{Bmatrix} u \\ v \\ w \end{Bmatrix} = \begin{Bmatrix} u_1 \\ u_2 \\ u_3 \end{Bmatrix} \tag{1.2}$$

1.2.2 Weak form for equilibrium – finite element discretization

The equilibrium equations (balance of linear momentum) are given in index form as

$$\sigma_{ji,j} + b_i = \rho \ddot{u}_i, \qquad i, j = 1, 2, 3 \tag{1.3}$$

where σ_{ij} are components of (Cauchy) stress, ρ is mass density, b_i are body force components and $(\dot{\ })$ denotes partial differentiation with respect to time. In the above, and in the sequel, we always use the convention that repeated indices in a term are summed over the range of the index. In addition, a partial derivative with respect to the coordinate x_i is indicated by a comma, and a superposed dot denotes partial differentiation with respect to time. Similarly, moment equilibrium (balance of angular momentum) yields symmetry of stress given indicially as

$$\sigma_{ij} = \sigma_{ji} \tag{1.4}$$

Equations (1.3) and (1.4) hold at all points x_i in the domain of the problem Ω. Stress boundary conditions are given by the traction condition

$$t_i = \sigma_{ji} n_j = \bar{t}_i \tag{1.5}$$

for all points which lie on the part of the boundary denoted as Γ_t.

A variational (weak) form of the equations may be written by using the procedures described in Chapter 3 of Volume 1 and yield the virtual work equations given by[1,8,9]

$$\int_\Omega \delta u_i \rho \ddot{u}_i \, \mathrm{d}\Omega + \int_\Omega \delta \varepsilon_{ij} \sigma_{ij} \, \mathrm{d}\Omega - \int_\Omega \delta u_i b_i \, \mathrm{d}\Omega - \int_{\Gamma_t} \delta u_i \bar{t}_i \, \mathrm{d}\Omega = 0 \tag{1.6}$$

In the above cartesian tensor form, virtual strains are related to virtual displacements as

$$\delta \varepsilon_{ij} = \tfrac{1}{2}(\delta u_{i,j} + \delta u_{j,i}) \tag{1.7}$$

In this book we will often use a transformation to matrix form where stresses are given in the order

$$\begin{aligned} \boldsymbol{\sigma} &= \begin{bmatrix} \sigma_{11} & \sigma_{22} & \sigma_{33} & \sigma_{12} & \sigma_{23} & \sigma_{31} \end{bmatrix}^{\mathrm{T}} \\ &= \begin{bmatrix} \sigma_{xx} & \sigma_{yy} & \sigma_{zz} & \sigma_{xy} & \sigma_{yz} & \sigma_{zx} \end{bmatrix}^{\mathrm{T}} \end{aligned} \tag{1.8}$$

and strains by

$$\begin{aligned} \boldsymbol{\varepsilon} &= \begin{bmatrix} \varepsilon_{11} & \varepsilon_{22} & \varepsilon_{33} & \gamma_{12} & \gamma_{23} & \gamma_{31} \end{bmatrix}^{\mathrm{T}} \\ &= \begin{bmatrix} \varepsilon_{xx} & \varepsilon_{yy} & \varepsilon_{zz} & \gamma_{xy} & \gamma_{yz} & \gamma_{zx} \end{bmatrix}^{\mathrm{T}} \end{aligned} \tag{1.9}$$

where symmetry of the tensors is assumed and 'engineering' shear strains are introduced as*

$$\gamma_{ij} = 2\varepsilon_{ij} \tag{1.10}$$

to make writing of subsequent matrix relations in a consistent manner.

The transformation to the six independent components of stress and strain is performed by using the index order given in Table 1.1. This ordering will apply to

* This form is necessary to allow the internal work always to be written as $\boldsymbol{\sigma}^{\mathrm{T}} \boldsymbol{\varepsilon}$.

Table 1.1 Index relation between tensor and matrix forms

Form	Index value					
Matrix	1	2	3	4	5	6
Tensor (1, 2, 3)	11	22	33	12	23	31
				21	32	13
Tensor (x, y, z)	xx	yy	zz	xy	yz	zx
				yx	zy	xz

many subsequent developments also. The order is chosen to permit reduction to two-dimensional applications by merely deleting the last two entries and treating the third entry as appropriate for plane or axisymmetric applications.

In matrix form, the virtual work equation is written as (see Chapter 3 of Volume 1)

$$\int_\Omega \delta\mathbf{u}^T \rho \ddot{\mathbf{u}}\, d\Omega + \int_\Omega \delta\boldsymbol{\varepsilon}^T \boldsymbol{\sigma}\, d\Omega - \int_\Omega \delta\mathbf{u}^T \mathbf{b}\, d\Omega - \int_{\Gamma_t} \delta\mathbf{u}^T \bar{\mathbf{t}}\, d\Gamma = 0 \tag{1.11}$$

Finite element approximations to displacements and virtual displacements are denoted by

$$\mathbf{u}(\mathbf{x}, t) = \mathbf{N}(\mathbf{x})\tilde{\mathbf{u}}(t) \quad \text{and} \quad \delta\mathbf{u}(\mathbf{x}) = \mathbf{N}(\mathbf{x})\,\delta\tilde{\mathbf{u}} \tag{1.12}$$

or in isoparametric form as

$$\mathbf{u}(\boldsymbol{\xi}, t) = \mathbf{N}(\boldsymbol{\xi})\tilde{\mathbf{u}}(t); \quad \delta\mathbf{u}(\boldsymbol{\xi}) = \mathbf{N}(\boldsymbol{\xi})\,\delta\tilde{\mathbf{u}} \quad \text{with} \quad \mathbf{x}(\boldsymbol{\xi}) = \mathbf{N}(\boldsymbol{\xi})\,\tilde{\mathbf{x}} \tag{1.13}$$

and may be used to compute virtual strains as

$$\delta\boldsymbol{\varepsilon} = \mathbf{S}\,\delta\mathbf{u} = (\mathbf{SN})\,\delta\tilde{\mathbf{u}} = \mathbf{B}\,\delta\tilde{\mathbf{u}} \tag{1.14}$$

in which the three-dimensional strain-displacement matrix is given by [see Eq. (6.11), Volume 1]

$$\mathbf{B} = \begin{bmatrix} \mathbf{N}_{,1} & \mathbf{0} & \mathbf{0} \\ \mathbf{0} & \mathbf{N}_{,2} & \mathbf{0} \\ \mathbf{0} & \mathbf{0} & \mathbf{N}_{,3} \\ \mathbf{N}_{,2} & \mathbf{N}_{,1} & \mathbf{0} \\ \mathbf{0} & \mathbf{N}_{,3} & \mathbf{N}_{,2} \\ \mathbf{N}_{,3} & \mathbf{0} & \mathbf{N}_{,1} \end{bmatrix} \tag{1.15}$$

In the above, $\tilde{\mathbf{u}}$ denotes time-dependent nodal displacement parameters and $\delta\tilde{\mathbf{u}}$ represents arbitrary virtual displacement parameters.

Noting that the virtual parameters $\delta\tilde{\mathbf{u}}$ are arbitrary we obtain for the discrete problem*

$$\mathbf{M}\ddot{\tilde{\mathbf{u}}} + \mathbf{P}(\boldsymbol{\sigma}) = \mathbf{f} \tag{1.16}$$

where

$$\mathbf{M} = \int_\Omega \mathbf{N}^T \rho \mathbf{N}\, d\Omega \tag{1.17}$$

$$\mathbf{f} = \int_\Omega \mathbf{N}^T \mathbf{b}\, d\Omega + \int_{\Gamma_t} \mathbf{N}^T \bar{\mathbf{t}}\, d\Gamma \tag{1.18}$$

* For simplicity we omit direct damping which leads to the term $\mathbf{C}\dot{\tilde{\mathbf{u}}}$ (see Chapter 17, Volume 1).

and

$$\mathbf{P}(\boldsymbol{\sigma}) = \int_\Omega \mathbf{B}^\mathrm{T} \boldsymbol{\sigma} \, \mathrm{d}\Omega \tag{1.19}$$

The term $\mathbf{P}$ is often referred to as the *stress divergence* or *stress force* term.

In the case of linear elasticity the stress is immediately given by the stress–strain relations (see Chapter 2, Volume 1) as

$$\boldsymbol{\sigma} = \mathbf{D}\boldsymbol{\varepsilon} \tag{1.20}$$

when effects of initial stress and strain are set to zero. In the above the $\mathbf{D}$ are the usual elastic moduli written in matrix form. If a displacement method is used the strains are obtained from the displacement field by using

$$\boldsymbol{\varepsilon} = \mathbf{B}\tilde{\mathbf{u}} \tag{1.21}$$

Equation (1.19) becomes

$$\mathbf{P}(\boldsymbol{\sigma}) = \left(\int_\Omega \mathbf{B}^\mathrm{T} \mathbf{D} \mathbf{B} \, \mathrm{d}\Omega \right) \tilde{\mathbf{u}} = \mathbf{K}\tilde{\mathbf{u}} \tag{1.22}$$

in which $\mathbf{K}$ is the linear stiffness matrix. In many situations, however, it is necessary to use non-linear or time-dependent stress–strain (constitutive) relations and in these cases we shall have to develop solution strategies directly from Eq. (1.19). This will be considered further in detail in later chapters. However, at this stage we simply need to note that

$$\boldsymbol{\sigma} = \boldsymbol{\sigma}(\boldsymbol{\varepsilon}) \tag{1.23}$$

quite generally and that the functional relationship can be very non-linear and occasionally non-unique. Furthermore, it will be necessary to use a mixed approach if constraints, such as near incompressibility, are encountered. We address this latter aspect in Sec. 1.2.4; however, before doing so we consider first the manner whereby solution of the transient equations may be computed by using step-by-step time integration methods discussed in Chapter 18 of Volume 1.

1.2.3 Non-linear formulation of transient and steady-state problems

To obtain a set of *algebraic equations* for transient problems we introduce a discrete approximation in time. We consider the GN22 method or the Newmark procedure as being applicable to the second-order equations (see Chapter 18, Volume 1). Dropping the tilde on discrete variables for simplicity we write the approximation to the solution as

$$\tilde{\mathbf{u}}(t_{n+1}) \approx \mathbf{u}_{n+1}$$

and now the equilibrium equation (1.16) at each discrete time t_{n+1} may be written in a *residual form* as

$$\boldsymbol{\Psi}_{n+1} = \mathbf{f}_{n+1} - \mathbf{M}\ddot{\mathbf{u}}_{n+1} - \mathbf{P}_{n+1} = \mathbf{0} \tag{1.24}$$

where

$$\mathbf{P}_{n+1} \equiv \int_{\Omega} \mathbf{B}^{\mathrm{T}} \boldsymbol{\sigma}_{n+1} \, \mathrm{d}\Omega = \mathbf{P}(\mathbf{u}_{n+1}) \tag{1.25}$$

Using the GN22 formulae, the discrete displacements, velocities, and accelerations are linked by [see Eq. (18.62), Volume 1]

$$\mathbf{u}_{n+1} = \mathbf{u}_n + \Delta t \dot{\mathbf{u}}_n + \tfrac{1}{2}(1 - \beta_2)\Delta t^2 \ddot{\mathbf{u}}_n + \tfrac{1}{2}\beta_2 \Delta t^2 \ddot{\mathbf{u}}_{n+1} \tag{1.26}$$

$$\dot{\mathbf{u}}_{n+1} = \dot{\mathbf{u}}_n + (1 - \beta_1)\Delta t \ddot{\mathbf{u}}_n + \beta_1 \Delta t \ddot{\mathbf{u}}_{n+1} \tag{1.27}$$

where $\Delta t = t_{n+1} - t_n$.

Equations (1.26) and (1.27) are simple, vector, linear relationships as the coefficient β_1 and β_2 are assigned *a priori* and it is possible to take the basic unknown in Eq. (1.24) as any one of the three variables at time step $n+1$ (i.e. $\mathbf{u}_{n+1}$, $\dot{\mathbf{u}}_{n+1}$ or $\ddot{\mathbf{u}}_{n+1}$).

A very convenient choice for explicit schemes is that of $\ddot{\mathbf{u}}_{n+1}$. In such schemes we take the constant β_2 as zero and note that this allows $\mathbf{u}_{n+1}$ to be evaluated directly from the initial values at time t_n without solving any simultaneous equations. Immediately, therefore, Eq. (1.24) will yield the values of $\ddot{\mathbf{u}}_{n+1}$ by simple inversion of matrix $\mathbf{M}$.

If the $\mathbf{M}$ matrix is diagonalized by any one of the methods which we have discussed in Volume 1, the solution for $\ddot{\mathbf{u}}_{n+1}$ is trivial and the problem can be considered solved. However, such explicit schemes are only conditionally stable as we have shown in Chapter 18 of Volume 1 and may require many time steps to reach a steady state solution. Therefore for transient problems and indeed for all static (steady state) problems, it is often more efficient to deal with implicit methods. Here, most conveniently, $\mathbf{u}_{n+1}$ can be taken as the basic variable from which $\dot{\mathbf{u}}_{n+1}$ and $\ddot{\mathbf{u}}_{n+1}$ can be calculated by using Eqs (1.26) and (1.27). The equation system (1.24) can therefore be written as

$$\boldsymbol{\Psi}(\mathbf{u}_{n+1}) \equiv \boldsymbol{\Psi}_{n+1} = \mathbf{0} \tag{1.28}$$

The solution of this set of equations will require an iterative process if the relations are non-linear. We shall discuss various non-linear calculation processes in some detail in Chapter 2; however, the Newton–Raphson method forms the basis of most practical schemes. In this method an iteration is as given below

$$\boldsymbol{\Psi}_{n+1}^{k+1} \approx \boldsymbol{\Psi}_{n+1}^{k} + \frac{\partial \boldsymbol{\Psi}^k}{\partial \mathbf{u}_{n+1}} \, d\mathbf{u}_n^k = \mathbf{0} \tag{1.29}$$

where $d\mathbf{u}_n^k$ is an increment to the solution* such that

$$\mathbf{u}_{n+1}^{k+1} = \mathbf{u}_{n+1}^{k} + d\mathbf{u}_n^k \tag{1.30}$$

For problems in which path dependence is involved it is necessary to keep track of the *total increment* during the iteration and write

$$\mathbf{u}_{n+1}^{k+1} = \mathbf{u}_n + \Delta \mathbf{u}_n^{k+1} \tag{1.31}$$

Thus the total increment can be accumulated by using the same solution increments as

$$\Delta \mathbf{u}_n^{k+1} = \mathbf{u}_{n+1}^{k+1} - \mathbf{u}_n = \Delta \mathbf{u}_n^k + d\mathbf{u}_n^k \tag{1.32}$$

* Note that an italic '*d*' is used for a solution increment and an upright 'd' for a differential.

in which a quantity without the superscript k denotes a converged value from a previous time step. The initial iterate may be taken as zero or, more appropriately, as the converged solution from the last time step. Accordingly,

$$\mathbf{u}_{n+1}^1 = \mathbf{u}_n \qquad \text{giving also} \qquad \Delta\mathbf{u}_n^1 = \mathbf{0} \tag{1.33}$$

A solution increment is now computed from Eq. (1.29) as

$$d\mathbf{u}_n^k = (\mathbf{K}_\mathrm{T}^k)^{-1}\mathbf{\Psi}_{n+1}^k \tag{1.34}$$

where the *tangent matrix* is computed as

$$\mathbf{K}_\mathrm{T}^k = -\frac{\partial \mathbf{\Psi}^k}{\partial \mathbf{u}_{n+1}}$$

From expressions (1.24) and (1.26) we note that the above equations can be rewritten as

$$\mathbf{K}_\mathrm{T}^k = \frac{\partial \mathbf{P}^k}{\partial \mathbf{u}_{n+1}} + \mathbf{M}\frac{\partial \ddot{\mathbf{u}}_{n+1}}{\partial \mathbf{u}_{n+1}} = \int_\Omega \mathbf{B}^\mathrm{T}\mathbf{D}_\mathrm{T}^k\mathbf{B}\,d\Omega + \frac{2}{\beta_2 \Delta t^2}\mathbf{M}$$

We note that the above relation is similar but not identical to that of linear elasticity. Here $\mathbf{D}_\mathrm{T}^k$ is the tangent modulus matrix for the stress–strain relation (which may or may not be unique but generally is related to deformations in a non-linear manner).

Iteration continues until a convergence criterion of the form

$$||\mathbf{\Psi}_{n+1}^k|| \leqslant \varepsilon ||\mathbf{\Psi}_{n+1}^1|| \tag{1.35}$$

or similar is satisfied for some small tolerance ε. A good practice is to assume the tolerance at half machine precision. Thus, if the machine can compute to about 16 digits of accuracy, selection of $\varepsilon = 10^{-8}$ is appropriate. Additional discussion on selection of appropriate convergence criteria is presented in Chapter 2.

Various forms of non-linear elasticity have in fact been used in the present context and here we present a simple approach in which we define a strain energy W as a function of $\boldsymbol{\varepsilon}$

$$W = W(\boldsymbol{\varepsilon}) = W(\varepsilon_{ij})$$

and we note that this definition gives us immediately

$$\boldsymbol{\sigma} = \frac{\partial W}{\partial \boldsymbol{\varepsilon}} \tag{1.36}$$

If the nature of the function W is known, we note that the tangent modulus $\mathbf{D}_\mathrm{T}^k$ becomes

$$\mathbf{D}_\mathrm{T}^k = \left(\frac{\partial^2 W}{\partial \boldsymbol{\varepsilon}\partial \boldsymbol{\varepsilon}}\right)_{n+1}^k \qquad \text{and} \qquad \mathbf{B} = \frac{\partial \boldsymbol{\varepsilon}}{\partial \mathbf{u}}$$

The algebraic non-linear solution in every time step can now be obtained by the process already discussed. In the general procedure during the time step, we have to take an initial value for $\mathbf{u}_{n+1}$, for example, $\mathbf{u}_{n+1}^1 = \mathbf{u}_n$ (and similarly for $\dot{\mathbf{u}}_{n+1}$ and $\ddot{\mathbf{u}}_{n+1}$) and then calculate at step 2 the value of $\mathbf{\Psi}_{n+1}^k$ at $k = 1$, and obtain

$d\mathbf{u}_{n+1}^1$ updating the value of $\mathbf{u}_{n+1}^k$ by Eq. (1.30). This of course necessitates calculation of stresses at t_{n+1} to obtain the necessary forces. It is worthwhile noting that the solution for steady state problems proceeds on identical lines with solution variable chosen as $\mathbf{u}_{n+1}$ but now we simply say $\ddot{\mathbf{u}}_{n+1} = \dot{\mathbf{u}}_{n+1} = \mathbf{0}$ as well as the corresponding terms in the governing equations.

1.2.4 Mixed or irreducible forms

The previous formulation was cast entirely in terms of the so-called displacement formulation which indeed was extensively used in the first volume. However, as we mentioned there, on some occasions it is convenient to use mixed finite element forms and these are especially necessary when constraints such as incompressibility arise. It has been frequently noted that certain constitutive laws, such as those of viscoelasticity and associative plasticity that we will discuss in Chapter 3, the material behaves in a nearly incompressible manner. For such problems a reformulation following the procedures given in Chapter 12 of Volume 1 is necessary. We remind the reader that on such occasions we have two choices of formulation. We can have the variables $\mathbf{u}$ and p (where p is the mean stress) as a two-field formulation (see Sec. 12.3 or 12.7 of Volume 1) or we can have the variables $\mathbf{u}$, p and ε_v (where ε_v is the volume change) as a three-field formulation (see Sec. 12.4, Volume 1). An alternative three-field form is the enhanced strain approach presented in Sec. 11.5.3 of Volume 1. The matter of which we use depends on the form of the constitutive equations. For situations where changes in volume affect only the pressure the two-field form can be easily used. However, for problems in which the response is coupled between the deviatoric and mean components of stress and strain the three-field formulations lead to much simpler forms from which to develop a finite element model. To illustrate this point we present again the mixed formulation of Sec. 12.4 in Volume 1 and show in detail how such coupled effects can be easily included without any change to the previous discussion on solving non-linear problems. The development also serves as a basis for the development of an extended form which permits the treatment of finite deformation problems. This extension will be presented in Sec. 10.4 of Chapter 10.

A three-field mixed method for general constitutive models

In order to develop a mixed form for use with constitutive models in which mean and deviatoric effects can be coupled we recall (Chapter 12 of Volume 1) that mean and deviatoric matrix operators are given by

$$\mathbf{m} = \begin{Bmatrix} 1 \\ 1 \\ 1 \\ 0 \\ 0 \\ 0 \end{Bmatrix}; \quad \mathbf{I}_d = \mathbf{I} - \frac{1}{3}\mathbf{m}\mathbf{m}^{\mathrm{T}}, \tag{1.37}$$

where $\mathbf{I}$ is the identity matrix.

As in Volume 1 we introduce independent parameters ε_v and p describing volumetric change and mean stress (pressure), respectively. The strains may now be expressed in a *mixed* form as

$$\boldsymbol{\varepsilon} = \mathbf{I}_d(\mathbf{S}\mathbf{u}) + \tfrac{1}{3}\mathbf{m}\,\varepsilon_{\mathrm{v}} \tag{1.38}$$

and the stresses in a mixed form as

$$\boldsymbol{\sigma} = \mathbf{I}_d\breve{\boldsymbol{\sigma}} + \mathbf{m}p \tag{1.39}$$

where $\breve{\boldsymbol{\sigma}}$ is the set of stresses deduced directly from the strains, incremental strains, or strain rates, depending on the particular constitutive model form. For the present we shall denote this stress by

$$\breve{\boldsymbol{\sigma}} = \boldsymbol{\sigma}(\boldsymbol{\varepsilon}) \tag{1.40}$$

where we note it is not necessary to split the model into mean and deviatoric parts.

The Galerkin (variational) equations for the case including transients are now given by

$$\begin{aligned}
&\int_\Omega \delta\mathbf{u}^{\mathrm{T}}\rho\ddot{\mathbf{u}}\,\mathrm{d}\Omega + \int_\Omega \delta(\mathbf{S}\mathbf{u})^{\mathrm{T}}\boldsymbol{\sigma}\,\mathrm{d}\Omega = \int_\Omega \delta\mathbf{u}^{\mathrm{T}}\mathbf{b}\,\mathrm{d}\Omega + \int_{\Gamma_t}\delta\mathbf{u}^{\mathrm{T}}\bar{\mathbf{t}}\,\mathrm{d}\Gamma\\
&\int_\Omega \delta\varepsilon_{\mathrm{v}}\left[\tfrac{1}{3}\mathbf{m}^{\mathrm{T}}\breve{\boldsymbol{\sigma}} - p\right]\mathrm{d}\Omega = 0\\
&\int_\Omega \delta p\left[\mathbf{m}^{\mathrm{T}}(\mathbf{S}\mathbf{u}) - \varepsilon_{\mathrm{v}}\right]\mathrm{d}\Omega = 0
\end{aligned} \tag{1.41}$$

Introducing finite element approximations to the variables as

$$\mathbf{u} \approx \hat{\mathbf{u}} = \mathbf{N}_u\tilde{\mathbf{u}}, \qquad p \approx \hat{p} = \mathbf{N}_p\tilde{\mathbf{p}} \qquad \text{and} \qquad \varepsilon_{\mathrm{v}} \approx \hat{\varepsilon}_{\mathrm{v}} = \mathbf{N}_{\mathrm{v}}\tilde{\boldsymbol{\varepsilon}}_{\mathrm{v}}$$

and similar approximations to virtual quantities as

$$\delta\mathbf{u} \approx \delta\hat{\mathbf{u}} = \mathbf{N}_u\delta\tilde{\mathbf{u}}, \qquad \delta p \approx \delta\hat{p} = \mathbf{N}_p\delta\tilde{\mathbf{p}} \qquad \text{and} \qquad \delta\varepsilon_{\mathrm{v}} \approx \delta\hat{\varepsilon}_{\mathrm{v}} = \mathbf{N}_{\mathrm{v}}\delta\tilde{\boldsymbol{\varepsilon}}_{\mathrm{v}}$$

the strain and virtual strain in an element become

$$\begin{aligned}
\boldsymbol{\varepsilon} &= \mathbf{I}_d\mathbf{B}\tilde{\mathbf{u}} + \tfrac{1}{3}\mathbf{m}\mathbf{N}_{\mathrm{v}}\tilde{\boldsymbol{\varepsilon}}_{\mathrm{v}},\\
\delta\boldsymbol{\varepsilon} &= \mathbf{I}_d\mathbf{B}\,\delta\tilde{\mathbf{u}} + \tfrac{1}{3}\mathbf{m}\mathbf{N}_{\mathrm{v}}\delta\tilde{\boldsymbol{\varepsilon}}_{\mathrm{v}}
\end{aligned} \tag{1.42}$$

in which $\mathbf{B}$ is the standard strain-displacement matrix given in Eq. (1.15). Similarly, the stresses in each element may be computed by using

$$\boldsymbol{\sigma} = \mathbf{I}_d\breve{\boldsymbol{\sigma}} + \mathbf{m}\mathbf{N}_p\tilde{\mathbf{p}} \tag{1.43}$$

where again $\breve{\boldsymbol{\sigma}}$ are stresses computed as in Eq. (1.40) in terms of the strains $\boldsymbol{\varepsilon}$.

Substituting the element stress and strain expressions from Eqs (1.42) and (1.43) into Eq. (1.41) we obtain the set of finite element equations

$$\begin{aligned}
\mathbf{P} + \mathbf{M}\ddot{\tilde{\mathbf{u}}} &= \mathbf{f}\\
\mathbf{P}_p - \mathbf{C}\tilde{\mathbf{p}} &= \mathbf{0}\\
-\mathbf{C}^{\mathrm{T}}\tilde{\boldsymbol{\varepsilon}}_{\mathrm{v}} + \mathbf{E}\tilde{\mathbf{u}} &= \mathbf{0}
\end{aligned} \tag{1.44}$$

where

$$\mathbf{P} = \int_{\Omega} \mathbf{B}^{\mathrm{T}} \boldsymbol{\sigma} \, \mathrm{d}\Omega, \qquad \mathbf{P}_p = \tfrac{1}{3} \int_{\Omega} \mathbf{N}_{\mathrm{v}}^{\mathrm{T}} \mathbf{m}^{\mathrm{T}} \breve{\boldsymbol{\sigma}} \, \mathrm{d}\Omega$$
$$\mathbf{C} = \int_{\Omega} \mathbf{N}_{\mathrm{v}}^{\mathrm{T}} \mathbf{N}_p \, \mathrm{d}\Omega, \qquad \mathbf{E} = \int_{\Omega} \mathbf{N}_p^{\mathrm{T}} \mathbf{m}^{\mathrm{T}} \mathbf{B} \, \mathrm{d}\Omega \tag{1.45}$$
$$\mathbf{f} = \int_{\Omega} \mathbf{N}_u^{\mathrm{T}} \mathbf{b} \, \mathrm{d}\Omega + \int_{\Gamma_t} \mathbf{N}_u^{\mathrm{T}} \bar{\mathbf{t}} \, \mathrm{d}\Gamma$$

If the pressure and volumetric strain approximations are taken locally in each element and $\mathbf{N}_{\mathrm{v}} = \mathbf{N}_p$ it is possible to solve the second and third equation of (1.44) in each element individually. Noting that the array $\mathbf{C}$ is now symmetric positive definite, we may always write these as

$$\begin{aligned} \tilde{\mathbf{p}} &= \mathbf{C}^{-1} \mathbf{P}_p \\ \tilde{\boldsymbol{\varepsilon}}_{\mathrm{v}} &= \mathbf{C}^{-1} \mathbf{E} \tilde{\mathbf{u}} = \mathbf{W} \tilde{\mathbf{u}} \end{aligned} \tag{1.46}$$

The mixed strain in each element may now be computed as

$$\boldsymbol{\varepsilon} = \left[\mathbf{I}_d \mathbf{B} + \frac{1}{3} \mathbf{m} \mathbf{B}_{\mathrm{v}} \right] \tilde{\mathbf{u}} = \begin{bmatrix} \mathbf{I}_d & \dfrac{1}{3} \mathbf{m} \end{bmatrix} \begin{bmatrix} \mathbf{B} \\ \mathbf{B}_{\mathrm{v}} \end{bmatrix} \tilde{\mathbf{u}} \tag{1.47}$$

where

$$\mathbf{B}_{\mathrm{v}} = \mathbf{N}_{\mathrm{v}} \mathbf{W} \tag{1.48}$$

defines a *mixed form* of the volumetric strain-displacement equations.

From the above results it is possible to write the vector $\mathbf{P}$ in the alternative forms[10,11]

$$\begin{aligned} \mathbf{P} &= \int_{\Omega} \mathbf{B}^{\mathrm{T}} \boldsymbol{\sigma} \, \mathrm{d}\Omega \\ &= \int_{\Omega} \mathbf{B}^{\mathrm{T}} \mathbf{I}_d \breve{\boldsymbol{\sigma}} \, \mathrm{d}\Omega + \int_{\Omega} \mathbf{B}^{\mathrm{T}} \mathbf{m} \mathbf{N}_p \, \mathrm{d}\Omega \, \mathbf{C}^{-1} \mathbf{P}_p \\ &= \int_{\Omega} \mathbf{B}^{\mathrm{T}} \mathbf{I}_d \breve{\boldsymbol{\sigma}} \, \mathrm{d}\Omega + \frac{1}{3} \int_{\Omega} \mathbf{W}^{\mathrm{T}} \mathbf{N}_{\mathrm{v}}^{\mathrm{T}} \mathbf{m}^{\mathrm{T}} \breve{\boldsymbol{\sigma}} \, \mathrm{d}\Omega \\ &= \int_{\Omega} \left[\mathbf{B}^{\mathrm{T}} \mathbf{I}_d + \frac{1}{3} \mathbf{B}_{\mathrm{v}}^{\mathrm{T}} \mathbf{m}^{\mathrm{T}} \right] \breve{\boldsymbol{\sigma}} \, \mathrm{d}\Omega \end{aligned} \tag{1.49}$$

The computation of $\mathbf{P}$ may then be represented in a matrix form as

$$\mathbf{P} = \int_{\Omega} \begin{bmatrix} \mathbf{B}^{\mathrm{T}} & \mathbf{B}_{\mathrm{v}}^{\mathrm{T}} \end{bmatrix} \begin{bmatrix} \mathbf{I}_d \\ \frac{1}{3} \mathbf{m}^{\mathrm{T}} \end{bmatrix} \breve{\boldsymbol{\sigma}} \, \mathrm{d}\Omega \tag{1.50}$$

in which we note the inclusion of the transpose of the matrices appearing in the expression for the mixed strain given in Eq. (1.47). Based on this result we observe that it is not necessary to compute the true mixed stress except when reporting final results where, for situations involving near incompressible behaviour, it is crucial to compute explicitly the mixed pressure to avoid any spurious volumetric stress effects.

The last step in the process is the computation of the tangent for the equations. This is straightforward using forms given by Eq. (1.40) where we obtain

$$d\breve{\boldsymbol{\sigma}} = \breve{\mathbf{D}}_\mathrm{T}\, d\boldsymbol{\varepsilon}$$

Use of Eq. (1.47) to express the incremental mixed strains then gives

$$\mathbf{K}_\mathrm{T} = \int_\Omega \begin{bmatrix} \mathbf{B}^\mathrm{T} & \mathbf{B}_\mathrm{v}^\mathrm{T} \end{bmatrix} \begin{bmatrix} \mathbf{I}_d \\ \frac{1}{3}\mathbf{m}^\mathrm{T} \end{bmatrix} \breve{\mathbf{D}}_\mathrm{T} \begin{bmatrix} \mathbf{I}_d & \frac{1}{3}\mathbf{m} \end{bmatrix} \begin{bmatrix} \mathbf{B} \\ \mathbf{B}_\mathrm{v} \end{bmatrix} \mathrm{d}\Omega \tag{1.51}$$

It should be noted that construction of a modified modulus term given by

$$\bar{\mathbf{D}}_\mathrm{T} = \begin{bmatrix} \mathbf{I}_d \\ \frac{1}{3}\mathbf{m}^\mathrm{T} \end{bmatrix} \breve{\mathbf{D}}_\mathrm{T} \begin{bmatrix} \mathbf{I}_d & \frac{1}{3}\mathbf{m} \end{bmatrix} = \begin{bmatrix} \mathbf{I}_d\breve{\mathbf{D}}_\mathrm{T}\mathbf{I}_d & \frac{1}{3}\mathbf{I}_d\breve{\mathbf{D}}_\mathrm{T}\mathbf{m} \\ \frac{1}{3}\mathbf{m}^\mathrm{T}\breve{\mathbf{D}}_\mathrm{T}\mathbf{I}_d & \frac{1}{9}\mathbf{m}^\mathrm{T}\breve{\mathbf{D}}_\mathrm{T}\mathbf{m} \end{bmatrix} \tag{1.52}$$

requires very few operations because of the sparsity and form of the arrays $\mathbf{I}_d$ and $\mathbf{m}$. Consequently, the multiplications by the coefficient matrices $\mathbf{B}$ and $\mathbf{B}_\mathrm{v}$ in this form is far more efficient than constructing a full $\bar{\mathbf{B}}$ as

$$\bar{\mathbf{B}} = \mathbf{I}_d\mathbf{B} + \tfrac{1}{3}\mathbf{m}\mathbf{B}_\mathrm{v} \tag{1.53}$$

and operating on $\breve{\mathbf{D}}_\mathrm{T}$ directly.

The above form for the mixed element generalizes the result in Volume 1 and is valid for use with many different linear and non-linear constitutive models. In Chapter 3 we consider stress–strain behaviour modelled by viscoelasticity, classical plasticity, and generalized plasticity formulations. Each of these forms can lead to situations in which a nearly incompressible response is required and for many examples included in this volume we shall use the above mixed formulation. Two basic forms are considered: four-noded quadrilateral or eight-noded brick isoparametric elements with constant interpolation in each element for one-term approximations to N_v and N_p by unity; and nine-noded quadrilateral or 27-noded brick isoparametric elements with linear interpolation for $\mathbf{N}_p$ and $\mathbf{N}_\mathrm{v}$.* Accordingly, in two dimensions we use

$$\mathbf{N}_p = \mathbf{N}_\mathrm{v} = \begin{bmatrix} 1 & \xi & \eta \end{bmatrix} \quad \text{or} \quad \begin{bmatrix} 1 & x & y \end{bmatrix}$$

and in three dimensions

$$\mathbf{N}_p = \mathbf{N}_\mathrm{v} = \begin{bmatrix} 1 & \xi & \eta & \zeta \end{bmatrix} \quad \text{or} \quad \begin{bmatrix} 1 & x & y & z \end{bmatrix}$$

The elements created by this process may be used to solve a wide range of problems in solid mechanics, as we shall illustrate in later chapters of this volume.

1.3 Non-linear quasi-harmonic field problems

In subsequent chapters we shall touch upon non-linear problems in the context of inelastic constitutive equations for solids, plates, and shells and in geometric effects

* Formulations using the eight-noded quadrilateral and twenty-noded brick serendipity elements may also be constructed; however, we showed in Chapter 11 of Volume 1 that these elements do not fully satisfy the mixed patch test.

arising from finite deformation. In Volume 3 non-linear effects will be considered for various fluid mechanics situations. However, non-linearity also may occur in many other problems and in these the techniques described in this chapter are still universally applicable. An example of such situations is the quasi-harmonic equation which is encountered in many fields of engineering. Here we consider a simple quasi-harmonic problem given by (e.g. heat conduction)

$$\rho c\dot{\phi} + \boldsymbol{\nabla}^{\mathrm{T}}\mathbf{q} - Q(\phi) = 0 \tag{1.54}$$

with suitable boundary conditions. Such a form may be used to solve problems ranging from temperature response in solids, seepage in porous media, magnetic effects in solids, and potential fluid flow. In the above, $\mathbf{q}$ is a flux and generally this can be written as

$$\mathbf{q} = \mathbf{q}(\phi, \boldsymbol{\nabla}\phi) = \mathbf{k}(\phi, \boldsymbol{\nabla}\phi)\boldsymbol{\nabla}\phi$$

or, after linearization,

$$d\mathbf{q} = \mathbf{k}^0\, d\phi + \mathbf{k}^1\, d(\boldsymbol{\nabla}\phi)$$

where

$$k_i^0 = \frac{\partial q_i}{\partial \phi} \qquad \text{and} \qquad k_{ij}^1 = \frac{\partial q_i}{\partial \phi_{,j}}$$

The source term $Q(\phi)$ also can introduce non-linearity.

A discretization based on Galerkin procedures gives after integration by parts of the $\mathbf{q}$ term the problem

$$\begin{aligned}\delta\Pi = &\int_\Omega \delta\phi\,\rho c\dot{\phi}\,\mathrm{d}\Omega - \int_\Omega (\boldsymbol{\nabla}\delta\phi)^{\mathrm{T}}\mathbf{q}\,\mathrm{d}\Omega \\ &- \int_\Omega \delta\phi\, Q(\phi)\,\mathrm{d}\Omega - \int_{\Gamma_q} \delta\phi\,\bar{q}_n\,\mathrm{d}\Gamma = 0\end{aligned} \tag{1.55}$$

and is still valid if $\mathbf{q}$ and/or Q (and indeed the boundary conditions) are dependent on ϕ or its derivatives. Introducing the interpolations

$$\phi = \mathbf{N}\tilde{\boldsymbol{\phi}} \quad \text{and} \quad \delta\phi = \mathbf{N}\,\delta\tilde{\boldsymbol{\phi}} \tag{1.56}$$

a discretized form is given as

$$\boldsymbol{\Psi} = \mathbf{f}(\tilde{\boldsymbol{\phi}}) - \mathbf{C}\dot{\tilde{\boldsymbol{\phi}}} - \mathbf{P}_q(\tilde{\boldsymbol{\phi}}) = \mathbf{0} \tag{1.57}$$

where

$$\begin{aligned}\mathbf{C} &= \int_\Omega \mathbf{N}^{\mathrm{T}}\rho c\mathbf{N}\,\mathrm{d}\Omega \\ \mathbf{P}_q &= -\int_\Omega \mathbf{N}^{\mathrm{T}}\mathbf{q}\,\mathrm{d}\Omega \\ \mathbf{f} &= \int_\Omega \mathbf{N}^{\mathrm{T}}Q(\phi)\,\mathrm{d}\Omega - \int_{\Gamma_q} \mathbf{N}^{\mathrm{T}}\bar{q}_n\,\mathrm{d}\Gamma\end{aligned} \tag{1.58}$$

Equation (1.57) may be solved following similar procedures described in Chapter 18, Volume 1. For instance, just as we did with GN22 we can now use GN11 as

$$\boldsymbol{\phi}_{n+1} = \boldsymbol{\phi}_n + (1-\theta)\dot{\boldsymbol{\phi}}_n + \theta\dot{\boldsymbol{\phi}}_{n+1} \tag{1.59}$$

Once again we have the choice of using $\boldsymbol{\phi}_{n+1}$ or $\dot{\boldsymbol{\phi}}_{n+1}$ as the primary solution variable. To this extent the process of solving transient problems follows absolutely the same lines as those described in the previous section (and indeed in the previous volume) and need not be further discussed. We note again that the use of $\boldsymbol{\phi}_{n+1}$ as the chosen variable will allow the solution method to be applied to static or steady state problems in which the first term of Eq. (1.54) becomes zero.

1.4 Some typical examples of transient non-linear calculations

In this section we report results of some transient problems of structural mechanics as well as field problems. As we mentioned earlier, we usually will not consider transient behaviour in latter parts of this book as the solution process for transients follow essentially the path described in Volume 1.

Transient heat conduction

The governing equation for this set of physical problems is discussed in the previous section, with ϕ being the temperature T now [Eq. (1.54)].

Non-linearity clearly can arise from the specific heat, c, thermal conductivity, k, and source, Q, being temperature-dependent or from a radiation boundary condition

$$k\frac{\partial T}{\partial n} = -\alpha(T - T_a)^n \tag{1.60}$$

with $n \neq 1$. Here α is a convective heat transfer coefficient and T_a is an ambient external temperature. We shall show two examples to illustrate the above.

The first concerns the *freezing* of ground in which the latent heat of freezing is represented by varying the material properties with temperature in a narrow zone, as shown in Fig. 1.1. Further, in the transition from the fluid to the frozen state a variation in conductivity occurs. We now thus have a problem in which both matrices $\mathbf{C}$ and $\mathbf{P}$ [Eq. (1.58)] are variable, and solution in Fig. 1.2 illustrates the progression of a freezing front which was derived by using the three-point (Lees) algorithm[12,13] with $\mathbf{C} = \mathbf{C}_n$ and $\mathbf{P} = \mathbf{P}_n$.

A computational feature of some significance arises in this problem as values of the specific heat become very high in the transition zone and, in time stepping can be missed if the temperature step *straddles* the freezing point. To avoid this difficulty and keep the heat balance correct the concept of enthalpy is introduced, defining

$$H = \int_0^T \rho c \, \mathrm{d}T \tag{1.61}$$

Now, whenever a change of temperature is considered, an appropriate value of ρc is calculated that gives the correct change of H.

The heat conduction problem involving phase change is of considerable importance in welding and casting technology. Some very useful finite element solutions of these problems have been obtained.[14] Further elaboration of the procedure described above is given in reference 15.

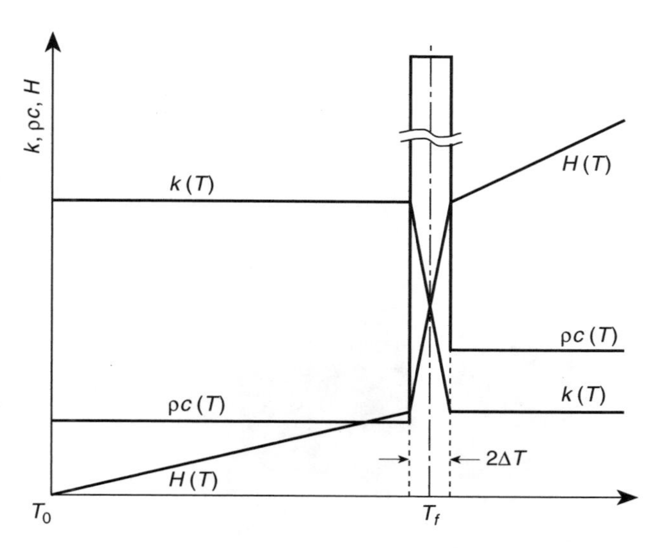

Fig. 1.1 Estimation of thermophysical properties in phase change problems. The latent heat effect is approximated by a large capacity over a small temperature interval $2\Delta T$.

The second non-linear example concerns the problem of *spontaneous ignition*.[16] We will discuss the steady state case of this problem in Chapter 3 and now will be concerned only with transient cases. Here the heat generated depends on the temperature

$$Q = \bar{\delta} e^{\mathrm{T}} \tag{1.62}$$

and the situation can become *physically unstable* with the computed temperature rising continuously to extreme values. In Fig. 1.3 we show a transient solution of a sphere at an initial temperature of $T = 290$ K immersed in a bath of 500 K. The solution is given for two values of the parameter $\bar{\delta}$ with $k = \rho c = 1$, and the non-linearities are now so severe that an iterative solution in each time increment is necessary. For the larger value of $\bar{\delta}$ the temperature increases to infinite value in a *finite time* and the time interval for the computation had to be changed continuously to account for this. The finite time for this point to be reached is known as the *induction time* and is shown in Fig. 1.3 for various values of $\bar{\delta}$.

The question of changing the time interval during the computation has not been discussed in detail, but clearly this must be done quite frequently to avoid large changes of the unknown function which will result in inaccuracies.

Structural dynamics

Here the examples concern dynamic structural transients with material and geometric non-linearity. A highly non-linear geometrical and material non-linearity generally occurs. Neglecting damping forces, Eq. (1.16) can be explicitly solved in an efficient manner.

If the explicit computation is pursued to the point when steady state conditions are approached, that is, until $\ddot{\mathbf{u}} = \dot{\mathbf{u}} \approx \mathbf{0}$ the solution to a static non-linear problem is obtained. This type of technique is frequently efficient as an alternative to the methods

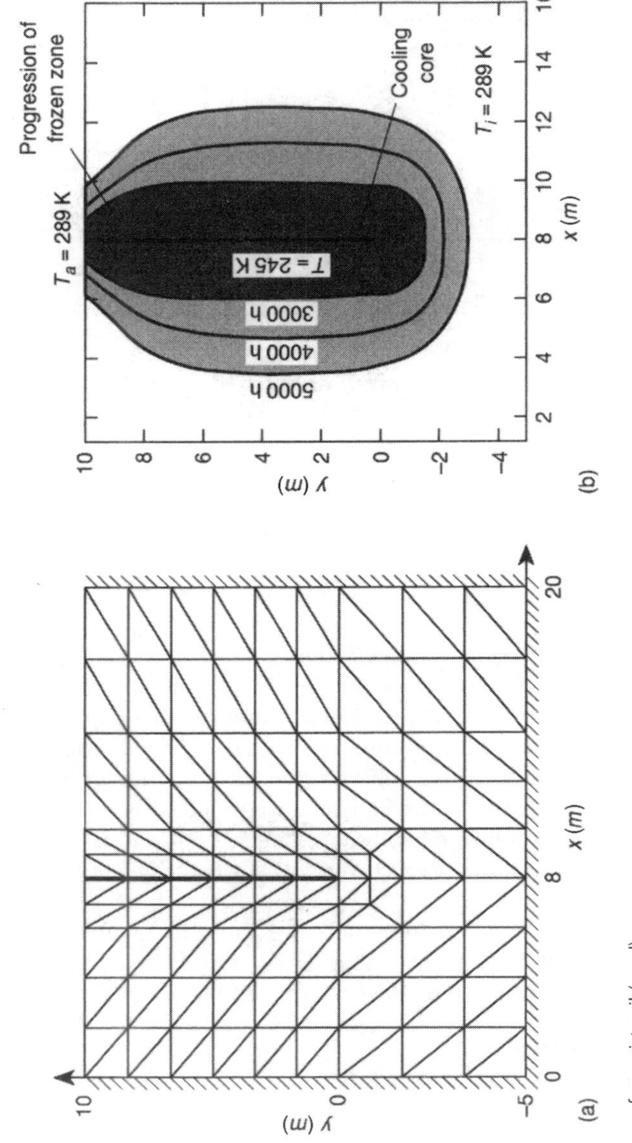

Fig. 1.2 Freezing of a moist soil (sand).

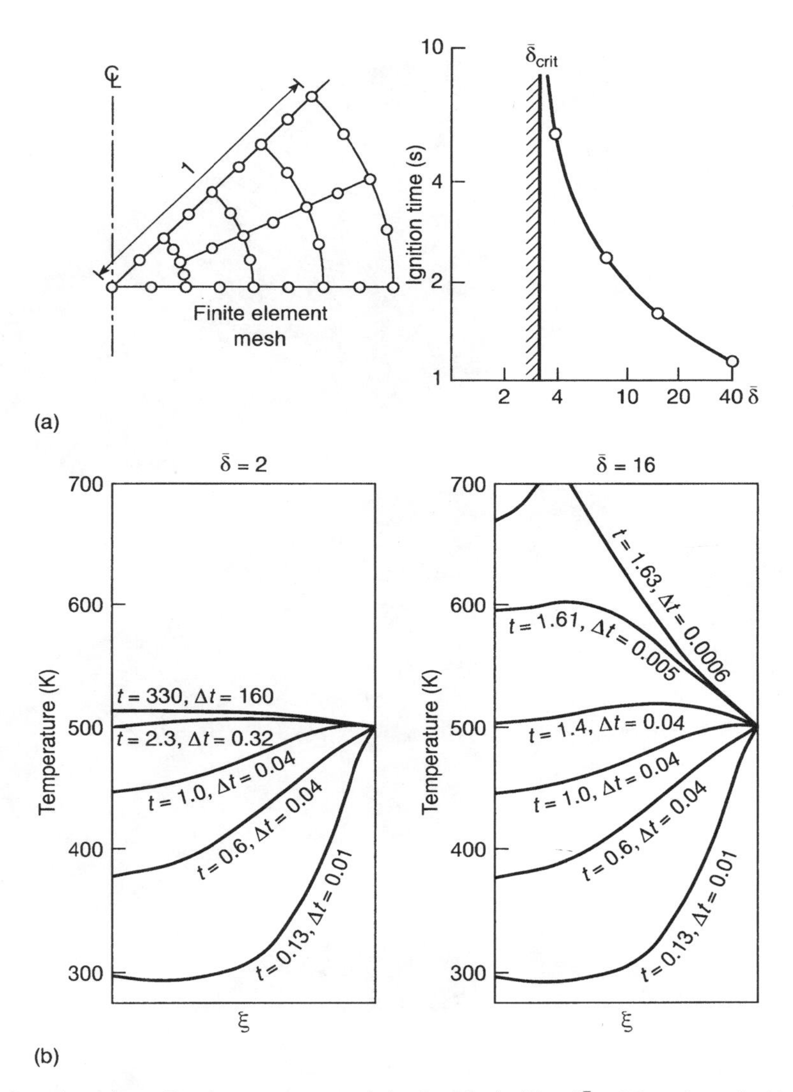

Fig. 1.3 Reactive sphere. Transient temperature behaviour for ignition ($\bar{\delta} = 16$) and non-ignition ($\bar{\delta} = 2$) cases: (a) induction time versus Frank–Kamenetskii parameter; temperature profiles; (b) temperature profiles for ignition ($\bar{\delta} = 16$) and non-ignition ($\bar{\delta} = 2$) transient behaviour of a reactive sphere.

described above and in Chapter 2 and has been applied successfully in the context of finite differences under the name of 'dynamic relaxation' for the solution of non-linear static problems.[17]

Two examples of explicit dynamic analysis will be given here. The first problem, illustrated in Plate 3, is a large three-dimensional problem and its solution was obtained with the use of an explicit dynamic scheme. In such a case implicit schemes would be totally inapplicable and indeed the explicit code provides a very efficient solution of the crash problem shown. It must, however, be recognized that such

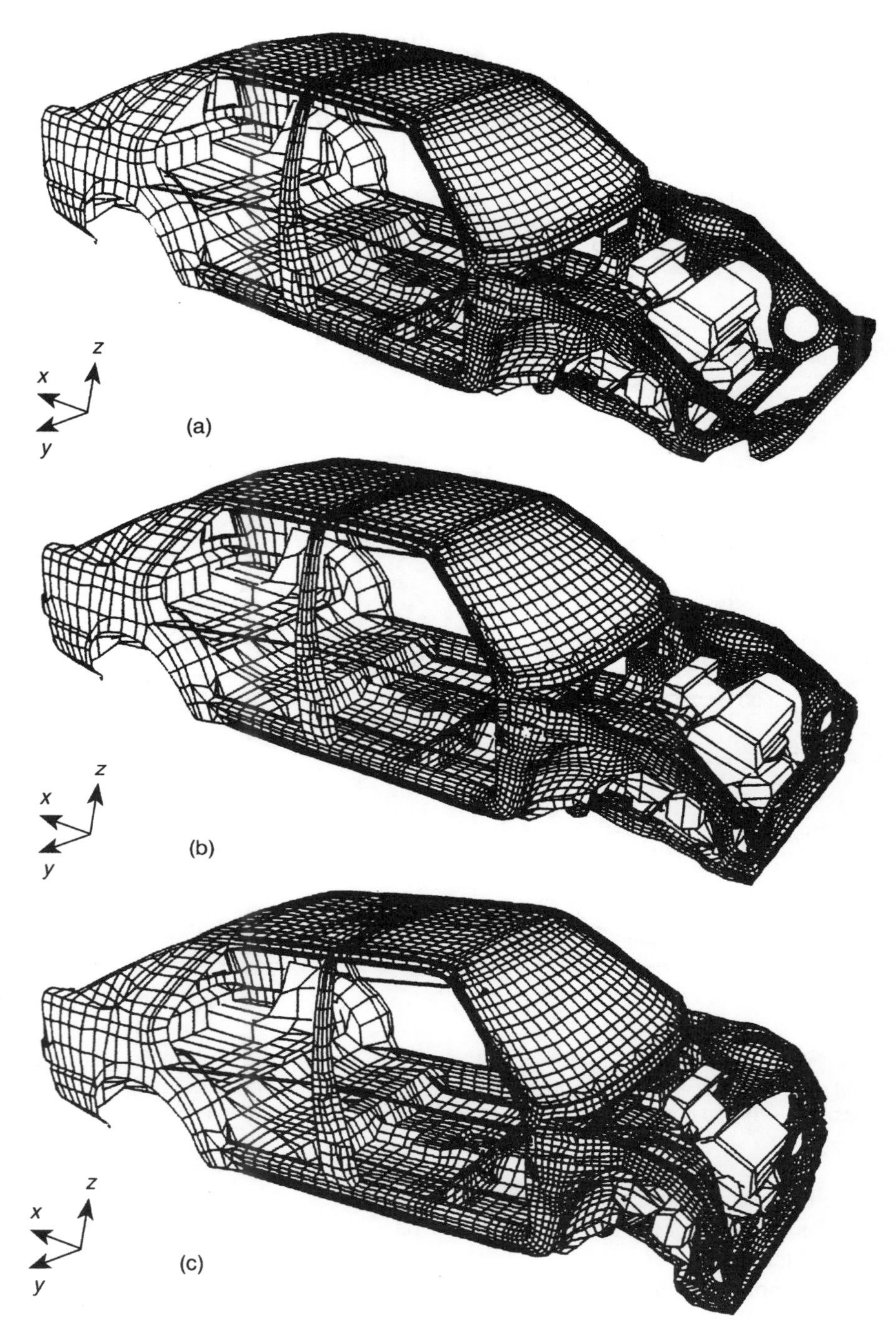

Fig. 1.4 Crash analysis: (a) mesh at $t = 0\,\text{ms}$; (b) mesh at $t = 20\,\text{ms}$; (c) mesh at $t = 40\,\text{ms}$.

final solutions are not necessarily unique. As a second example Figure 1.4 shows a typical crash analysis of a motor vehicle carried out by similar means.

Earthquake response of soil – structures

We have mentioned in Chapter 19, Volume 1, the essential problem involving interaction of the soil skeleton or matrix with the water contained in the pores. This problem is of extreme importance in earthquake engineering and here again solution

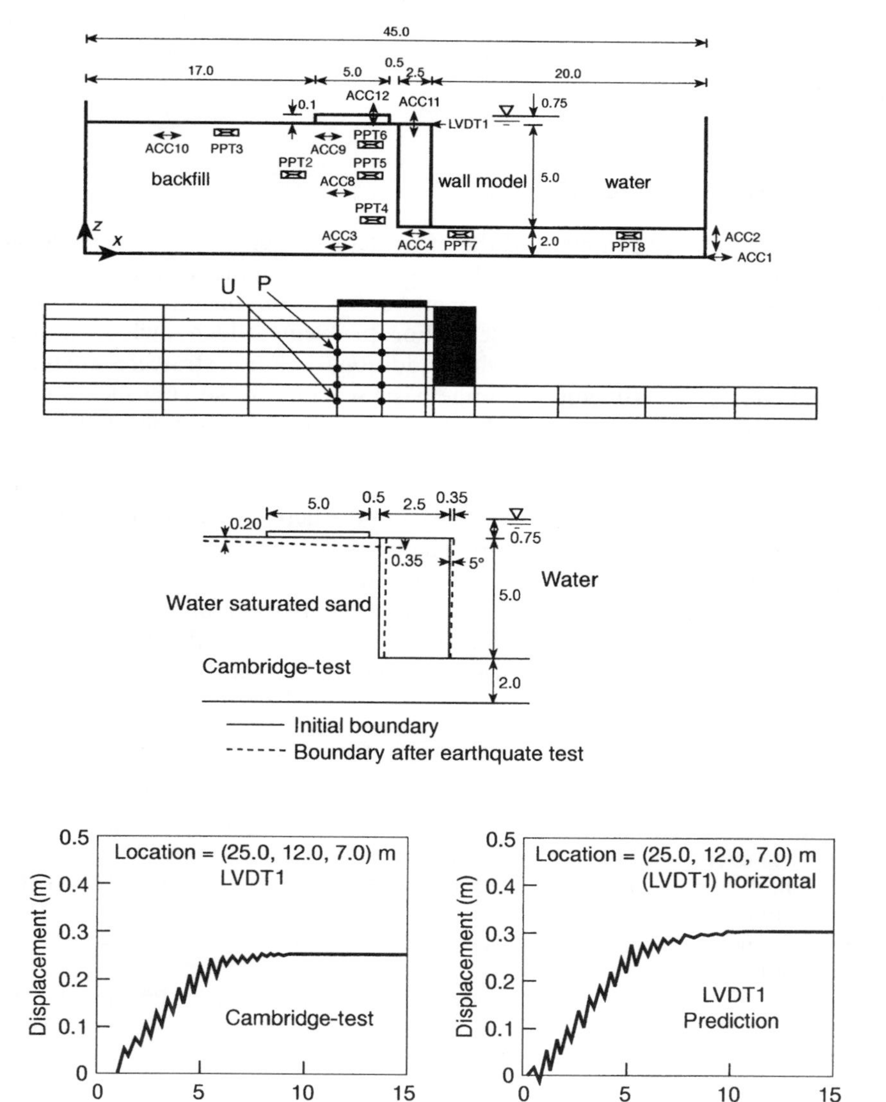

Fig. 1.5 Retaining wall subjected to earthquake excitation: comparison of experiment (centrifuge) and calculations.[18]

of transient non-linear equations is necessary. As in the mixed problem which we referred to earlier, the variables include displacement, and the pore pressure in the fluid p.

In Chapter 19 of Volume 1, we have in fact shown a comparison between some centrifuge results and computations showing the development of the pore pressure arising from a particular form of the constitutive relation assumed. Many such examples and indeed the full theory are given in a recent text,[18] and in Fig. 1.5 we show an example of comparison of calculations and a centrifuge model presented at a 1993 workshop known as VELACS[19]. This figure shows the displacements of a big retaining wall after the passage of an earthquake, which were measured in the centrifuge and also calculated.

1.5 Concluding remarks

In this chapter we have summarized the basic steps needed to solve a general small-strain solid mechanics problem as well as the quasi-harmonic field problem. Only a standard Newton–Raphson solution method has been mentioned to solve the resulting non-linear algebraic problem. For problems which include non-linear behaviour there are many situations where additional solution strategies are required. In the next chapter we will consider some basic schemes for solving such non-linear algebraic problems. In subsequent chapters we shall address some of these in the context of particular problems classes.

The reader will note that, except in the example solutions, we have not discussed problems in which large strains occur. We can note here, however, that the solution strategy described above remains valid. The parts which change are associated with the effects of finite deformation on computing stresses and thus the stress-divergence term and resulting tangent moduli. As these aspects involve more advanced concepts we have deferred the treatment of finite strain problems to the latter part of the volume where we will address basic formulations and applications.

References

1. O.C. Zienkiewicz and R.L. Taylor. *The Finite Element Method: The Basis*, Volume 1. Arnold, London, 5th edition, 2000.
2. S.P. Timoshenko and J.N. Goodier. *Theory of Elasticity*, McGraw-Hill, New York, 3rd edition, 1969.
3. I.S. Sokolnikoff, *The Mathematical Theory of Elasticity*, McGraw-Hill, New York, 2nd edition, 1956.
4. L.E. Malvern. *Introduction to the Mechanics of a Continuous Medium*, Prentice-Hall, Englewood Cliffs, NJ, 1969.
5. A.P. Boresi and K.P. Chong. *Elasticity in Engineering Mechanics*, Elsevier, New York, 1987.
6. P.C. Chou and N.J. Pagano. *Elasticity: Tensor, Dyadic and Engineering Approaches*, Dover Publications, Mineola, NY, 1992; reprinted from the 1967 Van Nostrand edition.
7. I.H. Shames and F.A. Cozzarelli. *Elastic and Inelastic Stress Analysis*, Taylor & Francis, Washington, DC, 1997; revised printing.

8. J.C. Simo, R.L. Taylor and K.S. Pister. Variational and projection methods for the volume constraint in finite deformation plasticity. *Comp. Meth. Appl. Mech. Eng.*, **51**, 177–208, 1985.
9. K. Washizu. *Variational Methods in Elasticity and Plasticity*, Pergamon Press, New York, 3rd edition, 1982.
10. T.J.R. Hughes. Generalization of selective integration procedures to anisotropic and non-linear media. *Int. J. Num. Meth. Eng.*, **15**, 1413–18, 1980.
11. J.C. Simo and T.J.R. Hughes. On the variational foundations of assumed strain methods. *J. Appl. Mech.*, **53**(1), 51–4, 1986.
12. M. Lees. A linear three level difference scheme for quasilinear parabolic equations. *Maths. Comp.*, **20**, 516–622, 1966.
13. G. Comini, S. Del Guidice, R.W. Lewis and O.C. Zienkiewicz. Finite element solution of non-linear conduction problems with special reference to phase change. *Int. J. Num. Meth. Eng.*, **8**, 613–24, 1974.
14. H.D. Hibbitt and P.V. Marcal. Numerical thermo-mechanical model for the welding and subsequent loading of a fabricated structure. *Computers and Structures*, **3**, 1145–74, 1973.
15. K. Morgan, R.W. Lewis and O.C. Zienkiewicz. An improved algorithm for heat convection problems with phase change. *Int. J. Num. Meth. Eng.*, **12**, 1191–95, 1978.
16. C.A. Anderson and O.C. Zienkiewicz. Spontaneous ignition: finite element solutions for steady and transient conditions. *Trans ASME, J. Heat Transfer*, 398–404, 1974.
17. J.R.H. Otter, E. Cassel and R.E. Hobbs. Dynamic relaxation. *Proc. Inst. Civ. Eng.*, **35**, 633–56, 1966.
18. O.C. Zienkiewicz, A.H.C. Chan, M. Pastor and B.A. Schrefler. *Computational Geomechanics: With Special Reference to Earthquake Engineering*, John Wiley, Chichester, Sussex, 1999.
19. K. Arulanandan and R.F. Scott (eds). *Proceedings of VELACS Symposium*, Balkema, Rotterdam, 1993.

2

Solution of non-linear algebraic equations

2.1 Introduction

In the solution of linear problems by a finite element method we always need to solve a set of simultaneous algebraic equations of the form

$$\mathbf{Ka} = \mathbf{f} \tag{2.1}$$

Provided the coefficient matrix is non-singular the solution to these equations is unique. In the solution of non-linear problems we will always obtain a set of algebraic equations; however, they generally will be non-linear, which we indicate as

$$\boldsymbol{\Psi}(\mathbf{a}) = \mathbf{f} - \mathbf{P}(\mathbf{a}) = \mathbf{0} \tag{2.2}$$

where $\mathbf{a}$ is the set of discretization parameters, $\mathbf{f}$ a vector which is independent of the parameters and $\mathbf{P}$ a vector dependent on the parameters. These equations may have multiple solutions [i.e. more than one set of $\mathbf{a}$ may satisfy Eq. (2.2)]. Thus, if a solution is achieved it may not necessarily be the solution sought. Physical insight into the nature of the problem and, usually, small-step incremental approaches from known solutions are essential to obtain realistic answers. Such increments are indeed always required if the constitutive law relating stress and strain changes is path dependent or if the load-displacement path has bifurcations or multiple branches at certain load levels.

The general problem should always be formulated as the solution of

$$\boldsymbol{\Psi}_{n+1} = \boldsymbol{\Psi}(\mathbf{a}_{n+1}) = \mathbf{f}_{n+1} - \mathbf{P}(\mathbf{a}_{n+1}) = \mathbf{0} \tag{2.3}$$

which starts from a nearby solution at

$$\mathbf{a} = \mathbf{a}_n, \qquad \boldsymbol{\Psi}_n = \mathbf{0}, \qquad \mathbf{f} = \mathbf{f}_n \tag{2.4}$$

and often arises from changes in the forcing function $\mathbf{f}_n$ to

$$\mathbf{f}_{n+1} = \mathbf{f}_n + \Delta\mathbf{f}_n \tag{2.5}$$

The determination of the change $\Delta\mathbf{a}_n$ such that

$$\mathbf{a}_{n+1} = \mathbf{a}_n + \Delta\mathbf{a}_n \tag{2.6}$$

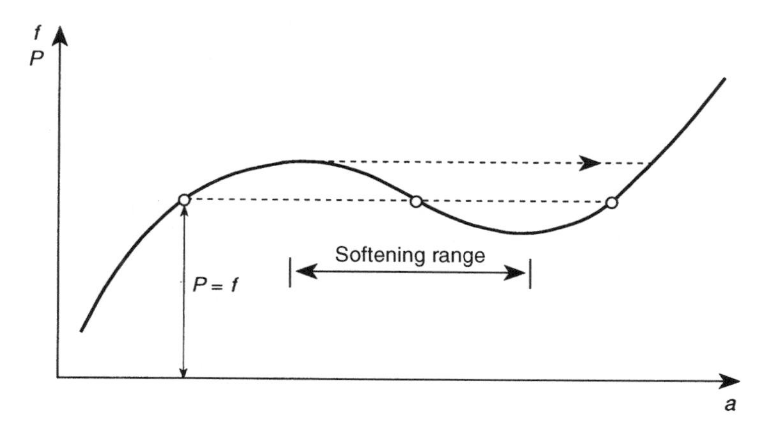

Fig. 2.1 Possibility of multiple solutions.

will be the objective and generally the increments of $\Delta\mathbf{f}_n$ will be kept reasonably small so that path dependence can be followed. Further, such incremental procedures will be useful in avoiding excessive numbers of iterations and in following the physically correct path. In Fig. 2.1 we show a typical non-uniqueness which may occur if the function $\boldsymbol{\Psi}$ decreases and subsequently increases as the parameter **a** uniformly increases. It is clear that to follow the path $\Delta\mathbf{f}_n$ will have both positive and negative signs during a complete computation process.

It is possible to obtain solutions in a single increment of **f** only in the case of mild non-linearity (and no path dependence), that is, with

$$\mathbf{f}_n = \mathbf{0}, \qquad \Delta\mathbf{f}_n = \mathbf{f}_{n+1} = \mathbf{f} \tag{2.7}$$

The literature on general solution approaches and on particular applications is extensive and, in a single chapter, it is not possible to encompass fully all the variants which have been introduced. However, we shall attempt to give a comprehensive picture by outlining first the *general* solution procedures.

In later chapters we shall focus on procedures associated with rate-independent material non-linearity (plasticity), rate-dependent material non-linearity (creep and visco-plasticity), some non-linear field problems, large displacments and other special examples.

2.2 Iterative techniques

2.2.1 General remarks

The solution of the problem posed by Eqs (2.3)–(2.6) cannot be approached directly and some form of iteration will always be required. We shall concentrate here on procedures in which repeated solution of linear equations (i.e. iteration) of the form

$$\mathbf{K}^i\, d\mathbf{a}_n^i = \mathbf{r}_{n+1}^i \tag{2.8}$$

in which a superscript i indicates the iteration number. In these a solution increment $d\mathbf{a}_n^i$ is computed.* Gaussian elimination techniques of the type discussed in Volume 1 can be used to solve the linear equations associated with each iteration. However, the application of an iterative solution method may prove to be more economical, and in later chapters we shall frequently refer to such possibilities although they have not been fully explored.

Many of the iterative techniques currently used to solve non-linear problems originated by intuitive application of physical reasoning. However, each of such techniques has a direct association with methods in numerical analysis, and in what follows we shall use the nomenclature generally accepted in texts on this subject.[1–5]

Although we state each algorithm for a set of non-linear algebraic equations, we shall illustrate each procedure by using a single scalar equation. This, though useful from a pedagogical viewpoint, is dangerous as convergence of problems with numerous degrees of freedom may depart from the simple pattern in a single equation.

2.2.2 The Newton–Raphson method

The Newton–Raphson method is the most rapidly convergent process for solutions of problems in which only one evaluation of $\boldsymbol{\Psi}$ is made in each iteration. Of course, this assumes that the initial solution is within the *zone of attraction* and, thus, divergence does not occur. Indeed, the Newton–Raphson method is the only process described here in which the asymptotic rate of convergence is quadratic. The method is sometimes simply called Newton's method but it appears to have been simultaneously derived by Raphson, and an interesting history of its origins is given in reference 6.

In this iterative method we note that, to the first order, Eq. (2.3) can be approximated as

$$\boldsymbol{\Psi}(\mathbf{a}_{n+1}^{i+1}) \approx \boldsymbol{\Psi}(\mathbf{a}_{n+1}^{i}) + \left(\frac{\partial \boldsymbol{\Psi}}{\partial \mathbf{a}}\right)_{n+1}^{i} d\mathbf{a}_n^i = \mathbf{0} \tag{2.9}$$

Here the iteration counter i usually starts by assuming

$$\mathbf{a}_{n+1}^{1} = \mathbf{a}_n \tag{2.10}$$

in which $\mathbf{a}_n$ is a converged solution at a previous load level or time step. The jacobian matrix (or in structural terms the stiffness matrix) corresponding to a tangent direction is given by

$$\mathbf{K}_{\mathrm{T}} = \frac{\partial \mathbf{P}}{\partial \mathbf{a}} = -\frac{\partial \boldsymbol{\Psi}}{\partial \mathbf{a}} \tag{2.11}$$

Equation (2.9) gives immediately the iterative correction as

$$\mathbf{K}_{\mathrm{T}}^{i}\, d\mathbf{a}_n^i = \boldsymbol{\Psi}_{n+1}^{i}$$

* Note the difference between a solution increment $d\mathbf{a}$ and a differential $\mathrm{d}\mathbf{a}$.

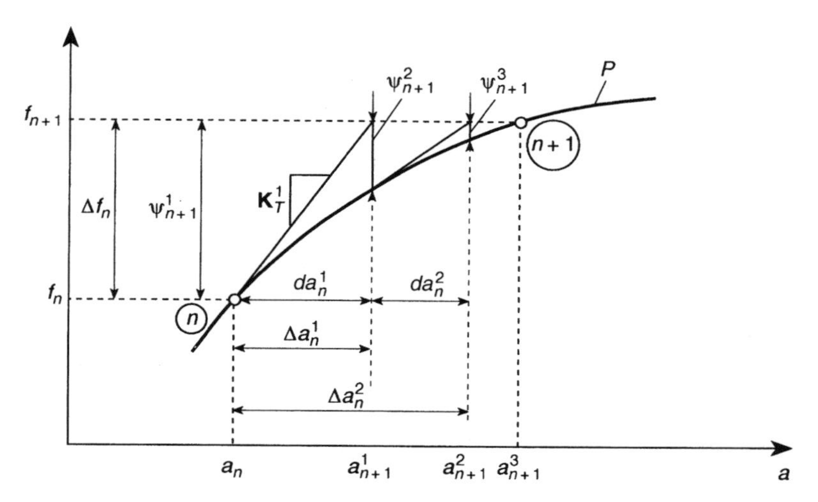

Fig. 2.2 The Newton–Raphson method.

or

$$d\mathbf{a}_n^i = \left(\mathbf{K}_\mathrm{T}^i\right)^{-1}\mathbf{\Psi}_{n+1}^i \tag{2.12}$$

A series of successive approximations gives

$$\begin{aligned}\mathbf{a}_{n+1}^{i+1} &= \mathbf{a}_{n+1}^i + d\mathbf{a}_{n+1}^i \\ &= \mathbf{a}_n + \Delta\mathbf{a}_n^i\end{aligned} \tag{2.13}$$

where

$$\Delta\mathbf{a}_n^i = \sum_{k=1}^{i} d\mathbf{a}_n^k \tag{2.14}$$

The process is illustrated in Fig. 2.2 and shows the very rapid convergence that can be achieved.

The need for the introduction of the total increment $\Delta\mathbf{a}_n^i$ is perhaps not obvious here but in fact it is essential if the solution process is path dependent, as we shall see in Chapter 3 for some non-linear constitutive equations of solids.

The Newton–Raphson process, despite its rapid convergence, has some negative features:

1. a new $\mathbf{K}_\mathrm{T}$ matrix has to be computed at each iteration;
2. if direct solution for Eq. (2.12) is used the matrix needs to be factored at each iteration;
3. on some occasions the tangent matrix is symmetric at a solution state but unsymmetric otherwise (e.g. in some schemes for integrating large rotation parameters[7] or non-associated plasticity). In these cases an unsymmetric solver is needed in general.

Some of these drawbacks are absent in alternative procedures, although generally then a quadratic asymptotic rate of convergence is lost.

2.2.3 Modified Newton–Raphson method

This method uses essentially the same algorithm as the Newton–Raphson process but replaces the variable jacobian matrix $\mathbf{K}_T^i$ by a constant approximation

$$\mathbf{K}_T^i \approx \bar{\mathbf{K}}_T \tag{2.15}$$

giving in place of Eq. (2.12),

$$d\mathbf{a}_n^i = \bar{\mathbf{K}}_T^{-1} \boldsymbol{\Psi}_{n+1}^i \tag{2.16}$$

Many possible choices exist here. For instance $\bar{\mathbf{K}}_T$ can be chosen as the matrix corresponding to the first iteration $\mathbf{K}_T^1$ [as shown in Fig. 2.3(a)] or may even be one corresponding to some previous time step or load increment $\mathbf{K}^0$ [as shown in Fig. 2.3(b)]. In the context of solving problems in solid mechanics the method is also known as the *stress transfer* or *initial stress method*. Alternatively, the approximation can be chosen every few iterations as $\bar{\mathbf{K}}_T = \mathbf{K}_T^j$ where $j \leqslant i$.

Obviously, the procedure generally will converge at a slower rate (generally a norm of the residual $\boldsymbol{\Psi}$ has linear asymptotic convergence instead of the quadratic one in the full Newton–Raphson method) but some of the difficulties mentioned above for the Newton–Raphson process disappear. However, some new difficulties can also arise as this method fails to converge when the tangent used has opposite 'slope' to the one at the current solution (e.g. as shown by regions with different slopes in Fig. 2.1). Frequently the 'zone of attraction' for the modified process is increased and previously divergent approaches can be made to converge, albeit slowly. Many variants of this process can be used and symmetric solvers often can be employed when a symmetric form of $\bar{\mathbf{K}}_T$ is chosen.

2.2.4 Incremental-secant or quasi-Newton methods

Once the first iteration of the preceding section has been established giving

$$d\mathbf{a}_n^1 = \bar{\mathbf{K}}_T^{-1} \boldsymbol{\Psi}_{n+1}^1 \tag{2.17}$$

a secant 'slope' can be found, as shown in Fig. 2.4, such that

$$d\mathbf{a}_n^1 = \left(\mathbf{K}_s^2\right)^{-1} \left(\boldsymbol{\Psi}_{n+1}^1 - \boldsymbol{\Psi}_{n+1}^2\right) \tag{2.18}$$

This 'slope' can now be used to establish $\mathbf{a}_n^2$ by using

$$d\mathbf{a}_n^2 = \left(\mathbf{K}_s^2\right)^{-1} \boldsymbol{\Psi}_{n+1}^2 \tag{2.19}$$

Quite generally, one could write in place of Eq. (2.19) for $i > 1$, now dropping subscripts,

$$d\mathbf{a}^i = \left(\mathbf{K}_s^i\right)^{-1} \boldsymbol{\Psi}^i \tag{2.20}$$

where $(\mathbf{K}_s^i)^{-1}$ is determined so that

$$d\mathbf{a}^{i-1} = \left(\mathbf{K}_s^i\right)^{-1} \left(\boldsymbol{\Psi}^{i-1} - \boldsymbol{\Psi}^i\right) = \left(\mathbf{K}_s^i\right)^{-1} \boldsymbol{\gamma}^{i-1} \tag{2.21}$$

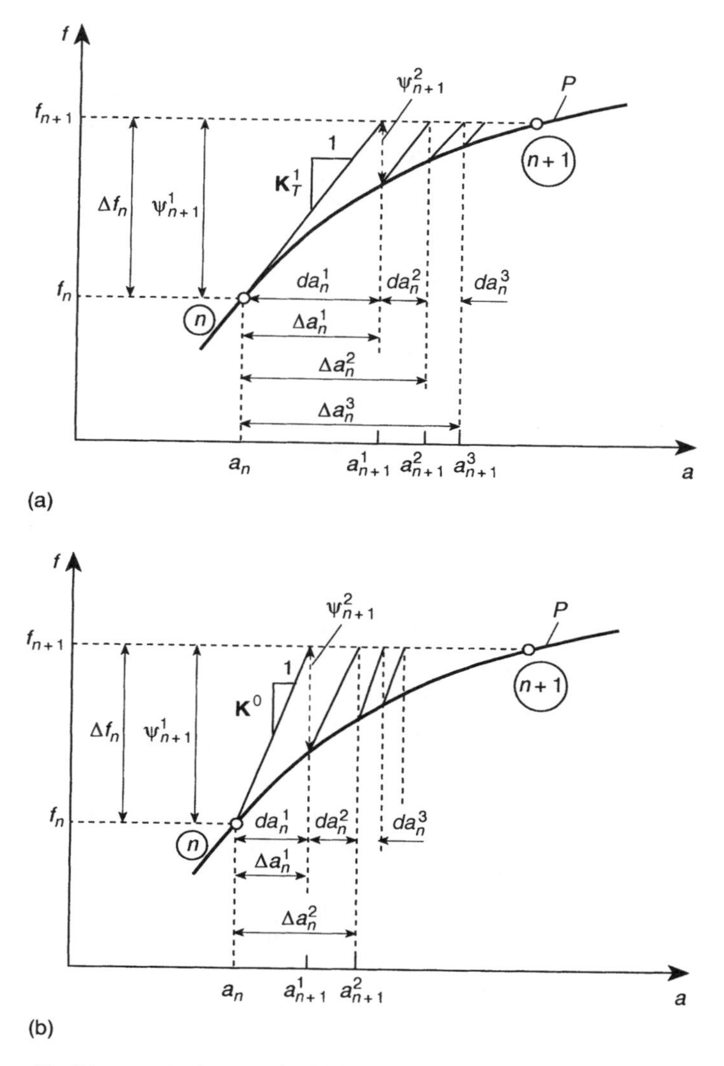

Fig. 2.3 The modified Newton–Raphson method: (a) with initial tangent in increment; (b) with initial problem tangent.

For the scalar system illustrated in Fig. 2.4 the determination of $\mathbf{K}_s^i$ is trivial and, as shown, the convergence is much more rapid than in the modified Newton–Raphson process (generally a super-linear asymptotic convergence rate is achieved for a norm of the residual).

For systems with more than one degree of freedom the determination of $\mathbf{K}_s^i$ or its inverse is more difficult and is not unique. Many different forms of the matrix $\mathbf{K}_s^i$ can satisfy relation (2.1) and, as expected, many alternatives are used in practice. All of these use some form of updating of a previously determined matrix or of its inverse in a manner that satisfies identically Eq. (2.21). Some such updates preserve the matrix symmetry whereas others do not. Any of the methods which begin with

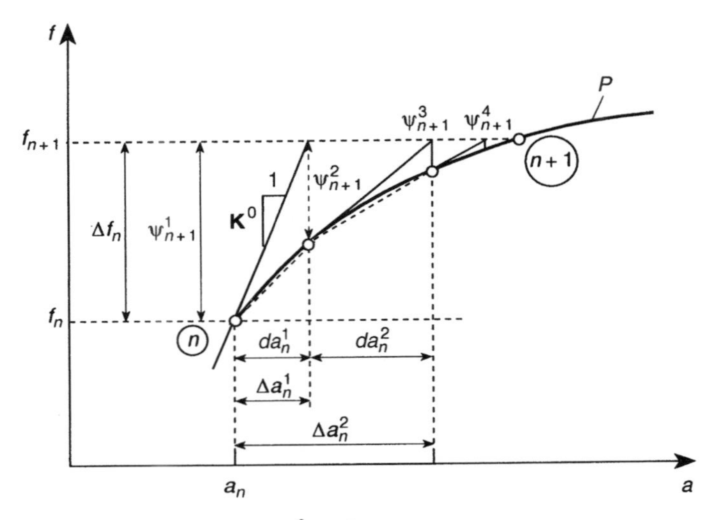

Fig. 2.4 The secant method starting from a $\mathbf{K}^0$ prediction.

a symmetric tangent can avoid the difficulty of non-symmetric matrix forms that arise in the Newton–Raphson process and yet achieve a faster convergence than is possible in the modified Newton–Raphson procedures.

Such secant update methods appear to stem from ideas introduced first by Davidon[8] and developed later by others. Dennis and More[9] survey the field extensively, while Matthies and Strang[10] appear to be the first to use the procedures in the finite element context. Further work and assessment of the performance of various update procedures is available in references 11–14.

The BFGS update[9] (named after Broyden, Fletcher, Goldfarb and Shanno) and the DFP update[9] (Davidon, Fletcher and Powell) preserve matrix symmetry and positive definiteness and both are widely used. We summarize below a step of the BFGS update for the inverse, which can be written as

$$(\mathbf{K}^i)^{-1} = (\mathbf{I} + \mathbf{w}_i \mathbf{v}_i^{\mathrm{T}})(\mathbf{K}^{i-1})^{-1}(\mathbf{I} + \mathbf{v}_i \mathbf{w}_i^{\mathrm{T}}) \tag{2.22}$$

where $\mathbf{I}$ is an identity matrix and

$$\begin{aligned} \mathbf{v}_i &= \left[1 - \frac{(d\mathbf{a}^{i-1})^{\mathrm{T}} \boldsymbol{\gamma}^{i-1}}{d(\mathbf{a}^i)^{\mathrm{T}} \boldsymbol{\Psi}^{i-1}}\right] \boldsymbol{\Psi}^{i-1} - \boldsymbol{\Psi}^i \\ \mathbf{w}_i &= \frac{1}{d\mathbf{a}^{(i-1)\mathrm{T}} \boldsymbol{\gamma}^{i-1}} d\mathbf{a}^{i-1} \end{aligned} \tag{2.23}$$

where $\boldsymbol{\gamma}$ is defined by Eq. (2.21). Some algebra will readily verify that substitution of Eqs (2.22) and (2.23) into Eq. (2.21) results in an identity. Further, the form of Eq. (2.22) guarantees preservation of the symmetry of the original matrix.

The nature of the update does not preserve any sparsity in the original matrix. For this reason it is convenient at every iteration to return to the original (sparse) matrix $\mathbf{K}_s^1$, used in the first iteration and to reapply the multiplication of Eq. (2.22) through

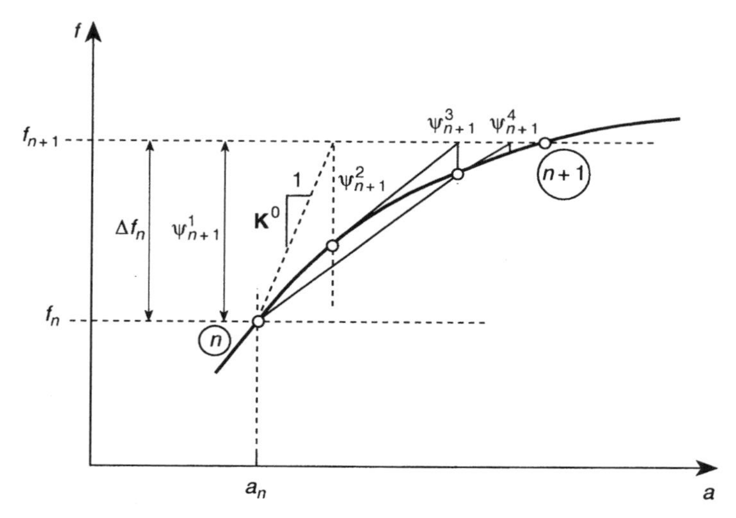

Fig. 2.5 Direct (or Picard) iteration.

all previous iterations. This gives the algorithm in the form

$$\mathbf{b}_1 = \prod_{j=2}^{i} \left(\mathbf{I} + \mathbf{v}_j \mathbf{w}_j^{\mathrm{T}}\right) \boldsymbol{\Psi}^i$$

$$\mathbf{b}_2 = \left(\mathbf{K}_{\mathrm{s}}^1\right)^{-1} \mathbf{b}_1 \tag{2.24}$$

$$d\mathbf{a}^i = \prod_{j=0}^{i-2} \left(\mathbf{I} + \mathbf{w}_{i-j} \mathbf{v}_{i-j}^{\mathrm{T}}\right) \mathbf{b}_2$$

This necessitates the storage of the vectors $\mathbf{v}_j$ and $\mathbf{w}_j$ for all previous iterations and their successive multiplications. Further details on the operations are described well in reference 10.

When the number of iterations is large ($i > 15$) the efficiency of the update decreases as a result of incipient instability. Various procedures are open at this stage, the most effective being the recomputation and factorization of a tangent matrix at the current solution estimate and restarting the process again.

Another possibility is to disregard *all* the previous updates and return to the original matrix $\mathbf{K}_{\mathrm{s}}^1$. Such a procedure was first suggested by Crisfield[11,15,16] in the finite element context and is illustrated in Fig. 2.5. It is seen to be convergent at a slightly slower rate but avoids totally the stability difficulties previously encountered and reduces the storage and number of operations needed. Obviously any of the secant update methods can be used here.

The procedure of Fig. 2.5 is identical to that generally known as direct (or Picard) iteration[1] and is particularly useful in the solution of non-linear problems which can be written as

$$\boldsymbol{\Psi}(\mathbf{a}) \equiv \mathbf{f} - \mathbf{K}(\mathbf{a})\mathbf{a} = \mathbf{0} \tag{2.25}$$

In such a case $\mathbf{a}_{n+1}^1 = \mathbf{a}_n$ is taken and the iteration proceeds as

$$\mathbf{a}_{n+1}^{i+1} = \left[\mathbf{K}(\mathbf{a}_{n+1}^i)\right]^{-1} \mathbf{f}_{n+1} \tag{2.26}$$

2.2.5 Line search procedures – acceleration of convergence

All the iterative methods of the preceding section have an identical structure described by Eqs (2.12)–(2.14) in which various approximations to the Newton matrix $\mathbf{K}_T^i$ are used. For all of these an iterative vector is determined and the new value of the unknowns found as

$$\mathbf{a}_{n+1}^{i+1} = \mathbf{a}_{n+1}^{i} + d\mathbf{a}_n^i \tag{2.27}$$

starting from

$$\mathbf{a}_{n+1}^{1} = \mathbf{a}_n$$

in which $\mathbf{a}_n$ is the known (converged) solution at the previous time step or load level. The objective is to achieve the reduction of $\boldsymbol{\Psi}_{n+1}^{i+1}$ to zero, although this is not always easily achieved by any of the procedures described even in the scalar example illustrated. To get a solution approximately satisfying such a scalar non-linear problem would have been in fact easier by simply evaluating the scalar $\boldsymbol{\Psi}_{n+1}^{i+1}$ for various values of $\mathbf{a}_{n+1}$ and by suitable interpolation arriving at the required answer. For multi-degree-of-freedom systems such an approach is obviously not possible unless some scalar norm of the residual is considered. One possible approach is to write

$$\mathbf{a}_{n+1}^{i+1,j} = \mathbf{a}_{n+1}^{i} + \eta_{i,j}\, d\mathbf{a}_n^i \tag{2.28}$$

and determine the *step size* $\eta_{i,j}$ so that a projection of the residual on the *search direction* $d\mathbf{a}_n^i$ is made zero. We could define this projection as

$$G_{i,j} \equiv \left(d\mathbf{a}_n^i\right)^{\mathrm{T}} \boldsymbol{\Psi}_{n+1}^{i+1,j} \tag{2.29}$$

where

$$\boldsymbol{\Psi}_{n+1}^{i+1,j} \equiv \boldsymbol{\Psi}\left(\mathbf{a}_{n+1}^{i} + \eta_{i,j}\, d\mathbf{a}_n^i\right), \qquad \eta_{i,0} = 1$$

Here, of course, other norms of the residual could be used.

This process is known as a *line search*, and $\eta_{i,j}$ can conveniently be obtained by using a *regula falsi* (or secant) procedure as illustrated in Fig 2.6. An obvious

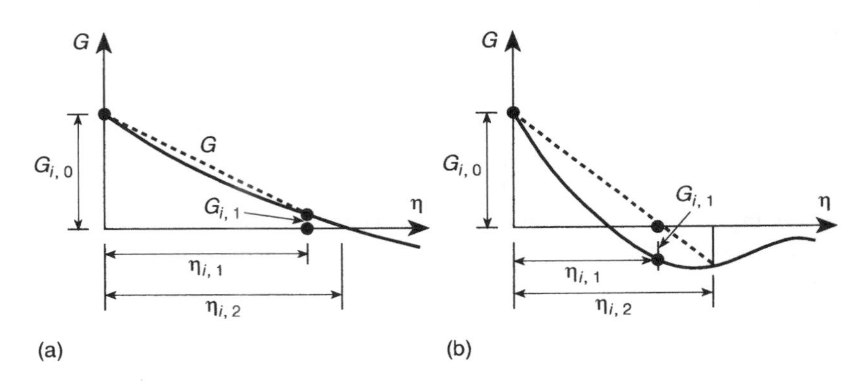

Fig. 2.6 *Regula falsi* applied to line search: (a) extrapolation; (b) interpolation.

disadvantage of a line search is the need for several evaluations of $\boldsymbol{\Psi}$. However, the acceleration of the overall convergence can be remarkable when applied to modified or quasi-Newton methods. Indeed, line search is also useful in the full Newton method by making the radius of attraction larger. A compromise frequently used[10] is to undertake the search only if

$$G_{i,0} > \varepsilon \left(d\mathbf{a}_n^i\right)^{\mathrm{T}} \boldsymbol{\Psi}_{n+1}^{i+1,j} \tag{2.30}$$

where the tolerance ε is set between 0.5 and 0.8. This means that if the iteration process directly resulted in a reduction of the residual to ε or less of its original value a line search is not used.

2.2.6 'Softening' behaviour and displacement control

In applying the preceding to load control problems we have implicitly assumed that the iteration is associated with positive increments of the forcing vector, **f**, in Eq. (2.5). In some structural problems this is a set of loads that can be assumed to be proportional to each other, so that one can write

$$\Delta\mathbf{f}_n = \Delta\lambda_n \mathbf{f}_0 \tag{2.31}$$

In many problems the situation will arise that no solution exists above a certain maximum value of **f** and that the real solution is a 'softening' branch, as shown in Fig. 2.1. In such cases $\Delta\lambda_n$ will need to be negative unless the problem can be recast as one in which the forcing can be applied by displacement control. In a simple case of a single load it is easy to recast the general formulation to increments of a single prescribed displacement and much effort has gone into such solutions.[11,17–23]

In all the successful approaches of incrementation of $\Delta\lambda_n$ the original problem of Eq. (2.3) is rewritten as the solution of

$$\boldsymbol{\Psi}_{n+1} \equiv \lambda_{n+1}\mathbf{f}_0 - \mathbf{P}(\mathbf{a}_{n+1}) = \mathbf{0}$$

with

$$\mathbf{a}_{n+1} = \mathbf{a}_n + \Delta\mathbf{a}_n \tag{2.32}$$

and

$$\lambda_{n+1} = \lambda_n + \Delta\lambda_n$$

being included as variables in any increment. Now an additional equation (constraint) needs to be provided to solve for the extra variable $\Delta\lambda_n$.

This additional equation can take various forms. Riks[19] assumes that in each increment

$$\Delta\mathbf{a}_n^{\mathrm{T}}\Delta\mathbf{a}_n + \Delta\lambda^2 \mathbf{f}_0^{\mathrm{T}}\mathbf{f}_0 = \Delta l^2 \tag{2.33}$$

where Δl is a prescribed 'length' in the space of $n+1$ dimensions. Crisfield[11,24] provides a more natural control on displacements, requiring that

$$\Delta\mathbf{a}_n^{\mathrm{T}}\Delta\mathbf{a}_n = \Delta l^2 \tag{2.34}$$

These so-called arc-length and spherical path controls are but some of the possible constraints.

Direct addition of the constraint Eqs (2.33) or (2.34) to the system of Eqs (2.32) is now possible and the previously described iterative methods could again be used. However, the 'tangent' equation system would always lose its symmetry so an alternative procedure is generally used.

We note that for a given iteration i we can write quite generally the solution as

$$\begin{aligned}\boldsymbol{\Psi}_{n+1}^{i} &= \lambda_{n+1}^{i}\mathbf{f}_0 - \mathbf{P}(\mathbf{a}_{n+1}^{i})\\ \boldsymbol{\Psi}_{n+1}^{i+1} &\approx \boldsymbol{\Psi}_{n+1}^{i} + d\lambda_{n+1}^{i}\mathbf{f}_0 - \mathbf{K}_{\mathrm{T}}^{i}\,d\mathbf{a}_n^{i}\end{aligned} \tag{2.35}$$

The solution increment for $\mathbf{a}$ may now be given as

$$\begin{aligned}d\mathbf{a}_n^{i} &= \left(\mathbf{K}_{\mathrm{T}}^{i}\right)^{-1}\left[\boldsymbol{\Psi}_{n+1}^{i} + d\lambda_n^{i}\mathbf{f}_0\right]\\ d\mathbf{a}_n^{i} &= d\breve{\mathbf{a}}_n^{i} + d\lambda_n^{i}\,d\hat{\mathbf{a}}_n^{i}\end{aligned} \tag{2.36}$$

where

$$\begin{aligned}d\breve{\mathbf{a}}_n^{i} &= \left(\mathbf{K}_{\mathrm{T}}^{i}\right)^{-1}\boldsymbol{\Psi}_{n+1}^{i}\\ d\hat{\mathbf{a}}_n^{i} &= \left(\mathbf{K}_{\mathrm{T}}^{i}\right)^{-1}\mathbf{f}_0\end{aligned} \tag{2.37}$$

Now an additional equation is cast using the constraint. Thus, for instance, with Eq. (2.34) we have

$$\left(\Delta\mathbf{a}_n^{i-1} + d\mathbf{a}_n^{i}\right)^{\mathrm{T}}\left(\Delta\mathbf{a}_n^{i-1} + d\mathbf{a}_n^{i}\right) = \Delta l^2 \tag{2.38}$$

where $\Delta\mathbf{a}_n^{i-1}$ is defined by Eq. (2.14). On substitution of Eq. (2.36) into Eq. (2.38) a quadratic equation is available for the solution of the remaining unknown $d\lambda_n^i$ (which may well turn out to be negative). Additional details may be found in references 11 and 24.

A procedure suggested by Bergan[20,23] is somewhat different from those just described. Here a fixed load increment $\Delta\lambda_n$ is first assumed and any of the previously introduced iterative procedures are used for calculating the increment $d\mathbf{a}_n^i$. Now a new increment $\Delta\lambda_n^*$ is calculated so that it minimizes a norm of the residual

$$\left[\left(\Delta\lambda_n^*\mathbf{f}_0 - \mathbf{P}_{n+1}^{i+1}\right)^{\mathrm{T}}\left(\Delta\lambda_n^*\mathbf{f}_0 - \mathbf{P}_{n+1}^{i+1}\right)\right] = \Delta l^2 \tag{2.39}$$

The result is thus computed from

$$\frac{\mathrm{d}\Delta l^2}{\mathrm{d}\Delta\lambda_n^*} = 0$$

and yields the solution

$$\Delta\lambda_n^* = \frac{\mathbf{f}_0^{\mathrm{T}}\mathbf{P}_{n+1}^{i+1}}{\mathbf{f}_0^{\mathrm{T}}\mathbf{f}_0} \tag{2.40}$$

This quantity may again well be negative, requiring a load decrease, and it indeed results in a rapid residual reduction in all cases, but precise control of displacement magnitudes becomes more difficult. The interpretation of the Bergan method in a one-dimensional example, shown in Fig. 2.7, is illuminating. Here it gives the exact answers – with a displacement control, the magnitude of which is determined by the initial $\Delta\lambda_n$ assumed to be the slope $\mathbf{K}_{\mathrm{T}}$ used in the first iteration.

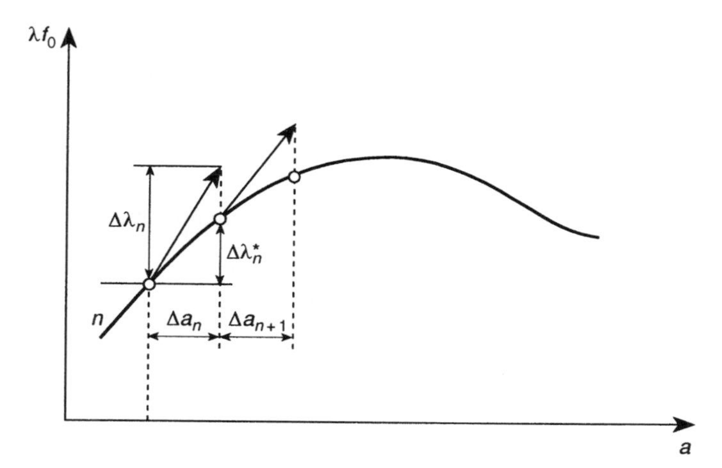

Fig. 2.7 One-dimensional interpretation of the Bergan procedure.

2.2.7 Convergence criteria

In all the iterative processes described the numerical solution is only approximately achieved and some tolerance limits have to be set to terminate the iteration. Since finite precision arithmetic is used in all computer calculations, one can never achieve a better solution than the round-off limit of the calculations.

Frequently, the criteria used involve a norm of the displacement parameter changes $||d\mathbf{a}_n^i||$ or, more logically, that of the residuals $||\boldsymbol{\Psi}_{n+1}^i||$. In the latter case the limit can often be expressed as some tolerance of the norm of forces $||\mathbf{f}_{n+1}||$. Thus, we may require that

$$||\boldsymbol{\Psi}_{n+1}^i|| \leqslant \varepsilon ||\mathbf{f}_{n+1}|| \tag{2.41}$$

where ε is chosen as a small number, and

$$||\boldsymbol{\Psi}|| = \left(\boldsymbol{\Psi}^{\mathrm{T}}\boldsymbol{\Psi}\right)^{1/2} \tag{2.42}$$

Other alternatives exist for choosing the comparison norm, and another option is to use the residual of the first iteration as a basis. Thus,

$$||\boldsymbol{\Psi}_{n+1}^i|| \leqslant \varepsilon ||\boldsymbol{\Psi}_{n+1}^1|| \tag{2.43}$$

The error due to the incomplete solution of the discrete non-linear equations is of course additive to the error of the discretization that we frequently measure in the energy norm (see Chapter 14 of Volume 1). It is possible therefore to use the same norm for bounding of the iteration process. We could, as a third option, require that the error in the energy norm satisfy

$$\begin{aligned} dE^i &= \left(d\mathbf{a}_{n+1}^{i,\mathrm{T}}\boldsymbol{\Psi}_{n+1}^i\right)^{1/2} \leqslant \varepsilon\left(d\mathbf{a}_{n+1}^{1,\mathrm{T}}\boldsymbol{\Psi}_{n+1}^1\right)^{1/2} \\ &\leqslant \varepsilon\, dE^1 \end{aligned} \tag{2.44}$$

In each of the above forms, problem types exist where the right-hand-side norm is zero. Thus a fourth form, which is quite general, is to compute the norm of the element residuals. If the problem residual is obtained as a sum over elements as

$$\boldsymbol{\Psi}_{n+1} = \sum_e \boldsymbol{\psi}^e_{n+1} \tag{2.45}$$

where e denotes an individual element and $\boldsymbol{\psi}^e$ the residual from each element, we can express the convergence criterion as

$$||\boldsymbol{\Psi}^i_{n+1}|| \leqslant \varepsilon ||\boldsymbol{\psi}^e_{n+1}|| \tag{2.46}$$

where

$$||\boldsymbol{\psi}^e_{n+1}|| = \sum_e ||(\boldsymbol{\psi}^e_{n+1})^i|| \tag{2.47}$$

Once a criterion is selected the problem still remains to choose an appropriate value for ε. In cases where a full Newton scheme is used (and thus asymptotic quadratic convergence should occur) the tolerance may be chosen at half the machine precision. Thus if the precision of calculations is about 16 digits one may choose $\varepsilon = 10^{-8}$ since quadratic convergence assures that the next residual (in the absence of round-off) would achieve full precision. For modified or quasi-Newton schemes such asymptotic rates are not assured, necessitating more iterations to achieve high precision. In these cases it is common practice by some to use much larger tolerance values (say 0.01 to 0.001). However, for problems where large numbers of steps are taken, instability in the solution may occur if the convergence tolerance is too large. We recommend therefore that whenever practical a tolerance of half machine precision be used.

2.2.8 General remarks – incremental and rate methods

The various iterative methods described provide an essential tool-kit for the solution of non-linear problems in which finite element discretization has been used. The precise choice of the optimal methodology is problem dependent and although many comparative solution cost studies have been published[10,15,25] the differences are often marginal. There is little doubt, however, that exact Newton–Raphson processes (with line search) should be used when convergence is difficult to achieve. Also the advantage of symmetric update matrices in the quasi-Newton procedures frequently make these a very economical candidate. When non-symmetric tangent moduli exist it may be better to consider one of the non-symmetric updates, for example, a Broyden method.[11,26]

We have not discussed in the preceding *direct iterative* methods such as the various conjugate direction methods[27–31] or *dynamic relaxation* methods in which an explicit dynamic transient analysis (see Chapter 18 of Volume 1) is carried out to achieve a steady-state solution.[32,33] These forms are often characterized by:

1. a diagonal or very sparse form of the matrix used in computing trial increments $d\mathbf{a}$ (and hence very low cost of an iteration) and
2. a significant number of total iterations and hence evaluations of the residual $\boldsymbol{\Psi}$.

These opposing trends imply that such methods offer the potential to solve large problems efficiently. However, to date such general solution procedures are effective only in certain problems.[34]

One final remark concerns the size of increments $\Delta\mathbf{f}$ or $\Delta\lambda$ to be adopted. First, it is clear that small increments reduce the total number of iterations required per computational step, and in many applications automatic guidance on the size of the increment to preserve a (nearly) constant number of iterations is needed. Here such processes as the use of the 'current stiffness parameter' introduced by Bergan[20] can be effective.

Second, if the behaviour is *path dependent* (e.g. as in plasticity-type constitutive laws) the use of small increments is desirable to preserve accuracy in solution changes. In this context, we have already emphasized the need for calculating such changes by using always the accumulated $\Delta\mathbf{a}_n^i$ change and not in adding changes arising from each iterative $d\mathbf{a}_n^i$ step in an increment.

Third, if only a single Newton–Raphson iteration is used in each increment of $\Delta\lambda$ then the procedure is equivalent to the solution of a standard rate problem incrementally by direct forward integration. Here we note that if Eq. (2.3) is rewritten as

$$\mathbf{P}(\mathbf{a}) = \lambda\mathbf{f}_0 \tag{2.48}$$

we can, on differentiation with respect to λ obtain

$$\frac{\mathrm{d}\mathbf{P}}{\mathrm{d}\mathbf{a}}\frac{\mathrm{d}\mathbf{a}}{\mathrm{d}\lambda} = \mathbf{f}_0 \tag{2.49}$$

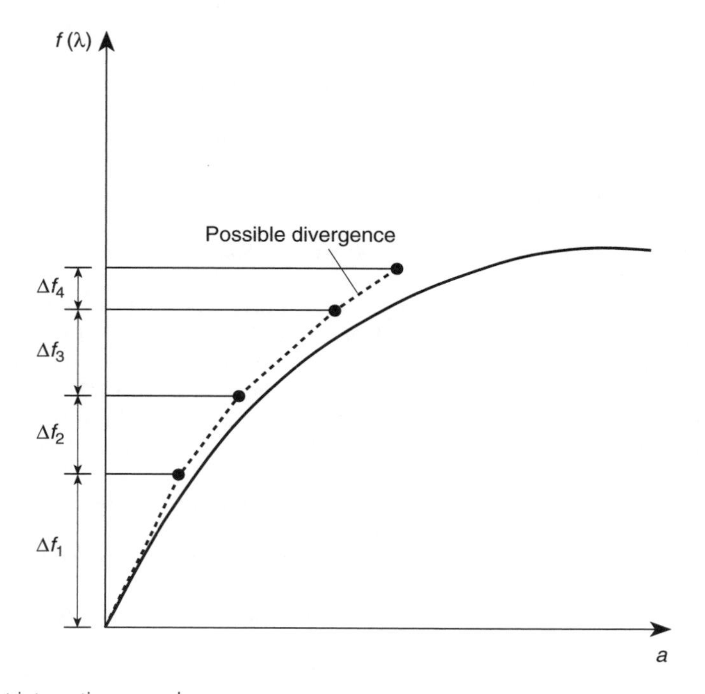

Fig. 2.8 Direct integration procedure.

and write this as

$$\frac{d\mathbf{a}}{d\lambda} = \mathbf{K}_T^{-1}\mathbf{f}_0 \tag{2.50}$$

Incrementally, this may be written in an explicit form by using an Euler method as

$$\Delta\mathbf{a}_n = \Delta\lambda\mathbf{K}_{Tn}^{-1}\mathbf{f}_0 \tag{2.51}$$

This direct integration is illustrated in Fig. 2.8 and can frequently be divergent as well as being only conditionally stable as a result of the Euler explicit method used. Obviously, other methods can be used to improve accuracy and stability. These include Euler implicit schemes and Runge–Kutta procedures.

References

1. A. Ralston. *A First Course in Numerical Analysis*, McGraw-Hill, New York, 1965.
2. L. Collatz. *The Numerical Treatment of Differential Equations*, Springer, Berlin, 1966.
3. G. Dahlquist and Å. Björck. *Numerical Methods*, Prentice-Hall, Englewood Cliffs, NJ, 1974.
4. H.R. Schwarz. *Numerical Analysis*, John Wiley, Chichester, Sussex, 1989.
5. J. Demmel. *Applied Numerical Linear Algebra*, SIAM, Philadelphia, PA, 1997.
6. N. Biĉaniĉ and K.W. Johnson. Who was 'Raphson'? *Int. J. Num. Meth. Eng.*, **14**, 148–52, 1979.
7. J.C. Simo and L. Vu-Quoc. A three-dimensional finite strain rod model. Part II: geometric and computational aspects. *Comp. Meth. Appl. Mech. Eng.*, **58**, 79–116, 1986.
8. W.C. Davidon. Variable metric method for minimization. Technical Report ANL-5990, Argonne National Laboratory, 1959.
9. J.E. Dennis and J. More. Quasi-Newton methods – motivation and theory. *SIAM Rev.*, **19**, 46–89, 1977.
10. H. Matthies and G. Strang. The solution of nonlinear finite element equations. *Int. J. Num. Meth. Eng.*, **14**, 1613–26, 1979.
11. M.A. Crisfield. *Non-linear Finite Element Analysis of Solids and Structures*. Volume 1, John Wiley, Chichester, Sussex, 1991.
12. M.A. Crisfield. *Non-linear Finite Element Analysis of Solids and Structures*. Volume 2, John Wiley, Chichester, Sussex, 1997.
13. K.-J. Bathe and A.P. Cimento. Some practical procedures for the solution of nonlinear finite element equations. *Comp. Meth. Appl. Mech. Eng.*, **22**, 59–85, 1980.
14. M. Geradin, S. Idelsohn and M. Hogge. Computational strategies for the solution of large nonlinear problems via quasi-Newton methods. *Comp. Struct.*, **13**, 73–81, 1981.
15. M.A. Crisfield. Finite element analysis for combined material and geometric nonlinearity. In W. Wunderlich, E. Stein and K.-J. Bathe (eds), *Nonlinear Finite Element Analysis in Structural Mechanics*, Springer-Verlag, Berlin, 1981.
16. M.A. Crisfield. A fast incremental/iterative solution procedure that handles 'snap through'. *Comp. Struct.*, **13**, 55–62, 1981.
17. T.H.H. Pian and P. Tong. Variational formulation of finite displacement analysis. *Symp. on High Speed Electronic Computation of Structures*, Liége, 1970.
18. O.C. Zienkiewicz. Incremental displacement in non-linear analysis. *Int. J. Num. Meth. Eng.*, **3**, 587–92, 1971.
19. E. Riks. An incremental approach to the solution of snapping and buckling problems. *International Journal of Solids and Structures*, **15**, 529–51, 1979.

20. P.G. Bergan. Solution algorithms for nonlinear structural problems. In *Int. Conf. on Engineering Applications of the Finite Element Method*, pages 13.1–13.39, Computas, 1979.
21. J.L. Batoz and G. Dhatt. Incremental displacement algorithms for nonlinear problems. *Int. J. Num. Meth. Eng.*, **14**, 1261–66, 1979.
22. E. Ramm. Strategies for tracing nonlinear response near limit points. In W. Wunderlich, E. Stein and K.-J. Bathe (eds), *Nonlinear Finite Element Analysis in Structural Mechanics*, pp. 63–89. Springer-Verlag, Berlin, 1981.
23. P. Bergan. Solution by iteration in displacement and load spaces. In W. Wunderlich, E. Stein and K.-J. Bathe (eds), *Nonlinear Finite Element Analysis in Structural Mechanics*, Springer-Verlag, Berlin, 1981.
24. M.A. Crisfield. Incremental/iterative solution procedures for nonlinear structural analysis. In C. Taylor, E. Hinton, D.R.J. Owen and E. Oñate (eds), *Numerical Methods for Nonlinear Problems*, Pineridge Press, Swansea, 1980.
25. A. Pica and E. Hinton. The quasi-Newton BFGS method in the large deflection analysis of plates. In C. Taylor, E. Hinton, D.R.J. Owen and E. Oñate (eds), *Numerical Methods for Nonlinear Problems*, Pineridge Press, Swansea, 1980.
26. C.G. Broyden. Quasi-Newton methods and their application to function minimization. *Math. Comp.*, **21**, 368–81, 1967.
27. M. Hestenes and E. Stiefel. Method of conjugate gradients for solving linear systems. *J. Res. Natl. Bur. Stand.*, **49**, 409–36, 1954.
28. R. Fletcher and C.M. Reeves. Function minimization by conjugate gradients. *The Computer Journal*, **7**, 149–54, 1964.
29. E. Polak. *Computational Methods in Optimization: A Unified Approach*, Academic Press, London, 1971.
30. B.M. Irons and A.F. Elsawaf. The conjugate Newton algorithm for solving finite element equations. In K.-J. Bathe, J.T. Oden and W. Wunderlich (eds), *Proc. U.S.–German Symp. on Formulations and Algorithms in Finite Element Analysis*, pp. 656–72, MIT Press, Cambridge, MA, 1977.
31. M. Papadrakakis and P. Ghionis. Conjugate gradient algorithms in nonlinear structural analysis problems. *Comp. Meth. Appl. Mech. Eng.*, **59**, 11–27, 1986.
32. J.R.H. Otter, E. Cassel and R.E. Hobbs. Dynamic relaxation. *Proc. Inst. Civ. Eng.*, **35**, 633–56, 1966.
33. O.C. Zienkiewicz and R. Löhner. Accelerated relaxation or direct solution? Future prospects for FEM. *Int. J. Num. Meth. Eng.*, **21**, 1–11, 1986.
34. M. Adams. Parallel multigrid solver algorithms and implementations for 3D unstructured finite element problems. *Supercomputing '99: High Performance Networking and Computing, http://www.sc99.org/proceedings*, Portland, OR, November 1999.

3

Inelastic and non-linear materials

3.1 Introduction

In Chapter 1 we presented a framework for solving general problems in solid mechanics. In this chapter we consider several classical models for describing the behaviour of engineering materials. Each model we describe is given in a *strain-driven* form in which a strain or strain increment obtained from each finite element solution step is used to compute the stress needed to evaluate the internal force, $\int \mathbf{B}^{\mathrm{T}} \boldsymbol{\sigma} \, \mathrm{d}\Omega$ as well as a tangent modulus matrix, or its approximation, for use in constructing the tangent stiffness matrix. Quite generally in the study of small deformation and inelastic materials (and indeed in some forms applied to large deformation) the strain (or strain rate) or the stress is assumed to split into an additive sum of parts. We can write this as

$$\boldsymbol{\varepsilon} = \boldsymbol{\varepsilon}^{\mathrm{e}} + \boldsymbol{\varepsilon}^{\mathrm{i}} \tag{3.1}$$

or

$$\boldsymbol{\sigma} = \boldsymbol{\sigma}^{\mathrm{e}} + \boldsymbol{\sigma}^{\mathrm{i}} \tag{3.2}$$

in which we shall generally assume that the elastic part is given by the linear model

$$\boldsymbol{\varepsilon}^{\mathrm{e}} = \mathbf{D}^{-1} \boldsymbol{\sigma} \tag{3.3}$$

in which $\mathbf{D}$ is the matrix of elastic moduli.

In the following sections we shall consider the problems of viscoelasticity, plasticity, and general creep in quite general form. By using these general types it is possible to present numerical solutions which accurately predict many physical phenomena. We begin with viscoelasticity, where we illustrate the manner in which we shall address the solution of problems given in a rate or differential form. This rate form of course assumes time dependence and all viscoelastic phenomena are indeed transient, with time playing an important part. We shall follow this section with a description of plasticity models in which times does not explicitly arise and the problems are time independent. However, we shall introduce for convenience a rate description of the behaviour. This is adopted to allow use of the same kind of algorithms for all forms discussed in this chapter.

3.2 Viscoelasticity – history dependence of deformation

Viscoelastic phenomena are characterized by the fact that the rate at which inelastic strains develop depends not only on the current state of stress and strain but, in general, on the *full history* of their development. Thus, to determine the increment of inelastic strain over a given time interval (or time step) it is necessary to know the state of stress and strain at all *preceding times*. In the computation process these can in fact be obtained and *in principle* the problem presents little theoretical difficulty. Practical limitations appear immediately, however, that each computation point must retain this history information – thus leading to very large storage demands. In the context of linear viscoelasticity, means of overcoming this limitation were introduced by Zienkiewicz *et al.*[1] and White.[2] Extensions to include thermal effects were also included in some of this early work.[3] Further considerations which extend this approach are also discussed in earlier editions of this book.[4,5]

3.2.1 Linear models for viscoelasticity

The representation of a constitutive equation for linear viscoelasticity may be given in the form of either a differential equation or an integral equation.[6,7] In a differential model the constitutive equation may be written as a linear elastic part with an added series of partial strains $\mathbf{q}$. Accordingly, we write

$$\boldsymbol{\sigma}(t) = \mathbf{D}_0\,\boldsymbol{\varepsilon}(t) + \sum_{m=1}^{M} \mathbf{D}_m \mathbf{q}^{(m)}(t) \tag{3.4}$$

where for a linear model the partial stresses are solutions of the first-order differential equations

$$\dot{\mathbf{q}}^{(m)} + \mathbf{T}_m \mathbf{q}^{(m)} = \dot{\boldsymbol{\varepsilon}} \tag{3.5}$$

with $\mathbf{T}_m$ a constant matrix of reciprocal *relaxation* times and $\mathbf{D}_0$, $\mathbf{D}_m$ constant moduli matrices. The presence of a split of stress as given by Eq. (3.2) is immediately evident in the above. Each of the forms in Eq. (3.5) represents an elastic response in series with a viscous response and is known as a *Maxwell model*. In terms of a spring–dashpot model, a representation for the Maxwell material is shown in Fig. 3.1(a) for a single stress component. Thus, the sum given by Eq. (3.4) describes a *generalized Maxwell solid* in which several elements are assembled in a parallel form and the $\mathbf{D}_0$ term becomes a spring alone.

In an integral form the stress–strain behaviour may be written in a convolution form as

$$\boldsymbol{\sigma} = \mathbf{D}(t)\boldsymbol{\varepsilon}(0) + \int_0^t \mathbf{D}(t - t')\,\frac{\partial \boldsymbol{\varepsilon}}{\partial t'}\,\mathrm{d}t' \tag{3.6}$$

where components of $\mathbf{D}(t)$ are *relaxation moduli* functions.

Inverse relations may be given where the differential model is expressed as

$$\boldsymbol{\varepsilon}(t) = \mathbf{J}_0\,\boldsymbol{\sigma}(t) + \sum_{m=1}^{M} \mathbf{J}_m \mathbf{r}^{(m)}(t) \tag{3.7}$$

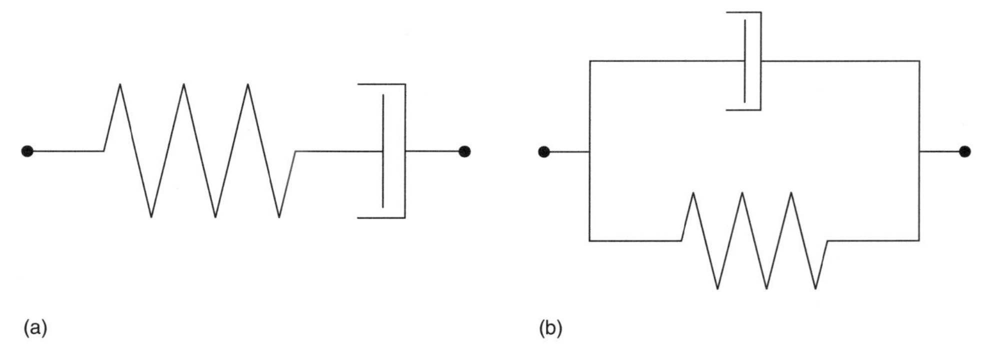

Fig. 3.1 Spring–dashpot models for linear viscoelasticity: (a) Maxwell element; (b) Kelvin element.

where for a linear model the partial stresses $\mathbf{r}$ are solutions of

$$\dot{\mathbf{r}}^{(m)} + \mathbf{V}_m \mathbf{r}^{(m)} = \boldsymbol{\sigma} \tag{3.8}$$

in which $\mathbf{V}_m$ are constant reciprocal *retardation* time parameters and $\mathbf{J}_0$, $\mathbf{J}_m$ constant compliances (i.e. reciprocal moduli). Each partial stress corresponds to a solution in which a linear elastic and a viscous response are combined in parallel to describe a *Kelvin model* as shown in Fig. 3.1(b). The total model thus is a *generalized Kelvin solid.*

In an integral form the strain–stress constitutive relation may be written as

$$\boldsymbol{\varepsilon} = \mathbf{J}(t)\boldsymbol{\sigma}(0) + \int_0^t \mathbf{J}(t-t')\,\frac{\partial \boldsymbol{\sigma}}{\partial t'}\,\mathrm{d}t' \tag{3.9}$$

where $\mathbf{J}(t)$ are known as *creep compliance* functions.

The parameters in the two forms of the model are related. For example, the creep compliances and relaxation moduli are related through

$$\mathbf{J}(t)\mathbf{D}(0) + \int_0^t \mathbf{J}(t-t')\,\frac{\partial \mathbf{D}}{\partial t'}\,\mathrm{d}t' = \mathbf{D}(t)\mathbf{J}(0) + \int_0^t \mathbf{D}(t-t')\,\frac{\partial \mathbf{J}}{\partial t'}\,\mathrm{d}t' = \mathbf{I} \tag{3.10}$$

as may easily be shown by applying, for example, Laplace transform theory to Eqs (3.6) and (3.9).

The above forms hold for isotropic and anisotropic linear viscoelastic materials. Solutions may be obtained by using standard numerical techniques to solve the constant coefficient differential or integral equations. Here we will proceed to describe a solution for the isotropic case where specific numerical schemes are presented. Generalization of the methods to the anisotropic case may be constructed by using a similar approach and is left as an exercise to the reader.

3.2.2 Isotropic models

To describe in more detail the ideas presented above we consider here isotropic models where we split the stress as

$$\boldsymbol{\sigma} = \mathbf{s} + \mathbf{m}p \qquad \text{with} \qquad p = \tfrac{1}{3}\mathbf{m}^{\mathrm{T}}\boldsymbol{\sigma} \tag{3.11}$$

where $\mathbf{s}$ is the stress deviator,* p is the mean (pressure) stress and, for a three-dimensional state of stress, $\mathbf{m}$ is given in Eq. (1.37). Similarly, a split of strain is expressed as

$$\boldsymbol{\varepsilon} = \mathbf{e} + \tfrac{1}{3}\mathbf{m}\theta \qquad \text{with} \qquad \theta = \mathbf{m}^{\mathrm{T}}\boldsymbol{\varepsilon} \tag{3.12}$$

where $\mathbf{e}$ is the strain deviator and θ is the volume change.

In the presentation given here, for simplicity we restrict the viscoelastic response to deviatoric parts and assume pressure–volume response is given by the linear elastic model

$$p = K\theta \tag{3.13}$$

where K is an elastic bulk modulus. A generalization to include viscoelastic behaviour in this component also may be easily performed by using the method described below for deviatoric components.

Differential equation model

The deviatoric part may be stated as differential equation models or in the form of integral equations as described above. In the differential equation model the constitutive equation may be written as

$$\mathbf{s} = 2G\left(\mu_0\mathbf{e} + \sum_{m=1}^{M}\mu_m\mathbf{q}^{(m)}\right) \tag{3.14}$$

in which μ_m are dimensionless parameters satisfying

$$\sum_{m=0}^{M}\mu_m = 1 \tag{3.15}$$

and dimensionless partial deviatoric strains $\mathbf{q}^{(m)}$ are obtained by solving

$$\dot{\mathbf{q}}^{(m)} + \frac{1}{\lambda_m}\mathbf{q}^{(m)} = \dot{\mathbf{e}} \tag{3.16}$$

in which λ_m are *relaxation times*. This form of the representation is again a *generalized Maxwell model* (a set of Maxwell models in parallel).

Each differential equation set may be solved numerically by using any of the finite-element-in-time methods described in Chapter 18 of Volume 1 (see Sec. 18.2). To solve numerically we first define a set of discrete points, t_k, at which we wish to obtain the solution. For a time t_{n+1} we assume the solution at all previous points up to t_n are known. Using a simple single-step method the solution for each partial stress is given by:

$$\left(1 + \frac{\theta\Delta t}{\lambda_m}\right)\mathbf{q}_{n+1}^{(m)} = \left(1 - \frac{(1-\theta)\Delta t}{\lambda_m}\right)\mathbf{q}_n^{(m)} + \mathbf{e}_{n+1} - \mathbf{e}_n \tag{3.17}$$

in which $\Delta t = t_{n+1} - t_n$.

* In Volume 1 $\boldsymbol{\sigma}^d$ was used to denote the deviatoric stress, and $\boldsymbol{\varepsilon}^d$ the deviatoric strain. Here we use the alternate notation $\mathbf{s}$ and $\mathbf{e}$ to avoid the extra superscript d.

We note that this form of the solution is given directly in a *strain-driven form*. Accordingly, given the strain from any finite element solution step we can immediately compute the stresses by using Eqs (3.13), (3.14) and (3.17) in Eqs (3.11) and (3.12). Inserting the above into a Newton-type solution strategy requires the computation of the tangent moduli. The tangent moduli for the viscoelastic model are deduced from

$$\mathbf{K}_{\mathrm{T}}|_{n+1} = \frac{\partial \boldsymbol{\sigma}_{n+1}}{\partial \boldsymbol{\varepsilon}_{n+1}} = \frac{\partial \mathbf{s}_{n+1}}{\partial \boldsymbol{\varepsilon}_{n+1}} + \mathbf{m}\,\frac{\partial p_{n+1}}{\partial \boldsymbol{\varepsilon}_{n+1}} \tag{3.18}$$

The tangent part for the volumetric term is elastic and given by

$$\mathbf{m}\,\frac{\partial p_{n+1}}{\partial \boldsymbol{\varepsilon}_{n+1}} = \mathbf{m}\,\frac{\partial p_{n+1}}{\partial \theta_{n+1}}\,\frac{\partial \theta_{n+1}}{\partial \boldsymbol{\varepsilon}_{n+1}} = K\mathbf{m}\mathbf{m}^{\mathrm{T}} \tag{3.19}$$

Similarly, the tangent part for the deviatoric term is deduced from Eq. (3.17) as

$$\frac{\partial \mathbf{s}_{n+1}}{\partial \boldsymbol{\varepsilon}_{n+1}} = \frac{\partial \mathbf{s}_{n+1}}{\partial \mathbf{e}_{n+1}}\,\frac{\partial \mathbf{e}_{n+1}}{\partial \boldsymbol{\varepsilon}_{n+1}} = 2G\left[\mu_0 + \sum_{m=1}^{M} \frac{\mu_m}{\left(1 + \dfrac{\theta \Delta t}{\lambda_m}\right)}\right]\mathbf{I}_d \tag{3.20}$$

where $\mathbf{I}_d$ is defined in Eq. (1.37). Using the above, tangent moduli are expressed as

$$\mathbf{K}_{\mathrm{T}}|_{n+1} = K\mathbf{m}\mathbf{m}^{\mathrm{T}} + 2G\left[\mu_0 + \sum_{m=1}^{M} \frac{\mu_m}{\left(1 + \dfrac{\theta \Delta t}{\lambda_m}\right)}\right]\mathbf{I}_d \tag{3.21}$$

and we note that the only difference from a linear elastic material is the replacement of the elastic shear modulus by the viscoelastic term

$$G \rightarrow G\left[\mu_0 + \sum_{m=1}^{M} \frac{\mu_m}{\left(1 + \dfrac{\theta \Delta t}{\lambda_m}\right)}\right]$$

This relation is independent of stress and strain and hence when it is used with a Newton scheme it converges in one iteration (i.e. the residual of a second iteration is numerically zero).

The set of first-order differential equations (3.16) may be integrated exactly for specified strains, **e**. The integral for each term is given by

$$\mathbf{q}^{(m)}(t) = \int_{-\infty}^{t} \exp\left[-(t - t')/\lambda_m\right] \frac{\partial \mathbf{e}}{\partial t'}\,\mathrm{d}t' \tag{3.22}$$

An advantage to the differential equation form, however, is that it may be extended to include *ageing* or other *nonlinear effects* by making the parameters time or solution dependent. The exact solution to the differential equations for such a situation will then involve integrating factors, leading to more involved expressions. In the following parts of this section we consider the integral equation form and its numerical solution for *linear* viscoelastic behaviour. Models and their solutions for more general cases are left as an exercise for the reader.

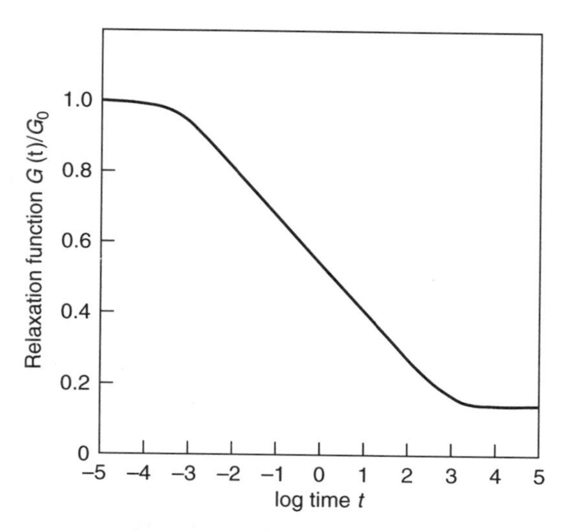

Fig. 3.2 Typical viscoelastic relaxation function.

Integral equation model

The integral equation form for the deviatoric stresses is expressed in terms of a relaxation modulus function which is defined by an idealized experiment in which, at time zero ($t = 0$), a specimen is subjected to suddenly applied and constant strain, $\mathbf{e}_0$, and the stress response, $\mathbf{s}(t)$, is measured. For a linear material a unique relation is obtained which is independent of the magnitude of the applied strain. This relation may be written as

$$\mathbf{s}(t) = 2G(t)\,\mathbf{e}_0 \tag{3.23}$$

where $G(t)$ is defined as the *shear relaxation modulus function*. A typical relaxation function is shown in Fig. 3.2. The function is shown on a logarithmic time scale since typical materials have time effects which cover wide ranges in time.

Using linearity and superposition for an arbitrary state of strain yields the integral equation specified as

$$\mathbf{s}(t) = \int_{-\infty}^{t} 2G(t - t')\,\frac{\partial \mathbf{e}}{\partial t'}\,\mathrm{d}t' \tag{3.24}$$

We note that the above form is a generalization to the Maxwell material. However, the integral equation form may be specialized to the generalized Maxwell model by assuming the shear relaxation modulus function in a Prony series form

$$G(t) = G\left[\mu_0 + \sum_{m=1}^{M} \mu_m \exp(-t/\lambda_m)\right] \tag{3.25}$$

where the μ_m satisfy Eq. (3.15).

Solution to integral equation with Prony series

The solution to the viscoelastic model is performed for a set of discrete points t_k. Thus, again assuming that all solutions are available up to time t_n, we desire to

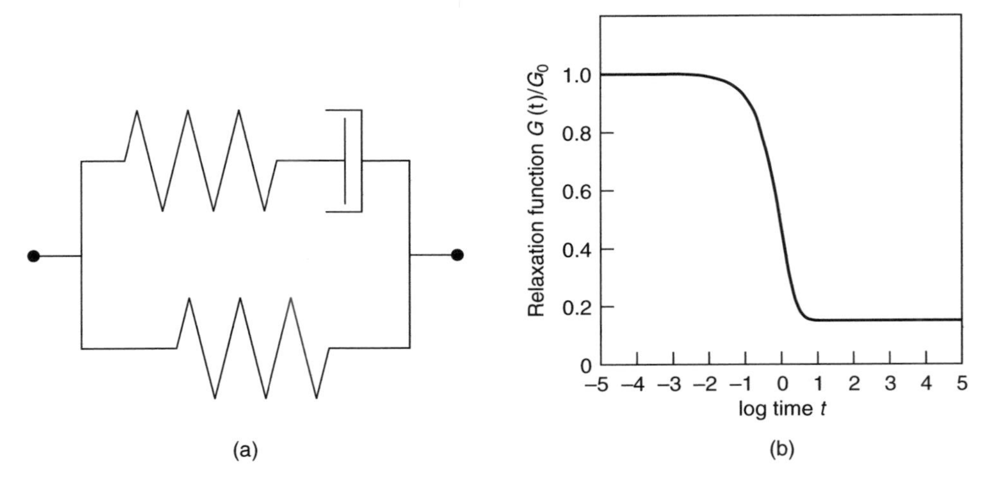

Fig. 3.3 Standard linear viscoelastic solid: (a) model for standard solid; (b) relaxation function.

compute the next step for time t_{n+1}. Solution of the general form would require summation over all previous time steps for each new time; however, by using the generalized Maxwell model we may reduce the solution to a recursion formula in which each new solution is computed by a simple update of the previous solution.

We will consider a special case of the generalized Maxwell material in which the number of terms M is equal to 1 [which defines a *standard linear solid*, Fig. 3.3(a)]. The addition of more terms is easily performed from the one-term solution. Accordingly, we take

$$G(t) = G[\mu_0 + \mu_1 \exp(-t/\lambda_1)] \tag{3.26}$$

where $\mu_0 + \mu_1 = 1$. For the standard solid only a limited range of time can be considered, as can be observed from Fig. 3.3(b) for the model given by

$$G(t) = G[0.15 + 0.85 \exp(-t)]$$

To consider a wider range it is necessary to use terms in which the λ_m cover the total time by using at least one term for each decade of time (a decade being one unit on the $\log_{10}$ time scale).

Substitution of Eq. (3.26) into Eq. (3.24) yields

$$\mathbf{s}(t) = 2G \int_{-\infty}^{t} [\mu_0 + \mu_1 \exp(-(t-t')/\lambda_1)] \frac{\partial \mathbf{e}}{\partial t'} \, \mathrm{d}t' \tag{3.27}$$

which may be split and expressed as

$$\begin{aligned} \mathbf{s}(t) &= 2G\mu_0 \boldsymbol{\varepsilon}(t) + 2G\mu_1 \int_{-\infty}^{t} \exp(-(t-t')/\lambda_1) \frac{\partial \mathbf{e}}{\partial t'} \, \mathrm{d}t' \\ &= 2G[\mu_0 \boldsymbol{\varepsilon}(t) + \mu_1 \mathbf{q}^{(1)}(t)] \end{aligned} \tag{3.28}$$

where we note that $\mathbf{q}^{(1)}$ is identical to the form given in Eq. (3.22). Thus use of a Prony series for $G(t)$ is identical to solving the differential equation model exactly.

In applications involving a linear viscoelastic model, it is usually assumed that the material is undisturbed until a time identified as zero. At time zero a strain may be suddenly applied and then varied over subsequent time. To evaluate a solution at time t_{n+1} the integral representation for the model may be simplified by dividing the integral into

$$\int_{-\infty}^{t_{n+1}} (\cdot)\, dt' = \int_{-\infty}^{0^-} (\cdot)\, dt' + \int_{0^-}^{0^+} (\cdot)\, dt' + \int_{0^+}^{t_n} (\cdot)\, dt' + \int_{t_n}^{t_{n+1}} (\cdot)\, dt' \tag{3.29}$$

In each analysis considered here the material is assumed to be unstrained before the time denoted as zero. Thus, the first term on the right-hand side is zero, the second term includes a jump term associated with $\mathbf{e}_0$ at time zero, and the last two terms cover the subsequent history of strain. The result of this separation when applied to Eq. (3.27) gives the recursion[3]

$$\mathbf{q}_{n+1}^{(1)} = \exp(-\Delta t/\lambda_1)\mathbf{q}_n^{(1)} + \Delta\mathbf{q}^{(1)} \tag{3.30}$$

where

$$\Delta\mathbf{q}^{(1)} = \int_{t_n}^{t_{n+1}} \exp[-(t_{n+1} - t')/\lambda_1] \frac{\partial \mathbf{e}}{\partial t'}\, dt' \tag{3.31}$$

and $\mathbf{q}_0^{(1)} = \mathbf{e}_0$.

To obtain a numerical solution, we approximate the strain rate in each time increment by a constant to obtain

$$\Delta\mathbf{q}_{n+1}^{(1)} = \frac{1}{\Delta t} \int_{t_n}^{t_{n+1}} \exp[-(t_{n+1} - t')/\lambda_1][\mathbf{e}_{n+1} - \mathbf{e}_n]\, dt' \tag{3.32}$$

The integral may now be evaluated directly over each time step as[3]

$$\Delta\mathbf{q}_{n+1}^{(1)} = \frac{\lambda_1}{\Delta t}[1 - \exp(-\Delta t/\lambda_1)](\mathbf{e}_{n+1} - \mathbf{e}_n) = \Delta q_{n+1}^{(1)}(\mathbf{e}_{n+1} - \mathbf{e}_n) \tag{3.33}$$

This approximation is singular for zero time steps; however, the limit value at $\Delta t = 0$ is one. Thus, for small time steps a series expansion may be used to yield accurate values, giving

$$\Delta q_{n+1}^{(1)} = 1 - \frac{1}{2}\left(\frac{\Delta t}{\lambda_1}\right) + \frac{1}{3!}\left(\frac{\Delta t}{\lambda_1}\right)^2 - \frac{1}{4!}\left(\frac{\Delta t}{\lambda_1}\right)^3 + \cdots \tag{3.34}$$

Using a few terms for very small time increment ratios yields numerically correct answers (to computer precision). Once the time increment ratio is larger than a certain small value the representation given in Eq. (3.33) is used directly.

The above form gives a recursion which is stable for small and large time steps and produces very smooth transitions under variable time steps.

A numerical approximation to Eq. (3.32) in which the integrand of Eq. (3.31) is evaluated at $t_{n+1/2}$ has also been used with success.[8] In the above recursion we note that a zero and infinite value of a time step produces a correct instantaneous and zero response, respectively, and thus is asymptotically accurate at both limits. The use of finite difference approximations on the differential equation form directly does not produce this property unless $\theta = 1$ and for this value is much less accurate than the solution given by Eq. (3.33).

Using the recursion formula, the constitutive equation now has the simple form

$$\mathbf{s}_{n+1} = 2G\,[\mu_0\,\mathbf{e}_{n+1} + \mu_1\,\mathbf{q}^{(1)}_{n+1}] \tag{3.35}$$

The process may also be extended to include effects of temperature on relaxation times for use with thermorheologically simple materials.[3]

The implementation of the above viscoelastic model into a Newton type solution process again requires the computation of a tangent tensor. Accordingly, for the deviatoric part we need to compute

$$\frac{\partial \mathbf{s}_{n+1}}{\partial \boldsymbol{\varepsilon}_{n+1}} = \frac{\partial \mathbf{s}_{n+1}}{\partial \mathbf{e}_{n+1}}\,\mathbf{I}_d \tag{3.36}$$

The partial derivative with respect to the deviatoric stress follows from Eq. (3.35) as

$$\frac{\partial \mathbf{s}}{\partial \mathbf{e}} = 2G\left[\mu_0\,\mathbf{I} + \mu_1\,\frac{\partial \mathbf{q}^{(1)}}{\partial \mathbf{e}}\right] \tag{3.37}$$

Using Eq. (3.33) the derivative of the last term becomes

$$\frac{\partial \mathbf{q}^{(1)}_{n+1}}{\partial \mathbf{e}_{n+1}} = \Delta q^{(1)}_{n+1}(\Delta t)\,\mathbf{I} \tag{3.38}$$

Thus, the tangent tensor is given by

$$\frac{\partial \mathbf{s}_{n+1}}{\partial \boldsymbol{\varepsilon}_{n+1}} = 2G\,[\mu_0 + \mu_1\,\Delta q^{(1)}_{n+1}(\Delta t)]\,\mathbf{I}_d \tag{3.39}$$

Again, the only modification from a linear elastic material is the substitution of the elastic shear modulus by

$$G \rightarrow G\,[\mu_0 + \mu_1\,\Delta q^{(1)}_{n+1}(\Delta t)] \tag{3.40}$$

We note that for zero Δt the full elastic modulus is recovered, whereas for very large increments the equilibrium modulus $\mu_0 G$ is used. Since the material is linear, use of this tangent modulus term again leads to convergence in one iteration (the second iteration produces a *numerically* *zero* residual).

The inclusion of more terms in the series reduces to evaluation of additional $\mathbf{q}^{(m)}_{n+1}$ integral recursions. Computer storage is needed to retain the $\mathbf{q}^{(m)}_n$ for each solution (quadrature) point in the problem and each term in the series.

Example: a thick-walled cylinder subjected to internal pressure

To illustrate the importance of proper element selection when performing analyses in which material behaviour approaches a near incompressible situation we consider the case of internal pressure on a thick-walled cylinder. The material is considered to be isotropic and modelled by viscoelastic response in deviatoric stress–strain only. Material properties are: modulus if elasticity, $E = 1000$; Poisson's ratio, $\nu = 0.3$; $\mu_1 = 0.99$; and $\lambda_1 = 1$. Thus, the viscoelastic relaxation function is given by

$$G(t) = \frac{1000}{2.6}\,[0.01 + 0.99\exp(-t)]$$

The ratio of the bulk modulus to shear modulus for instantaneous loading is given by $K/G(0) = 2.167$ and for long time loading by $K/G(\infty) = 216.7$ which indicates a

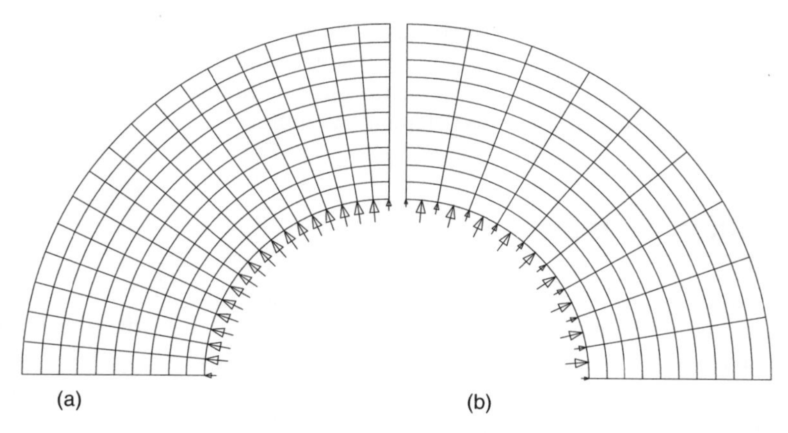

Fig. 3.4 Mesh and loads for internal pressure on a thick-walled cylinder: (a) four-noded quadrilaterals; (b) nine-noded quadrilaterals.

near incompressible behaviour for sustained loading cases (the effective Poisson ratio for infinite time is 0.498). The response for a suddenly applied internal pressure, $p = 10$, is computed to time 20 by using both displacement and the mixed element described in Chapter 1. Quadrilateral elements with four nodes (Q4) and nine nodes (Q9) are considered, and meshes with equivalent nodal forces are shown in Fig. 3.4. The exact solution to this problem is one-dimensional and, since all radial boundary conditions are traction ones, the stress distribution should be time independent. During the early part of the solution, when the response is still in the compressible range, the solutions from the two formulations agree well with this exact solution. However, during the latter part of the solution the answers from a displacement element diverge because of near incompressibility effects, whereas those from a mixed element do not. The distribution of quadrature point radial stresses at time $t = 20$ is shown in Fig. 3.5 where the highly oscillatory response of the displacement form is clearly evident. We note that extrapolation to reduced quadrature points

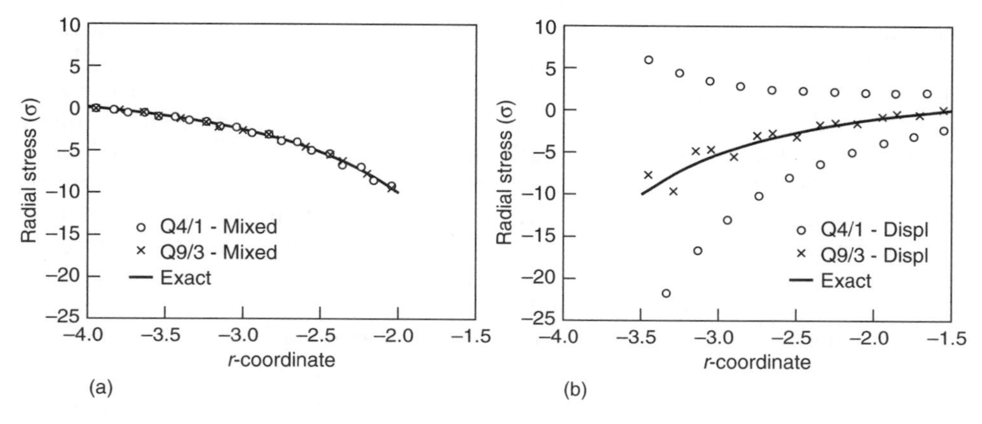

Fig. 3.5 Radial stress for internal pressure on a thick-walled cylinder: (a) mixed model; (b) displacement model.

would avoid these oscillations; however, use of fully reduced integration would lead to singularity in the stiffness matrix (as shown in Volume 1) and selective reduced integration is difficult to use with general non-linear material behaviour. Thus, for general applications the use of mixed elements is preferred.

3.2.3 Solution by analogies

The labour of step-by-step solutions for linear viscoelastic media can, on occasion, be substantially reduced. In the case of a homogeneous structure with linear isotropic viscoelasticity and constant Poisson ratio operator, the McHenry–Alfrey analogies allow single-step elastic solutions to be used to obtain stresses and displacements at a given time by the use of *equivalent loads*, *displacements* and *temperatures*.[9,10]

Some extensions of these analogies have been proposed by Hilton and Russell.[11] Further, when subjected to steady loads and when strains tend to a constant value at an infinite time, it is possible to determine the final stress distribution even in cases where the above analogies are not applicable. Thus, for instance, where the viscoelastic properties are temperature dependent and the structure is subject to a system of loads and temperatures which remain constant with time, long-term 'equivalent' elastic constants can be found and the problem solved as a single, non-homogeneous elastic one.[12]

The viscoelastic problem is a particular case of a creep phenomenon to which we shall return in Sect. 3.3 using some other classical non-linear models to represent material behaviour.

3.3 Classical time-independent plasticity theory

Classical 'plastic' behaviour of solids is characterized by a non-unique stress–strain relationship which is independent of the *rate* of loading but does depend on loading sequence that may be conveniently represented as a process evolving in time. Indeed, one definition of plasticity is the presence of irrecoverable strains on load removal. If uniaxial behaviour of a material is considered, as shown in Fig. 3.6(a), a non-linear relationship on loading alone does not determine whether *non-linear elastic or plastic behaviour* is exhibited. Unloading will immediately discover the difference, with an elastic material following the same path and a plastic material showing a *history-dependent* different path. We have referred to non-linearity elasticity already in Sect. 1.2 [see Eq. (1.36)] and will not give further attention to it here as the techniques used for plasticity problems or non-linear elasticity show great similarity. Representation of non-linear elastic behaviour for finite deformation applications is more complex as we shall show in Chapter 10.

Some materials show a nearly *ideal* plastic behaviour in which a limiting yield stress, Y (or σ_y), exists at which the strains are indeterminate. For all stresses below such yield, a linear (or non-linear) elastic relationship is assumed, Fig. 3.6(b) illustrates this. A further refinement of this model is one of a *hardening/softening plastic material* in which the yield stress depends on some parameter κ (such as the

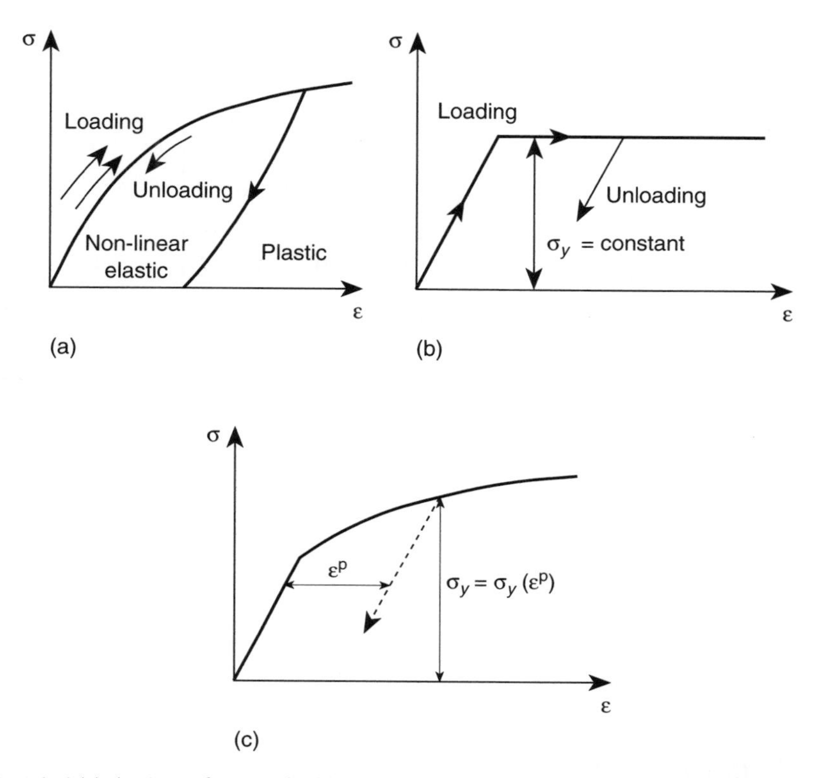

Fig. 3.6 Uniaxial behaviour of materials: (a) non-linear elastic and plastic behaviour; (b) ideal plasticity; (c) strain hardening plasticity.

accumulated plastic strain ε^p) [Fig. 3.6(c)]. It is with such kinds of plasticity that this section is concerned and for which much theory has been developed.[13,14]

In a multiaxial rather than a uniaxial state of stress the concept of yield needs to be generalized. It is important to note that in the following development of results in a matrix form all nine tensor components are used instead of the six 'engineering' component form used previously. To distinguish between the two we introduce an underbar on the symbol for all nine-component forms. Thus, we shall use:

$$\begin{aligned}
\boldsymbol{\sigma} &= \begin{bmatrix} \sigma_x & \sigma_y & \sigma_z & \sigma_{xy} & \sigma_{yz} & \sigma_{zx} \end{bmatrix}^{\mathrm{T}} \\
\underline{\boldsymbol{\sigma}} &= \begin{bmatrix} \sigma_x & \sigma_y & \sigma_z & \sigma_{xy} & \sigma_{yx} & \sigma_{yz} & \sigma_{zy} & \sigma_{zx} & \sigma_{xz} \end{bmatrix}^{\mathrm{T}} \\
\boldsymbol{\varepsilon} &= \begin{bmatrix} \varepsilon_x & \varepsilon_y & \varepsilon_z & \gamma_{xy} & \gamma_{yz} & \gamma_{zx} \end{bmatrix}^{\mathrm{T}} \\
\underline{\boldsymbol{\varepsilon}} &= \begin{bmatrix} \varepsilon_x & \varepsilon_y & \varepsilon_z & \varepsilon_{xy} & \varepsilon_{yx} & \varepsilon_{yz} & \varepsilon_{zy} & \varepsilon_{zx} & \varepsilon_{xz} \end{bmatrix}^{\mathrm{T}}
\end{aligned} \tag{3.41}$$

in which $\gamma_{ij} = 2\varepsilon_{ij}$. The transformations between the nine- and six-component forms needed later are obtained by using

$$\underline{\boldsymbol{\varepsilon}} = \mathbf{P}\boldsymbol{\varepsilon} \qquad \text{and} \qquad \boldsymbol{\sigma} = \mathbf{P}^{\mathrm{T}}\underline{\boldsymbol{\sigma}} \tag{3.42}$$

where

$$\mathbf{P}^{\mathrm{T}} = \frac{1}{2}\begin{bmatrix} 2 & 0 & 0 & 0 & 0 & 0 & 0 & 0 & 0 \\ 0 & 2 & 0 & 0 & 0 & 0 & 0 & 0 & 0 \\ 0 & 0 & 2 & 0 & 0 & 0 & 0 & 0 & 0 \\ 0 & 0 & 0 & 1 & 1 & 0 & 0 & 0 & 0 \\ 0 & 0 & 0 & 0 & 0 & 1 & 1 & 0 & 0 \\ 0 & 0 & 0 & 0 & 0 & 0 & 0 & 1 & 1 \end{bmatrix}$$

Accordingly, we first make all computations by using the nine 'tensor' components of stress and strain and only at the end do we reduce the computations to expressions in terms of the six independent 'engineering' quantities using **P**. This will permit final expressions for strain and equilibrium to be written in terms of **B** as in all previous developments. In addition we note that:

$$\mathbf{P}^{\mathrm{T}}\mathbf{I}\mathbf{P} = \mathbf{P}^{\mathrm{T}}\mathbf{P} = \mathbf{I}_0 \qquad \text{with} \qquad \mathbf{I}_0 = \frac{1}{2}\begin{bmatrix} 2 & & & & & \\ & 2 & & & & \\ & & 2 & & & \\ & & & 1 & & \\ & & & & 1 & \\ & & & & & 1 \end{bmatrix} \tag{3.43}$$

(see Section 12.2, Volume 1).

3.3.1 Yield functions

It is quite generally postulated, as an experimental fact, that yielding can occur only if the stress satisfies the general yield criterion

$$F(\underline{\boldsymbol{\sigma}}, \underline{\boldsymbol{\kappa}}, \kappa) = 0 \tag{3.44}$$

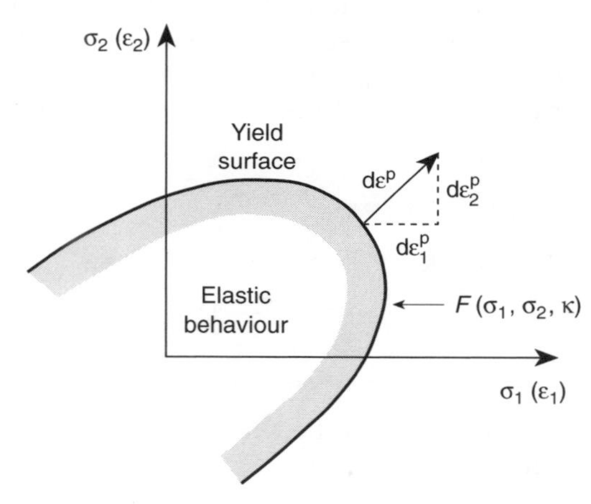

Fig. 3.7 Yield surface and normality criterion in two-dimensional stress space.

where $\underline{\boldsymbol{\sigma}}$ denotes a matrix form with all nine components of stress, $\underline{\boldsymbol{\kappa}}$ represents *kinematic hardening* parameters and κ an *isotropic hardening* parameter.[13] We shall discuss these particular sets of parameters later but, of course, many other types of parameters also can be used to define hardening.

This yield condition can be visualized as a surface in an n-dimensional space of stress with the position and size of the surface dependent on the instantaneous value of the parameters $\underline{\boldsymbol{\kappa}}$ and κ (Fig. 3.7).

3.3.2 Flow rule (normality principle)

Von Mises first suggested that basic behaviour defining the plastic strain increments is related to the yield surface.[15] Heuristic arguments for the validity of the relationship proposed have been given by various workers in the field[16–23] and at the present time the following hypothesis appears to be generally accepted for many materials; if $\underline{\boldsymbol{\varepsilon}}^{\mathrm{p}}$ denotes the components of the plastic strain tensor the rate of plastic strain is assumed to be given by*

$$\dot{\underline{\boldsymbol{\varepsilon}}}^{\mathrm{p}} = \dot{\lambda} F_{,\underline{\boldsymbol{\sigma}}} \tag{3.45}$$

where the notation

$$F_{,\underline{\boldsymbol{\sigma}}} \equiv \frac{\partial F}{\partial \underline{\boldsymbol{\sigma}}} \tag{3.46}$$

is introduced. In the above, $\dot{\lambda}$ is a proportionality constant, as yet undetermined, often referred to as the 'plastic consistency' parameter. During sustained plastic deformation we must have

$$\dot{F} = 0 \qquad \text{and} \qquad \dot{\lambda} > 0 \tag{3.47}$$

whereas during elastic loading/unloading $\dot{\lambda} = 0$ and $\dot{F} \neq 0$ leading to a general constraint condition in Kuhn–Tucker form[14]

$$\dot{F}\dot{\lambda} = 0 \tag{3.48}$$

The above rule is known as the *normality* principle because relation (3.45) can be interpreted as requiring the plastic strain rate components to be normal to the yield surface in the space of nine stress and strain dimensions.

Restrictions of the above rule can be removed by specifying separately a *plastic flow rule potential*

$$Q = Q(\underline{\boldsymbol{\sigma}}, \kappa) \tag{3.49}$$

which defines the plastic strain rate similarly to Eq. (3.45); that is, giving this as

$$\dot{\underline{\boldsymbol{\varepsilon}}}^{\mathrm{p}} = \dot{\lambda} Q_{,\underline{\boldsymbol{\sigma}}}, \qquad \dot{\lambda} \geqslant 0 \tag{3.50}$$

* Some authors prefer to write Eq. (3.45) in an incremental form

$$\mathrm{d}\underline{\boldsymbol{\varepsilon}}^{\mathrm{p}} = \mathrm{d}\lambda\, F_{,\underline{\boldsymbol{\sigma}}}$$

where then $\mathrm{d}\underline{\boldsymbol{\varepsilon}}^{\mathrm{p}} \equiv \dot{\underline{\boldsymbol{\varepsilon}}}^{\mathrm{p}}\, \mathrm{d}t$, and t is some pseudo-time variable. Here we prefer the rate form to permit use of common solution algorithms in which $d\underline{\boldsymbol{\varepsilon}}$ will denote an increment in a Newton-type solution. (Also note the difference in notation between a small increment 'd' and a differential 'd'.)

The particular case of $Q = F$ is known as *associative plasticity*. When this relation is not satisfied the plasticity is *non-associative*. In what follows this more general form will be considered initially (reductions to the associative case follow by simple substitution of $Q = F$).

The satisfaction of the normality rule for the associative case is essential for proving so called *upper and lower bound* theorems of plasticity as well as uniqueness. In the non-associative case the upper and lower bound do not exist and indeed it is not certain that the solutions are always unique. This does not prevent the validity of non-associated rules as it is well known that in frictional materials, for instance, uniqueness is seldom achieved but the existence of friction cannot be denied.

3.3.3 Hardening/softening rules

Isotropic hardening

The parameters $\underline{\boldsymbol{\kappa}}$ and κ must also be determined from rate equations and define hardening (or softening) of the plastic behaviour of the material. The evolution of κ, govern the *size* of the yield surface is commonly related to the rate of plastic work or directly to the consistency parameter. If related to the rate of plastic work κ has dimensions of stress and a relation of the type

$$\dot{\kappa} = \underline{\boldsymbol{\sigma}}^{\mathrm{T}}\, \dot{\underline{\boldsymbol{\varepsilon}}}^{\mathrm{p}} = Y(\kappa)\, \dot{\varepsilon}_u^{\mathrm{p}} \tag{3.51}$$

is used to match behaviour to a uniaxial tension or compression result. The slope

$$A = \frac{\partial Y}{\partial \kappa} \tag{3.52}$$

provides a modulus defining instantaneous *isotropic hardening*.

In the second approach κ is dimensionless (e.g., an accumulated plastic strain[14]) and is related directly to the consistency parameter using

$$\dot{\kappa} = \left[(\dot{\underline{\boldsymbol{\varepsilon}}}^{\mathrm{p}})^{\mathrm{T}}\, \dot{\underline{\boldsymbol{\varepsilon}}}^{\mathrm{p}}\right]^{1/2} = \dot{\lambda}[Q_{,\underline{\boldsymbol{\sigma}}}^{\mathrm{T}}\, Q_{,\underline{\boldsymbol{\sigma}}}]^{1/2} \tag{3.53}$$

A constitutive equation is then introduced to match uniaxial results. For example, a simple linear form is given by

$$\sigma_y(\kappa) = \sigma_{y0} + H_{i0}\kappa$$

where H_{i0} is a constant isotropic hardening modulus.

Kinematic hardening

A classical procedure to represent kinematic hardening was introduced by Prager[24] and modified by Ziegler.[25] Here the stress in each yield surface is replaced by a linear relation in terms of a 'back stress' $\underline{\boldsymbol{\kappa}}$ as

$$\underline{\varsigma} = \underline{\boldsymbol{\sigma}} - \underline{\boldsymbol{\kappa}} \tag{3.54}$$

with the yield function now given as

$$F(\underline{\boldsymbol{\sigma}} - \underline{\boldsymbol{\kappa}}, \kappa) = F(\underline{\varsigma}, \kappa) = 0 \tag{3.55}$$

during plastic behaviour. We note that with this approach derivatives of the yield surface differ only by a sign and are given by

$$F_{,\underline{\varsigma}} = F_{,\underline{\boldsymbol{\sigma}}} = -F_{,\underline{\boldsymbol{\kappa}}} \tag{3.56}$$

Accordingly, the yield surface will now *translate*, and if isotropic hardening is present will also expand or contract, during plastic loading.

A rate equation may be specified most directly by introducing a conjugate work variable $\underline{\boldsymbol{\beta}}$ from which the hardening parameter $\underline{\boldsymbol{\kappa}}$ is deduced by using a hardening potential $\mathcal{H}$. This may be stated as

$$\underline{\boldsymbol{\kappa}} = -\mathcal{H}_{,\underline{\boldsymbol{\beta}}} \tag{3.57}$$

which is completely analogous to use of an elastic energy to relate $\underline{\boldsymbol{\sigma}}$ and $\underline{\boldsymbol{\varepsilon}}^e$. A rate equation may be expressed now as

$$\dot{\underline{\boldsymbol{\beta}}} = \dot{\lambda} Q_{,\underline{\boldsymbol{\kappa}}} \tag{3.58}$$

It is immediately obvious that here also we have two possibilities. Using Q in the above expression defines a *non-associative* hardening, whereas replacing Q by F would give an *associative* hardening. Thus for a fully associative model we require that F be used to define both the plastic potential and the hardening. In such a case the relations of plasticity also may be deduced by using the *principle of maximum plastic dissipation.*[13,14,26,27] A quadratic form for the hardening potential may be adopted and written as

$$\mathcal{H} = \tfrac{1}{2}\underline{\boldsymbol{\beta}}^{\mathrm{T}} \underline{\mathbf{H}}_k \underline{\boldsymbol{\beta}} \tag{3.59}$$

in which $\underline{\mathbf{H}}_k$ is assumed to be an invertible set of constant hardening parameters. Now $\underline{\boldsymbol{\beta}}$ may be eliminated to give the simple rate form

$$\dot{\underline{\boldsymbol{\kappa}}} = -\dot{\lambda}\underline{\mathbf{H}}_k \frac{\partial Q}{\partial \underline{\boldsymbol{\kappa}}} = -\dot{\lambda}\underline{\mathbf{H}}_k Q_{,\underline{\boldsymbol{\kappa}}} \tag{3.60}$$

Use of a linear shift in relation (3.54) simplifies this, noting Eq. (3.56), to

$$\dot{\underline{\boldsymbol{\kappa}}} = \dot{\lambda}\underline{\mathbf{H}}_k Q_{,\underline{\varsigma}} \tag{3.61}$$

In our subsequent discussion we shall usually assume a general quadratic model for both elastic and hardening potentials. For a more general treatment the reader is referred to references 14 and 28.

Another approach to kinematic hardening was introduced by Armstrong and Frederick[29] and provides a means of retaining smoother transitions from elastic to inelastic behaviour during cyclic loading. Here the hardening is given as

$$\dot{\underline{\boldsymbol{\kappa}}} = \dot{\lambda}[\underline{\mathbf{H}}_k Q_{,\underline{\varsigma}} - H_{NL}\underline{\boldsymbol{\kappa}}] \tag{3.62}$$

Applications of this approach are presented by Chaboche[30,31] and numerical comparisons to a simpler approach using a generalized plasticity model[32,33] are given by Auricchio and Taylor.[34]

Many other approaches have been proposed to represent classical hardening behaviour and the reader is referred to the literature for additional information and discussion.[19–21,35–37] A physical procedure utilizing directly the finite element method is available to obtain both ideal plasticity and hardening. Here several ideal plasticity components, each with different yield stress, are put in series and it will be found that both hardening and softening behaviour can be obtained easily retaining the properties so far described. This approach was named by many authors as an 'overlay' model[38,39] and by others is described as a 'sublayer' model.

There are of course many other possibilities to define change in surfaces during the process of loading and unloading. Here frictional soils present one of the most difficult materials to model and for the non-associative case we find it convenient to use the generalized plasticity method described in Sect. 3.6.

3.3.4 Plastic stress–strain relations

To construct a constitutive model for plasticity, the strains are assumed to be divisible into elastic and plastic parts given as

$$\underline{\boldsymbol{\varepsilon}} = \underline{\boldsymbol{\varepsilon}}^{\mathrm{e}} + \underline{\boldsymbol{\varepsilon}}^{\mathrm{p}} \tag{3.63}$$

For linear elastic behaviour, the elastic strains are related to stresses by a symmetric 9×9 matrix of constants $\underline{\mathbf{D}}$. Differentiating Eq. (3.63) and incorporating the plastic relation (3.50) we obtain

$$\dot{\underline{\boldsymbol{\varepsilon}}} = \underline{\mathbf{D}}^{-1} \dot{\underline{\boldsymbol{\sigma}}} + \dot{\lambda} Q_{,\underline{\boldsymbol{\sigma}}} \tag{3.64}$$

The plastic strain (rate) will occur only if the 'elastic' stress changes

$$\dot{\underline{\boldsymbol{\sigma}}}^{\mathrm{e}} \equiv \underline{\mathbf{D}} \dot{\underline{\boldsymbol{\varepsilon}}} \tag{3.65}$$

tends to put the stress outside the yield surface, that is, is in the *plastic loading* direction. If, on the other hand, this stress change is such that *unloading* occurs then of course no plastic straining will be present, as illustrated for the one-dimensional case in Fig 3.6. The test of the above relation is therefore crucial in differentiating between loading and unloading operations and underlines the importance of the straining path in computing stress changes.

When plastic loading is occurring the stresses are on the yield surface given by Eq. (3.44). Differentiating this we can therefore write

$$\dot{F} = \frac{\partial F}{\partial \sigma_x} \dot{\sigma}_x + \frac{\partial F}{\partial \sigma_y} \dot{\sigma}_y + \cdots + \frac{\partial F}{\partial \kappa_x} \dot{\kappa}_x + \frac{\partial F}{\partial \kappa_y} \dot{\kappa}_y + \cdots + \frac{\partial F}{\partial \kappa} \dot{\kappa} = 0$$

or

$$\dot{F} = F_{,\underline{\boldsymbol{\sigma}}}^{\mathrm{T}} \dot{\underline{\boldsymbol{\sigma}}} + F_{,\underline{\boldsymbol{\kappa}}}^{\mathrm{T}} \dot{\underline{\boldsymbol{\kappa}}} - H_i \dot{\lambda} = 0 \tag{3.66}$$

in which we make the substitution

$$H_i \dot{\lambda} = -\frac{\partial F}{\partial \kappa} \dot{\kappa} = -F_{,\kappa} \dot{\kappa} \tag{3.67}$$

where H_i denotes an isotropic hardening modulus.

For the case where kinematic hardening is introduced, using Eq. (3.54) we can substitute Eq. (3.61) and modify Eq. (3.64) to

$$\underline{\mathbf{D}}\dot{\underline{\varepsilon}} = \dot{\underline{\varsigma}} + (\underline{\mathbf{D}} + \underline{\mathbf{H}}_k)\dot{\lambda} Q_{,\underline{\varsigma}} \tag{3.68}$$

Similarly, introducing Eq. (3.56) into Eq. (3.66) we obtain

$$\dot{F} = F_{,\underline{\varsigma}}^{\mathrm{T}} \dot{\underline{\varsigma}} - H_i \dot{\lambda} = 0 \tag{3.69}$$

Equations (3.68) and (3.69) now can be written in matrix form as

$$\begin{Bmatrix} \underline{\mathbf{D}}\dot{\underline{\varepsilon}} \\ 0 \end{Bmatrix} = \begin{bmatrix} \underline{\mathbf{I}} & (\underline{\mathbf{D}} + \underline{\mathbf{H}}_k) Q_{,\underline{\varsigma}} \\ F_{,\underline{\varsigma}}^{\mathrm{T}} & -H_i \end{bmatrix} \begin{Bmatrix} \dot{\underline{\varsigma}} \\ \dot{\lambda} \end{Bmatrix} \tag{3.70}$$

The indeterminate constant $\dot{\lambda}$ can now be eliminated (taking care not to multiply or divide by H_i or $\underline{\mathbf{H}}_k$ which are zero in ideal plasticity). To accomplish the elimination we solve the first set of Eq. (3.70) for $\dot{\underline{\varsigma}}$, giving

$$\dot{\underline{\varsigma}} = \underline{\mathbf{D}}\dot{\underline{\varepsilon}} - (\underline{\mathbf{D}} + \underline{\mathbf{H}}_k) Q_{,\underline{\varsigma}} \dot{\lambda}$$

and substitute into the second, yielding the expression

$$F_{,\underline{\varsigma}}^{\mathrm{T}} \underline{\mathbf{D}}\dot{\underline{\varepsilon}} - [H_i + F_{,\underline{\sigma}}^{\mathrm{T}} (\underline{\mathbf{D}} + \underline{\mathbf{H}}_k) Q_{,\underline{\varsigma}}] \dot{\lambda} = 0$$

Equation (3.64) now results in an explicit expansion that determines the *stress changes* in terms of imposed *strain changes*. Using Eq. (3.43) this may now be reduced to a form in which only six-independent components are present and expressed as*

$$\dot{\boldsymbol{\sigma}} = \mathbf{D}_{\mathrm{ep}}^{*} \dot{\boldsymbol{\varepsilon}} \tag{3.71}$$

and

$$\begin{aligned} \mathbf{D}_{\mathrm{ep}}^{*} &= \mathbf{P}^{\mathrm{T}} \underline{\mathbf{D}} \mathbf{P} - \frac{1}{H^*} \mathbf{P}^{\mathrm{T}} \underline{\mathbf{D}} Q_{,\underline{\varsigma}} F_{,\underline{\varsigma}}^{\mathrm{T}} \underline{\mathbf{D}} \mathbf{P} \\ &= \mathbf{D} - \frac{1}{H^*} \mathbf{P}^{\mathrm{T}} \underline{\mathbf{D}} Q_{,\underline{\varsigma}} F_{,\underline{\varsigma}}^{\mathrm{T}} \mathbf{D} \mathbf{P} \end{aligned} \tag{3.72}$$

where

$$H^* = H_i + F_{,\underline{\varsigma}}^{\mathrm{T}} (\underline{\mathbf{D}} + \underline{\mathbf{H}}_k) Q_{,\underline{\varsigma}}$$

The elasto-plastic matrix $\mathbf{D}_{\mathrm{ep}}^{*}$ takes the place of the elasticity matrix $\mathbf{D}_{\mathrm{T}}$ in a *continuum* rate formulation. We note that in the absence of kinematic hardening it is possible to make reductions to the six-component form for all the computations at the very beginning. However, the manner in which the back stress enters the computation is not the same as that for the plastic strain and would be necessary to scale the two differently to make the general reduction. Thus, for the developments reported here we prefer to carry out all calculations using the full nine-component form (or, in the case of plane stress, to follow a four-component form) and make final reductions using Eq. (3.72).

* We shall show this step in more detail below for the J_2 plasticity model. In general, however, the final result involves only the usual form of the **D** matrix and six independent components from the derivative of the yield function.

For a generalization of the above concepts to a yield surface possessing 'corners' where $Q_{,\underline{\sigma}}$ is indeterminate, the reader is referred to the work of Koiter[17] or the multiple surface treatments in Simo and Hughes.[14]

An alternative procedure exists here simply by smoothing the corners. We shall refer to it later in the context of the Mohr–Coulomb surface often used in geomechanics and the procedure can be applied to any form of yield surface.

The continuum elasto-plastic matrix is symmetric only when plasticity is associative *and* when kinematic hardening is symmetric. In general, non-associative materials present stability difficulties, and special care is needed to use them effectively. Similar difficulties occur if the hardening moduli are negative which, in fact, leads to a *softening* behaviour. This is addressed further in Secs 3.11 and 3.12.

The elasto-plastic matrix given above is defined even for ideal plasticity when H_i and $\underline{\mathbf{H}}_k$ are zero. Direct use of the continuum tangent in an incremental finite element context where the rates are approximated by

$$\dot{\boldsymbol{\varepsilon}}_{n+1}\Delta t \approx \Delta\boldsymbol{\varepsilon}_{n+1} \qquad \text{and} \qquad \dot{\boldsymbol{\sigma}}_{n+1}\Delta t \approx \Delta\boldsymbol{\sigma}_{n+1}$$

was first made by Yamada *et al.*[40] and Zienkiewicz *et al.*[41] However, this approach does not give quadratic convergence when used in the Newton–Raphson scheme. For the associative case we can introduce a *discrete time integration algorithm* in order to develop an exact (numerically consistent) tangent which does produce quadratic convergence when used in the Newton–Raphson iterative algorithm.

3.4 Computation of stress increments

We have emphasized that with the use of iterative procedures within a particular increment of loading, it is important to compute always the stresses as

$$\boldsymbol{\sigma}_{n+1}^k = \boldsymbol{\sigma}_n + \Delta\boldsymbol{\sigma}_n^k \tag{3.73}$$

corresponding to the total change in displacement parameters $\Delta\mathbf{a}_n^k$ and hence the total strain change

$$\Delta\boldsymbol{\varepsilon}_n^k = \mathbf{B}\Delta\mathbf{a}_n^k \qquad \Delta\mathbf{a}_n^k = \sum_{i=0}^{k} d\mathbf{a}_n^i \tag{3.74}$$

which has accumulated in all previous iterations within the step. This point is of considerable importance as constitutive models with path dependence (namely, plasticity-type models) have different responses for loading and unloading. If a decision on loading/unloading is based on the increment $d\mathbf{a}_n^k$ erroneous results will be obtained. Such decisions must *always* be performed with respect to the total increment $\Delta\mathbf{a}_n^k$.

In terms of the elasto-plastic modulus matrix given by Eq. (3.72) this means that the stresses have to be integrated as

$$\boldsymbol{\sigma}_{n+1}^k = \boldsymbol{\sigma}_n + \int_0^{\Delta\boldsymbol{\varepsilon}_n^k} \mathbf{D}_{\text{ep}}^* \, \mathrm{d}\boldsymbol{\varepsilon} \tag{3.75}$$

incorporating into $\mathbf{D}^*_{ep}$ the dependence on variables in a manner corresponding to a linear increase of $\Delta\boldsymbol{\varepsilon}^k_n$ (or $\Delta\mathbf{a}^k_n$). Here, of course, all other rate equations have to be suitably integrated, though this generally presents little additional difficulty.

Various procedures for integration of Eq. (3.75) have been adopted and can be classified into explicit and implicit categories.

3.4.1 Explicit methods

In explicit procedures either a direct integration process is used or some form of the Runge–Kutta process is adopted.[42] In the former the known increment $\Delta\boldsymbol{\varepsilon}^k_n$ is subdivided into m intervals and the integral of Eq. (3.75) is replaced by direct summation, writing

$$\Delta\boldsymbol{\sigma}^k_n = \frac{1}{m}\sum_{j=0}^{m-1} \mathbf{D}^*_{(n+j/m)}\,\Delta\boldsymbol{\varepsilon}^k_n \tag{3.76}$$

where $\mathbf{D}^*_{(n+j/m)}$ denotes the tangent matrix computed for stresses and hardening parameters updated from the previous increment in the sum.

This procedure, originally introduced in reference 43 and described in detail in references 44 and 45, is known as *subincrementation*. Its accuracy increases with the number of subincrements, m, used. In general it is difficult *a priori* to decide on this number, and accuracy of prediction is not easy to determine.

Such integration will generally result in the stress change departing from the yield surface by some margin. In problems such as those of ideal plasticity where the yield surface forms a meaningful limit a proportional scaling of stresses (or return map) has been practiced frequently to obtain stresses which are on the yield surface at all times.[45,46] In this process the effects of integrating the evolution equation for hardening must also be treated.

A more precise explicit procedure is provided by use of a Runge–Kutta method. Here, first an increment of $\Delta\boldsymbol{\varepsilon}/2$ is applied in a single-step explicit manner to obtain

$$\Delta\boldsymbol{\sigma}_{n+1/2} = \tfrac{1}{2}\,\mathbf{D}^*_n\,\Delta\boldsymbol{\varepsilon}_n \tag{3.77}$$

using the initial elasto-plastic matrix. This increment of stress (and corresponding $\boldsymbol{\kappa}_{n+1/2}$) is evaluated to compute $\mathbf{D}^*_{n+1/2}$ and finally we evaluate

$$\Delta\boldsymbol{\sigma}_n = \mathbf{D}^*_{n+1/2}\,\Delta\boldsymbol{\varepsilon}_n \tag{3.78}$$

This process has a second-order accuracy and, in addition, can give an estimate of errors incurred as

$$\Delta\boldsymbol{\sigma}_n - 2\Delta\boldsymbol{\sigma}_{n+1/2} \tag{3.79}$$

If such stress errors exceed a certain norm the size of the increment can be reduced. This approach is particularly useful for integration of non-associative models or models without yield functions where 'tangent' matrices are simply evaluated (see Sect. 3.6).

3.4.2 Implicit methods

The integration of Eq. (3.75) can, of course, be written in an implicit form. For instance, we could write in place of Eq. (3.75), during each iteration k, that

$$\Delta\boldsymbol{\sigma}_{n+1}^{k} = [(1-\theta)\,\mathbf{D}_n^* + \theta\,\mathbf{D}_{n+1}^{*,k}]\,\Delta\boldsymbol{\varepsilon}_{n+1}^{k} \tag{3.80}$$

where here $\mathbf{D}_n^*$ denotes the value of the tangential matrix at the beginning of the time step and $\mathbf{D}_{n+1}^{*,k}$ the current estimate to the tangential matrix at the end of the step.

This non-linear equation set could be solved by any of the procedures previously described; however, derivatives of the tangent matrix are quite complex and in any case a serious error is committed in the approximate form of Eq. (3.80). Further, there is no guarantee that the stresses do not depart from the yield surface.

Return map algorithm

In 1964 a very simple algorithm was introduced simultaneously by Maenchen and Sacks[47] and by Wilkins.[48] This algorithm uses a two-step process to compute the new stress and was originally implemented in an explicit time integration form, thus requiring no explicit construction of an elasto-plastic tangent matrix; however, later its versatility and robustness was demonstrated for implicit solutions.[49,50] The steps of the algorithm are:

1. Perform a predictor step in which the entire increment of strain (for the present discussion we omit the iteration counter k for simplicity)

$$\underline{\boldsymbol{\varepsilon}}_{n+1} = \underline{\boldsymbol{\varepsilon}}_n + \Delta\underline{\boldsymbol{\varepsilon}}_n$$

   is used to compute *trial* stresses (denoted by superscript TR) assuming elastic behaviour. Accordingly,

$$\underline{\boldsymbol{\sigma}}_{n+1}^{\mathrm{TR}} = \underline{\mathbf{D}}(\underline{\boldsymbol{\varepsilon}}_{n+1} - \underline{\boldsymbol{\varepsilon}}_n^{\mathrm{p}}) \tag{3.81}$$

   where only an elastic modulus $\underline{\mathbf{D}}$ is required.
2. Evaluate the yield function in terms of the trial stress and the values of the plastic parameters at the previous time:

$$F(\underline{\boldsymbol{\sigma}}^{\mathrm{TR}}, \underline{\boldsymbol{\kappa}}_n, \kappa_n) = \begin{cases} \leqslant 0, & \text{elastic} \\ >0, & \text{plastic} \end{cases} \tag{3.82}$$

   (a) For an elastic value of F set the current stress to the trial value, accordingly

$$\underline{\boldsymbol{\sigma}}_{n+1} = \underline{\boldsymbol{\sigma}}_{n+1}^{\mathrm{TR}}, \quad \underline{\boldsymbol{\kappa}}_{n+1} = \underline{\boldsymbol{\kappa}}_n \quad \text{and} \quad \kappa_{n+1} = \kappa_n$$

   (b) For a plastic state solve a discretized set of plasticity rate equations (namely, using any appropriate time integration method as described in Chapter 18 of Volume 1) such that the final value of F_{n+1} is zero.

A plastic correction can be most easily developed by returning to the original Eq. (3.64) and writing the relation for stress increment as

$$\Delta\underline{\boldsymbol{\sigma}}_n = \underline{\mathbf{D}}(\Delta\underline{\boldsymbol{\varepsilon}}_n - \Delta\underline{\boldsymbol{\varepsilon}}_n^{\mathrm{p}}) \tag{3.83}$$

Now integrating the plastic strain relation (3.50) using a form similar to that in Eq. (3.80) yields

$$\Delta \underline{\boldsymbol{\varepsilon}}_n^{\mathrm{p}} = \Delta\lambda[(1-\theta)Q_{,\underline{\boldsymbol{\sigma}}}|_n + \theta Q_{,\underline{\boldsymbol{\sigma}}}|_{n+1}] \tag{3.84}$$

where $\Delta\lambda$ represents an approximation to the change in consistency parameter over the time increment. Kinematic hardening is included by integrating Eq. (3.60) as

$$\Delta \underline{\boldsymbol{\kappa}}_n = -\Delta\lambda \underline{\mathbf{H}}_k[(1-\theta)Q_{,\underline{\boldsymbol{\kappa}}}|_n + \theta Q_{,\underline{\boldsymbol{\kappa}}}|_{n+1}] \tag{3.85}$$

Finally, during the plastic solution we enforce

$$F_{n+1} = 0 \tag{3.86}$$

thus ensuring that final values at t_{n+1} satisfy the yield condition exactly.

The above solution process is particularly simple for $\theta = 1$ (backward difference or Euler implicit) and now, eliminating $\Delta\underline{\boldsymbol{\varepsilon}}_n^{\mathrm{p}}$, we can write the above non-linear system in residual form

$$\begin{aligned}
\underline{\mathbf{R}}_\sigma^i &= \Delta\underline{\boldsymbol{\varepsilon}}_n - \underline{\mathbf{D}}^{-1}\Delta\underline{\boldsymbol{\sigma}}_n^i - \Delta\lambda Q_{,\underline{\boldsymbol{\sigma}}}|_{n+1}^i \\
\underline{\mathbf{R}}_\kappa^i &= -\underline{\mathbf{H}}_k^{-1}\Delta\underline{\boldsymbol{\kappa}}_n^i - \Delta\lambda Q_{,\underline{\boldsymbol{\kappa}}}|_{n+1}^i \\
r^i &= -F_{n+1}^i
\end{aligned}$$

and seek solutions which satisfy $\underline{\mathbf{R}}_\sigma^i = \mathbf{0}$, $\underline{\mathbf{R}}_\kappa^i = \mathbf{0}$ and $r^i = 0$. Any of the general iterative schemes described in Chapter 2 can now be used. In particular, the full Newton–Raphson process is convenient. Noting that $\Delta\underline{\boldsymbol{\varepsilon}}_n$ is treated here as a specified constant (actually, the $\Delta\underline{\boldsymbol{\varepsilon}}_n^k$ from the current finite element solution), we can write, on linearization

$$\begin{bmatrix} \underline{\mathbf{D}}^{-1} + \Delta\lambda Q_{,\underline{\boldsymbol{\sigma}\boldsymbol{\sigma}}} & \Delta\lambda Q_{,\underline{\boldsymbol{\sigma}\boldsymbol{\kappa}}} & Q_{,\underline{\boldsymbol{\sigma}}} \\ \Delta\lambda Q_{,\underline{\boldsymbol{\kappa}\boldsymbol{\sigma}}} & \underline{\mathbf{H}}_k^{-1} + \Delta\lambda Q_{,\underline{\boldsymbol{\kappa}\boldsymbol{\kappa}}} & Q_{,\underline{\boldsymbol{\kappa}}} \\ F_{,\underline{\boldsymbol{\sigma}}}^{\mathrm{T}} & F_{,\underline{\boldsymbol{\kappa}}}^{\mathrm{T}} & -H_i \end{bmatrix}_{n+1}^i \begin{Bmatrix} d\underline{\boldsymbol{\sigma}}^i \\ d\underline{\boldsymbol{\kappa}}^i \\ d\lambda^i \end{Bmatrix} = \begin{Bmatrix} \underline{\mathbf{R}}_\sigma^i \\ \underline{\mathbf{R}}_\kappa^i \\ r^i \end{Bmatrix} \tag{3.87}$$

where H_i is the same hardening parameter as that obtained in Eq. (3.67). Some complexity is introduced by the presence of the second derivatives of Q in Eq. (3.87) and the term may be omitted for simplicity (although at the expense of asymptotic quadratic convergence in the Newton–Raphson iteration). Analytical forms of such second derivatives are available for frequently used potential surfaces.[14,28,49–51] Appendix A also presents results for second derivatives of stress invariants.

It is important to note that the requirement that $F_{n+1} = -r^i$ [Eq. (3.87)] ensures that the r^i residual measures precisely the departure from the yield surface. This measure is not available for any of the tangential forms if $\mathbf{D}_{\mathrm{ep}}^*$ is adopted.

For the solution it is only necessary to compute $d\lambda^i$ and update as

$$\Delta\lambda^i = \sum_{j=0}^{i} d\lambda^j \tag{3.88}$$

This solution process can be done in precisely the same way as was done in establishing Eq. (3.72). Thus, a solution may be constructed by defining the following:

$$\underline{\mathbf{R}} = \begin{bmatrix} \underline{\mathbf{R}}_\sigma \\ \underline{\mathbf{R}}_\kappa \end{bmatrix}, \qquad \underline{\zeta} = \begin{bmatrix} \boldsymbol{\sigma} \\ \boldsymbol{\kappa} \end{bmatrix}$$

$$\underline{\nabla} F = \begin{bmatrix} F_{,\boldsymbol{\sigma}} \\ F_{,\boldsymbol{\kappa}} \end{bmatrix}, \qquad \underline{\nabla} Q = \begin{bmatrix} Q_{,\boldsymbol{\sigma}} \\ Q_{,\boldsymbol{\kappa}} \end{bmatrix} \tag{3.89}$$

$$\underline{\mathbf{A}} = \begin{bmatrix} \mathbf{D}^{-1} & \mathbf{0} \\ \mathbf{0} & \mathbf{H}_k^{-1} \end{bmatrix} + \Delta\lambda^i \begin{bmatrix} Q_{,\boldsymbol{\sigma\sigma}} & Q_{,\boldsymbol{\sigma\kappa}} \\ Q_{,\boldsymbol{\kappa\sigma}} & Q_{,\boldsymbol{\kappa\kappa}} \end{bmatrix}$$

and expressing Eq. (3.87) as

$$d\underline{\zeta}^i = \underline{\mathbf{A}}^{-1}\underline{\mathbf{R}}^i - \frac{1}{A^*}\underline{\mathbf{A}}^{-1}\underline{\nabla} Q^i\,[(\underline{\nabla} F^i)^{\mathrm{T}}\underline{\mathbf{A}}^{-1}\underline{\mathbf{R}}^i - r^i] \tag{3.90}$$

where

$$A^* = H_i + (\underline{\nabla} F^i)^{\mathrm{T}}\underline{\mathbf{A}}^{-1}\underline{\nabla} Q^i \tag{3.91}$$

Immediately, we observe that at convergence $\underline{\mathbf{R}}^i = \mathbf{0}$ and $r^i = 0$, thus, here we obtain a zero stress increment. At this point we have computed a stress state $\boldsymbol{\sigma}_{n+1}$ which satisfies the yield condition exactly. However, this stress, when substituted back into the finite element residual [e.g. Eq. (1.24) or (1.44)] may not satisfy the equilibrium condition and it is now necessary to compute a new iteration k and obtain a new strain increment $d\boldsymbol{\varepsilon}_n^k$ from which the process is repeated. We note that inserting this new increment into Eq. (3.87) will again result in a non-zero value for $\underline{\mathbf{R}}_\sigma$, but that $\underline{\mathbf{R}}_\kappa$ and r remain zero until subsequent iterations. Thus, Eq. (3.90) provides directly now the required tangent matrix $\tilde{\mathbf{D}}^*_{\mathrm{ep}}$ from

$$\begin{Bmatrix} d\boldsymbol{\sigma} \\ d\boldsymbol{\kappa} \end{Bmatrix} = \left[\underline{\mathbf{A}}^{-1} - \frac{1}{A^*}\underline{\mathbf{A}}^{-1}\underline{\nabla} Q(\underline{\nabla} F)^{\mathrm{T}}\underline{\mathbf{A}}^{-1}\right]\begin{Bmatrix} d\boldsymbol{\varepsilon} \\ \mathbf{0} \end{Bmatrix} = \begin{bmatrix} \tilde{\mathbf{D}}^*_{\mathrm{ep}} & \cdot \\ \cdot & \cdot \end{bmatrix}\begin{Bmatrix} d\boldsymbol{\varepsilon} \\ \mathbf{0} \end{Bmatrix} \tag{3.92}$$

Thus, we find the tangent matrix $\underline{\tilde{\mathbf{D}}}^*_{\mathrm{ep}}$ is obtained from the upper diagonal block of Eq. (3.92). We note that this development also follows exactly the procedure for computing $\mathbf{D}^*_{\mathrm{ep}}$ in Eq. (3.72). At this stage the terms may once again be reduced to their six-component form using $\mathbf{P}$ as indicated in Eq. (3.42).

Some remarks on the above algorithm are in order:

1. For non-associative plasticity (namely, $Q \neq F$) the return direction is *not* normal to the yield surface. In this case no solution may exist for some strain increments (in general, arbitrary selection of F and Q forms in non-associative does not assure stability) and the iteration process will not converge.
2. For associative plasticity the normality principle is valid, requiring a convex yield surface. In this case the above iteration process always converges for a hardening material.
3. Convergence of the finite element equations may not always occur if more than one quadrature point changes from elastic to plastic or from plastic to elastic in subsequent iterations.

Based on these comments it is evident that no universal method exists that can be used with the many alternatives which can occur in practice. In the next several sections we illustrate some formulations which employ the alternatives we have discussed above.

3.5 Isotropic plasticity models

We consider here some simple cases for isotropic plasticity-type models in which both a yield function and a flow rule are used. For an isotropic material linear elastic response may be expressed by moduli defined with two parameters. Here we shall assume these to be the bulk and shear moduli, as used previously in the viscoelastic section (Sec. 3.2). Accordingly, the stress at any discrete time t_{n+1} is computed from elastic strains in matrix form as

$$\begin{aligned}\underline{\boldsymbol{\sigma}}_{n+1} &= p_{n+1}\underline{\mathbf{m}} + \underline{\mathbf{s}}_{n+1} = K\underline{\mathbf{m}}\underline{\mathbf{m}}\underline{\boldsymbol{\varepsilon}}^{e}_{n+1} + 2G(\underline{\mathbf{I}} - \tfrac{1}{3}\underline{\mathbf{m}}\underline{\mathbf{m}}^{T})\underline{\boldsymbol{\varepsilon}}^{e}_{n+1} \\ &= \underline{\mathbf{D}}\left(\underline{\boldsymbol{\varepsilon}}_{n+1} - \underline{\boldsymbol{\varepsilon}}^{p}_{n+1}\right)\end{aligned} \tag{3.93}$$

where the elastic modulus matrix for an isotropic material is given in the simple form

$$\underline{\mathbf{D}} = K\underline{\mathbf{m}}\underline{\mathbf{m}}^{T} + 2G(\underline{\mathbf{I}} - \tfrac{1}{3}\underline{\mathbf{m}}\underline{\mathbf{m}}^{T}) \tag{3.94}$$

and $\underline{\mathbf{I}}$ is the 9×9 identity matrix and $\underline{\mathbf{m}}$ is the nine-component matrix

$$\underline{\mathbf{m}} = [1 \quad 1 \quad 1 \quad 0 \quad 0 \quad 0 \quad 0 \quad 0 \quad 0]^{T}$$

Using Eqs (3.42) and (3.43) immediately reduces the above to

$$\mathbf{D} = K\mathbf{m}\mathbf{m}^{T} + 2G(\mathbf{I}_0 - \tfrac{1}{3}\mathbf{m}\mathbf{m}^{T}) \tag{3.95}$$

The above relation yields the stress at the current time provided we know the current total strain and the current plastic strain values. The total strain is available from the finite element equations using the current value of nodal displacements, and the plastic strain is assumed to be computed with use of one of the algorithms given above. In the discussion to follow we consider relations for various classical yield surfaces.

3.5.1 Isotropic yield surfaces

The general procedures outlined in the previous section allow determination of the tangent matrices for almost any yield surface applicable in practice. For an isotropic material all functions can be represented in terms of the three stress invariants:*

$$\begin{aligned}I_1 &= \sigma_{ii} = \underline{\mathbf{m}}^{T}\underline{\boldsymbol{\sigma}} \\ 2J_2 &= s_{ij}s_{ji} = \underline{\mathbf{s}}^{T}\underline{\mathbf{s}} = |\underline{\mathbf{s}}|^2 \\ 3J_3 &= s_{ij}s_{jk}s_{ki} = \det \underline{\mathbf{s}}\end{aligned} \tag{3.96}$$

where we can observe that definition of all the invariants is most easily performed in indicial notation.

* Appendix A presents a summary of invariants and their derivatives.

One useful form of these invariants for use in yield functions is given by[43]

$$\begin{aligned} 3\sigma_m &= I_1 \\ \bar{\sigma} &= \sqrt{J_2} \\ 3\theta &= \sin^{-1}\left(-\frac{3\sqrt{3}J_3^{1/3}}{2\bar{\sigma}}\right) \quad \text{with} \quad -\frac{\pi}{6} \leqslant \theta \leqslant \frac{\pi}{6} \end{aligned} \tag{3.97}$$

Using these definitions the surface for several classical yield conditions can be given as:

1. Tresca:

$$F = 2\bar{\sigma}\cos\theta - Y(\kappa) = 0 \tag{3.98}$$

2. Huber–von Mises:

$$F = \sqrt{2}\bar{\sigma} - \sqrt{\frac{2}{3}}Y(\kappa) = |\underline{\mathbf{s}}| - \sqrt{\frac{2}{3}}Y(\kappa) = 0 \tag{3.99}$$

   Both conditions 1 and 2 are well verified in metal plasticity. For soils, concrete and other 'frictional' materials the Mohr–Coulomb or Drucker–Prager surfaces is frequently used.[52]

3. Mohr–Coulomb:

$$F = \sigma_m \sin\phi + \bar{\sigma}\left(\cos\theta - \frac{1}{\sqrt{3}}\sin\phi\sin\theta\right) - c\cos\phi = 0 \tag{3.100}$$

   where $c(\kappa)$ and $\phi(\kappa)$ are the cohesion and the angle of friction, respectively, which can depend on an isotropic strain hardening parameter κ.

4. Drucker–Prager:

$$F = 3\alpha'(\kappa)\sigma_m + \bar{\sigma} - K(\kappa) = 0 \tag{3.101}$$

   where

$$\alpha' = \frac{2\sin\phi}{\sqrt{3}(3-\sin\phi)} \qquad K = \frac{6\cos\phi}{\sqrt{3}(3-\sin\phi)}$$

   and again c and ϕ can depend on a strain hardening parameter.

These forms lead to a convenient definition of the gradients $F_{,\underline{\sigma}}$ or $Q_{,\underline{\sigma}}$, irrespective of whether the surface is used as a yield condition or a flow potential. Thus we can always write

$$F_{,\underline{\sigma}} = F_{,\sigma_m}\frac{\partial\sigma_m}{\partial\underline{\sigma}} + F_{,\bar{\sigma}}\frac{\partial\bar{\sigma}}{\partial\underline{\sigma}} + F_{,\theta}\frac{\partial\theta}{\partial\underline{\sigma}} \tag{3.102}$$

and upon noting that

$$\begin{aligned} \frac{\partial\bar{\sigma}}{\partial\underline{\sigma}} &= \frac{\partial\bar{\sigma}}{\partial J_2}\frac{\partial J_2}{\partial\underline{\sigma}} = \frac{1}{2\sqrt{J_2}}\frac{\partial J_2}{\partial\underline{\sigma}} \\ \frac{\partial\theta}{\partial\underline{\sigma}} &= \frac{\partial\theta}{\partial J_2}\frac{\partial J_2}{\partial\underline{\sigma}} + \frac{\partial\theta}{\partial J_3}\frac{\partial J_3}{\partial\underline{\sigma}} = \tan 3\theta\left[\frac{1}{9}J_3\frac{\partial J_3}{\partial\underline{\sigma}} - \frac{1}{6}J_2\frac{\partial J_2}{\partial\underline{\sigma}}\right] \end{aligned} \tag{3.103}$$

Table 3.1 Invariant derivatives for various yield conditions

Yield condition	$F_{,\sigma_m}$	$\sqrt{J_2}F_{,J_2}$	$J_2 F_{,J_3}$
Tresca	0	$2\cos\theta(1+\tan\theta\tan 3\theta)$	$\dfrac{\sqrt{3}\sin\theta}{\cos 3\theta}$
Huber–von Mises	0	$\sqrt{3}$	0
Mohr–Coulomb	$\sin\phi$	$\frac{1}{2}\cos\theta\left[1 = \tan\theta\sin 3\theta + \frac{1}{\sqrt{3}}\sin\phi(\tan 3\theta - \tan\theta)\right]$	$\dfrac{\sqrt{3}\sin\theta + \sin\phi\cos\theta}{2\cos 3\theta}$
Drucker–Prager	$3\alpha'$	1	0

Alternatively, we can always write:

$$F_{,\underline{\sigma}} = F_{,\sigma_m}\frac{\partial\sigma_m}{\partial\underline{\sigma}} + F_{,J_2}\frac{\partial J_2}{\partial\underline{\sigma}} + F_{,J_3}\frac{\partial J_3}{\partial\underline{\sigma}} \tag{3.104}$$

which can be put into a matrix form as shown in Appendix A.

The values of the three derivatives with respect to the invariants are shown in Table 3.1 for the various yield surfaces mentioned. The form of the various yield surfaces given above is shown with respect to the principal stress space in Fig. 3.8, though many more elaborate ones have been developed, particularly for soil (geomechanics) problems.[53–55]

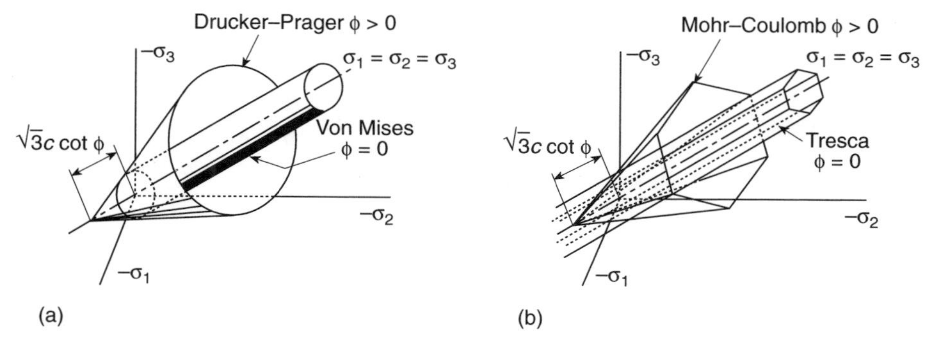

Fig. 3.8 Isotropic yield surfaces in principal stress space: (a) Drucker–Prager and von Mises; (b) Mohr–Coulomb and Trexa.

3.5.2 J_2 model with isotropic and kinematic hardening (Prandtl–Reuss equations)

As noted in Table 3.1 a particularly simple form results if we assume the yield function involves only the second invariant of the deviatoric stresses J_2. Here we present a more detailed discussion of results obtained by using an associated form and the return map algorithm. Since the yield function involves deviatoric quantities only we can initially make all the calculations in terms of these. Accordingly, the elastic deviatoric stress–strain relation is given as

$$\underline{s} = 2G\underline{e}^{e} = 2G(\underline{e} - \underline{e}^{p}) \tag{3.105}$$

Continuum rate form

Before constructing the return map solution we first consider the form of the plasticity equations in rate form for this simple model. The plastic deviatoric strain rates are deduced from

$$\dot{\underline{\mathbf{e}}}^{\mathrm{p}} = \dot{\lambda} \frac{\partial F}{\partial \underline{\mathbf{s}}} = \dot{\lambda} F_{,\underline{\mathbf{s}}} \tag{3.106}$$

Including the effects of isotropic and kinematic hardening the Huber–von Mises yield function may be expressed as

$$F = |\underline{\mathbf{s}} - \underline{\boldsymbol{\kappa}}| - \sqrt{\tfrac{2}{3}}\, Y(\kappa) = 0 \tag{3.107}$$

in which $\underline{\boldsymbol{\kappa}}$ are back stresses from kinematic hardening and κ is an isotropic hardening parameter. We assume linear isotropic hardening given by*

$$Y(\kappa) = Y_0 + H_i \kappa \tag{3.108}$$

Here a rate of κ is computed from a norm of the plastic strains, by using Eq. (3.53), as

$$\dot{\kappa} = \sqrt{\tfrac{2}{3}}\, \dot{\lambda} \tag{3.109}$$

in which the factor $\sqrt{2/3}$ is introduced to match uniaxial behaviour given by Eq. (3.108).

On differentiation of F it will be found that

$$\frac{\partial F}{\partial \underline{\mathbf{s}}} = -\frac{\partial F}{\partial \underline{\boldsymbol{\kappa}}} = \underline{\mathbf{n}} \qquad \text{where} \qquad \underline{\mathbf{n}} = \frac{\underline{\mathbf{s}} - \underline{\boldsymbol{\kappa}}}{|\underline{\mathbf{s}} - \underline{\boldsymbol{\kappa}}|} \tag{3.110}$$

Using the above, the plastic strains are given by

$$\dot{\underline{\mathbf{e}}}^{\mathrm{p}} = \dot{\lambda} \underline{\mathbf{n}} \tag{3.111}$$

and, when substituted into a rate-of-stress relation, yield

$$\dot{\underline{\mathbf{s}}} = 2G \left[\dot{\underline{\mathbf{e}}} - \dot{\lambda} \underline{\mathbf{n}} \right] \tag{3.112}$$

A rate form for the kinematic hardening is taken as

$$\dot{\underline{\boldsymbol{\kappa}}} = \tfrac{2}{3} H_k \dot{\lambda} \underline{\mathbf{n}} \tag{3.113}$$

The rate of the yield function becomes

$$\dot{F} = \underline{\mathbf{n}}^{\mathrm{T}} (\dot{\underline{\mathbf{s}}} - \dot{\underline{\boldsymbol{\kappa}}}) - \tfrac{2}{3} H_i \dot{\lambda} \tag{3.114}$$

and when combined with the other rate equations gives the expression for the plastic consistency parameter as (noting that with the nine-component form $\underline{\mathbf{n}}^{\mathrm{T}} \underline{\mathbf{n}} = 1$)

$$\dot{\lambda} = \frac{G}{G^*} \underline{\mathbf{n}}^{\mathrm{T}} \dot{\underline{\mathbf{e}}} \tag{3.115}$$

* More general forms of hardening may be approximated by piecewise linear segments, thus making the present formulation quite general.

where

$$G^* = G + \tfrac{1}{3}(H_i + H_k) \qquad (3.116)$$

Substitution of Eq. (3.115) into Eq. (3.112), and using Eq. (3.42) to reduce to the six-component form, gives the rate form for stress–strain deviators as

$$\dot{\mathbf{s}} = 2G\left[\mathbf{I}_0 - \frac{2G}{G^*}\mathbf{n}\mathbf{n}^{\mathrm{T}}\right]\dot{\mathbf{e}} \qquad (3.117)$$

We note that for perfect plasticity $H_i = H_k = 0$ leading to $2G/G^* = 1$ and, thus, the elastic–plastic tangent for this special case is also here obtained.

Use of Eq. (3.117) in the rate form of Eq. (3.93) gives the final continuum elastic–plastic tangent

$$\mathbf{D}^*_{\mathrm{ep}} = K\mathbf{m}\mathbf{m}^{\mathrm{T}} + 2G\left[\mathbf{I}_0 - \frac{1}{3}\mathbf{m}\mathbf{m}^{\mathrm{T}} - \frac{2G}{G^*}\mathbf{n}\mathbf{n}^{\mathrm{T}}\right] \qquad (3.118)$$

This then establishes the well-known Prandtl–Reuss stress–strain relations generalized for linear isotropic and kinematic hardening.

Incremental return map form

The return map form for the equations is established by using a backward (Euler implicit) difference form as described previously (see Sec. 3.4.2). Omitting the subscript on the $n+1$ quantities the plastic strain equation becomes, using Eqs (3.106) and (3.110),

$$\underline{\mathbf{e}}^{\mathrm{p}} = \underline{\mathbf{e}}^{\mathrm{p}}_n + \Delta\lambda\underline{\mathbf{n}} \qquad (3.119)$$

and the accumulated (effective) plastic strain

$$\kappa = \kappa_n + \sqrt{\tfrac{2}{3}}\Delta\lambda \qquad (3.120)$$

Thus, now the discrete constitutive equation is

$$\underline{\mathbf{s}} = 2G(\underline{\mathbf{e}} - \underline{\mathbf{e}}^{\mathrm{p}}) \qquad (3.121)$$

the kinematic hardening is

$$\underline{\boldsymbol{\kappa}} = \underline{\boldsymbol{\kappa}}_n + \tfrac{2}{3}H_k\Delta\lambda\underline{\mathbf{n}} \qquad (3.122)$$

and the yield function is

$$F = |\underline{\mathbf{s}} - \underline{\boldsymbol{\kappa}}| - \sqrt{\tfrac{2}{3}}Y_n - \tfrac{2}{3}H_i\Delta\lambda \qquad (3.123)$$

where $Y_n = Y_0 + \sqrt{2/3}\,\kappa_n$.

The trial stress, which establishes whether plastic behaviour occurs, is given by

$$\underline{\mathbf{s}}^{\mathrm{TR}} = 2G(\underline{\mathbf{e}} - \underline{\mathbf{e}}^{\mathrm{p}}_n) \qquad (3.124)$$

which for situations where plasticity occurs permits the final stress to be given as

$$\underline{\mathbf{s}} = \underline{\mathbf{s}}^{\mathrm{TR}} - 2G\Delta\lambda\underline{\mathbf{n}} \qquad (3.125)$$

Using the definition of $\underline{\mathbf{n}}$, we may now combine the stress and kinematic hardening relations as

$$|\underline{\mathbf{s}} - \underline{\boldsymbol{\kappa}}|\,\underline{\mathbf{n}} = |\underline{\mathbf{s}}^{\mathrm{TR}} - \underline{\boldsymbol{\kappa}}_n|\,\underline{\mathbf{n}}^{\mathrm{TR}} - \left(2G + \tfrac{2}{3}H_k\right)\Delta\lambda\,\underline{\mathbf{n}} \tag{3.126}$$

and noting from this that we must have

$$\underline{\mathbf{n}}^{\mathrm{TR}} = \underline{\mathbf{n}} \tag{3.127}$$

we may solve the yield function directly for the consistency parameter as[14,34]

$$\Delta\lambda = \frac{|\underline{\mathbf{s}}^{\mathrm{TR}} - \underline{\boldsymbol{\kappa}}_n| - \sqrt{2/3}\,Y_n}{2G^*} \tag{3.128}$$

where G^* is given by Eq. (3.116).

We can also easily establish the relations for the consistent tangent matrix for this J_2 model. From Eqs (3.121) and (3.119) we obtain the incremental expression

$$d\underline{\mathbf{s}} = 2G\left[d\underline{\mathbf{e}} - \underline{\mathbf{n}}\,d\lambda - \Delta\lambda\,d\underline{\mathbf{n}}\right] \tag{3.129}$$

The increment of relation (3.127) gives[14]

$$d\underline{\mathbf{n}} = d\underline{\mathbf{n}}^{\mathrm{TR}} = \frac{2G}{|\underline{\mathbf{s}} - \underline{\boldsymbol{\kappa}}|}\left[\underline{\mathbf{I}} - \underline{\mathbf{n}}\,\underline{\mathbf{n}}^{\mathrm{T}}\right] d\underline{\mathbf{e}} \tag{3.130}$$

and from Eq. (3.128) we have

$$d\lambda = \frac{G}{G^*}\,\underline{\mathbf{n}}^{\mathrm{T}}\,d\underline{\mathbf{e}} \tag{3.131}$$

Substitution into Eq. (3.129) gives the consistent tangent matrix

$$d\underline{\mathbf{s}} = 2G\left[\left(1 - \frac{2G\Delta\lambda}{|\underline{\mathbf{s}}^{\mathrm{TR}} - \underline{\boldsymbol{\kappa}}_n|}\right)\underline{\mathbf{I}} - \left(\frac{G}{G^*} - \frac{2G\Delta\lambda}{|\underline{\mathbf{s}}^{\mathrm{TR}} - \underline{\boldsymbol{\kappa}}_n|}\right)\underline{\mathbf{n}}\,\underline{\mathbf{n}}^{\mathrm{T}}\right] d\underline{\mathbf{e}} \tag{3.132}$$

This may now be expressed in terms of the total strains, combined with the elastic volumetric term and reduced to six-components to give

$$\mathbf{D}^*_{\mathrm{ep}} = K\mathbf{m}\mathbf{m}^{\mathrm{T}} + 2G\left[\left(1 - \frac{2G\Delta\lambda}{|\underline{\mathbf{s}}^{\mathrm{TR}} - \underline{\boldsymbol{\kappa}}_n|}\right)\mathbf{I}_0 - \left(\frac{G}{G^*} - \frac{2G\Delta\lambda}{|\underline{\mathbf{s}}^{\mathrm{TR}} - \underline{\boldsymbol{\kappa}}_n|}\right)\mathbf{n}\mathbf{n}^{\mathrm{T}}\right] \tag{3.133}$$

We here note also that when $\Delta\lambda = 0$ the tangent for the return map becomes the continuum tangent, thus establishing consistency of form.

3.5.3 J_2 plane stress

The discussion in the previous part of this section may be applied to solve problems in plane strain, axisymmetry, and general three-dimensional behaviour. In plane strain and axisymmetric problems it is only necessary to note that some strain components are zero. For problems in plane stress, however, it is necessary to modify the algorithm to achieve an efficient solution process. In a plane stress process only the four stresses σ_x, σ_y, τ_{xy} and τ_{yx} need be considered. When considering deviatoric components, however, there are five components, s_x, s_y, s_z, s_{xy} and s_{yx}. The deviators may be expressed in terms of the independent stresses as

$$\underline{\mathbf{s}} = \begin{Bmatrix} s_x \\ s_y \\ s_z \\ s_{xy} \\ s_{yx} \end{Bmatrix} = \frac{1}{3} \begin{bmatrix} 2 & -1 & 0 & 0 \\ -1 & 2 & 0 & 0 \\ -1 & -1 & 0 & 0 \\ 0 & 0 & 3 & 0 \\ 0 & 0 & 0 & 3 \end{bmatrix} \begin{Bmatrix} \sigma_x \\ \sigma_y \\ \tau_{xy} \\ \tau_{yx} \end{Bmatrix} = \mathbf{P}_s \underline{\boldsymbol{\sigma}} \tag{3.134}$$

The Huber–von Mises yield function may be written as

$$F = \left[(\underline{\boldsymbol{\sigma}} - \underline{\boldsymbol{\kappa}})^{\mathrm{T}} \mathbf{P}_s^{\mathrm{T}} \mathbf{P}_s (\underline{\boldsymbol{\sigma}} - \underline{\boldsymbol{\kappa}}) \right]^{1/2} - \sqrt{\tfrac{2}{3}}\, Y(\kappa) \leqslant 0 \tag{3.135}$$

Expanding Eq. (3.135) gives the plane stress yield function

$$F = \left[\varsigma_x^2 - \varsigma_x \varsigma_y + \varsigma_y^2 + 1.5 (\varsigma_{xy}^2 + \varsigma_{yx}^2) \right]^{1/2} - Y(\kappa) \leqslant 0 \tag{3.136}$$

where

$$\varsigma_x = \sigma_x - \kappa_x, \qquad \varsigma_y = \sigma_y - \kappa_y, \qquad \varsigma_{xy} = \tau_{xy} - \kappa_{xy}, \qquad \varsigma_{yx} = \tau_{yx} - \kappa_{yx} \tag{3.137}$$

define stresses which are shifted by the kinematic hardening back stress. Plastic strain rates may now be computed by direct differentiation of the yield function, giving

$$\underline{\dot{\boldsymbol{\varepsilon}}}^{\mathrm{p}} = \dot{\lambda} F_{,\underline{\boldsymbol{\sigma}}} = \begin{Bmatrix} \dot{\varepsilon}_x \\ \dot{\varepsilon}_y \\ \dot{\varepsilon}_{xy} \\ \dot{\varepsilon}_{yx} \end{Bmatrix} = \frac{\dot{\lambda}}{2|\varsigma|} \begin{bmatrix} 2 & -1 & 0 & 0 \\ -1 & 2 & 0 & 0 \\ 0 & 0 & 3 & 0 \\ 0 & 0 & 0 & 3 \end{bmatrix} \begin{Bmatrix} \sigma_x \\ \sigma_y \\ \tau_{xy} \\ \tau_{yx} \end{Bmatrix} = \frac{\dot{\lambda}}{|\underline{\varsigma}|} \mathbf{A}_s \underline{\boldsymbol{\sigma}} \tag{3.138}$$

where $\mathbf{A}_s = \mathbf{P}_s^{\mathrm{T}} \mathbf{P}_s$. Similarly, the rate of the back stress for the kinematic hardening case is given by

$$\frac{1}{H_k} \underline{\dot{\boldsymbol{\kappa}}} = \frac{\dot{\lambda}}{|\underline{\varsigma}|} \mathbf{A}_s \underline{\boldsymbol{\sigma}} \tag{3.139}$$

The elastic components are computed by using the plane stress relation. Accordingly, for a plastic step the constitution is given by

$$\underline{\dot{\boldsymbol{\sigma}}} = \underline{\mathbf{D}} (\underline{\dot{\boldsymbol{\varepsilon}}} - \underline{\dot{\boldsymbol{\varepsilon}}}^{\mathrm{p}}) \tag{3.140}$$

where for isotropic behaviour

$$\underline{\mathbf{D}} = \frac{E}{1 - \nu^2} \begin{bmatrix} 1 & \nu & 0 & 0 \\ \nu & 1 & 0 & 0 \\ 0 & 0 & 1 - \nu & 0 \\ 0 & 0 & 0 & 1 - \nu \end{bmatrix} \tag{3.141}$$

with E the modulus of elasticity and ν the Poisson's ratio.

We note that for a J_2 model the volumetric plastic strain must always be zero; consequently, we can complete the determination of plastic strains at any instant by using

$$\varepsilon_z^{\mathrm{p}} = -\varepsilon_x^{\mathrm{p}} - \varepsilon_y^{\mathrm{p}} \tag{3.142}$$

This may be combined with the elastic strain given by

$$\varepsilon_x^{\mathrm{e}} = -\frac{\nu}{E}\left(\sigma_x + \sigma_y\right) \tag{3.143}$$

to compute the total strain ε_z and, thus, the thickness change. The solution process now follows the procedures given for the general return mapping case. A procedure which utilizes a spectral transformation on the elastic and plastic parts is given in references 14 and 50. The process given there is more elegant but lacks the clarity of working directly with the stress and plastic strain increments.

3.6 Generalized plasticity – non-associative case

Plastic behaviour characterized by irreversibility of stress paths and the development of permanent strain changes after a stress cycle can be described in a variety of ways. One form of such description has been given in Sec. 3.3. Another general method is presented here.

3.6.1 Non-associative case – frictional materials

This approach assumes *a priori* the existence of a rate process which may be written directly as

$$\dot{\underline{\boldsymbol{\sigma}}} = \underline{\mathbf{D}}^{*}\,\dot{\underline{\boldsymbol{\varepsilon}}} \tag{3.144}$$

in which the matrix $\underline{\mathbf{D}}^{*}$ depends not only on the stress $\underline{\boldsymbol{\sigma}}$ and the state of parameters $\underline{\boldsymbol{\kappa}}$, but also on the direction of the applied stress (or strain) rate $\dot{\underline{\boldsymbol{\sigma}}}$ (or $\dot{\underline{\boldsymbol{\varepsilon}}}$).[56] A slightly less ambitious description arises if we accept the dependence of $\mathbf{D}^{*}$ only on two directions – those of loading and unloading. If in the general stress space we specify a 'loading' direction by a unit vector $\underline{\mathbf{n}}$ given at every point (and also depending on the state parameters $\underline{\boldsymbol{\kappa}}$), as shown in Fig. 3.9, we can describe plastic loading and unloading by the sign of the projection $\underline{\mathbf{n}}^{\mathrm{T}}\,\dot{\underline{\boldsymbol{\sigma}}}$. Thus

$$\underline{\mathbf{n}}^{\mathrm{T}}\,\dot{\underline{\boldsymbol{\sigma}}} \quad \begin{cases} >0 & \text{for loading} \\ <0 & \text{for unloading} \end{cases} \tag{3.145}$$

while $\underline{\mathbf{n}}^{\mathrm{T}}\,\dot{\underline{\boldsymbol{\sigma}}} = 0$ is a neutral direction in which only elastic straining occurs. One can now write quite generally that

$$\dot{\underline{\boldsymbol{\sigma}}} = \begin{cases} \underline{\mathbf{D}}_{\mathrm{L}}^{*}\,\dot{\underline{\boldsymbol{\varepsilon}}} & \text{for loading} \\ \underline{\mathbf{D}}_{\mathrm{U}}^{*}\,\dot{\underline{\boldsymbol{\varepsilon}}} & \text{for unloading} \end{cases} \tag{3.146}$$

where the matrices $\underline{\mathbf{D}}_{\mathrm{L}}^{*}$ and $\underline{\mathbf{D}}_{\mathrm{U}}^{*}$ depend only on the state described by $\underline{\boldsymbol{\sigma}}$ and $\underline{\boldsymbol{\kappa}}$.

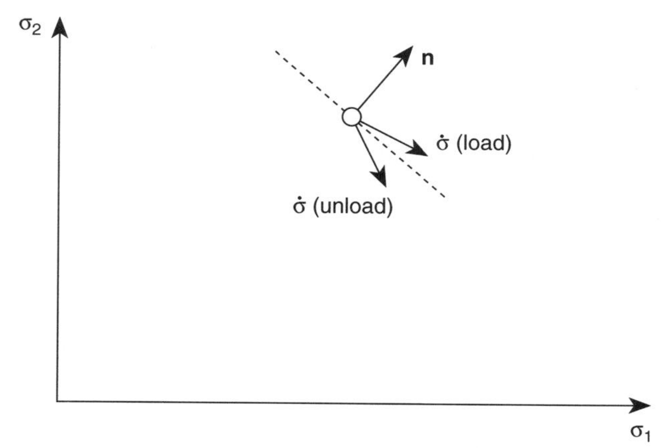

Fig. 3.9 Loading and unloading directions in stress space.

The specification of $\underline{\mathbf{D}}_L^*$ and $\underline{\mathbf{D}}_U^*$ must be such that in the neutral direction of the stress increment $\dot{\underline{\boldsymbol{\sigma}}}$ the strain rates corresponding to this are equal. Thus we require

$$\dot{\underline{\boldsymbol{\varepsilon}}} = (\underline{\mathbf{D}}_L^*)^{-1}\dot{\underline{\boldsymbol{\sigma}}} = \underline{\mathbf{D}}_U^{*-1}\dot{\underline{\boldsymbol{\sigma}}} \qquad \text{when } \underline{\mathbf{n}}^T\dot{\underline{\boldsymbol{\sigma}}} = 0 \tag{3.147}$$

A general way to achieve this end is to write

$$(\underline{\mathbf{D}}_L^*)^{-1} \equiv \underline{\mathbf{D}}^{-1} + \frac{1}{H_L}\underline{\mathbf{n}}_{gL}\underline{\mathbf{n}}^T \quad \text{and} \quad (\underline{\mathbf{D}}_U^*)^{-1} \equiv \underline{\mathbf{D}}^{-1} + \frac{1}{H_U}\underline{\mathbf{n}}_{gU}\underline{\mathbf{n}}^T \tag{3.148}$$

where $\underline{\mathbf{D}}$ is the elastic matrix, $\underline{\mathbf{n}}_{gL}$ and $\underline{\mathbf{n}}_{gU}$ are arbitrary unit stress vectors for loading and unloading directions, and H_L and H_U are appropriate plastic moduli which in general depend on $\underline{\boldsymbol{\sigma}}$ and $\underline{\boldsymbol{\kappa}}$.

The value of the tangent matrices $\underline{\mathbf{D}}_L^*$ and $\underline{\mathbf{D}}_U^*$ can be obtained by direct inversion if $H_{L/U} \neq 0$, but more generally can be deduced following procedures given in Sect. 3.3.4 or can be written directly using the *Sherman–Morrison–Woodbury formula*[57] as:

$$\underline{\mathbf{D}}_L^* = \underline{\mathbf{D}} - \frac{1}{H_L^*}\underline{\mathbf{D}}\,\underline{\mathbf{n}}_{gL}\underline{\mathbf{n}}^T\underline{\mathbf{D}} \qquad H_L^* = H_L + \underline{\mathbf{n}}^T\underline{\mathbf{D}}\,\underline{\mathbf{n}}_{gL} \tag{3.149}$$

This form resembles Eq. (3.72) and indeed its derivation is almost identical. We note further that $(\underline{\mathbf{D}}_L^*)^{-1}$ is now well behaved for H_L zero and a form identical to that of perfect plasticity is represented. Of course, a similar process is used to obtain $\underline{\mathbf{D}}_U^*$.

This simple and general description of *generalized plasticity* was introduced by Mróz and Zienkiewicz.[58,59] It allows:

1. the full model to be specified by a direct prescription of $\underline{\mathbf{n}}$, $\underline{\mathbf{n}}_g$ and H for loading and unloading at any point of the stress space;
2. existence of plasticity in both loading and unloading directions;
3. relative simplicity for description of experimental results when these are complex and when the existence of a yield surface of the kind encountered in ideal plasticity is uncertain.

For the above reasons the generalized plasticity forms have proved useful in describing the complex behaviour of soils.[60–64] Here other descriptions using various interpolations of $\underline{\mathbf{n}}$ and moduli form a unique yield surface, known as *bounding surface plasticity* models, are indeed particular forms of the above generalization and have proved to be useful.[65]

Classical plasticity is indeed a special case of the generalized models. Here the yield surface may be used to define a unit normal vector as

$$\underline{\mathbf{n}} = \frac{1}{[F^{\mathrm{T}}_{,\underline{\sigma}} F_{,\underline{\sigma}}]^{1/2}} F_{,\underline{\sigma}} \tag{3.150}$$

and the plastic potential may be used to define

$$\mathbf{n}_g = \frac{1}{[Q^{\mathrm{T}}_{,\underline{\sigma}} Q_{,\underline{\sigma}}]^{1/2}} Q_{,\underline{\sigma}} \tag{3.151}$$

where once again some care must be exercised in defining the matrix notation. Substitution of such values for the unit vectors into Eq. (3.149) will of course retrieve the original form of Eq. (3.72). However, interpretation of generalized plasticity in classical terms is more difficult.

The success of generalized plasticity in practical applications has allowed many complex phenomena of soil dynamics to be solved.[66,67] We shall refer to such applications later but in Fig. 3.10 we show how complex cyclic response with plastic loading and unloading can be followed.

While we have specified initially the loading and unloading directions in terms of the total stress rate $\dot{\underline{\sigma}}$ this definition ceases to apply when strain softening occurs and the plastic modulus H becomes negative. It is therefore more convenient to check the loading or unloading direction by the elastic stress increment $\dot{\underline{\sigma}}^{\mathrm{e}}$ of

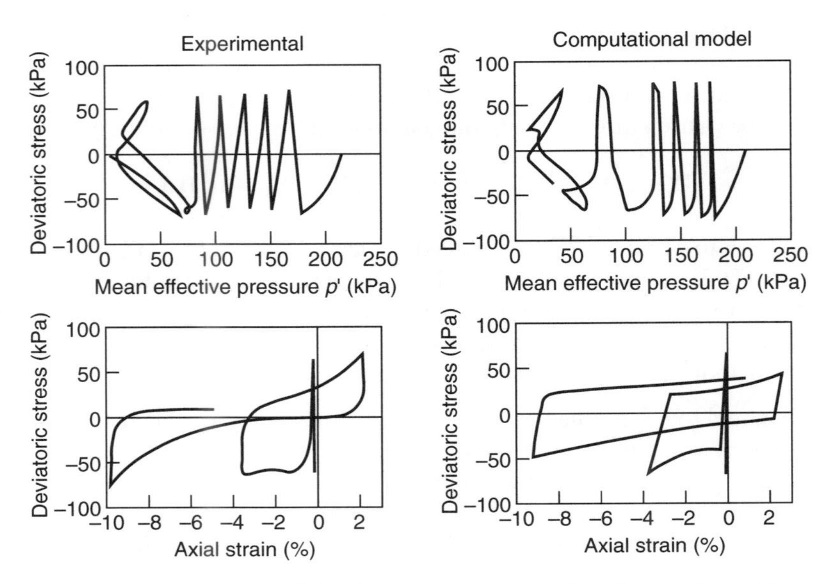

Fig. 3.10 A generalized plasticity model describing a very complex path, and comparison with experimental data. Undrained two-way cyclic loading of Nigata sand.[68] (Note that in an undrained soil test the fluid restrains all volumetric strains, and pore pressures develop; see Sec. 19.3.5 of Volume 1).

Eq. (3.65) and to specify

$$\underline{\mathbf{n}}^{\mathrm{T}}\dot{\underline{\boldsymbol{\sigma}}}^{e}\begin{cases} >0 & \text{for loading} \\ <0 & \text{for unloading} \end{cases} \tag{3.152}$$

This, of course, becomes identical to the previous definition of loading and unloading in the case of hardening.

3.6.2 Associative case – J_2 generalized plasticity

Another modification to the classical rate-independent approach is one in which the transition from an elastic to a fully plastic solution is accomplished with a smooth transition. This approach is useful in improving the match with experimental data for cyclic loading. A particularly simple form applicable to the J_2 model was introduced by Lubliner.[32,33] In this approach, the yield function is modified to a rate form directly and is expressed as

$$h(F)\dot{F} - \dot{\lambda} = 0 \tag{3.153}$$

where $h(F)$ is given by the function

$$h(F) = \frac{F}{(\beta - F)\delta + H\beta} \tag{3.154}$$

in which $H = H_i + H_k$, and δ, β are two positive parameters with dimension of stress. In particular, β is a distance between a *limit plastic state* and the current radius of the yield surface, and δ is a parameter controlling the approach to the limit state with increasing accumulated plastic strain.

On discretization and combination with the return map algorithm a rate-independent process is evident and again only minor modifications to the algorithm presented previously is necessary. A full description of the steps involved is given by Auricchio and Taylor.[34] Their paper also includes a development for the non-linear kinematic hardening model given in Eq. (3.62). In the case where the yield function is associative (i.e. $F = Q$) the use of the non-linear kinematic hardening model leads to an unsymmetric tangent stiffness when used with the return map algorithm. On the other hand, the generalized plasticity model is fully symmetric for this case.

In the next section we present further discussion on use of generalized plasticity to model the behaviour of frictional materials. In general, these involve use of non-associative models where the return map algorithm cannot be used effectively.

3.7 Some examples of plastic computation

The finite element discretization technique in plasticity problems follows precisely the same procedures as those of corresponding elasticity problems. Any of the elements already discussed can be used for problems in plane stress; however, for plane strain, axisymmetry, and three-dimensional problems it is usually necessary to use elements which perform well in *constrained* situations such as encountered for near incompressibility. For this latter class of problems use of mixed elements is generally recommended, although elements and constitutive forms that permit use of reduced integration may also be used.

The use of mixed elements is especially important in metal plasticity as the Huber–von Mises flow rule does not permit any volume changes. As the extent of plasticity spreads at the collapse load the deformation becomes nearly incompressible, and with conventional (fully integrated) displacement elements the system *locks* and a true collapse load cannot be obtained.[69,70]

Finally, we should remark that the possibility of solving plastic problems is not limited to a displacement and mixed formulation alone. Equilibrium fields and, indeed, most of the formulations described in Chapters 11 and 12 of Volume 1 form a suitable vehicle,[71–73] but owing to their convenient and easy interpretation displacement and mixed forms are most commonly used.

3.7.1 Perforated plate – plane stress solutions

Figure 3.11 shows the configuration and the division into simple triangular and quadrilateral elements. In this example plane stress conditions are assumed and

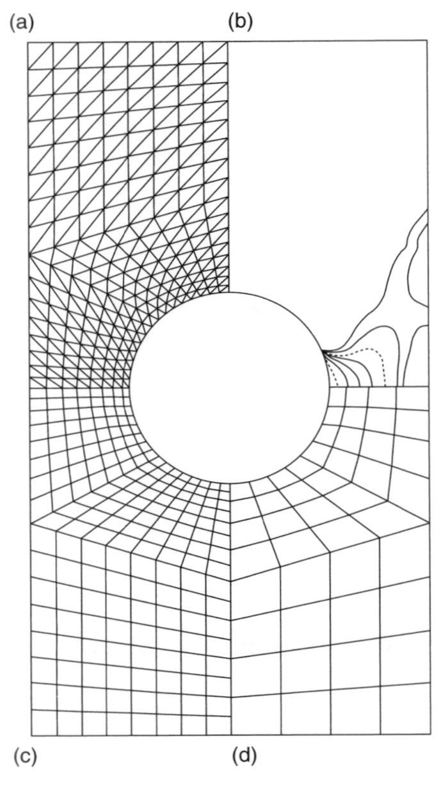

Fig. 3.11 Perforated plane stress tension strip: mesh used and development of plastic zones at loads of 0.55, 0.66, 0.75, 0.84, 0.92, 0.98, 1.02 times σ_y. (a) T3 triangles; (b) plastic zone spread; (c) Q4 quadrilaterals; (d) Q9 quadrilaterals.

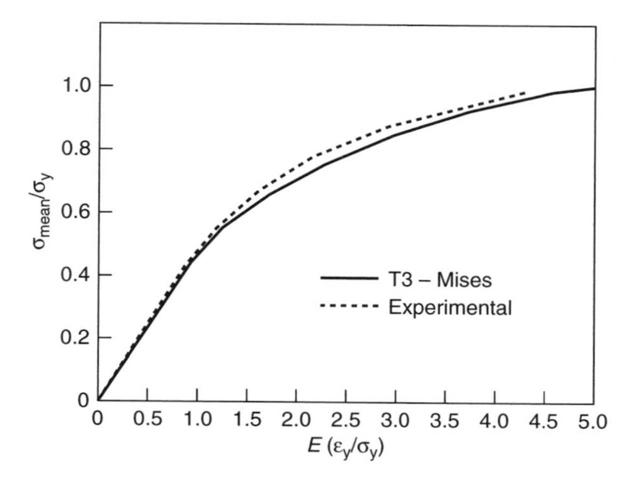

Fig. 3.12 Perforated plane stress tension strip: load deformation for strain hardening case ($H = 225\,\text{kg/mm}^2$).

solution is obtained for both ideal plasticity and strain hardening. This problem was studied experimentally by Theocaris and Marketos[74] and was first analysed using finite element methods by Marcal and King[75] and Zienkiewicz *et al.*[41] (See reference 5 for discussion on these early solutions.) The von Mises criterion is used and, in the case of strain hardening, a constant slope of the uniaxial hardening curve, H, is taken. Data for the problem, from reference 74, are $E = 7000\,\text{kg/mm}^2$, $H = 225\,\text{kg/mm}^2$ and $\sigma_y = 24.3\,\text{kg/mm}^2$. Poisson's ratio is not given but is here taken as in reference 41 as $\nu = 0.3$. To match a configuration considered in the experimental study a strip with 200 mm width and 360 mm length containing a central hole of 200 mm diameter. Using symmetry only one quadrant is discretized as shown in Fig. 3.11. Displacement boundary restraints are imposed for normal components on symmetry boundaries and the top boundary. Sliding is permitted, to impose the necessary zero tangential traction boundary condition. Loading is applied by a uniform non-zero normal displacement with equal increments. Displacement elements of type T3, Q4, and Q9 are used with the same nodal layout. Results for the three elements are nearly the same, with the extent of plastic zones indicated for various loads in Fig. 3.11 obtained using the Q4 element. The load–deformation characteristics of the problem are shown in Fig. 3.12 and compared to experimental results. The strain ε_y is the peak value occurring at the hole boundary. This plane stress problem is relatively insensitive to element type and load increment size. Indeed, doubling the number of elements resulted in small changes of all essential quantities.

3.7.2 Perforated plate – plane strain solutions

The problem described above is now analysed assuming a plane strain situation. Data are the same as for the plane stress case except the lateral boundaries are also restrained to create a zero normal displacement boundary condition. This increases

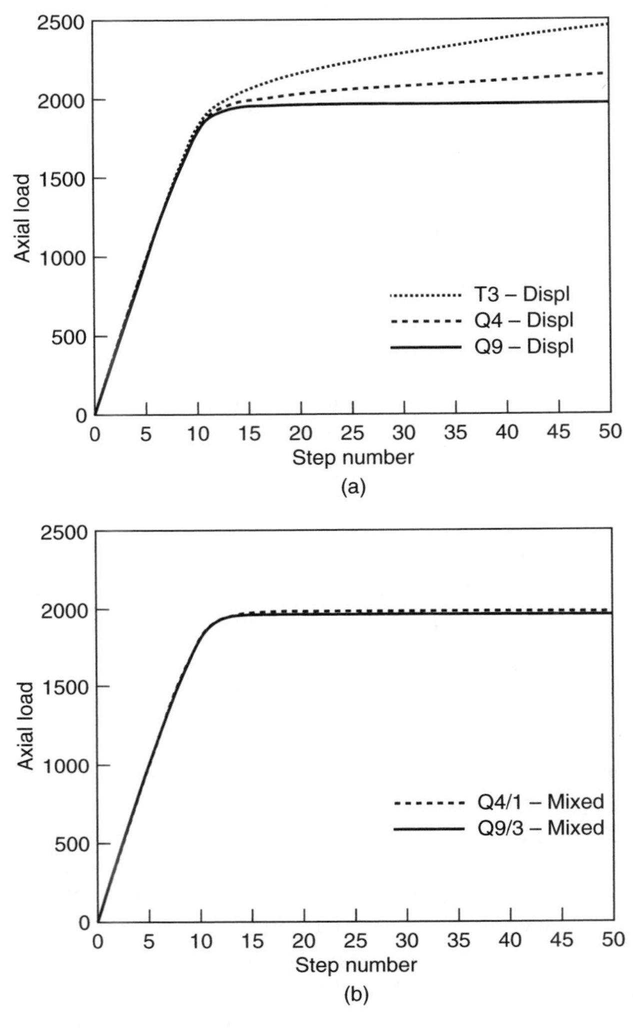

Fig. 3.13 Limit load behaviour for plane strain perforated strip: (a) displacement (displ.) formulation results; (b) mixed formulation results.

the confinement on the mesh and shows more clearly the locking condition cited previously. In Fig. 3.13 we plot the resultant axial load for each load step in the solution. Figure 3.13(a) shows results for the displacement model using T3, Q4, and Q9 elements and it is evident that the T3 and Q4 elements result in an erroneous increasing resultant load after the fully plastic state has developed. The Q9 element shows a clear limit state and indicates that higher order elements are less prone to locking (even though we have shown that for the fully incompressible state the Q9 displacement element will lock!). Figure 3.13(b) presents the same results for the Q4/1 and Q9/3 mixed elements and both give a clear limit load after the fully plastic state is reached.

3.7.3 Steel pressure vessel

This final example, for which test results obtained by Dinno and Gill[76] are available, illustrates a practical application, and the objectives are twofold. First, we show that this problem which can really be described as a thin shell can be adequately represented by a limit number (53) of isoparametric quadratic elements. Indeed, this model simulates well both the overall behaviour and the local stress concentration effects [Fig. 3.14(a)]. Second, this problem is loaded by an internal pressure and a solution is performed up to the 'collapse' point (where, because there is no hardening, the strains increase without limit) by incrementing the pressure rather than displacement. A comparison of calculated and measured deflections in Fig. 3.14(b) shows how well the objectives are achieved.

3.8 Basic formulation of creep problems

The phenomenon of 'creep' is manifested by a time-dependent deformation under a constant stress. Indeed the viscoelastic behaviour described in Sect. 3.2 is a particular model for linear creep. Here we shall deal with some non-linear models. Thus, in addition to an instantaneous strain, the material develops creep strains, $\boldsymbol{\varepsilon}^c$, which generally increase with duration of loading. The constitutive law of creep will usually be of a form in which the *rate of creep strain* is defined as some function of stresses and the total creep strains ($\boldsymbol{\varepsilon}^c$), that is,

$$\dot{\boldsymbol{\varepsilon}}^c \equiv \frac{\partial \boldsymbol{\varepsilon}^c}{\partial t} = \boldsymbol{\beta}(\boldsymbol{\sigma}, \boldsymbol{\varepsilon}^c) \tag{3.155}$$

If we consider the instantaneous strains are elastic ($\boldsymbol{\varepsilon}^e$), the total strain can be written again in an additive form as

$$\boldsymbol{\varepsilon} = \boldsymbol{\varepsilon}^e + \boldsymbol{\varepsilon}^c \tag{3.156}$$

with

$$\boldsymbol{\varepsilon}^e = \mathbf{D}^{-1}\boldsymbol{\sigma} \tag{3.157}$$

where we neglect any initial (thermal) strains or initial (residual) stresses. A special case of this form was considered for linear viscoelasticity in Sec. 3.2. Here we consider a more general non-linear approach commonly used in modelling behaviour of metals at elevated temperatures and in modelling creep in cementitious materials.

We can again use any of the time integration schemes considered above and approximate the constitutive equations in a form similar to that used in plasticity as

$$\begin{aligned} \boldsymbol{\sigma}_{n+1} &= \mathbf{D}(\boldsymbol{\varepsilon}_{n+1} - \boldsymbol{\varepsilon}^c_{n+1}) \\ \boldsymbol{\varepsilon}^c_{n+1} &= \boldsymbol{\varepsilon}^c_n + \Delta t\,\boldsymbol{\beta}_{n+\theta} \end{aligned} \tag{3.158}$$

where $\boldsymbol{\beta}_{n+\theta}$ is calculated as

$$\boldsymbol{\beta}_{n+\theta} = (1-\theta)\,\boldsymbol{\beta}_n + \theta\,\boldsymbol{\beta}_{n+1}$$

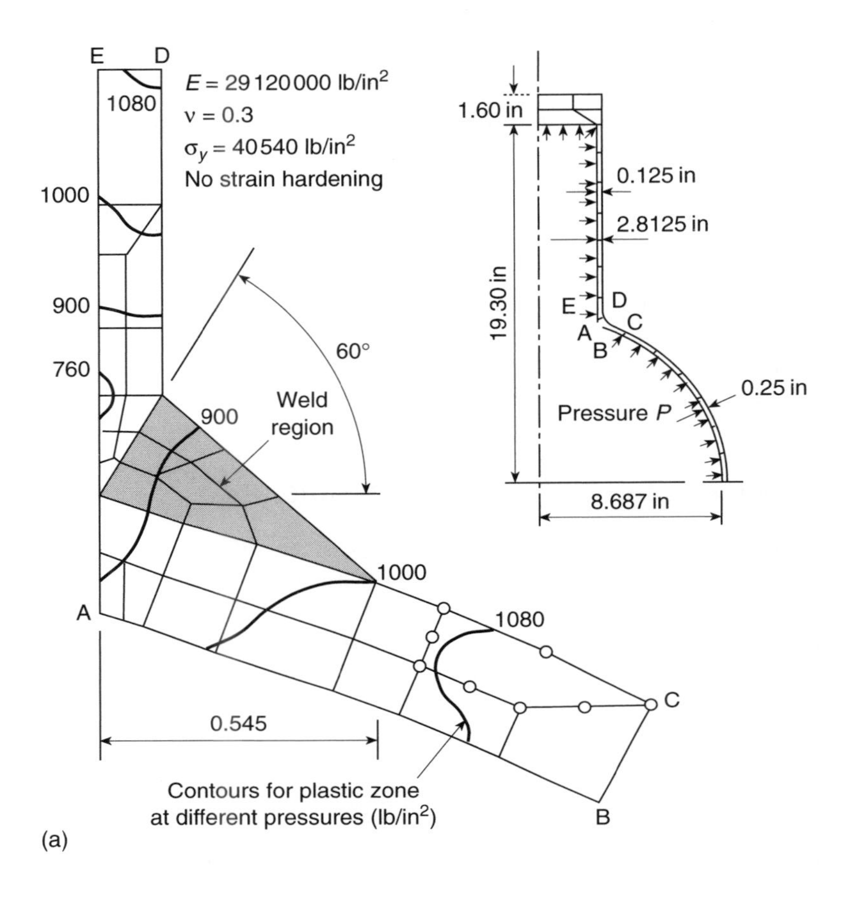

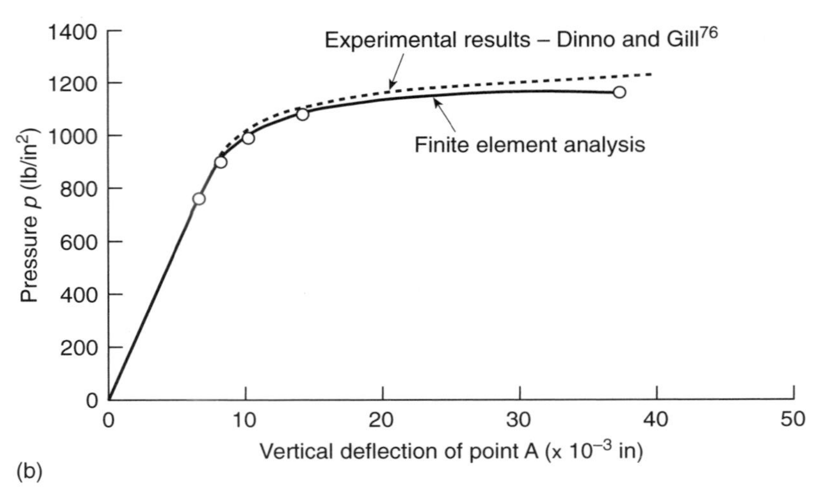

Fig. 3.14 Steel pressure vessel: (a) element subdivision and spread of plastic zones; (b) vertical deflection at point A with increasing pressure.

On eliminating $\Delta\varepsilon^c$ we have simply a non-linear equation

$$\mathbf{R}_{n+1} \equiv \boldsymbol{\varepsilon}_{n+1} - \mathbf{D}^{-1}\boldsymbol{\sigma}_{n+1} - \boldsymbol{\varepsilon}_n^c - \Delta t \boldsymbol{\beta}_{n+\theta} = \mathbf{0} \tag{3.159}$$

The system of equations can be solved iteratively using, say the Newton–Raphson procedure. Starting from some initial guess, say $\boldsymbol{\sigma}_{n+1} = \boldsymbol{\sigma}_n$ and an increment of strain is given by the finite element process, the general iterative/incremental solution can be written as

$$\mathbf{R}^{i+1} = \mathbf{0} = \mathbf{R}^i - (\mathbf{D}^{-1} + \Delta t \mathbf{C}_{n+1})\, d\boldsymbol{\sigma}_{n+1}^i \tag{3.160}$$

where

$$\mathbf{C}_{n+1} = \left.\frac{\partial \boldsymbol{\beta}}{\partial \boldsymbol{\sigma}}\right|_{n+\theta} = \theta \left.\frac{\partial \boldsymbol{\beta}}{\partial \boldsymbol{\sigma}}\right|_{n+1} \tag{3.161}$$

Solving this set of equations until the residual $\mathbf{R}$ is zero we obtain a set of stresses $\boldsymbol{\sigma}_{n+1}$ and tangent matrix

$$\mathbf{D}_{n+1}^* \equiv \left[\mathbf{D}^{-1} + \Delta t \mathbf{C}_{n+1}\right]^{-1} \tag{3.162}$$

which may once again be used to perform any needed iterations on the finite element equilibrium equations. The iterative computation that follows is very similar to that used in plasticity, but here Δt is an actual time and the solution becomes *rate dependent*.

While in plasticity we have generally used implicit (backward difference) procedures; here many simple alternatives are possible. In particular, two schemes with a single iterative step are popular.

3.8.1 Fully explicit solutions

'Initial strain' procedure: $\theta = 0$

Here, from Eqs (3.161) and (3.162) we see that

$$\mathbf{C}_{n+1} = \mathbf{0} \qquad \text{and} \qquad \mathbf{D}_{n+1}^* = \mathbf{D} \tag{3.163}$$

Thus, from Eq. (3.159) we obtain

$$\boldsymbol{\sigma}_{n+1} = \mathbf{D}[\boldsymbol{\varepsilon}_{n+1} - \boldsymbol{\varepsilon}_n^c - \Delta t \boldsymbol{\beta}_n] \tag{3.164}$$

which may be used in Eq. (1.19) of Chapter 1 to satisfy a discretized equilibrium equation. We note that this form will lead to a standard elastic stiffness matrix. This, of course, is equivalent to evaluating the increment of creep strain from the initial stress values at each time t_n and is exceedingly simple to calculate. While the process has been popular since the earliest days of finite elements[77–79] it is obviously less accurate for a finite step than other alternatives. Of course accuracy will improve if small time steps are used in such calculations. Further, if the time step is too large, unstable results will be obtained. Thus it is necessary for

$$\Delta t \leqslant \Delta t_{\text{crit}} \tag{3.165}$$

where Δt_{crit} is determined in a suitable manner (see Chapter 18, Volume 1).

A 'rule of thumb' that proves quite effective in practice is that the increment of creep strain should not exceed one half the total elastic strain[80]

$$\Delta t \left[\boldsymbol{\beta}_n^{\mathrm{T}} \boldsymbol{\beta}_n\right]^{1/2} \leqslant \tfrac{1}{2} \left[(\boldsymbol{\varepsilon}^{\mathrm{e}})^{\mathrm{T}} \boldsymbol{\varepsilon}^{\mathrm{e}}\right]^{1/2} \tag{3.166}$$

Fully explicit process with modified stiffness: $\frac{1}{2} \leqslant \theta \leqslant 1$

Here the main difference from the first explicit process is that the matrix $\mathbf{C}$ is not equal to zero but within a single step is taken as a constant, that is,

$$\mathbf{C}_{n+1} \approx \theta \left.\frac{\partial \boldsymbol{\beta}}{\partial \boldsymbol{\sigma}}\right|_n \tag{3.167}$$

This is equivalent to a modified Newton–Raphson scheme in which the tangent is held constant at its initial value in the step. Now

$$\mathbf{D}^*_{n+1} = \left[\mathbf{D}^{-1} + \Delta t \mathbf{C}_{n+1}\right]^{-1}$$

This process is more expensive than the simple explicit one previously mentioned, as the finite element tangent matrix has to be formed and solved for every time step. Further, such matrices can be non-symmetric, adding to computational expense.

Neither of the simplified iteration procedures described above give any attention to errors introduced in the estimates of the creep strain. However, for accuracy the iterative process with $\theta \geqslant \frac{1}{2}$ is recommended. Such full iterative procedures were introduced by Cyr and Teter,[81] and later by Zienkiewicz and co-workers.[82,83]

We shall note that the process has much similarity with iterative solutions of plastic problems of Sec. 3.3 in the case of viscoplasticity, which we shall discuss in the next section.

3.9 Viscoplasticity – a generalization

3.9.1 General remarks

The purely plastic behaviour of solids postulated in Sec. 3.3 is probably a fiction as the maximum stress that can be carried is invariably associated with the rate at which this is applied. A purely elasto-plastic behaviour in a uniaxial loading is described in a model of Fig. 3.15(a) in which the plastic strain rate is zero for stresses below yield, that is,

$$\dot{\varepsilon}^{\mathrm{p}} = 0 \qquad \text{if } |\sigma - \sigma_y| < 0 \quad \text{and} \quad |\sigma| > 0$$

and $\dot{\varepsilon}^{\mathrm{p}}$ is indeterminate when $\sigma - \sigma_y = 0$.

An elasto-viscoplastic material, on the other hand, can be modelled as shown in Fig. 3.15(b), where a dashpot is placed in parallel with the plastic element. Now stresses can exceed σ_y for strain rates other than zero.

The viscoplastic (or creep) strain rate is now given by a general expression

$$\dot{\varepsilon}^{\mathrm{vp}} = \gamma \langle \phi(\sigma - \sigma_y) \rangle \tag{3.168}$$

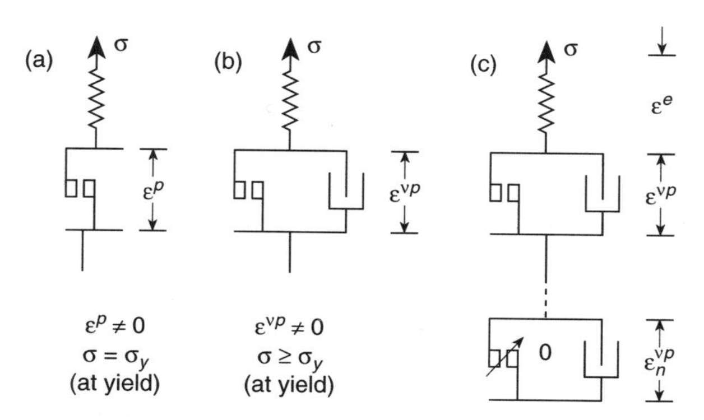

Fig. 3.15 (a) Elastoplastic; (b) elasto-viscoplastic; (c) series of elasto-viscoplastic models.

where the arbitrary function ϕ is such that

$$\begin{aligned} \langle \phi(\sigma - \sigma_y) \rangle &= 0 && \text{if } |\sigma - \sigma_y| \leqslant 0 \\ \langle \phi(\sigma - \sigma_y) \rangle &= \phi(\sigma - \sigma_y) && \text{if } |\sigma - \sigma_y| > 0 \end{aligned} \tag{3.169}$$

The model suggested is, in fact, of a creep-type category described in the previous sections and often is more realistic than that of classical plasticity.

A viscoplastic model for a general stress state is given here and follows precisely the arguments of the plasticity section. In a three-dimensional context ϕ becomes a function of the yield condition $F(\boldsymbol{\sigma}, \boldsymbol{\kappa}, \kappa)$ defined in Eq. (3.44). If this is less than zero, no 'plastic' flow will occur. To include the viscoplastic behaviour we modify Eq. (3.44) as

$$\gamma \langle \phi(F) \rangle - \dot{\lambda} = 0 \tag{3.170}$$

and use Eq. (3.45) to define the plastic strain. Equation (3.175) implies

$$\langle \phi(F) \rangle = \begin{cases} 0 & \text{if } F \leqslant 0 \\ \phi(F) & \text{if } F > 0 \end{cases} \tag{3.171}$$

and γ is some 'viscosity' parameter. Once again *associated* or *non-associated* flows can be invoked, depending on whether $F = Q$ or not. Further, any of the yield surfaces described in Sec. 3.3.1 and hardening forms described in Sec. 3.3.3 can be used to define the appropriate flow in detail. For simplicity, $\phi(F) = F^m$ where m is a positive power often used to define the viscoplastic rate effects in Eq. (3.170).[84]

The concept of viscoplasticity in one of its earliest versions was introduced by Bingham in 1922[85] and a survey of such modelling is given in references 86 and 88. The computational procedure of using the viscoplastic model can follow any of the general methods described in Sec. 3.4. Early applications commonly used the straightforward Euler (explicit) method.[89–93] The stability requirements for this approach have been considered for several types of yield conditions by Cormeau.[94] A *tangential* process can again be used, but unless the viscoplastic flow is associated ($F = Q$), non-symmetric systems of equations have to be solved at each step. Use of an explicit method will yield solution for the associative and non-associative cases and the system matrix remains symmetric. This process is thus similar to that of a modified Newton–Raphson method (initial stress method) and is quite efficient. Indeed within the stability limit it has been shown that use of an over relaxation method leads to rapid convergence.

3.9.2 Iterative solution

The complete iterative solution scheme for viscoplasticity is identical to that used in plasticity except for the use of Eq. (3.170) instead of Eq. (3.44). To underline this similarity we consider the constitutive model without hardening and use the return map implicit algorithm. The linearized relations are identical except for the treatment of relation (3.175). The form becomes

$$\begin{bmatrix} \mathbf{D}^{-1} + \Delta\lambda Q_{,\sigma\sigma} & Q_{,\sigma} \\ \phi' F_{,\sigma}^{\mathrm{T}} & -H_i - \dfrac{1}{\gamma\Delta t} \end{bmatrix}_{n+1}^{i} \begin{Bmatrix} d\boldsymbol{\sigma}^i \\ d\lambda^i \end{Bmatrix} = \begin{Bmatrix} \mathbf{R}_\sigma^i \\ r^i \end{Bmatrix} \tag{3.172}$$

where the discrete residual for Eq. (3.171) is given by

$$r_n = -\phi(F)_n + \frac{1}{\gamma\Delta t}\Delta\lambda_n \tag{3.173}$$

and

$$\phi' = \frac{\mathrm{d}\phi}{\mathrm{d}F}$$

Now the equations are almost identical to those of plasticity [see Eq. (3.87)], with differences appearing only in the ϕ' and $1/(\gamma\Delta t)$ terms.

Again, a consistent tangent can be obtained by elimination of the $d\lambda^i$ and a general iterative scheme is once more available.

Indeed, as expected, $\gamma\Delta t = \infty$ will now correspond to the exact plasticity solution. This will always be reached by any solution tending to steady state. However, for transient situations this is not the case and use of finite values for $\gamma\Delta t$ will invariably lead to some rate effects being present in the solution.

The viscoplastic laws can easily be generalized to include a series of components, as shown in Fig. 3.15(c). Now we write

$$\dot{\boldsymbol{\varepsilon}}^{\mathrm{v}} = \dot{\boldsymbol{\varepsilon}}_1^{\mathrm{v}} + \dot{\boldsymbol{\varepsilon}}_2^{\mathrm{v}} + \cdots \tag{3.179}$$

and again the standard formulation suffices. If, as shown in the last element of Fig. 3.15(c), the plastic yield is set to zero, a 'pure' creep situation arises in which flow occurs at all stress levels. If a finite value is in a term a corresponding rate equation for the associated $\dot{\lambda}_j$ must be used. This is similar to the Koiter treatment for multi-surface plasticity.[17,95]

The use of the Duvaut and Lions[88] approach modifies the return map algorithm for a rate-independent plasticity solution. Once this solution is available a reduction in the value of $\Delta\lambda$ is computed to account for rate effects. The interested reader should consult references 14 and 28 for additional information on this approach.

3.9.3 Creep of metals

If an associated form of viscoplasticity using the von Mises yield criterion of Eq. (3.99) is considered the viscoplastic strain rate can be written as

$$\dot{\boldsymbol{\varepsilon}}^{\mathrm{vp}} = \dot{\lambda}\frac{\partial|\mathbf{s}|}{\partial\boldsymbol{\sigma}} = \dot{\lambda}\mathbf{n} \tag{3.175}$$

with the rate expressed again as

$$\dot{\lambda} = \gamma \langle \bar{\sigma} - \sigma_y \rangle \tag{3.176}$$

If σ_y, the yield stress, is set to zero we can write the above as

$$\dot{\boldsymbol{\varepsilon}}^{\mathrm{vp}} = \gamma \bar{\sigma}^m \mathbf{n} \tag{3.177}$$

and we obtain the well-known Norton–Soderberg creep law. In this, generally the parameter γ is a function of time, temperature, and the total creep strain (e.g. the analogue to the plastic strain $\boldsymbol{\varepsilon}^{\mathrm{p}}$). For a survey of such laws the reader can consult specialized references.[96,97]

An example initially solved using a large number of triangular elements[83] is presented in Fig. 3.16, where a much smaller number of isoparametric quadrilaterals are used in a general viscoplastic program.[93]

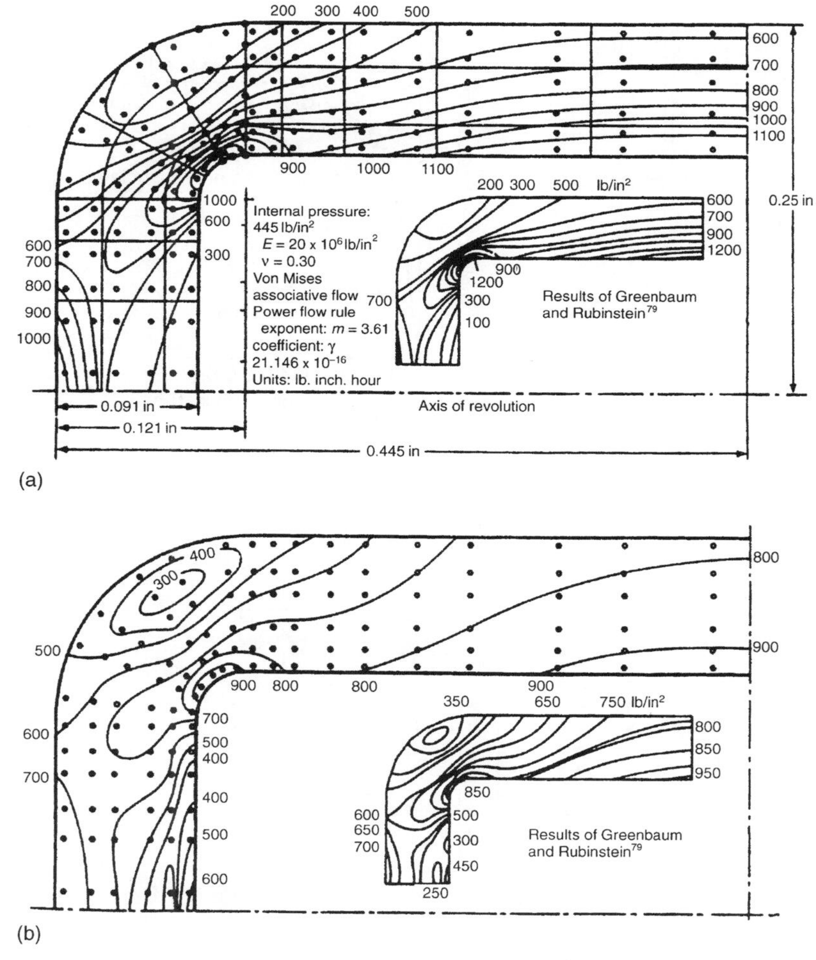

Fig. 3.16 Creep in a pressure vessel: (a) mesh end effective stress contours at start of pressurization; (b) effective stress contours 3 h after pressurization.

3.9.4 Soil mechanics applications

As we have already mentioned, the viscoplastic model provides a simple and effective tool for the solution of plasticity problems in which transient effects are absent. This includes many classical problems which have been solved in references 93 and 98, and the reader is directed there for details. In this section some problems of soil mechanics are discussed in which the facility of the process for solving non-associated behaviour is demonstrated.[99] The whole subject of the behaviour of soils and similar porous media is one in which much yet needs to be done to formulate good constitutive models. For a fuller discussion the reader is referred to texts, conferences, and papers on the subject.[100,101]

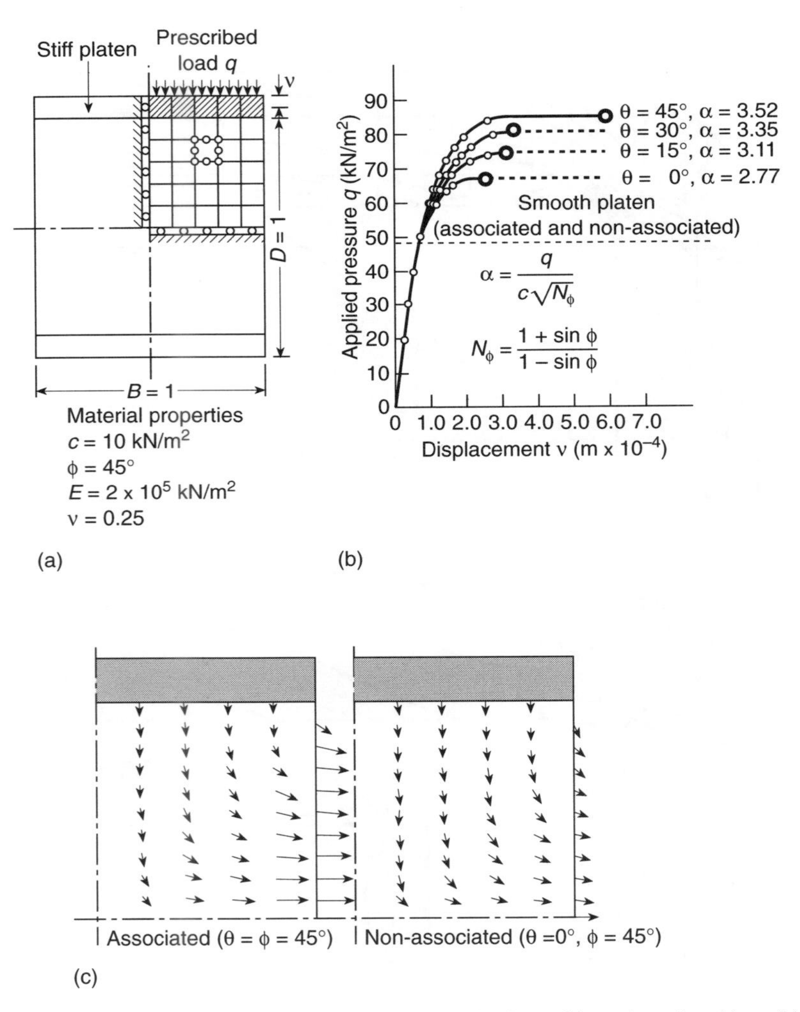

Fig. 3.17 Uniaxial, axisymmetric compression between rough plates: (a) mesh and problem; (b) pressure displacement result; (c) plastic flow velocity patterns.

One particular controversy centres on the 'associated' versus 'non-associated' nature of soil behaviour. In the example of Fig. 3.17, dealing with an axisymmetric sample, the effect of these different assumptions is investigated.[86] Here a Mohr–Coulomb law is used to describe the yield surface, and a similar form, but with a different friction angle, $\bar{\phi}$, is used in the plastic potential, thus reducing the plastic potential to the Tresca form of Fig. 3.8 when $\bar{\phi} = 0$ and suppressing volumetric strain changes. As can be seen from the results, only moderate changes in collapse load occur, although very appreciable differences in plastic flow patterns exist.

Figure 3.18 shows a similar study carried out for an embankment. Here, despite quite different flow patterns, a prediction of collapse load was almost unaffected by the flow rate law assumed.

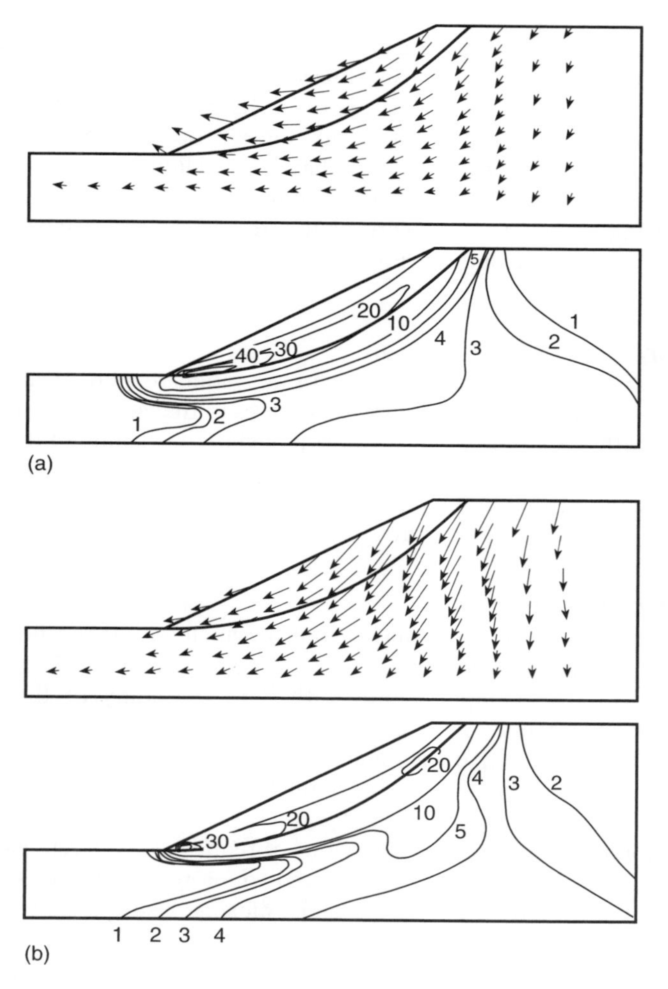

Fig. 3.18 Embankment under action of gravity, relative plastic flow velocities at collapse, and effective shear strain rate contours at collapse: (a) associative behaviour; (b) non-associative (zero volume change) behaviour.

The non-associative plasticity, in essence caused by frictional behaviour, may lead to non-uniqueness of solution. The equivalent viscoplastic form is, however, always unique and hence viscoplasticity is on occasion used as a *regularizing* procedure.

3.10 Some special problems of brittle materials

3.10.1 The no-tension material

A hypothetical material capable of sustaining only compressive stresses and straining without resistance in tension is in many respects similar to an ideal plastic material. While in practice such an ideal material does not exist, it gives a reasonable approximation of the behaviour of randomly jointed rock and other granular materials. While an explicit stress–strain relation cannot be generally written, it suffices to carry out the analysis elastically and wherever tensile stresses develop to reduce these to zero. The initial stress (modified Newton–Raphson) process here is natural and indeed was developed in this context.[102] The steps of calculation are obvious but it is important to remember that the *principal tensile stresses* have to be eliminated.

The 'constitutive' law as stated above can at best approximate to the true situation, no account being taken of the closure of fissures on reapplication of compressive stresses. However, these results certainly give a clear insight into the behaviour of real rock structures.

An underground power station

Figure 3.19(a) and (b) shows an application of this model to a practical problem.[102] In Fig. 3.19(a) an elastic solution is shown for stresses in the vicinity of an underground power station with 'rock bolt' pre-stressing applied in the vicinity of the opening. The zones in which tension exists are indicated. In Fig. 3.19(b) a *no-tension* solution is given for the same problem, indicating the rather small general redistribution and the zones where 'cracking' has occurred.

Reinforced concrete

A variant on this type of material may be one in which a finite tensile strength exists but when this is once exceeded the strength drops to zero (on fissuring). Such an analysis was used by Valliappan and Nath[103] in the study of the behaviour of reinforced concrete beams. Good correlation with experimental results for under-reinforced beams (in which development of compressive yield is not important) have been obtained. The beam is one for which test results were obtained by Krahl *et al.*[104] Figure 3.20 shows some relevant results.

Much development work on the behaviour of reinforced concrete has taken place, with various plasticity forms being introduced to allow for compressive failure and procedures that take into account the crack-closing history. References 105 and 106 list some of the basic approaches on this subject.

The subject of analysis of reinforced concrete has proved to be of great importance in recent years and publications in this field are proliferating. Publications 107 to 110 guide the reader to current practice in this field.

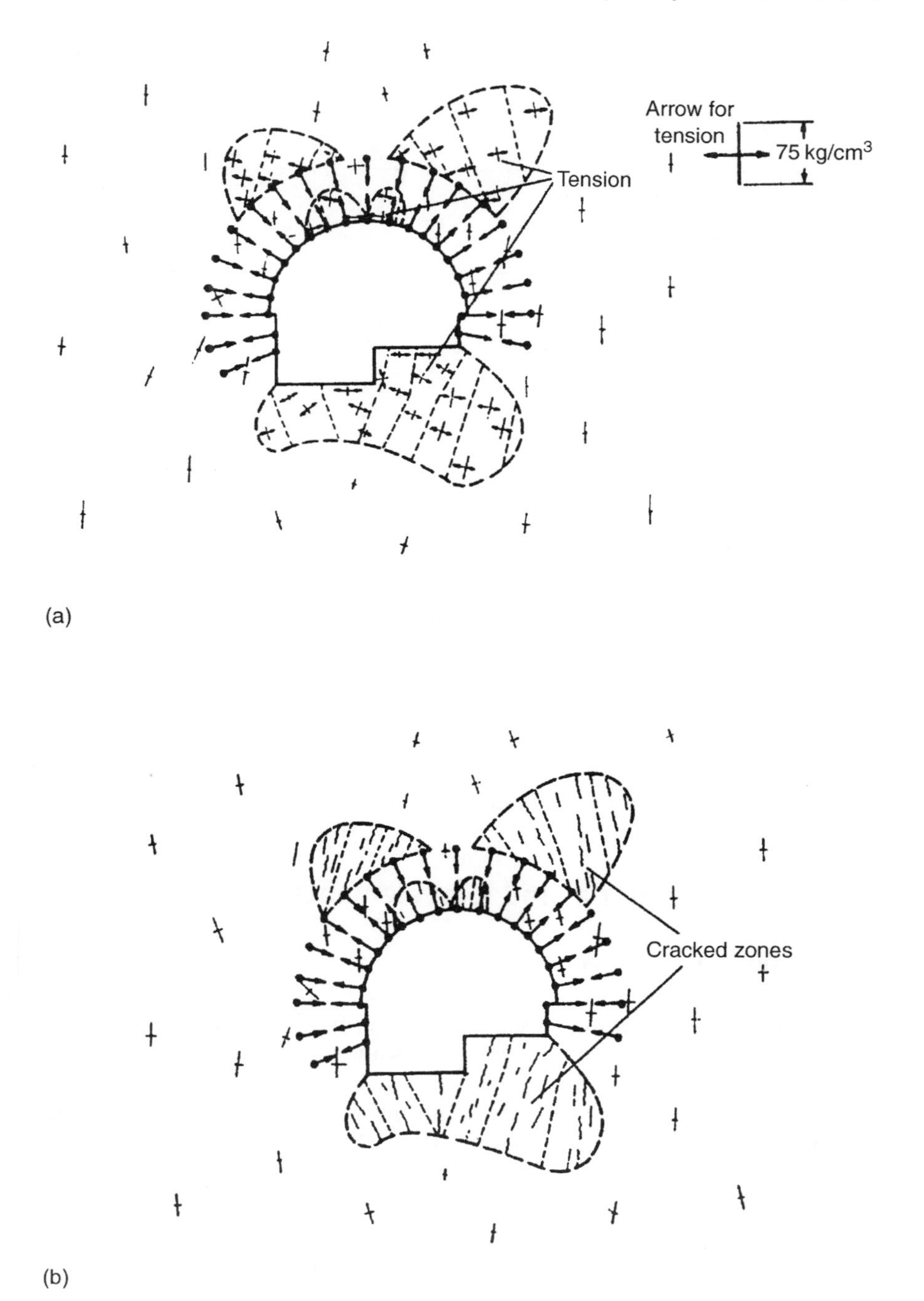

Fig. 3.19 Underground power station: gravity and prestressing loads. (a) Elastic stresses; (b) 'no-tension' stresses.

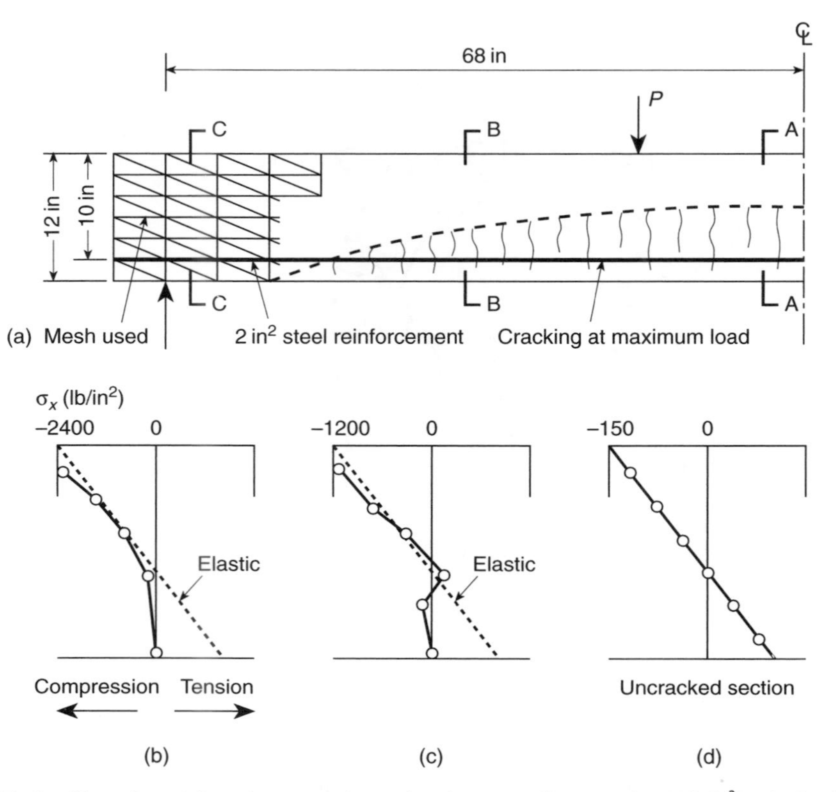

Fig. 3.20 Cracking of a reinforced concrete beam (maximum tensile strength 200 lb/in^2). Distribution of stresses at various sections.[103] (a) Mesh used; (b) section AA; (c) section BB; (d) section CC.

3.10.2 'Laminar' material and joint elements

Another idealized material model is one that is assumed to be built up of a large number of elastic and inelastic laminae. When under compression, these can transmit shear stress parallel to their direction – providing this does not exceed the frictional resistance. No tensile stresses can, however, be transmitted in the normal direction to the laminae.

This idealized material has obvious uses in the study of rock masses with parallel joints but has much wider applicability. Figure 3.21 shows a two-dimensional situation involving such a material. With a local coordinate axis x' oriented in the direction of the laminae we can write for a simple Coulomb friction joint

$$\begin{aligned} |\tau_{x'y'}| &< \mu\sigma_{y'} \quad \text{if } \sigma_{y'} \leqslant 0 \\ \sigma_{y'} &= 0 \quad\quad\;\; \text{if } \varepsilon_{y'} > 0 \end{aligned} \tag{3.178}$$

for stresses at which purely elastic behaviour occurs. In the above, μ is the friction coefficient applicable between the laminae.

If elastic stresses exceed the limits imposed the stresses have to be reduced to the limiting values given above. The application of the initial stress process in this

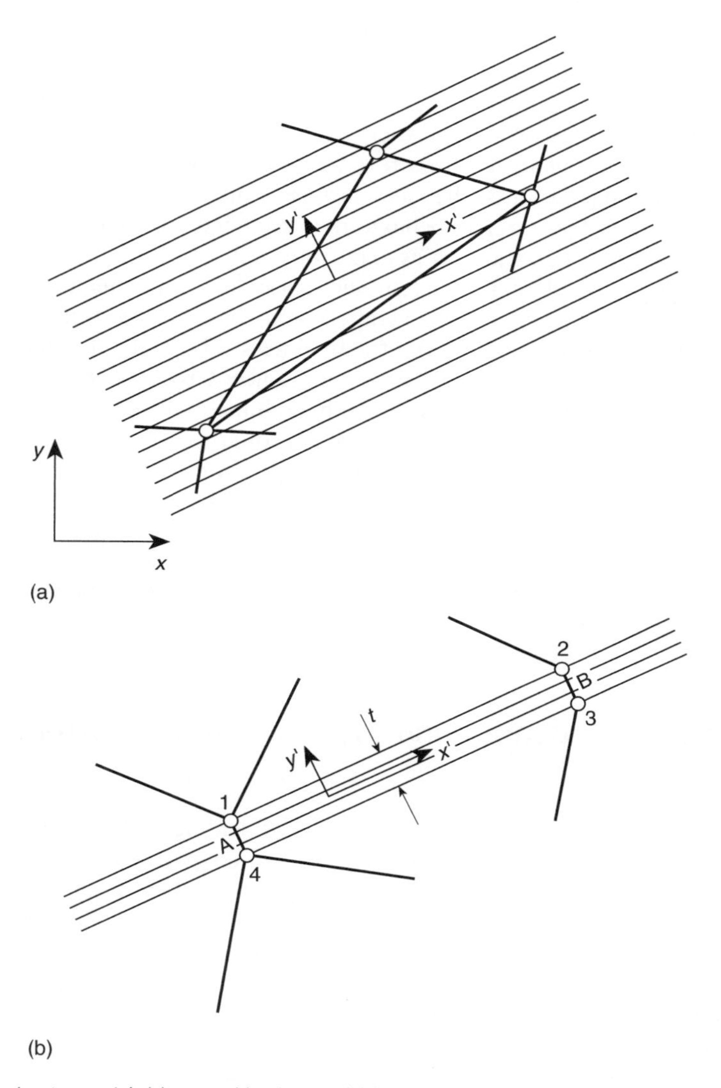

Fig. 3.21 'Laminar' material: (a) general laminarity; (b) laminar in narrow joint.

context is again self-evident, and the problem is very similar to that implied in the no-tension material of the previous section. At each step of elastic calculation, first the existence of tensile stresses $\sigma_{y'}$ is checked and, if these develop, a corrective initial stress reducing these and the shearing stresses to zero is applied. If $\sigma_{y'}$ stresses are compressive, the absolute magnitude of the shearing stresses $\tau_{x'y'}$ are checked again; if these exceed the value given by Eq. (3.178) they are reduced to their proper limit.

However, such a procedure poses the question of the manner in which the stresses are reduced, as two components have to be considered. It is, therefore, preferable to use the statements of relations (3.178) as definitions of plastic yield surfaces (F). The assumption of additional plastic potentials (Q) will now define the flow, and we note that associated behaviour, with Eq. (3.178) used as the potential, will imply a simultaneous separation and sliding of the laminae (as the corresponding strain rates $\dot{\gamma}_{x'y'}$ and $\dot{\varepsilon}_{y'}$ are finite). Non-associated plasticity (or viscoplasticity) techniques have therefore to be used. Once again, if stress reversal is possible it is necessary to note the opening of the laminae, that is, the yield surface is made strain dependent.

In some instances the laminar behaviour is confined to a narrow joint between relatively homogeneous elastic masses. This may well be of a nature of a geological fault or a major crushed rock zone. In such cases it is convenient to use narrow, generally rectangular elements whose geometry may be specified by mean coordinates of two ends A and B [Fig. 3.21(b)] and the thickness. The element still has, however, separate points of continuity (1–4) with the adjacent rock mass.[111,112] Such joint elements can be simple rectangles, as shown here, but equally can take more complex shapes if represented by using isoparametric coordinates.

Laminations may not be confined to one direction only – and indeed the interlaminar material itself may possess a plastic limit. The use of such multilaminate models in the context of rock mechanics has proved very effective;[113] with a random distribution of laminations we return, of course, to a typical soil-like material, and the possibilities of extending such models to obtain new and interesting constitutive relations have been highlighted by Pande and Sharma.[114]

3.11 Non-uniqueness and localization in elasto-plastic deformations

In the preceding sections the general processes of dealing with complex, non-linear constitutive relations have been examined and some particular applications were discussed. Clearly, the subject is large and of great practical importance; however, presentation in a single chapter is not practical or possible. For different materials alternate forms of constitutive relations can be proposed and experimentally verified. Once such constitutive relations are available the processes of this chapter serve as a guide for constructing effective numerical solution strategies. Indeed, it is possible to build standard computing systems applicable to a wide variety of material properties in which new specifications of behaviour may be inserted.

What must be restated once more is that, in non-linear problems:

1. non-uniqueness of solution may arise;
2. convergence can never be, *a priori*, guaranteed;
3. the cost of solution is invariably much greater than it is in linear solutions.

Here, of course, the item of most serious concern is the first one, that is, that of non-uniqueness, which could lead to a physically irrelevant solution even if numerical convergence occurred and possibly large computational expense was incurred. Such non-uniqueness may be due to several reasons in elasto-plastic computations:

1. the existence of corners in the yield (or potential) surfaces at which the gradients are not uniquely defined;

2. the use of a *non-associated* formulation[18,115,116] (to which we have already referred in Sec. 3.9.4);
3. the development of strain softening and localization.[117,118]

The first problem is the least serious and can readily be avoided by modifying the yield (or potential) surface forms to avoid corners. A simple modification of the Mohr–Coulomb (or Tresca) surface expressions [Eq. (3.100)] is easily achieved by writing[53]

$$F = \sigma_m \sin\phi - c\cos\phi + \frac{\bar{\sigma}}{g(\theta)} \tag{3.179}$$

where

$$g(\theta) = \frac{2K}{(1+K) - (1-K)\sin 3\theta}$$

and

$$K = \frac{3 - \sin\phi}{3 + \sin\phi}$$

Figure 3.22 shows how the angular section of the Mohr–Coulomb surface in the Π plane (constant σ_m) now becomes rounded. Similar procedures have been suggested by others.[119] An alternative to smoothing is to introduce a multisurface model and use a solution process which gives unique results for a corner.[14,28]

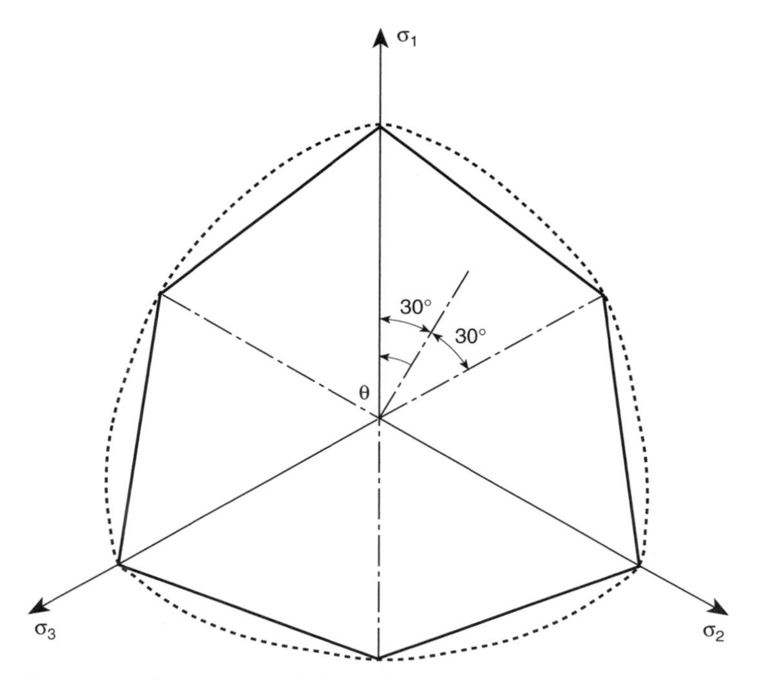

Fig. 3.22 Π plane section of Mohr–Coulomb yield surface in principal stress space, with $\phi = 25°$ (solid line); smooth approximation of Eq. (3.183) (dotted line).

Much more serious are the second and third possible causes of non-uniqueness mentioned above. Here, theoretical non-uniqueness can be avoided by considering the plastic deformation to be a limit state of viscoplastic behaviour in a manner we have already referred to in Sec. 3.9. Such a process, mathematically known as *regularization*, has allowed us to obtain many realistic solutions for both non-associative and strain softening behaviour in problems which are subjected to steady-state or quasi-static loading, as already shown. For fully transient cases, however, the process is quite delicate and much care is needed to obtain a valid regularization.

However, on occasion (though not invariably), both forms of behaviour can lead to *localization* phenomena where strain (and displacement) discontinuities develop.[116–128] The non-uniqueness can be particularly evident in strain softening plasticity. We illustrate this in an example of Fig. 3.23 where a bar of length L, divided into elements of length h, is subject to a uniformly increasing extension u. The material is initially elastic with a modulus E and after exceeding a stress of σ_y, the yield stress softens (plastically) with a negative modulus H.

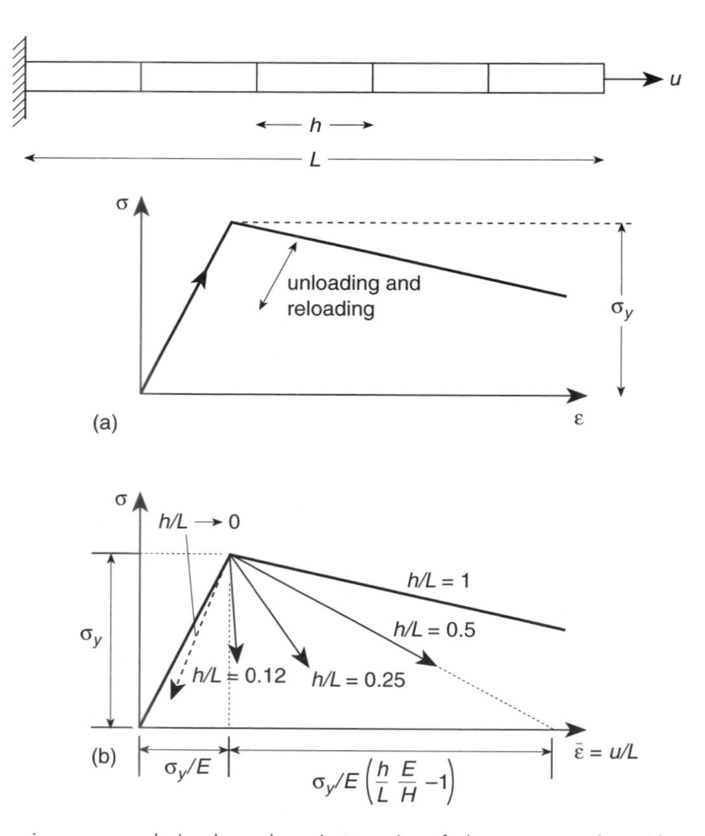

Fig. 3.23 Non-uniqueness: mesh size dependence in extension of a homogeneous bar with a strain softening material. (Peak value of yield stress, σ_y perturbed in a single element). (a) Stress σ versus strain ε for material; (b) stress $\bar{\sigma}$ versus average strain $\bar{\varepsilon}$ ($= u/L$) assuming yielding in a single element of length h.

The strain–stress relation is thus [Fig. (3.23(a)]

$$\sigma = E\varepsilon \qquad \text{if} \quad \varepsilon < \sigma_y/E = \varepsilon_y \tag{3.180}$$

and for increasing ε only,

$$\sigma = \sigma_y - H\left(\varepsilon - \varepsilon_y\right) \qquad \text{if} \quad \varepsilon > \varepsilon_y \tag{3.181}$$

For unloading from any plastic point the material behaves elastically as shown.

One possible solution is, of course, that in which all elements yield identically. Plotting the applied stress versus the elongation strain $\bar{\varepsilon} = u/L$ the material behaviour curve is simply obtained identically as shown in Fig. 3.23(b) ($h/L = 1$). However, it is equally possible that after reaching the maximum stress σ_y only one element (probably one with infinitesimally smaller yield stress owing to computer round-off) continues into the plastic range while all the others unload elastically. The total elongation strain is now given by

$$\bar{\varepsilon} = \frac{u}{L} = \frac{\sigma}{E} - \frac{h(\sigma - \sigma_y)}{LH} \tag{3.182}$$

and as h tends to 0 then $\bar{\varepsilon}$ tends to σ/E. Clearly, a multitude of solutions is possible for any arbitrary element subdivision and in this trivial example a unique finite element solution is impossible (with localization to a single element always occurring). Further, the above simple 'thought experiment' points to another unacceptable paradox implying the inadmissibility of the softening model specified with constant softening modulus. The difficulties are as follows.

1. The behaviour seems to depend on the size (h) of the subdivision chosen (also called a *mesh sensitive* result). Clearly this is unacceptable physically.
2. If the element size falls below a value given by $h = HL/E$ only a catastrophic, brittle, behaviour is possible without involving an unacceptable energy gain.

Similar difficulties can arise with non-associated plasticity which exhibits occasionally an effectively strain softening behaviour in some circumstances (see reference 129).

The computational difficulties can be overcome to some extent by introducing visco-plasticity as a start to any computation. Such *regularization* was introduced as early as 1974[93] and was considered seriously by De Borst and co-authors.[130] However, most of the difficulties remain as steady state is approached.

The problem remains a serious line of research but two possible alternative treatments have emerged. The first of these is physically difficult to accept but is very effective in practice. This is the concept of properties which are labelled as *non-local*. In such an approach the softening modulus is made dependent on the element size. Many authors have contributed here, with the earliest being Bazant and co-workers.[124,125] Other relevant references are 130 and 131.

The second approach, that of a *concentrated discontinuity*, is more elegant but, we believe, computationally more difficult. It was first suggested by Simo, Oliver and Armero in 1993[132] and extended in later publications.[133–136]

Both approaches allow strain and indeed displacement discontinuities to develop following the brittle failure behaviour on which we have already remarked. In the

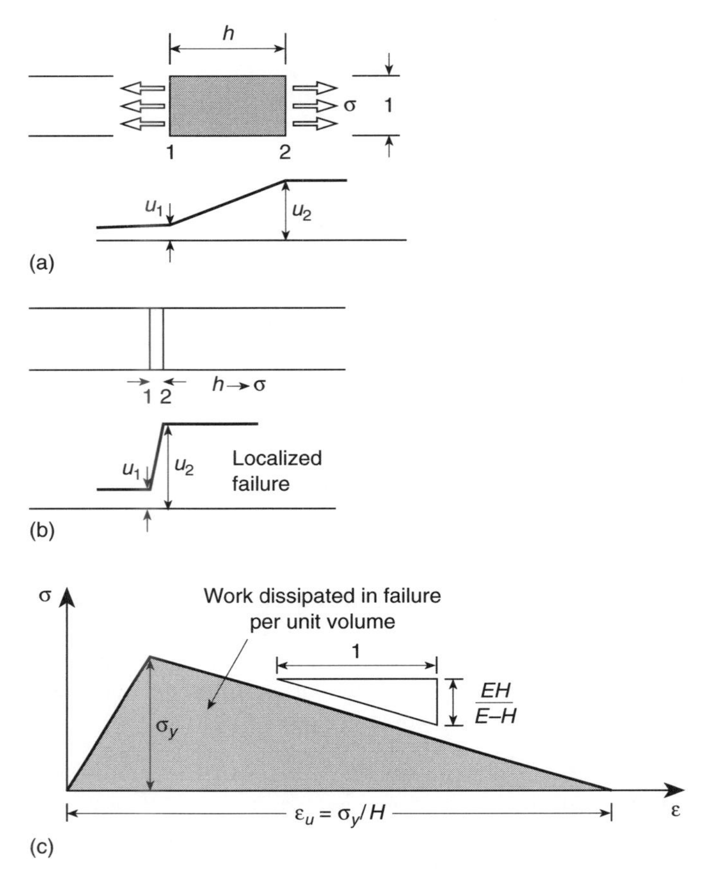

Fig. 3.24 Illustration of a non-local approach (work dissipation in failure is assumed to be constant for all elements): (a) an element in which localization is considered; (b) localization; (c) stress–strain curve showing work dissipated in failure.

numerical application this limit is *approached* as element size decreases or alternatively when stress singularities, such as corners, trigger this type of behaviour.

In the second approach, continuous plastic behaviour is not permitted and all action is concentrated on discontinuity lines which have to be suitably placed.

A particular form of the non-local approach is illustrated in Fig. 3.24. Here we examine in detail a unit width of an element in which the displacement discontinuity is approximated. In the examples which we shall consider later this discontinuity is a slip one with the 'failure' being modelled as shown. However an identical approach has been used to model strain softening behaviour of concrete in cracking.[124,125]

The most basic form of non-local behaviour assumes that the work (or energy) expended in achieving the discontinuity must be the same whatever the dimension h of the element. This work is equal to

$$\frac{1}{2}\sigma_y \varepsilon_y h \approx \frac{1}{2}\sigma_y^2 \frac{h}{H} = \frac{1}{2}\sigma_y \Delta U \tag{3.183}$$

If this work is to be identical in all highly strained elements we will require that

$$\frac{H}{h} = \text{constant} \tag{3.184}$$

Such a requirement is easy to apply in an adaptive refinement process.

At this stage we can comment on the concentrated discontinuity approach of Bazant and co-workers.[124,125] In this we shall simply assume that the displacement increment of Eq. (3.183), that is, ΔU is permitted to occur only on a discontinuity line and that its magnitude is strictly related to the energy density previously specified in Eq. (3.183). In the next section we shall show how a very effective treatment and capture of discontinuity can be made adaptively.

3.12 Adaptive refinement and localization (slip-line) capture

3.12.1 Introduction

The simple discussion of localization phenomena given in the previous section is sufficient we believe to convince the reader that with softening plastic behaviour localization and indeed rapid failure will occur inevitably. Similar behaviour will often be observed with ideal plasticity especially if large deformations are present (see Chap. 10). Here, however, 'brittle type' of failure will be replaced by collapse in which displacement can continue to increase without any increase of load. It is well known that during such continuing displacement

1. the elastic strains will remain unchanged;
2. all displacement is confined to *plastic mechanisms*. Such mechanisms will (frequently) involve discontinuous displacements, such as sliding, and will therefore involve localization.

To control and minimize the errors of the analysis it will be necessary to estimate errors and adaptively remesh in each step of an elasto-plastic computation. This, of course, implies a difficult and costly process. Nevertheless, many attempts to use adaptive refinement were made and references 122, 130, and 137–143 provide a list of some successful attempts.

3.12.2 Adaptive refinement based on energy norm error estimates

In Chapters 14 and 15 of Volume 1 we provide a comprehensive survey of error estimation and h refinement in adaptivity. Most of the procedures there described could be applied with success to elasto-plastic analysis. One in particular, the *recovery*

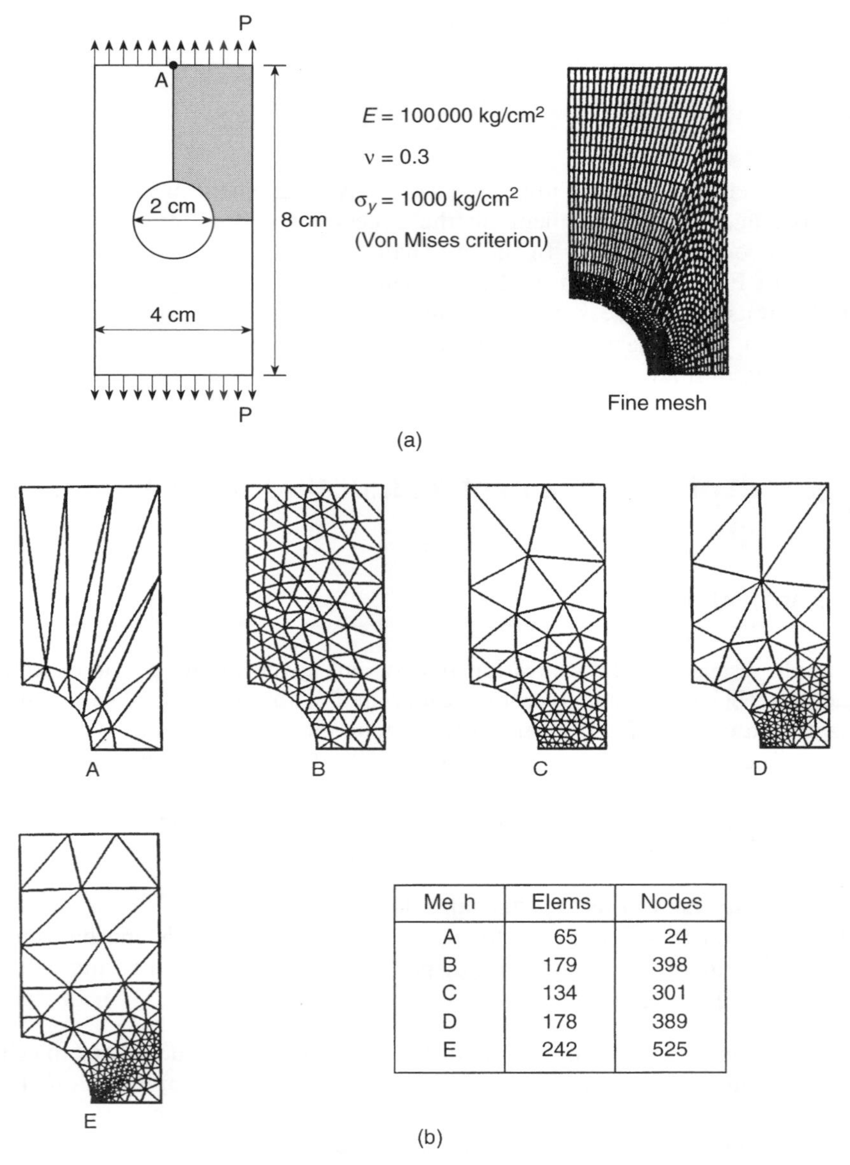

Me h	Elems	Nodes
A	65	24
B	179	398
C	134	301
D	178	389
E	242	525

Fig. 3.25 Adaptive refinement applied to the problem of a perforated strip: (a) the geometry of the strip and a very fine mesh are used to obtain an 'exact' solution; (b) various stages of refinement aiming to achieve a 5% energy norm, relative, error at each load increment (quadratic elements T6/3B/3D were used); (c) local displacement results.

procedures for stress and strain, can be used very efficiently. Indeed, in references 144 and 145 the SPR and REP methods (see Chapter 14, Volume 1) are used successfully to estimate the errors. In Fig. 3.25 we show an analysis of a tension strip using procedures of Volume 1.

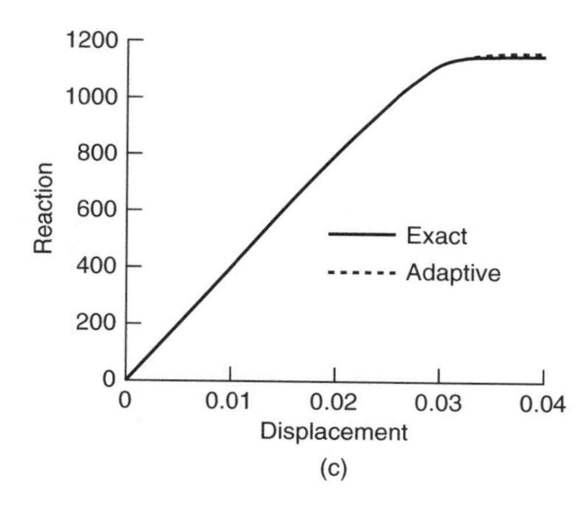

Fig. 3.25 Continued.

It will be observed that as the load increases the refined mesh tends to capture a solution in which displacements are localized. Note that in this solution T6/3B/3D elements were used as these have an excellent performance in incompressibility and may be incorporated easily into an adaptive process. As in the problem discussed in Volume 1, we have attempted here to keep the error to 5% of the energy norm in each refinement.

3.12.3 Alternate refinement using error indicators: discontinuity capture

In the illustrative example of the previous section we have shown how a refinement on the test of a specified energy norm error can indicate and capture discontinuity and slip lines. Nevertheless the process is not economical and may require the use of a very large number of elements. More direct processes have been developed for adaptive refinement in high-speed fluid dynamics where shocks presenting very similar discontinuity properties form. We shall describe the refinement procedures in Volume 3 to which the interested reader may proceed. However, we give a brief summary below.

The processes developed are based on the recognition that in certain directions the unknown function which we are attempting to model exhibits higher gradient or curvatures. High degree of refinement can be achieved economically in high gradient areas with elongated elements. In such areas the smaller side of elements (h_{min}) is placed across the discontinuity, and the larger side (h_{max}) in the direction parallel to the discontinuity. We show such a directionality in Fig. 3.26. For determination of gradients and curvatures, we shall require a scalar function to be considered. The scalar variable which we frequently use in plasticity problems could be the absolute displacement value

$$U = (\mathbf{u}^{\mathrm{T}}\mathbf{u})^{1/2} \tag{3.185}$$

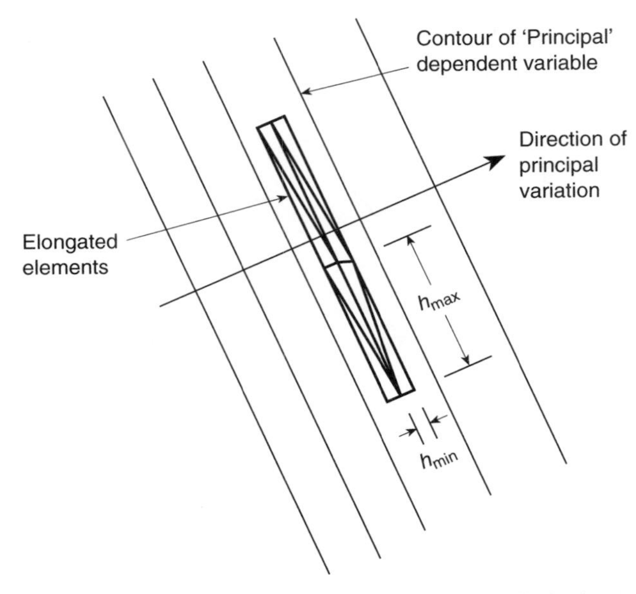

Fig. 3.26 Elongation of elements used to model the nearly one-dimensional behaviour and the discontinuity.

The original refinement indicator of the type we are describing attempts to achieve an equal interpolation error ensuring that in the major and minor direction the equality

$$h_{\min}^2 \left. \frac{\partial^2 U}{\partial x_1^2} \right|_{\max} = h_{\max}^2 \left. \frac{\partial^2 U}{\partial x_2^2} \right|_{\min} = \text{constant} \tag{3.186}$$

By fixing the value of the constant in the above equation and evaluating the approximate curvatures and the ratio of stretching $h_{\max}/h_{\min}$ of the function U immediately we have sufficient data to design a new mesh from the existing one. The procedures for such mesh generation are given in references 146 and 147, although other methods can be adopted.

As an alternative to the above-mentioned procedure we can aim at limiting the first derivative of U by making

$$h \frac{\partial U}{\partial x_1} = \text{constant} \tag{3.187}$$

For this procedure it is not easy to evaluate the stretching ratio. However, the first-derivative condition is useful for guiding the refinement.

A plastic localization calculation based on Eq. (3.186) is shown in Fig. 3.27. This is an early example taken from reference 138. Here, purely plastic flow is shown, ignoring the elastic effects and the refinement is based on the second derivatives. Such a flow formulation (rigid plastic flow) is frequently used in metal forming calculations, to which we return in Volume 3.

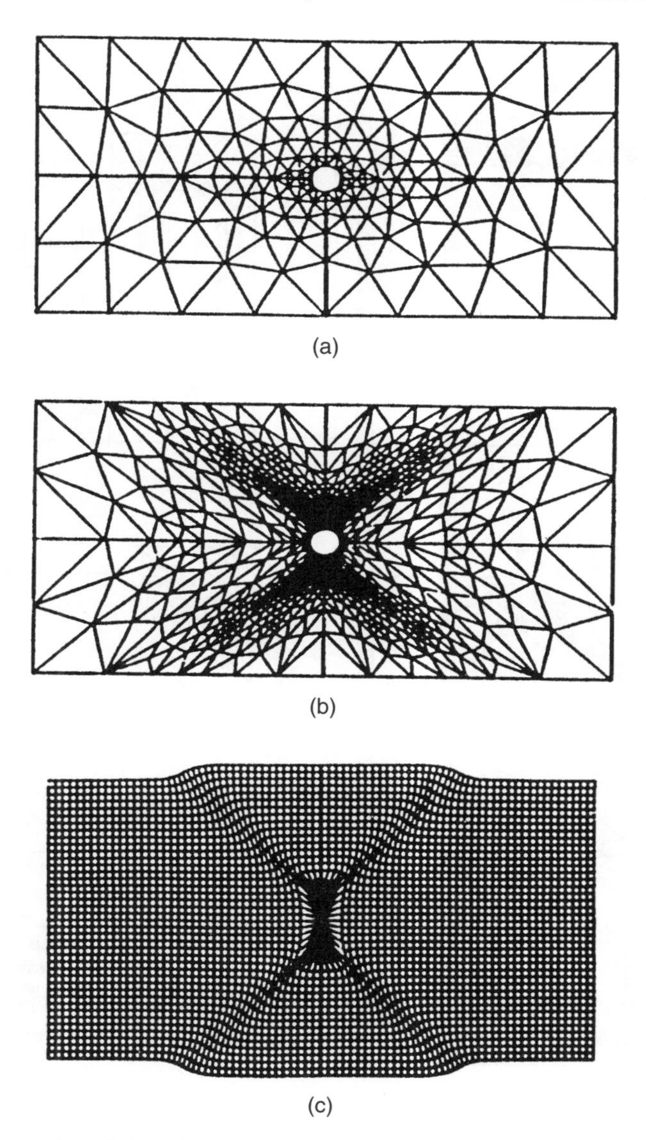

Fig. 3.27 Adaptive analysis of plastic flow deformation in a perforated plate: (a) initial mesh, 273 degrees of freedom; (b) final adapted mesh; (c) displacement of an initially uniform grid embedded in the material.

In the next example we shall use adaptivity based on the first derivatives. Figure 3.28 illustrates a load on a rigid footing over a vertical cut. Here a T6/3C element (triangular with quadratic displacement and linear pressure) is used. Both coarse and adaptively refined meshes give nearly exact answers for the case of ideal plasticity.

For strain softening with a plastic modulus $H = -5000$ answers appear to be mesh dependent. Here, we show how answers become almost mesh independent if H is

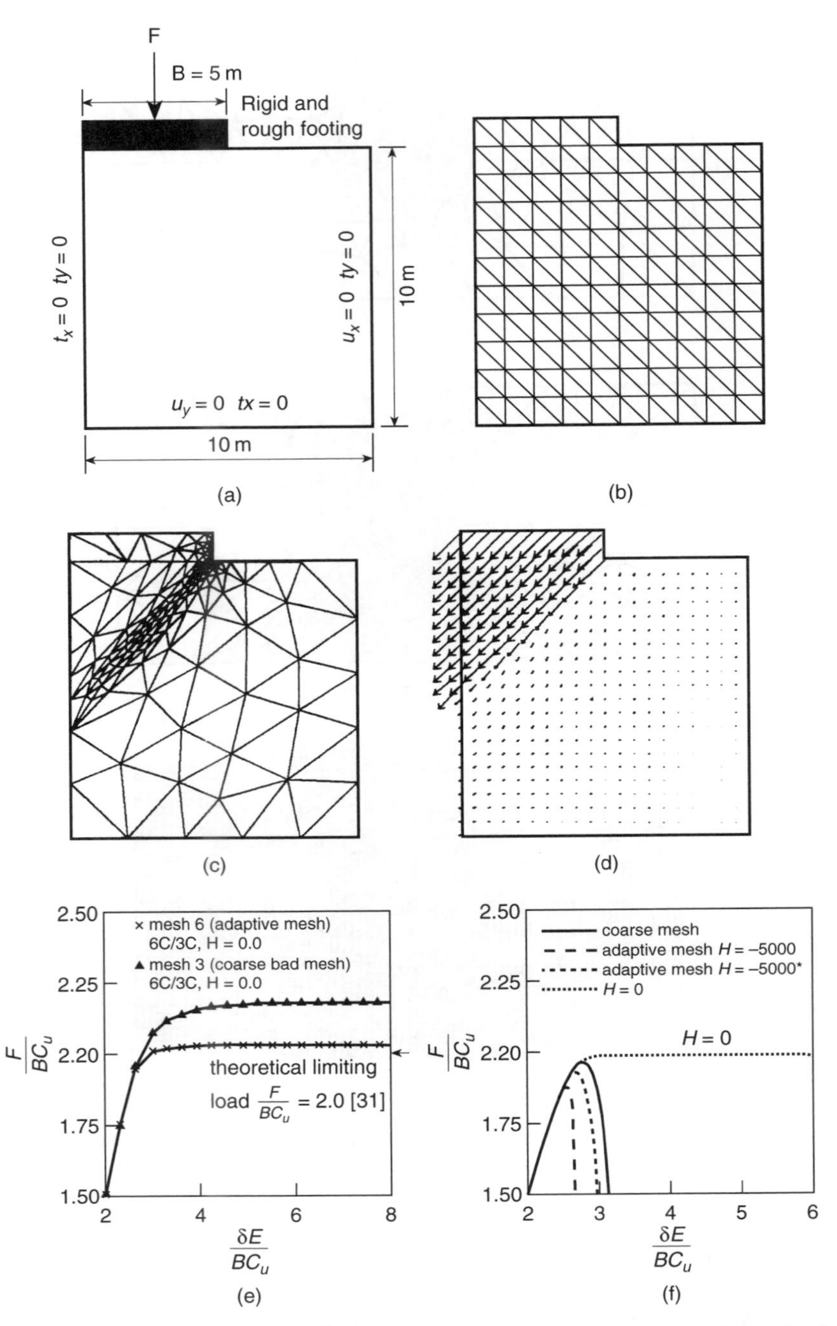

Fig. 3.28 Failure of a rigid footing on a vertical cut. Ideal von Mises plasticity and quadratic triangles with linear variation for pressure (T6C/3C) elements are assumed. (a) Geometrical data; (b) coarse mesh; (c) final adapted mesh; (d) displacements after failure; (e) displacement–load diagrams for adaptive mesh and ideal plasticity ($H = 0$); (f) softening behaviour. Coarse mesh and adapted mesh results are with a constant H of -5000 and a variable H starting from -5000 at coarse mesh size.

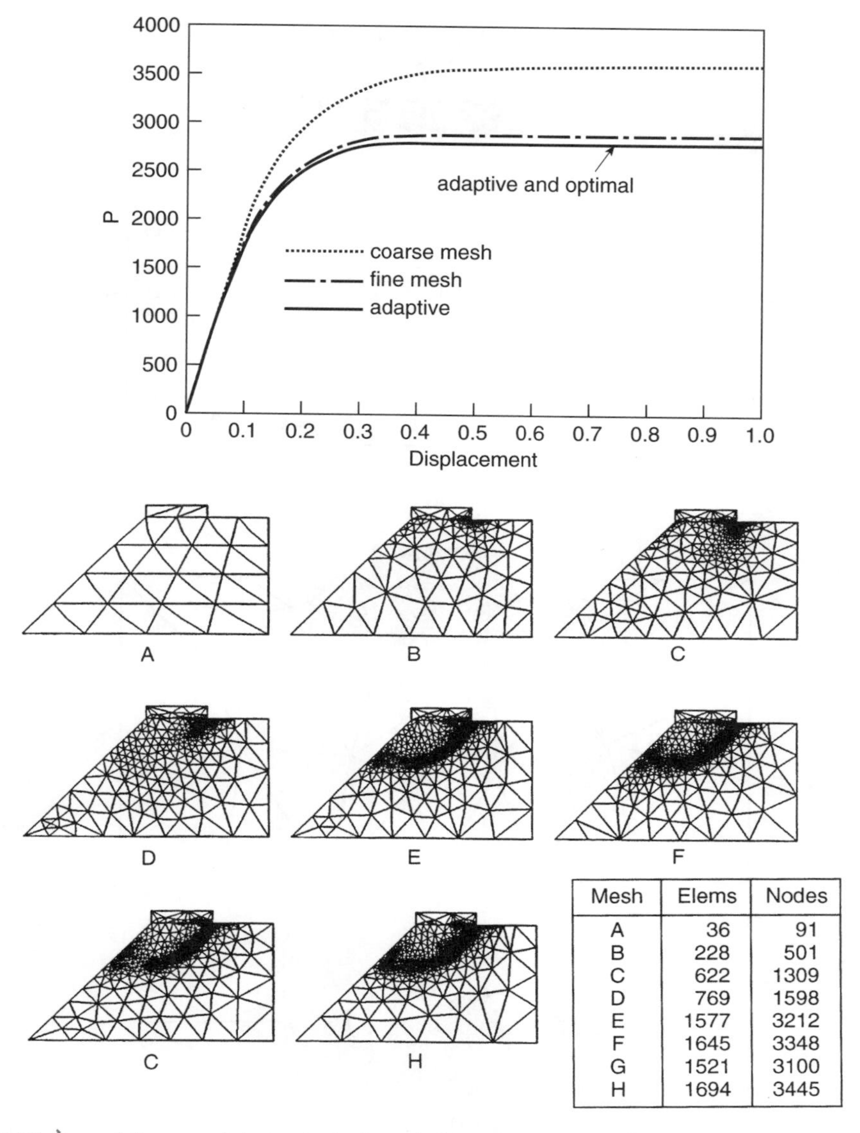

Mesh	Elems	Nodes
A	36	91
B	228	501
C	622	1309
D	769	1598
E	1577	3212
F	1645	3348
G	1521	3100
H	1694	3445

Fig. 3.29 A $p-\delta$ diagram of elasto-plastic slope aiming at 2.5% error in ultimate load (15% incremental energy error) with use of quadratic triangular elements (T6/3B). Mesh A: $u = 0.0$ (coarse mesh). Mesh B: $u = 0.025$. Mesh C: $u = 0.15$. Mesh D: $u = 0.3$. Mesh E: $u = 0.45$. Mesh F: $u = 0.6$. Mesh G: $u = 0.75$. Mesh H: $u = 0.9$. The last mesh (Mesh H, named as the 'optimal mesh') is used for the solution of the problem from the first load step, without further refinement.

varied with element size in the manner discussed in Sec. 3.11 [see Eq. (3.184)]. Figures 3.29 and 3.30 show, respectively, the behaviour of a rigid footing placed on an embankment and on a flat foundation with eccentric loading. All cases illustrate the excellent discontinuity capturing properties of the adaptive refinement.

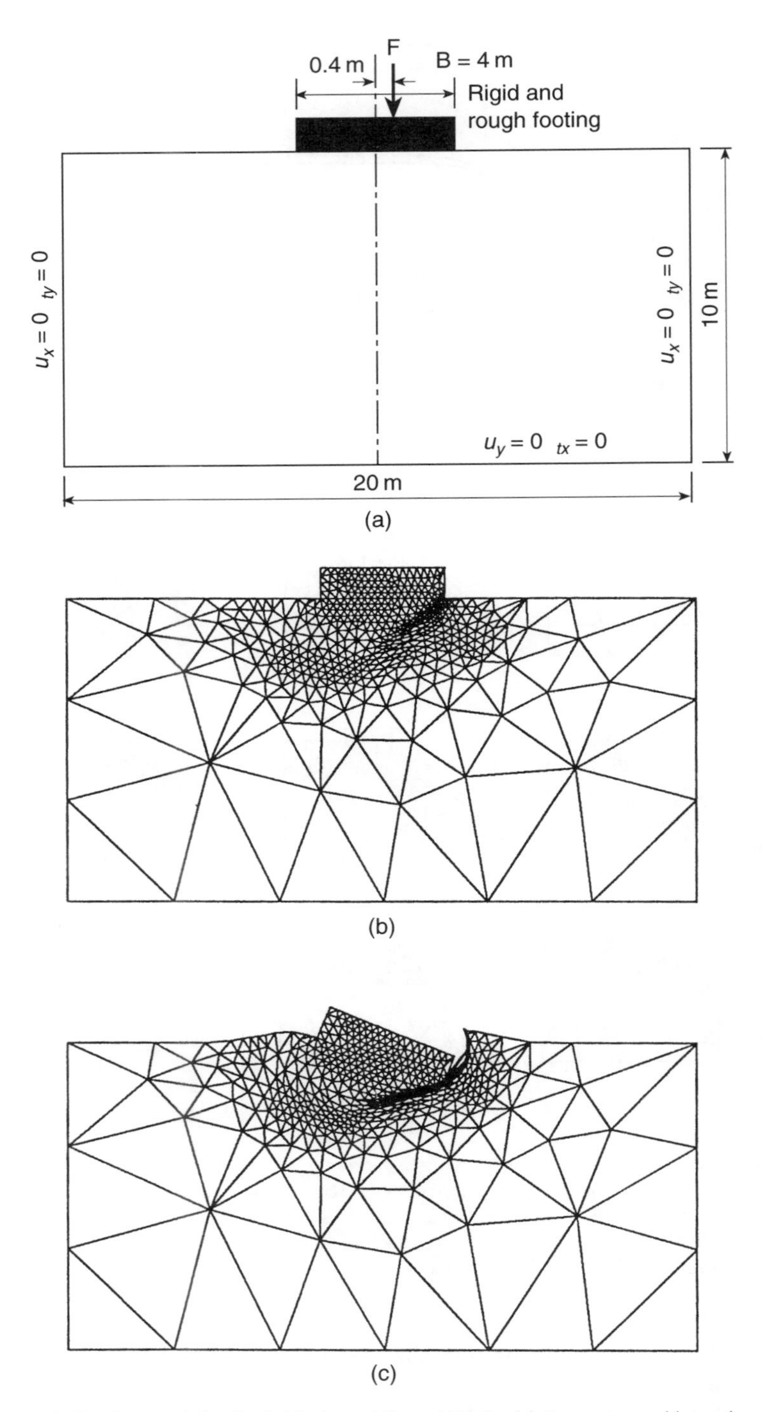

Fig. 3.30 Foundation (eccentric loading); ideal von Mises plasticity. (a) Geometry and boundary conditions; (b) adaptive mesh; (c) deformed mesh using T6C/1D elements ($H = 0$, $\nu = 0.49$).

3.13 Non-linear quasi-harmonic field problems

Non-linearity may arise in many problems beyond those of solid mechanics, but the techniques described in this chapter are still universally applicable. Here we shall look again at one class of problems which is governed by the quasi-harmonic field equations of Chapter 1.

In some formulations it is assumed that

$$\mathbf{q} = -k(\phi)\,\boldsymbol{\nabla}\phi \tag{3.188}$$

which gives, then (with use of definitions from Sec. 1.2.4),

$$\mathbf{P}_q = \mathbf{H}(\boldsymbol{\phi})\,\boldsymbol{\phi} \tag{3.189}$$

where now $\mathbf{H}$ has the familiar form

$$\mathbf{H} = \int_\Omega (\boldsymbol{\nabla}\mathbf{N})^{\mathrm{T}} k(\phi)\,\boldsymbol{\nabla}\mathbf{N}\,\mathrm{d}\Omega \tag{3.190}$$

In this form the general non-linear problem may be solved by direct iteration methods; however, as these often fail to converge it is frequently necessary to use a scheme for which a tangential matrix to $\boldsymbol{\Psi}$ is required, as presented in Sec. 2.2.4 [see Eq. (2.26)]. The tangent for the form given by Eq. (3.188) is generally unsymmetric; however, special forms can be devised which lead to symmetry.[148] In many physical problems, however, the values of k in Eq. (3.188) depend on the absolute value of the gradient of $\boldsymbol{\nabla}\phi$, that is,

$$\begin{aligned} V &= \sqrt{(\boldsymbol{\nabla}\phi)^{\mathrm{T}}\boldsymbol{\nabla}\phi} \\ \bar{k}' &= \frac{\mathrm{d}k}{\mathrm{d}V} \end{aligned} \tag{3.191}$$

In such cases, we can write

$$\mathbf{H}_{\mathrm{T}} = \frac{\partial \mathbf{H}(\boldsymbol{\phi})\,\boldsymbol{\phi}}{\partial \boldsymbol{\phi}} = \mathbf{H} + \mathbf{A} \tag{3.192}$$

where

$$\mathbf{A} = \int_\Omega (\boldsymbol{\nabla}\mathbf{N})^{\mathrm{T}} \left[(\boldsymbol{\nabla}\phi)^{\mathrm{T}}\bar{k}'\,\boldsymbol{\nabla}\phi\right] \boldsymbol{\nabla}\mathbf{N},\mathrm{d}\Omega \tag{3.193}$$

and symmetry is preserved.

Situations of this kind arise in seepage flow where the permeability is dependent on the absolute value of the flow velocity,[149,150] in magnetic fields,[151–154] where magnetic response is a function of the absolute field strength, in slightly compressible fluid flow, and indeed in many other physical situations.[155] Figure 3.31 from reference 151 illustrates a typical non-linear magnetic field solution.

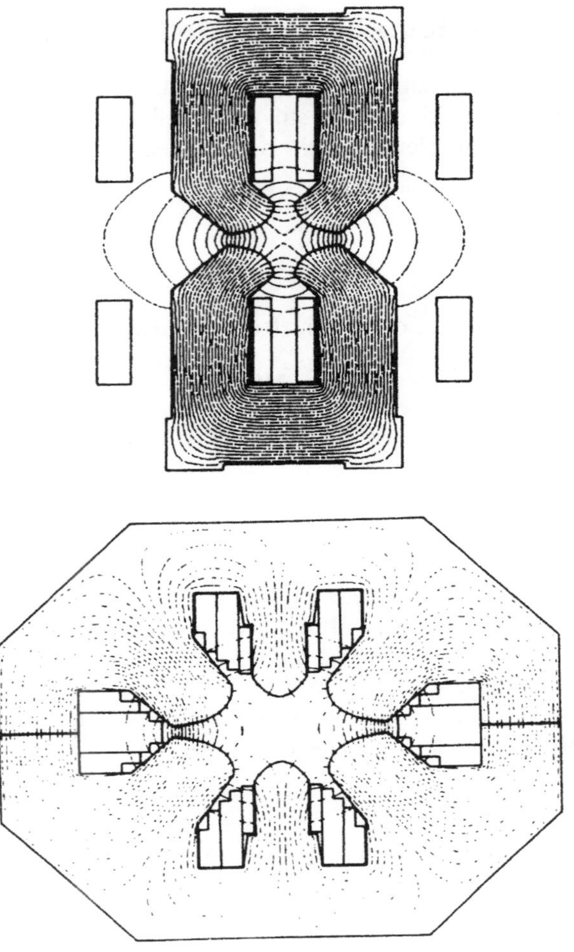

Fig. 3.31 Magnetic field in a six-pole magnet with non-linearity owing to saturation.[151]

While many more interesting problems could be quoted we conclude with one in which the only non-linearity is that due to the heat generation term Q [see Chapter 1, Eq. (1.54)]. This particular problem of spontaneous ignition, in which Q depends exponentially on the temperature, serves to illustrate the point about the possibility of multiple solutions and indeed the non-existence of any solution in certain non-linear cases.[156]

Taking $k = 1$ and $Q = \bar{\delta} \exp \phi$, we examine an elliptic domain in Fig. 3.32. For various values of δ, a Newton–Raphson iteration is used to obtain a solution, and we find that no convergence (and indeed *no solution*) exists when $\bar{\delta} > \bar{\delta}_{\text{crit}}$ exists; above the critical value of $\bar{\delta}$ the temperature rises indefinitely and *spontaneous ignition* of the material occurs. For values below this, *two* solutions are possible and the starting point of the iteration determines which one is in fact obtained.

This last point illustrates that an insight into the problem is, in non-linear solutions, even more important than elsewhere.

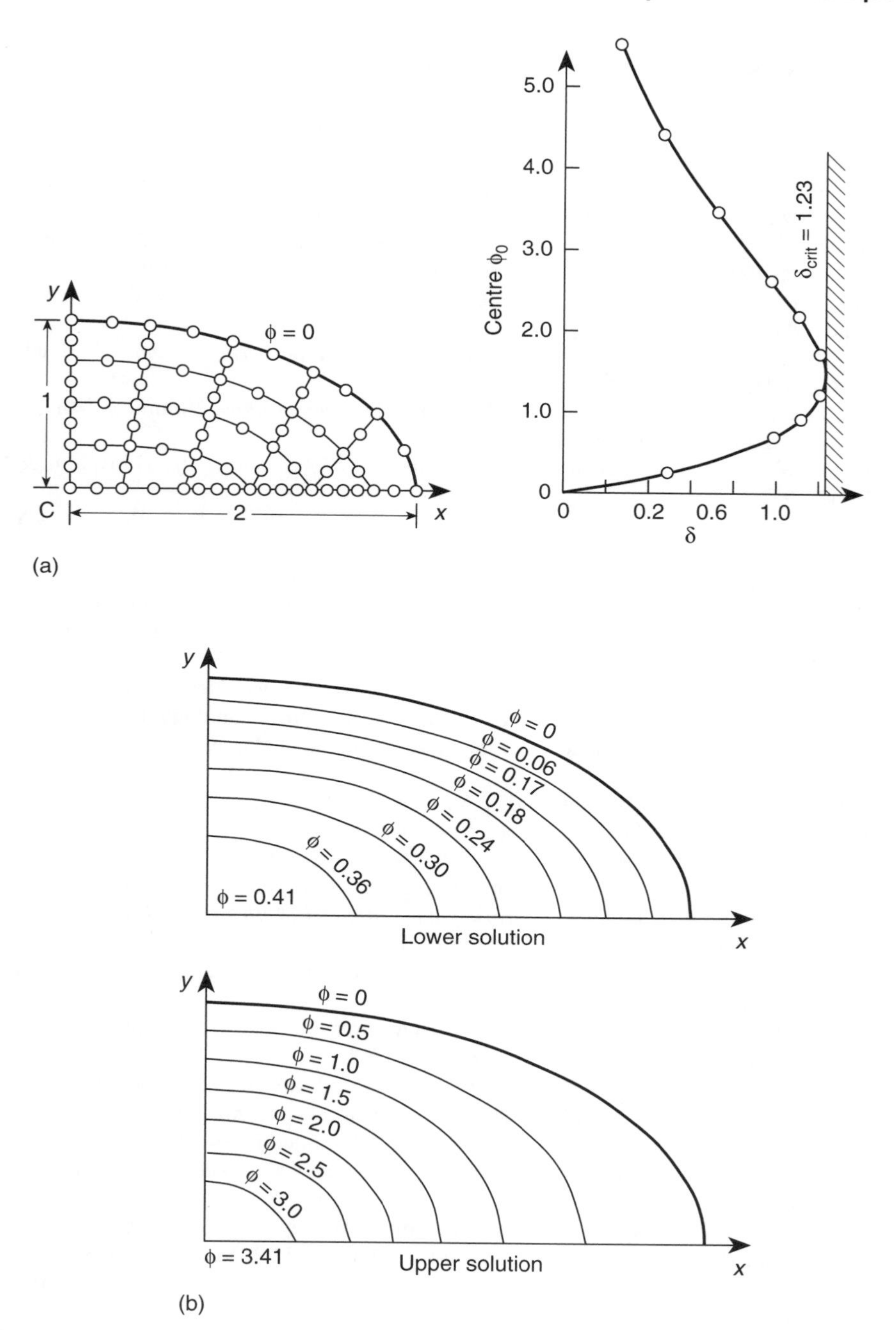

Fig. 3.32 A non-linear heat-generation problem illustrating the possibility of multiple or no solutions depending on the heat generation parameter $\bar{\delta}$; spontaneous combustion.[156] (a) Solution mesh and variation of temperature at point C; (b) two possible temperature distributions for $\bar{\delta} = 0.75$.

References

1. O.C. Zienkiewicz, M. Watson and I.P. King. A numerical method of visco-elastic stress analysis. *Int. J. Mech. Sci.*, **10**, 807–27, 1968.
2. J.L. White. Finite elements in linear viscoelastic analysis. In L. Berke, R.M. Bader, W.J. Mykytow, J.S. Przemienicki and M.H. Shirk (eds), *Proc. 2nd Conf. Matrix Methods in Structural Mechanics*, Volume AFFDL-TR-68-150, pages 489–516, Air Force Flight Dynamics Laboratory, Wright Patterson Air Force Base, OH, October 1968.
3. R.L. Taylor, K.S. Pister and G.L. Goudreau. Thermomechanical analysis of viscoelastic solids. *Int. J. Num. Meth. Eng.*, **2**, 45–79, 1970.
4. O.C. Zienkiewicz. *The Finite Element Method in Engineering Science*, 2nd edition, McGraw-Hill, London, 1971.
5. O.C. Zienkiewicz and R.L. Taylor. *The Finite Element Method*, Volume 2, 4th edition, McGraw-Hill, London, 1991.
6. B. Gross. *Mathematical Structure of the Theories of Viscoelasticity*, Herrmann & Cie, Paris, 1953.
7. R.M. Christensen. *Theory of Viscoelasticity: An Introduction*, Academic Press, New York, 1971.
8. L.R. Herrmann and F.E. Peterson. A numerical procedure for viscoelastic stress analysis. In *Proceedings 7th ICRPG Mechanical Behavior Working Group*, Orlando, FL, 1968.
9. D. McHenry. A new aspect of creep in concrete and its application to design. *Proc. ASTM*, **43**, 1064, 1943.
10. T. Alfrey. *Mechanical Behavior of High Polymer*, Interscience, New York, 1948.
11. H.H. Hilton and H.G. Russell. An extension of Alfrey's analogy to thermal stress problems in temperature dependent linear visco-elastic media. *J. Mech. Phys. Solids*, **9**, 152–64, 1961.
12. O.C. Zienkiewicz, M. Watson and Y.K. Cheung. Stress analysis by the finite element method – thermal effects. In *Proc. Conf. on Prestressed Concrete Pressure Vessels*, Institute of Civil Engineers, London, 1967.
13. J. Lubliner. *Plasticity Theory*, Macmillan, New York, 1990.
14. J.C. Simo and T.J.R. Hughes. *Interdisciplinary Applied Mathematics: Volume 7. Computational Inelasticity*, Springer-Verlag, Berlin, 1998.
15. R. von Mises. Mechanik der Plastischen Formänderung der Kristallen. *Z. angew. Math. Mech.*, **8**, 161–85, 1928.
16. D.C. Drucker. A more fundamental approach to plastic stress–strain solutions. In *Proc. 1st U.S. Nat. Cong. Appl. Mech.*, pp. 487–91, 1951.
17. W.T. Koiter. Stress–strain relations, uniqueness and variational theorems for elastic–plastic materials with a singular yield surface. *Q. J. Appl. Math.*, **11**, 350–54, 1953.
18. R. Hill. *The Mathematical Theory of Plasticity*, Clarenden Press, Oxford, 1950.
19. W. Johnson and P.W. Mellor. *Plasticity for Mechanical Engineers*, Van Nostrand, New York, 1962.
20. W. Prager. *An Introduction to Plasticity*, Addison-Wesley, Reading, MA, 1959.
21. W.F. Chen. *Plasticity in Reinforced Concrete*, McGraw-Hill, New York, 1982.
22. D.C. Drucker. Conventional and unconventional plastic response and representations. *Appl. Mech. Rev.*, **41**, 151–67, 1988.
23. G.C. Nayak and O.C. Zienkiewicz. Elasto-plastic stress analysis. Generalization for various constitutive relations including strain softening. *Int. J. Num. Meth. Eng.*, **5**, 113–35, 1972.
24. W. Prager. A new method of analysing stress and strains in work-hardening plastic solids. *J. Appl. Mech.*, **23**, 493–96, 1956.
25. H. Ziegler. A modification of Prager's hardening rule. *Q. Appl. Math.*, **17**, 55–65, 1959.

26. J. Mandel. Contribution theorique a l'etude de l'ecrouissage et des lois de l'ecoulement plastique. In *Proc. 11th Int. Cong. of Appl. Mech.*, pp. 502–9, 1964.
27. J. Lubliner. A maximum-dissipation principle in generalized plasticity. *Acta Mechanica*, **52**, 225–37, 1984.
28. J.C. Simo. Topics on the numerical analysis and simulation of plasticity. In P.G. Ciarlet and J.L. Lions (eds), *Handbook of Numerical Analysis*, Volume III, pp. 183–499, Elsevier, Amsterdam, 1999.
29. P.J. Armstrong and C.O. Frederick. A mathematical representation of the multi-axial Bauschinger, effect. Technical Report RD/B/N731, CEGB, Berkeley Nuclear Laboratories, R&D Department, 1965.
30. J.L. Chaboche. Constitutive equations for cyclic plasticity and cyclic visco-plasticity. *Int. J. Plasticity*, **5**, 247–302, 1989.
31. J.L. Chaboche. On some modifications of kinematic hardening to improve the description of ratcheting effects. *Int. J. Plasticity*, **7**, 661–78, 1991.
32. J.L. Lubliner. A simple model of generalized plasticity. *International Journal of Solids and Structures*, **28**, 769, 1991.
33. J.L. Lubliner. A new model of generalized plasticity. *International Journal of Solids and Structures*, **30**, 3171–84, 1993.
34. F. Auricchio and R.L. Taylor. Two material models for cyclic plasticity: nonlinear kinematic hardening and generalized plasticity. *Int. J. Plasticity*, **11**, 65–98, 1995.
35. J.F. Besseling. A theory of elastic, plastic and creep deformations of an initially isotropic material. *J. Appl. Mech.*, **25**, 529–36, 1958.
36. Z. Mróz. An attempt to describe the behaviour of metals under cyclic loads using a more general work hardening model. *Acta Mech.*, **7**, 199, 1969.
37. O.C. Zienkiewicz, C.T. Chang, N. Bićanić and E. Hinton. Earthquake response of earth and concrete in the partial damage range. In *Proc. 13th Int. Cong. on Large Dams*, pp. 1033–47, New Delhi, 1979.
38. O.C. Zienkiewicz, G.C. Nayak and D.R.J. Owen. Composite and 'Overlay' models in numerical analysis of elasto-plastic continua. In A. Sawczuk (ed.), *Foundations of Plasticity*, pp. 107–22, Noordhoff, Dordrecht, 1972.
39. D.R.J. Owen, A. Prakash and O.C. Zienkiewicz. Finite element analysis of non-linear composite materials by use of overlay systems. *Computers and Structures*, **4**, 1251–67, 1974.
40. Y. Yamada, N. Yishimura and T. Sakurai. Plastic stress–strain matrix and its application for the solution of elasto-plastic problems by the finite element method. *Int. J. Mech. Sci.*, **10**, 343–54, 1968.
41. O.C. Zienkiewicz, S. Valliappan and I.P. King. Elasto-plastic solutions of engineering problems. Initial stress, finite element approach. *Int. J. Num. Meth. Eng.*, **1**, 75–100, 1969.
42. F.B. Hildebrand. *Introduction to Numerical Analysis*, 2nd edition, Dover Publishers, 1987.
43. G.C. Nayak and O.C. Zienkiewicz. Note on the 'alpha'-constant stiffness method for the analysis of nonlinear problems. *Int. J. Num. Meth. Eng.*, **4**, 579–82, 1972.
44. G.C. Nayak. *Plasticity and Large Deformation Problems by the Finite Element Method.* PhD thesis, Department of Civil Engineering, University of Wales, Swansea, 1971.
45. E. Hinton and D.R.J. Owen. *Finite Element Programming*, Academic Press, New York, 1977.
46. R.D. Krieg and D.N. Krieg. Accuracy of numerical solution methods for the elastic, perfectly plastic model. *J. Pressure Vessel Tech., Trans. ASME*, **99**, 510–15, 1977.
47. G. Maenchen and S. Sacks. The tensor code. In B. Alder (ed.), *Methods in Computational Physics*, Volume 3, pp. 181–210, Academic Press, New York, 1964.
48. M.L. Wilkins. Calculation of elastic–plastic flow. In B. Alder (ed.), *Methods in Computational Physics*, Volume 3, pp. 211–63. Academic Press, New York, 1964.

49. J.C. Simo and R.L. Taylor. Consistent tangent operators for rate-independent elastoplasticity. *Comp. Meth. Appl. Mech. Eng.*, **48**, 101–18, 1985.
50. J.C. Simo and R.L. Taylor. A return mapping algorithm for plane stress elastoplasticity. *Int. J. Num. Meth. Eng.*, **22**, 649–70, 1986.
51. I.H. Shames and F.A. Cozzarelli. *Elastic and Inelastic Stress Analysis*, revised edition, Taylor & Francis, Washington, DC, 1997.
52. D.C. Drucker and W. Prager. Soil mechanics and plastic analysis or limit design. *Q. J. Appl. Math.*, **10**, 157–65, 1952.
53. O.C. Zienkiewicz and G.N. Pande. Some useful forms of isotropic yield surfaces for soil and rock mechanics. In G. Gudehus (ed.), *Finite Elements in Geomechanics*, pp. 171–90. John Wiley, Chichester, Sussex, 1977.
54. O.C. Zienkiewicz, V.A. Norris and D.J. Naylor. Plasticity and viscoplasticity in soil mechanics with special reference to cyclic loading problems. In *Proc. Int. Conf. on Finite Elements in Nonlinear Solid and Structural Mechanics*, Volume 2, Tapir Press, Trondheim, 1977.
55. O.C. Zienkiewicz, V.A. Norris, L.A. Winnicki, D.J. Naylor and R.W. Lewis. A unified approach to the soil mechanics problems of offshore foundations. In O.C. Zienkiewicz, R.W. Lewis and K.G. Stagg (eds), *Numerical Methods in Offshore Engineering*, chapter 12, John Wiley, Chichester, Sussex, 1978.
56. F. Darve. An incrementally nonlinear constitutive law of second order and its application to localization. In C.S. Desai and R.H. Gallegher (eds), *Mechanics of Engineering Materials*, pp. 179–96. John Wiley, Chichester, Sussex, 1984.
57. G.H. Golub and C.F. Van Loan. *Matrix Computations*, 3rd edition, The Johns Hopkins University Press, Baltimore, MD, 1996.
58. Z. Mróz and O.C. Zienkiewicz. Uniform formulation of constitutive equations for clays and sands. In *Mechanics of Engineering Materials*, pp. 415–50, John Wiley, Chichester, Sussex, 1984.
59. O.C. Zienkiewicz and Z. Mróz. Generalized plasticity formulation and applications to geomechanics. In *Mechanics of Engineering Materials*, pp. 655–80. John Wiley, Chichester, Sussex, 1984.
60. O.C. Zienkiewicz, K.H. Leung and M. Pastor. Simple model for transient soil loading in earthquake analysis: Part I – basic model and its application. *International Journal for Numerical Analysis Methods in Geomechanics*, **9**, 453–76, 1985.
61. O.C. Zienkiewicz, K.H. Leung and M. Pastor. Simple model for transient soil loading in earthquake analysis: Part II – non-associative models for sands. *International Journal for Numerical Analysis Methods in Geomechanics*, **9**, 477–98, 1985.
62. O.C. Zienkiewicz, S. Qu, R.L. Taylor and S. Nakazawa. The patch test for mixed formulations. *Int. J. Num. Meth. Eng.*, **23**, 1873–83, 1986.
63. M. Pastor and O.C. Zienkiewicz. A generalized plasticity, hierarchical model for sand under monotonic and cyclic loading. In G.N. Pande and W.F. Van Impe, (eds), *Proc. Int. Symp. on Numerical Models in Geomechanics*, pp. 131–50, Jackson, England, 1986. Ghent.
64. M. Pastor and O.C. Zienkiewicz. Generalized plasticity and modelling of soil behaviour. *International Journal for Numerical Analysis Methods in Geomechanics*, **14**, 151–90, 1990.
65. Y. Defalias and E.P. Popov. A model of nonlinear hardening material for complex loading. *Acta Mechanica*, **21**, 173–92, 1975.
66. O.C. Zienkiewicz, A.H.C. Chan, M. Pastor, D.K. Paul and T. Shiomi. Static and dynamic behaviour of soils: a rational approach to quantitative solutions, I. *Proceedings of the Royal Society of London*, **429**, 285–309, 1990.
67. O.C. Zienkiewicz, Y.M. Xie, B.A. Schrefler, A. Ledesma and N. Bîĉanîĉ. Static and dynamic behaviour of soils: a rational approach to quantitative solutions, II. *Proceedings of the Royal Society of London*, **429**, 311–21, 1990.

68. T. Tatsueka and K. Ishihera. Yielding of sand in triaxial compression. *Soil Foundation*, **14**, 63–76, 1974.
69. J.C. Nagtegaal, D.M. Parks and J.R. Rice. On numerical accurate finite element solutions in the fully plastic range. *Comp. Meth. Appl. Mech. Eng.*, **4**, 153–77, 1974.
70. R.M. McMeeking and J.R. Rice. Finite element formulations for problems of large elastic–plastic deformation. *Int. J. Solids Struct.*, **11**, 601–16, 1975.
71. E.F. Rybicki and L.A. Schmit. An incremental complementary energy method of non-linear stress analysis. *Journal of AIAA*, **8**, 1105–12, 1970.
72. R.H. Gallagher and A.K. Dhalla. Direct flexibility finite element elasto-plastic analysis. In *Proc. 1st SMIRT Conf.*, Volume 6, Berlin, 1971.
73. J.A. Stricklin, W.E. Haisler and W. von Reisman. Evaluation of solution procedures for material and/or geometrically non-linear structural analysis. *Journal of AIAA*, **11**, 292–9, 1973.
74. P.S. Theocaris and E. Marketos. Elastic–plastic analysis of perforated thin strips of a strain-hardening material. *J. Mech. Phys. Solids*, **12**, 377–90, 1964.
75. P.V. Marcal and I.P. King. Elastic–plastic analysis of two-dimensional stress systems by the finite element method. *Int. J. Mech. Sci.*, **9**, 143–55, 1967.
76. K.S. Dinno and S.S. Gill. An experimental investigation into the plastic behaviour of flush nozzles in spherical pressure vessels. *Int. J. Mech. Sci.*, **7**, 817, 1965.
77. A. Mendelson, M.H. Hirschberg and S.S. Manson. A general approach to the practical solution of creep problems. *Trans. ASME, J. Basic Eng.*, **81**, 85–98, 1959.
78. O.C. Zienkiewicz and Y.K. Cheung. *The Finite Element Method in Structural Mechanics*, McGraw-Hill, London, 1967.
79. G.A. Greenbaum and M.F. Rubinstein. Creep analysis of axisymmetric bodies using finite elements. *Nucl. Eng. Des.*, **7**, 379–97, 1968.
80. P.V. Marcal. Selection of creep strains for explicit calculations, private communication, 1972.
81. N.A. Cyr and R.D. Teter. Finite element elastic plastic creep analysis of two-dimensional continuum with temperature dependent material properties. *Computers and Structures*, **3**, 849–63, 1973.
82. O.C. Zienkiewicz. Visco-plasticity, plasticity, creep and visco-plastic flow (problems of small, large and continuing deformation). In *Computational Mechanics*, TICOM Lecture Notes on Mathematics 461. Springer-Verlag, Berlin, 1975.
83. M.B. Kanchi, D.R.J. Owen and O.C. Zienkiewicz. The visco-plastic approach to problems of plasticity and creep involving geometrically non-linear effects. *Int. J. Num. Meth. Eng.*, **12**, 169–81, 1978.
84. T.J.R. Hughes and R.L. Taylor. Unconditionally stable algorithms for quasi-static elasto/visco-plastic finite element analysis. *Computers and Structures*, **8**, 169–73, 1978.
85. E.C. Bingham. *Fluidity and Plasticity*, pp. 215–18. McGraw-Hill, New York, 1922.
86. P. Perzyna. Fundamental problems in viscoplasticity. *Advances in Applied Mechanics*, **9**, 243–377, 1966.
87. P. Perzyna. Thermodynamic theory of viscoplasticity. In *Advances in Applied Mechanics*, Volume 11, Academic Press, New York, 1971.
88. G. Duvaut and J.-L. Lions. *Inequalities in Mechanics and Physics*, Springer-Verlag, Berlin, 1976.
89. O.C. Zienkiewicz and I.C. Cormeau. Visco-plasticity solution by finite element process. *Arch. Mech.*, **24**, 873–88, 1972.
90. J. Zarka. Généralisation de la théorie du potential multiple en viscoplasticité. *J. Mech. Phys. Solids*, **20**, 179–95, 1972.
91. Q.A. Nguyen and J. Zarka. Quelques méthodes de resolution numérique en elastoplasticité classique et en elasto-viscoplasticité. *Sciences et Technique de l'Armement*, **47**, 407–36, 1973.

92. O.C. Zienkiewicz and I.C. Cormeau. Visco-plasticity and plasticity: an alternative for finite element solution of material non-linearities. In *Proc. Colloque Methodes Calcul. Sci. Tech.*, pp. 171–99, IRIA, Paris, 1973.
93. O.C. Zienkiewicz and I.C. Cormeau. Visco-plasticity, plasticity and creep in elastic solids – a unified numerical solution approach. *Int. J. Num. Meth. Eng.*, **8**, 821–45, 1974.
94. I.C. Cormeau. Numerical stability in quasi-static elasto-visco-plasticity. *Int. J. Num. Meth. Eng.*, **9**, 109–28, 1975.
95. W.T. Koiter. General theorems for elastic–plastic solids. In I.N. Sneddon and R. Hill (eds), *Progress in Solid Mechanics*, Volume 6, pp. 167–221. North-Holland, Amsterdam, 1960.
96. F.A. Leckie and J.B. Martin. Deformation bounds for bodies in a state of creep. *Trans. ASME, J. Appl. Mech.*, **34**, 411–17, 1967.
97. I. Finnie and W.R. Heller. *Creep of Engineering Materials*, McGraw-Hill, New York, 1959.
98. O.C. Zienkiewicz, A.H.C. Chan, M. Pastor and B.A. Schrefler. *Computational Geomechanics: With Special Reference to Earthquake Engineering*, John Wiley, Chichester, Sussex, 1999.
99. O.C. Zienkiewicz, C. Humpheson and R.W. Lewis. Associated and non-associated visco-plasticity and plasticity in soil mechanics. *Geotechnique*, **25**, 671–89, 1975.
100. G.N. Pande and O.C. Zienkiewicz. (eds), *Soil Mechanics – Transient and Cyclic Loads*, John Wiley, Chichester, Sussex, 1982.
101. C.S. Desai and R.H. Gallagher. (eds), *Mechanics of Engineering Materials*, John Wiley, Chichester, Sussex, 1984.
102. O.C. Zienkiewicz, S. Valliappan and I.P. King. Stress analysis of rock as a 'no tension' material. *Geotechnique*, **18**, 56–66, 1968.
103. S. Valliappan and P. Nath. Tensile crack propagation in reinforced concrete beams by finite element techniques. In *Int. Conf. on Shear, Torsion and Bond in Reinforced Concrete*, Coimbatore, India, January 1969.
104. N.W. Krahl, W. Khachaturian and C.P. Seiss. Stability of tensile cracks in concrete beams. *Proc. Am. Soc. Civ. Eng.*, **93**(ST1), 235–54, 1967.
105. D.V. Phillips and O.C. Zienkiewicz. Finite element non-linear analysis of concrete structures. *Proc. Inst. Civ. Eng.*, Part 2, **61**, 59–88, 1976.
106. O.C. Zienkiewicz, D.V. Phillips and D.R.J. Owens. Finite element analysis of some concrete non-linearities: theories and examples. In *IABSE Symp. on Concrete Structures to Triaxial Stress*, Bergamo, 17–19 May 1974.
107. J.G. Rots and J. Blaauwendraad. Crack models for concrete: Discrete or smeared? Fixed, multidirectional or rotating? *Heron*, **34**, 1–59, 1989.
108. J. Isenberg. (ed.), *Finite Element Analysis of Concrete Structures*, New York, 1993. American Society of Civil Engineers. ISBN 0-87262-983-X.
109. G. Hoffstetter and H.A. Mang. *Computational Mechanics of Reinforced Concrete Structures*, Vieweg & Sohn, Braunschweig, 1995. ISBN 3-528-06390-4.
110. R. De Borst, N. Biĉaniĉ, H. Mang and G. Meschke. (eds), *Computational Modelling of Concrete Structures*, Vol. 1 & 2, Rotterdam, 1998. Balkema. ISBN 90-5410-9467.
111. R.E. Goodman, R.L. Taylor and T. Brekke. A model for the mechanics of jointed rock. *Proc. Am. Soc. Civ. Eng.*, **94**(SM3), 637–59, 1968.
112. O.C. Zienkiewicz and B. Best. Some non-linear problems in soil and rock mechanics – finite element solutions. In *Conf. on Rock Mechanics*, University of Queensland, Townsville, 1969.
113. O.C. Zienkiewicz and G.N. Pande. Time dependent multilaminate models for rock – a numerical study of deformation and failure of rock masses. *International Journal for Numerical Analysis Methods in Geomechanics*, **1**, 219–47, 1977.

114. G.N. Pande and K.G. Sharma. Multi-laminate model of clays – a numerical evaluation of the influence of rotation of the principal stress axes. *International Journal for Numerical Analysis Methods in Geomechanics*, **7**, 397–418, 1983.
115. J. Mandel. Conditions de stabilité et postulat de Drucker. In *Proc. IUTAM Symp. on Rheology and Soil Mechanics*, pp. 58–68, Springer-Verlag, Berlin, 1964.
116. J.W. Rudnicki and J.R. Rice. Conditions for the localization of deformations in pressure sensitive dilatant materials. *J. Mech. Phys. Solids*, **23**, 371–94, 1975.
117. J.R. Rice. The localization of plastic deformation. In W.T. Koiter (ed.), *Theoretical and Applied Mechanics*, pp. 207–20, North-Holland, Amsterdam, 1977.
118. S.T. Pietruszczak and Z. Mróz. Finite element analysis of strain softening materials. *Int. J. Num. Meth. Eng.*, **10**, 327–34, 1981.
119. S.W. Sloan and J.R. Booker. Removal of singularities in Tresea and Mohr–Coulomb yield functions. *Comm. App. Num. Meth.*, **2**, 173–79, 1986.
120. N. Biĉaniĉ, E. Pramono, S. Sture and K.J. Willam. On numerical prediction of concrete fracture localizations. In *Proc. NUMETA Conf.*, pp. 385–92, Balkema, Rotterdam, 1985.
121. J.C. Simo. Strain softening and dissipation: a unification of approaches. In *Proc. NUMETA Conf.*, pp. 440–61, Balkema, Rotterdam, 1985.
122. M. Ortiz, Y. Leroy and A. Needleman. A finite element method for localized failure analysis. *Comp. Meth. Appl. Mech. Eng.*, **61**, 189–214, 1987.
123. A. Needleman. Material rate dependence and mesh sensitivity in localization problems. *Comp. Meth. Appl. Mech. Eng.*, **67**, 69–85, 1988.
124. Z.P. Bazant and F.B. Lin. Non-local yield limit degradation. *Int. J. Num. Meth. Eng.*, **26**, 1805–23, 1988.
125. Z.P. Bazant and G. Pijaudier-Cabot. Non-linear continuous damage, localization instability and convergence. *J. Appl. Mech.*, **55**, 287–93, 1988.
126. J. Mazars and Z.P. Basant. (eds), *Cracking and Damage*, Elsevier, Dordrecht, 1989.
127. Y. Leroy and M. Ortiz. Finite element analysis of strain localization in frictional materials. *International Journal for Numerical Analysis Methods in Geomechanics*, **13**, 53–74, 1989.
128. T. Belytschko, J. Fish and A. Bayless. The spectral overlay on finite elements for problems with high gradients. *Comp. Meth. Appl. Mech. Eng.*, **81**, 71–89, 1990.
129. G.N. Pande and S. Pietruszczak. Symmetric tangent stiffness formulation for non-associated plasticity. *Computers and Geotechniques*, **2**, 89–99, 1986.
130. R. De Borst, L.J. Sluys, H.B. Hühlhaus and J. Pamin. Fundamental issues in finite element analysis of localization of deformation. *Engineering Computations*, **10**, 99–121, 1993.
131. T. Belytschko and M. Tabbara. *h*-adaptive finite element methods for dynamic problems, with emphasis on localization. *Int. J. Num. Meth. Eng.*, **36**, 4245–65, 1994.
132. J. Simo, J. Oliver and F. Armero. An analysis of strong discontinuities induced by strain–softening in rate-independent inelastic solids. *Comp. Mech.*, **12**, 277–96, 1993.
133. J. Simo and J. Oliver. A new approach to the analysis and simulations of strong discontinuities. In Z.P. Bazant *et al.* (eds), *Fracture and Damage in Quasi-brittle Structures*, pp. 25–39, E & FN Spon, London, 1994.
134. J. Oliver. Modelling strong discontinuities in solid mechanics via strain softening constitutive equations: part 1, fundamentals. *Int. J. Num. Meth. Eng.*, **39**, 3575–600, 1996.
135. J. Oliver. Modelling strong discontinuities in solid mechanics via strain softening constitutive equations: part 2, numerical simulation. *Int. J. Num. Meth. Eng.*, **39**, 3601–23, 1996.
136. J. Oliver, M. Cervera and O. Manzoli. Strong discontinuities and continuum plasticity models: the strong discontinuity approach. *Int. J. Plasticity*, **15**, 319–51, 1999.
137. O.C. Zienkiewicz, H.C. Huang and M. Pastor. Localization problems in plasticity using the finite elements with adaptive remeshing. *International Journal for Numerical Analysis Methods in Geomechanics*, **19**, 127–48, 1995.

138. O.C. Zienkiewicz, M. Pastor and M. Huang. Softening, localization and adaptive remeshing: capture of discontinuous solutions. *Comp. Mech.*, **17**, 98–106, 1995.
139. J. Yu, D. Peric and D.R.J. Owen. Adaptive finite element analysis of strain localization problem for elasto plastic Cosserat continuum. In D.R.J. Owen *et al.* (eds), *Computational Plasticity III: Models, Software and Applications*, pp. 551–66, Pineridge Press, Swansea, 1992.
140. P. Steinmann and K. Willam. Adaptive techniques for localization analysis. In *Proc. ASME – WAM 92*, ASME, New York, 1992.
141. M. Pastor, J. Peraire and O.C. Zienkiewicz. Adaptive remeshing for shear band localization problems. *Archive of Applied Mechanics*, **61**, 30–91, 1991.
142. M. Pastor, C. Rubio, R Mira, J. Peraire and O.C. Zienkiewicz. Numerical analysis of localization. In G.N. Pande and S. Pietruszczak, (eds), *Numerical Models in Geomechanics*, pp. 339–48, A.A. Balkema, Rotterdam, 1991.
143. R. Larsson, K. Runcsson and N.S. Ottosen. Discontinuous displacement approximations for capturing localization. *Int. J. Num. Meth. Eng.*, **36**, 2087–2105, 1993.
144. O.C. Zienkiewicz and J.Z. Zhu. The superconvergent patch recovery (SPR) and *a posteriori* error estimates. Part 1: the recovery technique. *Int. J. Num. Meth. Eng.*, **33**, 1331–64, 1992.
145. O.C. Zienkiewicz, B. Boroomand and J.Z. Zhu. Recovery procedures in error estimation and adaptivity: adaptivity in linear problems. In P. Ladevèze and J.T. Oden, (eds), *Advances in Adaptive Computational Mechanics in Mechanics*, pp. 3–23. Elsevier, 1998.
146. J. Peraire, M. Vahdati, K. Morgan and O.C. Zienkiewicz. Adaptive remeshing for compressible flow computations. *Journal of Computational Physics*, **72**, 449–66, 1987.
147. O.C. Zienkiewicz and J. Wu. Automatic directional refinement in adaptive analysis of compressible flows. *Int. J. Num. Meth. Eng.*, **37**, 2189–219, 1994.
148. M. Muscat. *The Flow of Homogeneous Fluids through Porous Media*, Edwards, London, 1964.
149. R.E. Volker. Non-linear flow in porous media by finite elements. *Proc. Am. Soc. Civ. Eng.*, **95**(HY6), 1969.
150. H. Ahmed and D.K. Suneda. Non-linear flow in porous media by finite elements. *Proc. Am. Soc. Civ. Eng.*, **95**(HY6), 1847–59, 1969.
151. A.M. Winslow. Numerical solution of the quasi-linear Poisson equation in a non-uniform triangle 'mesh'. *J. Comp. Phys.*, **1**, 149–72, 1966.
152. M.V.K. Chari and P. Silvester. Finite element analysis of magnetically saturated D.C. motors. In *IEEB Winter Meeting on Power*, New York, 1971.
153. J.F. Lyness, D.R.J. Owen and O.C. Zienkiewicz. The finite element analysis of engineering systems governed by a non-linear quasi-harmonic equation. *Computers and Structures*, **5**, 65–79, 1975.
154. O.C. Zienkiewicz, J.F. Lyness and D.R.J. Owen. Three-dimensional magnetic field determination using a scalar potential: a finite element solution. *Trans. IEEE Magn.*, **13**, 1649–56, 1977.
155. D. Gelder. Solution of the compressible flow equations. *Int. J. Num. Meth. Eng.*, **3**, 35–43, 1971.
156. C.A. Anderson and O.C. Zienkiewicz. Spontaneous ignition: finite element solutions for steady and transient conditions. *Trans ASME, J. Heat Transfer*, pp. 398–404, 1974.

4

Plate bending approximation: thin (Kirchhoff) plates and C_1 continuity requirements

4.1 Introduction

The subject of bending of plates and indeed its extension to shells was one of the first to which the finite element method was applied in the early 1960s. At that time the various difficulties that were to be encountered were not fully appreciated and for this reason the topic remains one in which research is active to the present day. Although the subject is of direct interest only to applied mechanicians and structural engineers there is much that has more general applicability, and many of the procedures which we shall introduce can be directly translated to other fields of application.

Plates and shells are but a particular form of a three-dimensional solid, the treatment of which presents no theoretical difficulties, at least in the case of elasticity. However, the thickness of such structures (denoted throughout this and later chapters as t) is very small when compared with other dimensions, and complete three-dimensional numerical treatment is not only costly but in addition often leads to serious numerical ill-conditioning problems. To ease the solution, even long before numerical approaches became possible, several classical assumptions regarding the behaviour of such structures were introduced. Clearly, such assumptions result in a series of approximations. Thus numerical treatment will, in general, concern itself with the approximation to an already approximate theory (or mathematical model), the validity of which is restricted. On occasion we shall point out the shortcomings of the original assumptions, and indeed modify these as necessary or convenient. This can be done simply because now we are granted more freedom than that which existed in the 'pre-computer' era.

The *thin plate* theory is based on the assumptions formalized by Kirchhoff in 1850,[1] and indeed his name is often associated with this theory, though an early version was presented by Sophie Germain in 1811.[2–4] A relaxation of the assumptions was made by Reissner in 1945[5] and in a slightly different manner by Mindlin[6] in 1951. These modified theories extend the field of application of the theory to *thick plates* and we shall associate this name with the Reissner–Mindlin postulates.

It turns out that the thick plate theory is simpler to implement in the finite element method, though in the early days of analytical treatment it presented more difficulties. As it is more convenient to introduce first the thick plate theory and by imposition of additional assumptions to limit it to thin plate theory we shall follow this path in the present chapter. However, when discussing numerical solutions we shall reverse the process and follow the historical procedure of dealing with the thin plate situations first in this chapter. The extension to thick plates and to what turns out always to be a *mixed* formulation will be the subject of Chapter 5.

In the thin plate theory it is possible to represent the state of deformation by one quantity w, the lateral displacement of the middle plane of the plate. Clearly, such a formulation is *irreducible*. The achievement of this irreducible form introduces second derivatives of w in the strain definition and continuity conditions between elements have now to be imposed not only on this quantity but also on its derivatives (C_1 continuity). This is to ensure that the plate remains continuous and does not 'kink'.* Thus at nodes on element interfaces it will always be necessary to use both the value of w and its slopes (first derivatives of w) to impose continuity.

Determination of suitable shape functions is now much more complex than those needed for C_0 continuity. Indeed, as complete slope continuity is required on the interfaces between various elements, the mathematical and computational difficulties often rise disproportionately fast. It is, however, relatively simple to obtain shape functions which, while preserving continuity of w, may violate its slope continuity between elements, though normally not at the node where such continuity is imposed.† If such chosen functions satisfy the 'patch test' (see Chapter 10, Volume 1) then convergence will still be found. The first part of this chapter will be concerned with such 'non-conforming' or 'incompatible' shape functions. In later parts new functions will be introduced by which continuity can be restored. The solution with such 'conforming' shape functions will now give bounds to the energy of the correct solution, but, on many occasions, will yield inferior accuracy to that achieved with non-conforming elements. Thus, for practical usage the methods of the first part of the chapter are often recommended.

The shape functions for rectangular elements are the simplest to form for thin plates and will be introduced first. Shape functions for triangular and quadrilateral elements are more complex and will be introduced later for solutions of plates of *arbitrary* shape or, for that matter, for dealing with shell problems where such elements are essential.

The problem of thin plates is associated with *fourth-order differential equations* leading to a potential energy function which contains *second derivatives* of the unknown function. It is characteristic of a large class of physical problems and, although the chapter concentrates on the structural problem, the reader will find that the procedures developed also will be equally applicable to any problem which is of fourth order.

The difficulty of imposing C_1 continuity on the shape functions has resulted in many alternative approaches to the problems in which this difficulty is side-stepped. Several possibilities exist. Two of the most important are:

1. independent interpolation of rotations $\boldsymbol{\theta}$ and displacement w, imposing continuity as a special constraint, often applied at discrete points only;

* If 'kinking' occurs the second derivative or curvature becomes infinite and squares of infinite terms occur in the energy expression.

† Later we show that even slope discontinuity at the node may be used.

2. the introduction of lagrangian variables or indeed other variables to avoid the necessity of C_1 continuity.

Both approaches fall into the class of mixed formulations and we shall discuss these briefly at the end of the chapter. However, a fuller statement of mixed approaches will be made in the next chapter where both thick and thin approximations will be dealt with simultaneously.

4.2 The plate problem: thick and thin formulations

4.2.1 Governing equations

The mechanics of plate action is perhaps best illustrated in one dimension, as shown in Fig. 4.1. Here we consider the problem of cylindrical bending of plates.[2] In this problem the plate is assumed to have infinite extent in one direction (here assumed the y direction) and to be loaded and supported by conditions independent of y. In this case we may analyse a strip of unit width subjected to some *stress resultants* M_x, P_x, and S_x, which denote x-direction bending moment, axial force and transverse

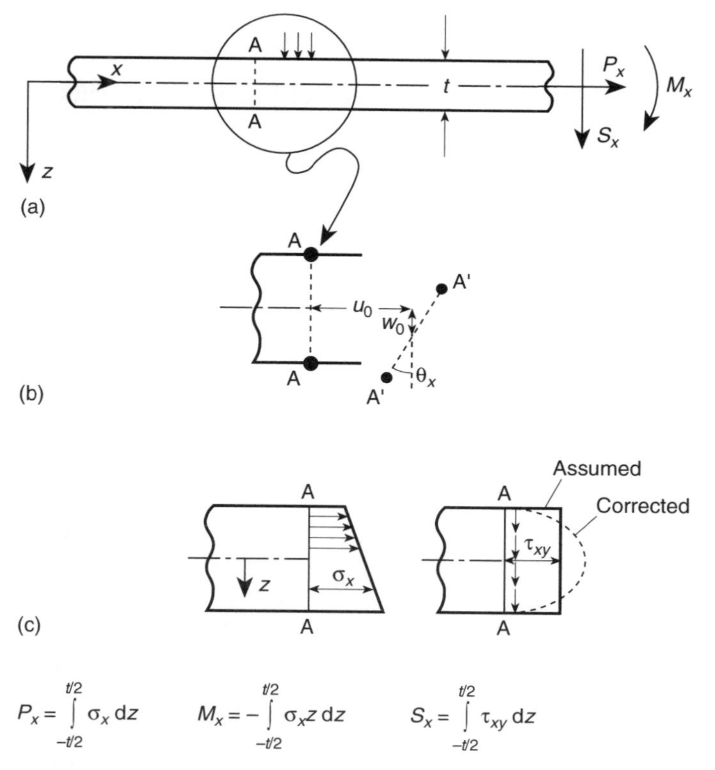

Fig. 4.1 Displacements and stress resultants for a typical beam.

shear force, respectively. For cross-sections that are originally normal to the middle plane of the plate we can use the approximation that at some distance from points of support or concentrated loads plane sections will remain plane during the deformation process. The postulate that sections normal to the middle plane remain plane during deformation is thus the *first* and most important assumption of the theory of plates (and indeed shells). To this is added the *second* assumption. This simply observes that the direct stresses in the normal direction, z, are small, that is, of the order of applied lateral load intensities, q, and hence direct strains in that direction can be neglected. This 'inconsistency' in approximation is compensated for by assuming plane stress conditions in each lamina.

With these two assumptions it is easy to see that the total state of deformation can be described by displacements u_0 and w_0 of the middle surface ($z = 0$) and a rotation θ_x of the normal. Thus the local displacements in the directions of the x and z axes are taken as

$$u(x, z) = u_0(x) - z\theta_x(x) \qquad \text{and} \qquad w(x, z) = w_0(x) \tag{4.1}$$

Immediately the strains in the x and z directions are available as

$$\begin{aligned} \varepsilon_x &= \frac{\partial u}{\partial x} = \frac{\partial u_0}{\partial x} - z\frac{\partial \theta_x}{\partial x} \\ \varepsilon_z &= 0 \\ \gamma_{xz} &= \frac{\partial u}{\partial z} + \frac{\partial w}{\partial x} = -\theta_x + \frac{\partial w_0}{\partial x} \end{aligned} \tag{4.2}$$

For the cylindrical bending problem a state of linear elastic, plane stress for each lamina yields the stress–strain relations

$$\sigma_x = \frac{E}{1 - \nu^2}\varepsilon_x \qquad \text{and} \qquad \tau_{xz} = G\gamma_{xz}$$

The stress resultants are obtained as

$$\begin{aligned} P_x &= \int_{-t/2}^{t/2} \sigma_x \, \mathrm{d}z = B\frac{\partial u_0}{\partial x} \\ S_x &= \int_{-t/2}^{t/2} \tau_{xz} \, \mathrm{d}z = \kappa G t\left(\frac{\partial w_0}{\partial x} - \theta_x\right) \\ M_x &= -\int_{-t/2}^{t/2} \sigma_x z \, \mathrm{d}z = D\frac{\partial \theta_x}{\partial x} \end{aligned} \tag{4.3}$$

where B is the in-plane plate stiffness and D the bending stiffness computed from

$$B = \frac{Et}{1 - \nu^2} \qquad \text{and} \qquad D = \frac{Et^3}{12(1 - \nu^2)} \tag{4.4}$$

with ν Poisson's ratio, E and G direct and shear elastic moduli, respectively.*

* A constant κ has been added here to account for the fact that the shear stresses are not constant across the section. A value of $\kappa = 5/6$ is exact for a rectangular, homogeneous section and corresponds to a parabolic shear stress distribution.

Three equations of equilibrium complete the basic formulation. These equilibrium equations may be computed directly from a differential element of the plate or by integration of the local equilibrium equations. Using the latter approach and assuming zero body and inertial forces we have for the axial resultant

$$\begin{aligned}\int_{-t/2}^{t/2}\left[\frac{\partial\sigma_x}{\partial x}+\frac{\partial\tau_{xz}}{\partial z}\right]\mathrm{d}z&=0\\ \frac{\partial}{\partial x}\int_{-t/2}^{t/2}\sigma_x\,\mathrm{d}z+\tau_{xz}|_{t/2}-\tau_{xz}|_{-t/2}&=0\\ \frac{\partial P_x}{\partial x}&=0\end{aligned}\tag{4.5}$$

where the shear stress on the top and bottom of the plate are assumed to be zero. Similarly, the shear resultant follows from

$$\begin{aligned}\int_{-t/2}^{t/2}\left[\frac{\partial\tau_{xz}}{\partial x}+\frac{\partial\sigma_z}{\partial z}\right]\mathrm{d}z&=0\\ \frac{\partial}{\partial x}\int_{-t/2}^{t/2}\tau_{xz}\,\mathrm{d}z+\sigma_z|_{t/2}-\sigma_z|_{-t/2}&=0\\ \frac{\partial S_x}{\partial x}+q_z&=0\end{aligned}\tag{4.6}$$

where the transverse loading q_z arises from the resultant of the normal traction on the top and/or bottom surfaces. Finally, the moment equilibrium is deduced from

$$\begin{aligned}-\int_{-t/2}^{t/2}z\left[\frac{\partial\sigma_x}{\partial x}+\frac{\partial\tau_{xz}}{\partial z}\right]\mathrm{d}z&=0\\ -\frac{\partial}{\partial x}\int_{-t/2}^{t/2}z\sigma_x\,\mathrm{d}z+\int_{-t/2}^{t/2}\tau_{xz}\,\mathrm{d}z&=0\\ \frac{\partial M_x}{\partial x}+S_x&=0\end{aligned}\tag{4.7}$$

In the elastic case of a plate it is easy to see that the in-plane displacements and forces, u_0 and P_x, decouple from the other terms and the problem of lateral deformations can be dealt with separately. We shall thus only consider bending in the present chapter, returning to the combined problem, characteristic of shell behaviour, in later chapters.

Equations (4.1)–(4.7) are typical for thick plates, and the thin plate theory adds an additional assumption. This simply neglects the shear deformation and puts $G=\infty$. Equation (4.3) thus becomes

$$\frac{\partial w_0}{\partial x}-\theta_x=0\tag{4.8}$$

This thin plate assumption is equivalent to stating that the normals to the middle plane remain normal to it during deformation and is the same as the well-known Bernoulli–Euler assumption for thin beams. The thin, constrained theory is very

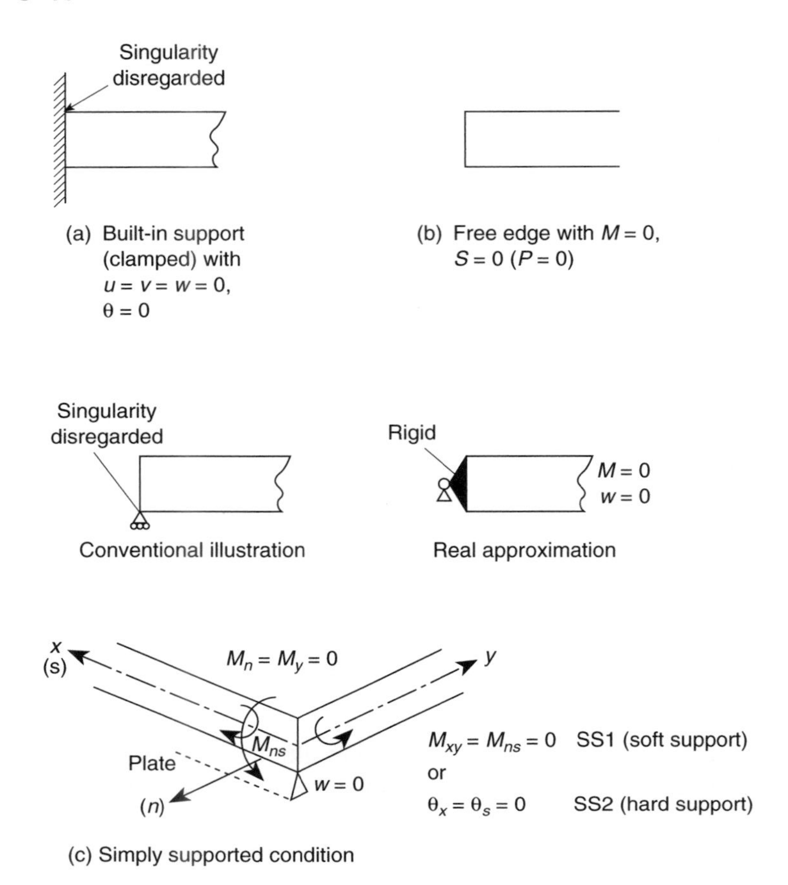

Fig. 4.2 Support (end) conditions for a plate or a beam. Note: the conventionally illustrated simple support leads to infinite displacement – reality is different.

widely used in practice and proves adequate for a large number of structural problems, though, of course, should not be taken literally as the true behaviour near supports or where local load action is important and is three dimensional.

In Fig. 4.2 we illustrate some of the boundary conditions imposed on plates (and beams) and immediately note that the diagrammatic representations of simple support as a knife edge would lead to infinite displacements and stresses. Of course, if a rigid bracket is added in the manner shown this will alter the behaviour to that which we shall generally assume.

The one-dimensional problem of plates and the introduction of thick and thin assumptions translate directly to the general theory of plates. In Fig. 4.3 we illustrate the extensions necessary and write, in place of Eq. (4.1) (assuming u_0 and v_0 to be zero)

$$u = -z\theta_x(x, y) \quad v = -z\theta_y(x, y) \quad w = w_0(x, y) \tag{4.9}$$

where we note that displacement parameters are now functions of x and y.

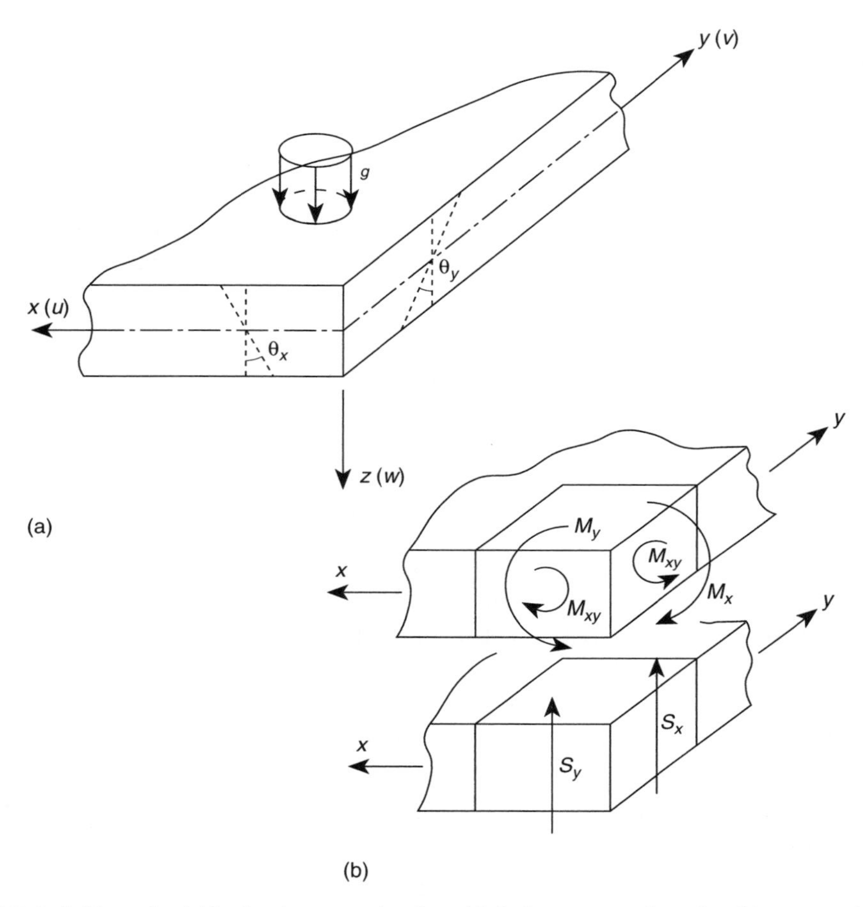

Fig. 4.3 Definitions of variables for plate approximations: (a) displacements and rotation; (b) stress resultants.

The strains may now be separated into bending (in-plane components) and transverse shear groups and we have, in place of Eq. (4.2),

$$\boldsymbol{\varepsilon} = \begin{Bmatrix} \varepsilon_x \\ \varepsilon_y \\ \gamma_{xy} \end{Bmatrix} = -z \begin{bmatrix} \dfrac{\partial}{\partial x} & 0 \\ 0 & \dfrac{\partial}{\partial y} \\ \dfrac{\partial}{\partial y} & \dfrac{\partial}{\partial x} \end{bmatrix} \begin{Bmatrix} \theta_x \\ \theta_y \end{Bmatrix} \equiv -z\mathbf{L}\boldsymbol{\theta} \tag{4.10}$$

and

$$\boldsymbol{\gamma} = \begin{Bmatrix} \gamma_{xz} \\ \gamma_{yz} \end{Bmatrix} = \begin{Bmatrix} \dfrac{\partial w}{\partial x} \\ \dfrac{\partial w}{\partial y} \end{Bmatrix} - \begin{Bmatrix} \theta_x \\ \theta_y \end{Bmatrix} = \boldsymbol{\nabla} w - \boldsymbol{\theta} \tag{4.11}$$

We note that now in addition to normal bending moments M_x and M_y, now defined by expression (4.3) for the x and y directions, respectively, a twisting moment arises defined by

$$M_{xy} = -\int_{-t/2}^{t/2} \tau_{xy}\,\mathrm{d}z \tag{4.12}$$

Introducing appropriate constitutive relations, all moment components can be related to displacement derivatives. For isotropic elasticity we can thus write, in place of Eq. (4.3),

$$\mathbf{M} = \begin{Bmatrix} M_x \\ M_y \\ M_{xy} \end{Bmatrix} = \mathbf{D}\mathbf{L}\boldsymbol{\theta} \tag{4.13}$$

where, assuming plane stress behaviour in each layer,

$$\mathbf{D} = D\begin{bmatrix} 1 & \nu & 0 \\ \nu & 1 & 0 \\ 0 & 0 & (1-\nu)/2 \end{bmatrix} \tag{4.14}$$

where ν is Poisson's ratio and D is defined by the second of Eqs (4.4). Further, the shear force resultants are

$$\mathbf{S} = \begin{Bmatrix} S_x \\ S_y \end{Bmatrix} = \boldsymbol{\alpha}(\boldsymbol{\nabla} w - \boldsymbol{\theta}) \tag{4.15}$$

For isotropic elasticity (though here we deliberately have not related G to E and ν to allow for possibly different shear rigidities)

$$\boldsymbol{\alpha} = \kappa G t \mathbf{I} \tag{4.16}$$

where $\mathbf{I}$ is a 2×2 identity matrix.

Of course, the constitutive relations can be simply generalized to anisotropic or inhomogeneous behaviour such as can be manifested if several layers of materials are assembled to form a *composite*. The only apparent difference is the structure of the $\mathbf{D}$ and $\boldsymbol{\alpha}$ matrices, which can always be found by simple integration.

The governing equations of thick and thin plate behaviour are completed by writing the equilibrium relations. Again omitting the 'in-plane' behaviour we have, in place of Eq. (4.6),

$$\left[\frac{\partial}{\partial x}, \quad \frac{\partial}{\partial y}\right]\begin{Bmatrix} S_x \\ S_y \end{Bmatrix} + q \equiv \boldsymbol{\nabla}^{\mathrm{T}}\mathbf{S} + q = 0 \tag{4.17}$$

and, in place of Eq. (4.7),

$$\begin{bmatrix} \dfrac{\partial}{\partial x} & 0 & \dfrac{\partial}{\partial y} \\ 0 & \dfrac{\partial}{\partial y} & \dfrac{\partial}{\partial x} \end{bmatrix}\begin{Bmatrix} M_x \\ M_y \\ M_{xy} \end{Bmatrix} + \begin{Bmatrix} S_x \\ S_y \end{Bmatrix} \equiv \mathbf{L}^{\mathrm{T}}\mathbf{M} + \mathbf{S} = \mathbf{0} \tag{4.18}$$

Equations (4.13)–(4.18) are the basis from which the solution of both thick and thin plates can start. For thick plates any (or all) of the independent variables can be approximated independently, leading to a mixed formulation which we shall discuss in Chapter 5 and also briefly in Sec. 4.16 of this chapter.

For thin plates in which the shear deformations are suppressed Eq. (4.15) is rewritten as

$$\boldsymbol{\nabla} w - \boldsymbol{\theta} = \mathbf{0} \tag{4.19}$$

and the strain-displacement relations (4.10) become

$$\boldsymbol{\varepsilon} = -z\mathbf{L}\boldsymbol{\nabla} w = -z \begin{Bmatrix} \dfrac{\partial^2 w}{\partial x^2} \\ \dfrac{\partial^2 w}{\partial y^2} \\ 2\dfrac{\partial^2 w}{\partial x \partial y} \end{Bmatrix} = -z\boldsymbol{\kappa} \tag{4.20}$$

where $\boldsymbol{\kappa}$ is the matrix of changes in curvature of the plate. Using the above form for the thin plate, both irreducible and mixed forms can now be written. In particular, it is an easy matter to eliminate $\mathbf{M}$, $\mathbf{S}$ and $\boldsymbol{\theta}$ and leave only w as the variable.

Applying the operator $\boldsymbol{\nabla}^{\mathrm{T}}$ to expression (4.17), inserting Eqs (4.13) and (4.17) and finally replacing $\boldsymbol{\theta}$ by the use of Eq. (4.19) gives a scalar equation

$$(\mathbf{L}\boldsymbol{\nabla})^{\mathrm{T}}\mathbf{D}\mathbf{L}\boldsymbol{\nabla} w - q = 0 \tag{4.21}$$

where, using Eq. (4.20),

$$(\mathbf{L}\boldsymbol{\nabla}) = \left[\frac{\partial^2}{\partial x^2}, \; \frac{\partial^2}{\partial y^2}, \; 2\frac{\partial^2}{\partial x \partial y} \right]^{\mathrm{T}}$$

In the case of isotropy with constant bending stiffness D this becomes the well-known biharmonic equation of plate flexure

$$D\left(\frac{\partial^4 w}{\partial x^4} + 2\frac{\partial^4 w}{\partial x^2 \partial y^2} + \frac{\partial^4 w}{\partial y^4} \right) - q = 0 \tag{4.22}$$

4.2.2 The boundary conditions

The boundary conditions which have to be imposed on the problem (see Figs 4.2 and 4.4) include the following classical conditions.

1. *Fixed boundary*, where displacements on restrained parts of the boundary are given specified values.* These conditions are expressed as

$$w = \bar{w}; \quad \theta_n = \bar{\theta}_n \quad \text{and} \quad \theta_s = \bar{\theta}_s$$

* Note that in thin plates the specification of w along s automatically specifies θ_s by Eq. (4.19), but this is not the case in thick plates where the quantities are independently prescribed.

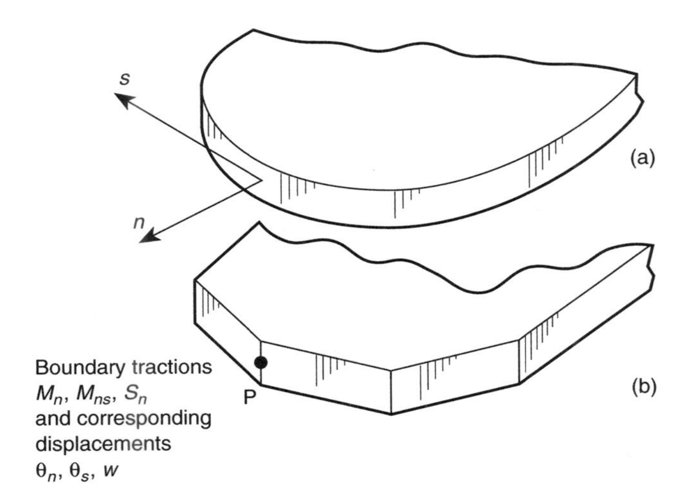

Fig. 4.4 Boundary traction and conjugate displacement. Note: the simply supported condition requiring $M_n = 0$, $\theta_s = 0$ and $w = 0$ is identical at a corner node to specifying $\theta_n = \theta_s = 0$, that is, a clamped support. This leads to a paradox if a curved boundary (a) is modelled as a polygon (b).

Here n and s are directions normal and tangential to the boundary curve of the middle surface. A clamped edge is a special case with zero values assigned.

2. *Traction boundary*, where stress resultants M_n, M_{ns} and S_n (conjugate to the displacements θ_n, θ_s and w) are given prescribed values. A free edge is a special case with zero values assigned.
3. *'Mixed' boundary conditions*, where both traction and displacements can be specified. Typical here is the simply supported edge (see Fig. 4.2). For this, clearly, $M_n = 0$ and $w = 0$, but it is less clear whether M_{ns} or θ_s needs to be given. Specification of $M_{ns} = 0$ is *physically* a more acceptable condition and does not lead to difficulties. This should always be adopted for thick plates.

In thin plates θ_s is automatically specified from w and we shall find certain difficulties, and indeed anomalies, associated with this assumption.[7,8] For instance, in Fig. 4.4 we see how a specification of $\theta_s = 0$ at corner nodes implicit in thin plates formally leads to the prescription of all boundary parameters, which is identical to boundary conditions of a clamped plate for this point.

4.2.3 The irreducible, thin plate approximation

The thin plate approximation when cast in terms of a single variable w is clearly irreducible and is in fact typical of a displacement formulation. The equations (4.17) and (4.18) can be written together as

$$(\mathbf{L}\boldsymbol{\nabla})^{\mathrm{T}}\mathbf{M} - q = 0 \tag{4.23}$$

and the constitutive relation (4.13) can be recast by using Eq. (4.19) as

$$\mathbf{M} = \mathbf{D}\mathbf{L}\boldsymbol{\nabla} w \tag{4.24}$$

The derivation of the finite element equations can be obtained either from a weak form of Eq. (4.23) obtained by weighting with an arbitrary function (say $v = \mathbf{N}\tilde{\mathbf{v}}$) and integration by parts (done twice) or, more directly, by application of the virtual work equivalence. Using the latter approach we may write the internal virtual work for the plate as

$$\delta\Pi_{\text{int}} = \int_\Omega (\delta\boldsymbol{\varepsilon})^{\mathrm{T}}\mathbf{D}\boldsymbol{\varepsilon}\,\mathrm{d}\Omega = \int_\Omega \delta w(\mathbf{L}\boldsymbol{\nabla})^{\mathrm{T}}\mathbf{D}(\mathbf{L}\boldsymbol{\nabla})w\,\mathrm{d}\Omega \tag{4.25}$$

where Ω denotes the area of the plate reference (middle) surface and $\mathbf{D}$ is the plate stiffness, which for isotropy is given by Eq. (4.14).

Similarly the external work is given by[2]

$$\delta\Pi_{\text{ext}} = \int_\Omega \delta w\, q\,\mathrm{d}\Omega + \int_{\Gamma_n} \delta\theta_n \bar{M}_n\,\mathrm{d}\Gamma + \int_{\Gamma_t} \delta\theta_s \bar{M}_{ns}\,\mathrm{d}\Gamma + \int_{\Gamma_s} \delta w\,\bar{S}_n\,\mathrm{d}\Gamma \tag{4.26}$$

where $\bar{M}_n$, $\bar{M}_{ns}$, $\bar{S}_n$ are specified values and Γ_n, Γ_t and Γ_s are parts of the boundary where each component is specified. For thin plates with straight edges Eq. (4.19) gives immediately $\theta_s = \partial w/\partial s$ and thus the last two terms above may be combined as

$$\int_{\Gamma_t} \delta\theta_s \bar{M}_{ns}\,\mathrm{d}\Gamma + \int_{\Gamma_s} \delta w\,\bar{S}_n\,\mathrm{d}\Gamma = \int_{\Gamma_s} \delta w\left(\bar{S}_n - \frac{\partial \bar{M}_{ns}}{\partial s}\right)\mathrm{d}\Gamma + \sum_i \delta w_i R_i \tag{4.27}$$

where R_i are concentrated forces arising at locations where corners exist (see Fig 4.2).[2]

Substituting into Eqs (4.25) and (4.26) the discretization

$$w = \mathbf{N}\mathbf{a} \tag{4.28}$$

where $\mathbf{a}$ are appropriate parameters, we can obtain for a linear case standard displacement approximation equations

$$\mathbf{K}\mathbf{a} = \mathbf{f} \tag{4.29}$$

with

$$\mathbf{K}\mathbf{a} = \left(\int_\Omega \mathbf{B}^{\mathrm{T}}\mathbf{D}\mathbf{B}\,\mathrm{d}\Omega\right)\mathbf{a} \equiv \int_\Omega \mathbf{B}^{\mathrm{T}}\mathbf{M}\,\mathrm{d}\Omega \tag{4.30}$$

and

$$\mathbf{f} = \int_\Omega \mathbf{N}^{\mathrm{T}} q\,\mathrm{d}\Omega + \mathbf{f}_{\mathrm{b}} \tag{4.31}$$

where $\mathbf{f}_{\mathrm{b}}$ is the boundary contribution to be discussed later and

$$\mathbf{M} = \mathbf{D}\mathbf{B}\mathbf{a} \tag{4.32}$$

with

$$\mathbf{B} = (\mathbf{L}\boldsymbol{\nabla})\mathbf{N} \tag{4.33}$$

It is of interest, and indeed important to note, that when tractions are prescribed to non-zero values the force term $\mathbf{f}_{\mathrm{b}}$ includes all prescribed values of M_n, M_{ns} and S_n irrespective of whether the thick or thin formulation is used. The reader can verify that this term is

$$\mathbf{f}_{\mathrm{b}} = \int_\Gamma (\mathbf{N}_n^{\mathrm{T}}\bar{M}_n + \mathbf{N}_s^{\mathrm{T}}\bar{M}_{ns} + \mathbf{N}^{\mathrm{T}}\bar{S}_n)\,\mathrm{d}\Gamma \tag{4.34a}$$

where $\bar{M}_n$, $\bar{M}_{ns}$ and $\bar{S}_n$ are prescribed values and for thin plates [though, of course, relation (4.34a) is valid for thick plates also]:

$$\mathbf{N}_n = \frac{\partial \mathbf{N}}{\partial n} \quad \text{and} \quad \mathbf{N}_s = \frac{\partial \mathbf{N}}{\partial s} \tag{4.34b}$$

The reader will recognize in the above the well-known ingredients of a displacement formulation (see Chapter 2 of Volume 1, and Chapter 1 of this volume) and the procedures are almost automatic once **N** is chosen.

4.2.4 Continuity requirement for shape functions (C_1 continuity)

In Sections 4.3–4.13 we will be concerned with the above formulation [starting from Eqs (4.24) and (4.26)], and the presence of the second derivatives indicates quite clearly that we shall need C_1 continuity of the shape functions for the irreducible, thin plate, formulation. This continuity is difficult to achieve and reasons for this are given below.

To ensure the continuity of both w and its normal slope across an interface we must have both w and $\partial w/\partial n$ uniquely defined by values of nodal parameters along such an interface. Consider Fig. 4.5 depicting the side 1–2 of a rectangular element. The normal direction n is in fact that of y and we desire w and $\partial w/\partial y$ to be uniquely determined by values of w, $\partial w/\partial x$, $\partial w/\partial y$ at the nodes lying along this line.

Following the principles expounded in Chapter 8 of Volume 1, we would write along side 1–2,

$$w = A_1 + A_2 x + A_3 y + \cdots \tag{4.35}$$

and

$$\frac{\partial w}{\partial y} = B_1 + B_2 x + B_3 y + \cdots \tag{4.36}$$

with a number of constants in each expression just sufficient to determine a unique solution for the nodal parameters associated with the line.

Thus, for instance, if only two nodes are present a cubic variation of w should be permissible noting that $\partial w/\partial x$ and w are specified at each node. Similarly, only a linear, or two-term, variation of $\partial w/\partial y$ would be permissible.

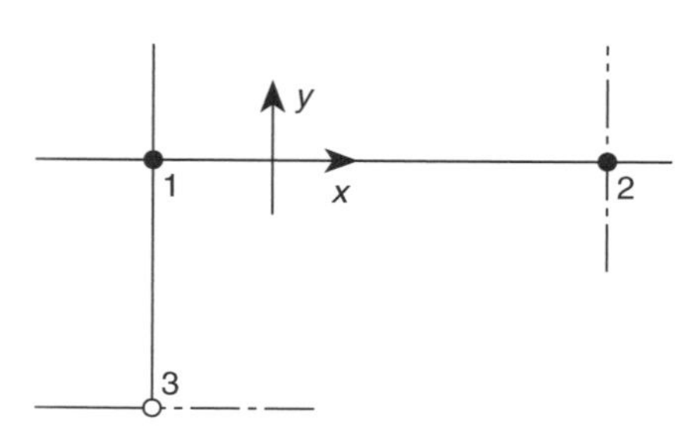

Fig. 4.5 Continuity requirement for normal slopes.

Note, however, that a similar exercise could be performed along the side placed in the y direction preserving continuity of $\partial w/\partial x$ along this. Along side 1–2 we thus have $\partial w/\partial y$, depending on nodal parameters of line 1–2 only, and along side 1–3 we have $\partial w/\partial x$, depending on nodal parameters of line 1–3 only. Differentiating the first with respect to x, on line 1–2 we have $\partial^2 w/\partial x \partial y$, depending on nodal parameters of line 1–2 only, and similarly, on line 1–3 we have, $\partial^2 w/\partial y \partial x$, depending on nodal parameters of line 1–3 only.

At the common point, 1, an inconsistency arises immediately as we cannot automatically have there the necessary identity for continuous functions

$$\frac{\partial^2 w}{\partial x \partial y} \equiv \frac{\partial^2 w}{\partial y \partial x} \tag{4.37}$$

for arbitrary values of the parameters at nodes 2 and 3. *It is thus impossible to specify simple polynomial expressions for shape functions ensuring full compatibility when only* w *and its slopes are prescribed at corner nodes.*[9]

Thus if any functions satisfying the compatibility are found with the three nodal variables, they must be such that at corner nodes these functions are not continuously differentiable and the cross-derivative is not unique. Some such functions are discussed in the second part of this chapter.[10–16]

The above proof has been given for a rectangular element. Clearly, the arguments can be extended for any two arbitrary directions of interface at the corner node 1.

A way out of this difficulty appears to be obvious. We could specify the cross-derivative as one of the nodal parameters. This, for an assembly of rectangular elements, is convenient and indeed permissible. Simple functions of that type have been suggested by Bogner *et al.*[17] and used with some success. Unfortunately, the extension to nodes at which a number of element interfaces meet with different angles (Fig. 4.6) is not, in general, permissible. Here, the continuity of cross-derivatives in several sets of orthogonal directions implies, in fact, a specification of *all second derivatives at a node.*

This, however, violates physical requirements if the plate stiffness varies abruptly from element to element, for then equality of moments normal to the interfaces cannot be maintained. However, this process has been used with some success in homogeneous plate situations[18–25] although Smith and Duncan[18] comment adversely on the effect of imposing such *excessive continuities* on several orders of higher derivatives.

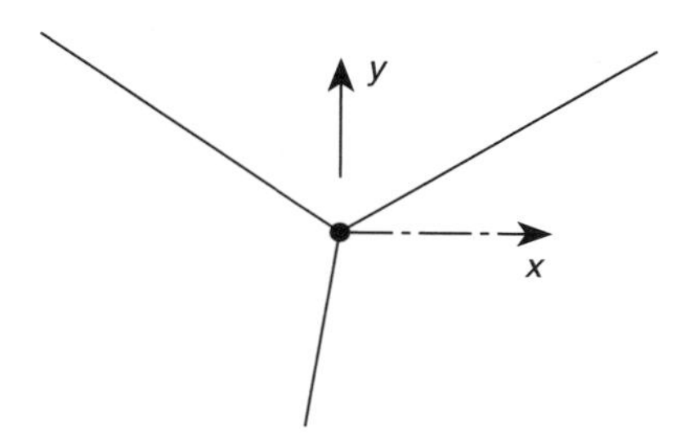

Fig. 4.6 Nodes where elements meet in arbitrary directions.

The difficulties of finding compatible displacement functions have led to many attempts at ignoring the complete slope continuity while still continuing with the other necessary criteria. Proceeding perhaps from a naive but intuitive idea that the imposition of slope continuity at nodes only must, in the limit, lead to a complete slope continuity, several successful, 'non-conforming', elements have been developed.[11,26–40]

The convergence of such elements is not obvious but can be proved either by application of the patch test or by comparison with finite difference algorithms. We have discussed the importance of the patch test extensively in Chapter 11 of Volume 1 and additional details are available in references 41–43.

In plate problems the importance of the patch test in both design and testing of elements is paramount and this test should never be omitted. In the first part of this chapter, dealing with non-conforming elements, we shall repeatedly make use of it. Indeed, we shall show how some of the most successful elements currently used have developed via this analytical interpretation.[44–49]

Non-conforming shape functions

4.3 Rectangular element with corner nodes (12 degrees of freedom)

4.3.1 Shape functions

Consider a rectangular element of a plate *ijkl* coinciding with the xy plane as shown in Fig. 4.7. At each node, n, displacements $\mathbf{a}_n$ are introduced. These have three components: the first a displacement in the z direction, w_n, the second a rotation about the x axis, $(\hat{\theta}_x)_n$ and the third a rotation about the y axis $(\hat{\theta}_y)_n$.*

The nodal displacement vectors are defined below as $\mathbf{a}_i$. The element displacement will, as usual, be given by a listing of the nodal displacements, now totalling twelve:

$$\mathbf{a}^e = \begin{Bmatrix} \mathbf{a}_i \\ \mathbf{a}_j \\ \mathbf{a}_k \\ \mathbf{a}_l \end{Bmatrix} \qquad \mathbf{a}_i = \begin{Bmatrix} w_i \\ \hat{\theta}_{xi} \\ \hat{\theta}_{yi} \end{Bmatrix} \tag{4.38}$$

A polynomial expression is conveniently used to define the shape functions in terms of the 12 parameters. Certain terms must be omitted from a complete fourth-order

* Note that we have changed here the convention from that of Fig. 4.3 in this chapter. This allows transformations needed for shells to be carried out in an easier manner. However, when manipulating the equations of Chapter 5 we shall return to the orginal definitions of Fig. 4.3. Similar difficulties are discussed by Hughes,[50] and a simple transformation is as follows:

$$\hat{\boldsymbol{\theta}} = \mathbf{T}\boldsymbol{\theta} \quad \text{where} \quad \mathbf{T} = \begin{bmatrix} 0 & 1 \\ -1 & 0 \end{bmatrix}$$

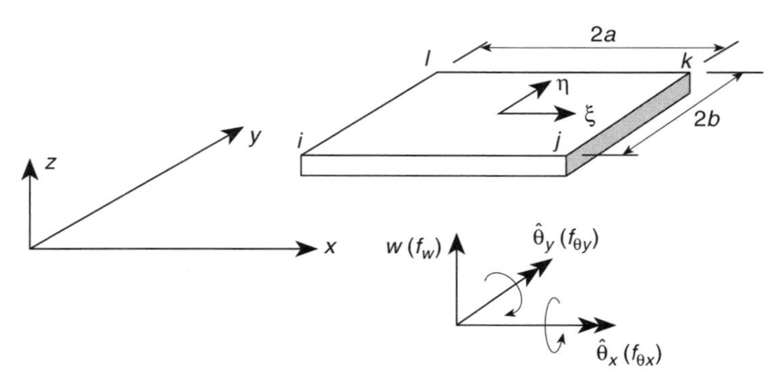

Fig. 4.7 A rectangular plate element.

polynomial. Writing

$$
\begin{aligned}
w &= \alpha_1 + \alpha_2 x + \alpha_3 y + \alpha_4 x^2 + \alpha_5 xy + \alpha_6 y^2 + \alpha_7 x^3 + \alpha_8 x^2 y \\
&\quad + \alpha_9 xy^2 + \alpha_{10} y^3 + \alpha_{11} x^3 y + \alpha_{12} xy^3 \\
&\equiv \mathbf{P}\boldsymbol{\alpha}
\end{aligned}
\tag{4.39}
$$

has certain advantages. In particular, along any x constant or y constant line, the displacement w will vary as a cubic. The element boundaries or interfaces are composed of such lines. As a cubic is uniquely defined by four constants, the two end values of slopes and the two displacements at the ends will therefore define the displacements along the boundaries uniquely. As such end values are common to adjacent elements continuity of w will be imposed along any interface.

It will be observed that the gradient of w normal to any of the boundaries also varies along it in a cubic way. (Consider, for instance, values of the normal $\partial w/\partial y$ along a line on which x is constant.) As on such lines only two values of the normal slope are defined, the cubic is not specified uniquely and, in general, a discontinuity of normal slope will occur. The function is thus 'non-conforming'.

The constants α_1 to α_{12} can be evaluated by writing down the 12 simultaneous equations linking the values of w and its slopes at the nodes when the coordinates take their appropriate values. For instance,

$$
\begin{aligned}
w_i &= \alpha_1 + \alpha_2 x_i + \alpha_3 y_i + \cdots \\
\left(\frac{\partial w}{\partial y}\right)_i &= \hat{\theta}_{xi} = \alpha_3 + \alpha_5 x_i + \cdots \\
-\left(\frac{\partial w}{\partial x}\right)_i &= \hat{\theta}_{yi} = -\alpha_2 - \alpha_5 y_i - \cdots
\end{aligned}
$$

Listing all 12 equations, we can write, in matrix form,

$$
\mathbf{a}^e = \mathbf{C}\boldsymbol{\alpha} \tag{4.40}
$$

where $\mathbf{C}$ is a 12×12 matrix depending on nodal coordinates, and $\boldsymbol{\alpha}$ is a vector of the 12 unknown constants. Inverting we have

$$\boldsymbol{\alpha} = \mathbf{C}^{-1}\mathbf{a}^{e} \tag{4.41}$$

This inversion can be carried out by the computer or, if an explicit expression for the stiffness, etc., is desired, it can be performed algebraically. This was in fact done by Zienkiewicz and Cheung.[26]

It is now possible to write the expression for the displacement within the element in a standard form as

$$\mathbf{u} \equiv w = \mathbf{N}\mathbf{a}^{e} = \mathbf{P}\mathbf{C}^{-1}\mathbf{a}^{e} \tag{4.42}$$

where

$$\mathbf{P} = (1, x, y, x^2, xy, y^2, x^3, x^2y, xy^2, y^3, x^3y, xy^3)$$

The form of the $\mathbf{B}$ is obtained directly from Eqs (4.28) and (4.33). We thus have

$$\mathbf{L}\boldsymbol{\nabla}w = \begin{Bmatrix} +2\alpha_4 & +6\alpha_7 x & +2\alpha_8 y + 6\alpha_{11}xy \\ +2\alpha_6 & +2\alpha_9 x & +6\alpha_{10} y + 6\alpha_{12}xy \\ +2\alpha_5 & +4\alpha_8 x & +4\alpha_9 y + 6\alpha_{11}x^2 + 6\alpha_{12}y^2 \end{Bmatrix}$$

We can write the above as

$$\mathbf{L}\boldsymbol{\nabla}w = \mathbf{Q}\boldsymbol{\alpha} = \mathbf{Q}\mathbf{C}^{-1}\mathbf{a}^{e} = \mathbf{B}\mathbf{a}^{e} \quad \text{and thus} \quad \mathbf{B} = \mathbf{Q}\mathbf{C}^{-1} \tag{4.43}$$

in which

$$\mathbf{Q} = \begin{bmatrix} 0 & 0 & 0 & 2 & 0 & 0 & 6x & 2y & 0 & 0 & 6xy & 0 \\ 0 & 0 & 0 & 0 & 0 & 2 & 0 & 0 & 2x & 6y & 0 & 6xy \\ 0 & 0 & 0 & 0 & 2 & 0 & 0 & 4x & 4y & 0 & 6x^2 & 6y^2 \end{bmatrix} \tag{4.44}$$

It is of interest to remark now that the displacement function chosen does in fact permit a state of constant strain (curvature) to exist and therefore satisfies one of the criteria of convergence stated in Volume 1.*

An explicit form of the shape function $\mathbf{N}$ was derived by Melosh[36] and can be written simply in terms of normalized coordinates. Thus, we can write for any node

$$\mathbf{N}_i^{\mathrm{T}} = \tfrac{1}{8}(1+\xi_0)(1+\eta_0)\begin{Bmatrix} 2 + \xi_0 + \eta_0 - \xi^2 - \eta^2 \\ b\eta_i(1-\eta^2) \\ -a\xi_i(1-\xi^2) \end{Bmatrix} \tag{4.45}$$

with normalized coordinates defined as:

$$\xi = \frac{x - x_c}{a} \quad \text{where} \quad \xi_0 = \xi\xi_i$$

$$\eta = \frac{y - y_c}{b} \quad \text{where} \quad \eta_0 = \eta\eta_i$$

* If α_7 to α_{12} are zero, then the 'strain' defined by second derivatives is constant. By Eq. (4.40), the corresponding $\mathbf{a}^{e}$ can be found. As there is a unique correspondence between $\mathbf{a}^{e}$ and $\boldsymbol{\alpha}$ such a state is therefore unique. All this presumes that $\mathbf{C}^{-1}$ does in fact exist. The algebraic inversion shows that the matrix $\mathbf{C}$ is never singular.

This form avoids the explicit inversion of **C**; however, for simplicity we pursue the direct use of polynomials to deduce the stiffness and load matrices.

4.3.2 Stiffness and load matrices

Standard procedures can now be followed, and it is almost superfluous to recount the details. The stiffness matrix relating the nodal *forces* (given by lateral force and two moments at each node) to the corresponding nodal displacement is

$$\mathbf{K}^e = \int_{\Omega^e} \mathbf{B}^T \mathbf{D} \mathbf{B} \, dx \, dy \tag{4.46}$$

or, substituting Eq. (4.43) into this expression,

$$\mathbf{K}^e = \mathbf{C}^{-T} \left(\int_{-b}^{b} \int_{-a}^{a} \mathbf{Q}^T \mathbf{D} \mathbf{Q} \, dx \, dy \right) \mathbf{C}^{-1} \tag{4.47}$$

The terms not containing x and y have now been moved from the operation of integrating. The term within the integration sign can be multiplied out and integrated explicitly without difficulty if **D** is constant.

The external forces at nodes arising from distributed loading can be assigned 'by inspection', allocating specific areas as contributing to any node. However, it is more logical and accurate to use once again the standard expression (4.31) for such an allocation.

The contribution of these forces to each of the nodes is

$$\mathbf{f}_i = \begin{Bmatrix} f_{w_i} \\ f_{\theta_{xi}} \\ f_{\theta_{yi}} \end{Bmatrix} = \int_{-b}^{b} \int_{-a}^{a} \mathbf{N}^T q \, dx \, dy \tag{4.48}$$

or, by Eq. (4.42),

$$\mathbf{f}_i = -\mathbf{C}^{-T} \int_{-b}^{b} \int_{-a}^{a} \mathbf{P}^T q \, dx \, dy \tag{4.49}$$

The integral is again evaluated simply. It will now be noted that, in general, all three components of external force at any node will have non-zero values. This is a result that the simple allocation of external loads would have missed. The nodal load vector for uniform loading q is given by

$$\mathbf{f}_1 = \tfrac{1}{12} qab \begin{Bmatrix} 3 \\ b \\ -a \end{Bmatrix}, \quad \mathbf{f}_2 = \tfrac{1}{12} qab \begin{Bmatrix} 3 \\ -b \\ -a \end{Bmatrix}, \quad \mathbf{f}_3 = \tfrac{1}{12} qab \begin{Bmatrix} 3 \\ b \\ a \end{Bmatrix}, \quad \mathbf{f}_4 = \tfrac{1}{12} qab \begin{Bmatrix} 3 \\ -b \\ a \end{Bmatrix} \tag{4.50}$$

The vector of nodal plate forces due to initial strains and initial stresses can be found in a similar way. It is necessary to remark in this connection that initial strains, such as may be due to a temperature rise, is seldom confined in its effects on curvatures. Usually, direct (in-plane) strains in the plate are introduced additionally, and the complete problem can be solved only by consideration of the plane stress problem as well as that of bending.

4.4 Quadrilateral and parallelogram elements

The rectangular element developed in the preceding section passes the patch test[41] and is always convergent. However, it cannot be easily generalized into a quadrilateral shape. Transformation of coordinates of the type described in Chapter 9 of Volume 1 can be performed but unfortunately now it will be found that the constant curvature criterion is violated. As expected, such elements behave badly but by arguments given in Chapter 9 of Volume 1 convergence may still occur providing the patch test is passed in the curvilinear coordinates. Henshell *et al.*[40] studied the performance of such an element (and also some of a higher order) and concluded that reasonable accuracy is attainable. Their paper gives all the details of transformations required for an isoparametric mapping and the resulting need for numerical integration.

Only for the case of a parallelogram is it possible to achieve states of constant curvature exclusively using functions of ξ and η and the patch test is satisfied. For a parallelogram the local coordinates can be related to the global ones by the explicit expression (Fig. 4.8)

$$
\begin{aligned}
\xi &= \frac{x - y \cot \alpha}{a} \\
\eta &= \frac{y \csc \alpha}{b}
\end{aligned}
\tag{4.51}
$$

and all expressions for the stiffness and loads can therefore also be derived directly. Such an element is suggested in the discussion in reference 26, and the stiffness matrices have been worked out by Dawe.[28] A somewhat different set of shape functions was suggested by Argyris.[29]

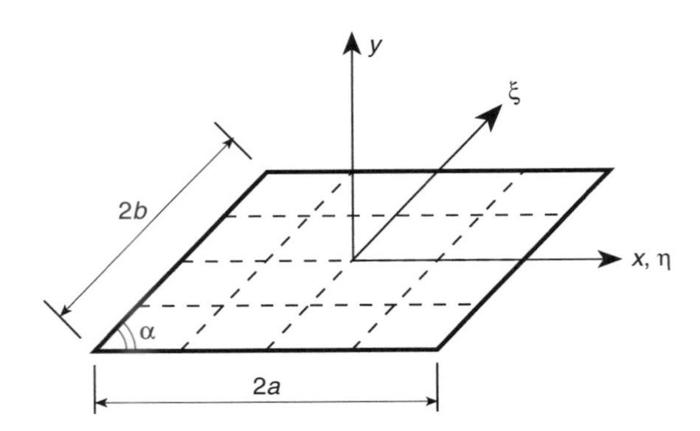

Fig. 4.8 Parallelogram element and skew coordinates.

4.5 Triangular element with corner nodes (9 degrees of freedom)

At first sight, it would seem that once again a simple polynomial expansion could be used in a manner identical to that of the previous section. As only nine independent

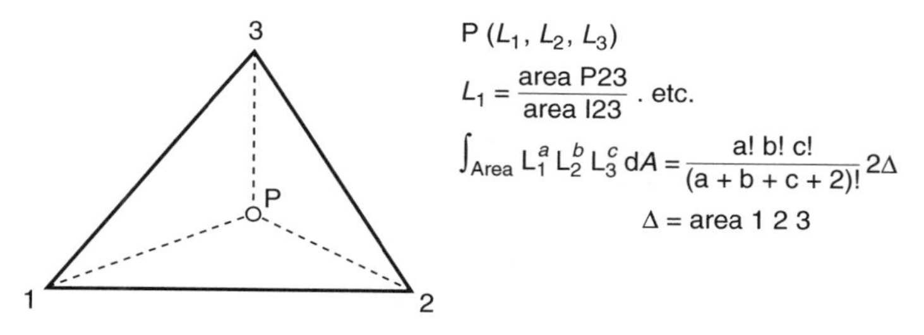

Fig. 4.9 Area coordinates.

movements are imposed, only nine terms of the expansion are permissible. Here an immediate difficulty arises as the full cubic expansion contains 10 terms [Eq. (4.39) with $\alpha_{11} = \alpha_{12} = 0$] and any omission has to be made arbitrarily. To retain a certain symmetry of appearance all 10 terms could be retained and two coefficients made equal (for example $\alpha_8 = \alpha_9$) to limit the number of unknowns to nine. Several such possibilities have been investigated but a further, much more serious, problem arises. The matrix corresponding to $\mathbf{C}$ of Eq. (4.40) becomes singular for certain orientations of the triangle sides. This happens, for instance, when two sides of the triangle are parallel to the x and y axes respectively.

An 'obvious' alternative is to add a central node to the formulation and eliminate this by static condensation. This would allow a complete cubic to be used, but again it was found that an element derived on this basis does not converge to correct answers.

Difficulties of asymmetry can be avoided by the use of area coordinates described in Sec. 8.8 of Volume 1. These are indeed nearly always a natural choice for triangles, see (Fig. 4.9).

4.5.1 Shape functions

As before we shall use polynomial expansion terms, and it is worth remarking that these are given in area coordinates in an unusual form. For instance,

$$\alpha_1 L_1 + \alpha_2 L_2 + \alpha_3 L_3 \tag{4.52}$$

gives the three terms of a *complete* linear polynomial and

$$\alpha_1 L_1^2 + \alpha_2 L_2^2 + \alpha_3 L_3^2 + \alpha_4 L_1 L_2 + \alpha_5 L_2 L_3 + \alpha_6 L_3 L_1 \tag{4.53}$$

gives all six terms of a quadratic (containing within it the linear terms).* The 10 terms of a cubic expression are similarly formed by the products of all possible cubic

* However, it is also possible to write a complete quadratic as

$$\alpha_1 L_1 + \alpha_2 L_2 + \alpha_3 L_3 + \alpha_4 L_1 L_2 + \alpha_5 L_2 L_3 + \alpha_6 L_3 L_1$$

and so on, for higher orders. This has the advantage of explicitly stating all retained terms of polynomials of lower order.

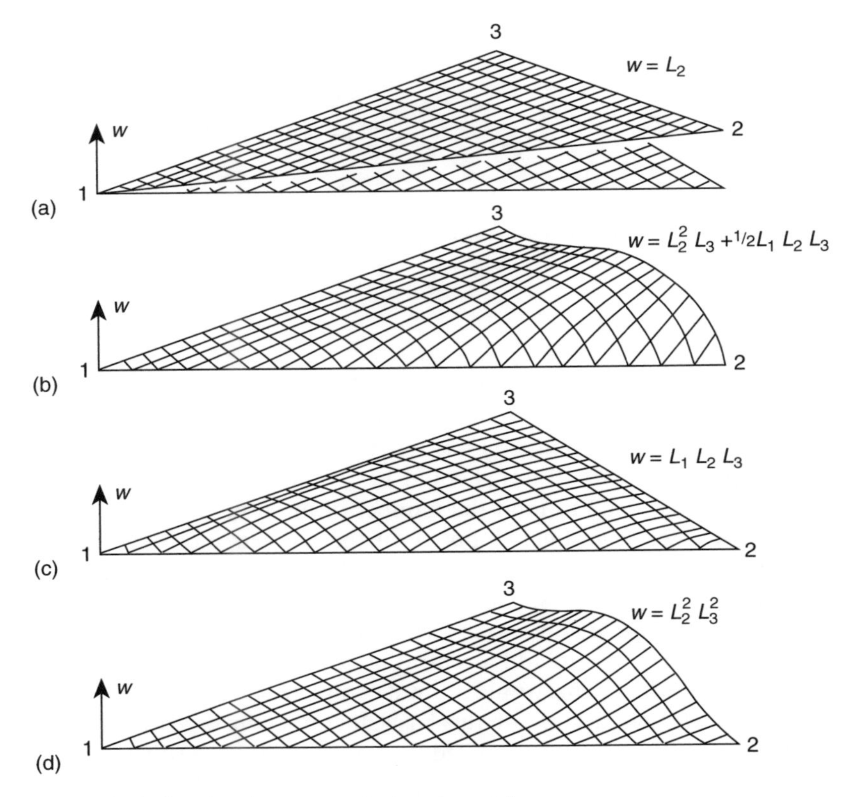

Fig. 4.10 Some basic functions in area coordinate polynomials.

combinations, that is,

$$L_1^3,\ L_2^3,\ L_3^3,\ L_1^2L_2,\ L_1^2L_3,\ L_2^2L_3,\ L_2^2L_1,\ L_3^2L_1,\ L_3^2L_2,\ L_1L_2L_3 \qquad (4.54)$$

For a 9 degree-of-freedom element any of the above terms can be used in a suitable combination, remembering, however, that only nine independent functions are needed and that constant curvature states have to be obtained. Figure 4.10 shows some functions that are of importance. The first [Fig. 4.10(a)] gives one of three functions representing a simple, unstrained rotation of the plate. Obviously, these must be available to produce the rigid body modes. Further, functions of the type $L_1^2 L_2$, of which there are six in the cubic expression, will be found to take up a form similar (though not identical) to Fig. 4.10(b).

The cubic function $L_1 L_2 L_3$ is shown in Fig 4.10(c), illustrating that this is a purely internal (bubble) mode with zero values and slopes at all three corner nodes (though slopes are not zero along edges). This function could thus be useful for a nodeless or internal variable but will not, in isolation, be used as it cannot be prescribed in terms of corner variables. It can, however, be added to any other basic shape in any proportion, as indicated in Fig. 4.10(b).

The functions of the second kind are of special interest. They have zero values of w at all corners and indeed always have zero slope in the direction of one side. A linear combination of two of these (for example $L_2^2 L_1$ and $L_2^2 L_3$) are capable of providing any desired slopes in the x and y directions at one node while maintaining all other nodal slopes at zero.

For an element envisaged with 9 degrees of freedom we must ensure that all six quadratic terms are present. In addition we select three of the cubic terms. The quadratic terms ensure that a constant curvature, necessary for patch test satisfaction, is possible. Thus, the polynomials we consider are

$$\mathbf{P} = \left[L_1,\ L_2,\ L_3,\ L_1L_2,\ L_2L_3,\ L_3L_1,\ L_1^2L_2,\ L_2^2L_3,\ L_3^2L_1\right]$$

and we write the interpolation as

$$w = \mathbf{P}\boldsymbol{\alpha} \tag{4.55}$$

where $\boldsymbol{\alpha}$ are parameters to be expressed in terms of nodal values. The nine nodal values are denoted as

$$(\,\hat{w}_i,\ \ \hat{\theta}_{xi},\ \ \hat{\theta}_{yi}\,) = \left(\hat{w}_i,\ \ \frac{\partial \hat{w}}{\partial y}\bigg|_i,\ \ -\frac{\partial \hat{w}}{\partial x}\bigg|_i\right); \quad i = 1, 2, 3$$

Upon noting that

$$\begin{Bmatrix} \dfrac{\partial}{\partial x} \\ \dfrac{\partial}{\partial y} \end{Bmatrix} = \begin{bmatrix} \dfrac{\partial L_1}{\partial x} & \dfrac{\partial L_2}{\partial x} & \dfrac{\partial L_3}{\partial x} \\ \dfrac{\partial L_1}{\partial y} & \dfrac{\partial L_2}{\partial y} & \dfrac{\partial L_3}{\partial y} \end{bmatrix} \begin{Bmatrix} \dfrac{\partial}{\partial L_1} \\ \dfrac{\partial}{\partial L_2} \\ \dfrac{\partial}{\partial L_3} \end{Bmatrix} = \frac{1}{2\Delta} \begin{bmatrix} b_1 & b_2 & b_3 \\ c_1 & c_2 & c_3 \end{bmatrix} \begin{Bmatrix} \dfrac{\partial}{\partial L_1} \\ \dfrac{\partial}{\partial L_2} \\ \dfrac{\partial}{\partial L_3} \end{Bmatrix} \tag{4.56}$$

where

$$\begin{aligned} 2\Delta &= b_1c_2 - b_2c_1 \\ b_i &= y_j - y_k \\ c_i &= x_k - x_j \end{aligned}$$

with i, j, k a cyclic permutation of indices (see Chapter 9 of Volume 1), we now determine the shape function by a suitable inversion [see Sec. 4.3.1, Eq. (4.42)], and write for node i

$$\mathbf{N}_i^{\mathrm{T}} = \begin{Bmatrix} 3L_i^2 - 2L_i^3 \\ L_i^2\,(b_jL_k - b_kL_j) + \frac{1}{2}\,(b_j - b_k)\,L_1L_2L_3 \\ L_i^2\,(c_jL_k - c_kL_j) + \frac{1}{2}\,(c_j - c_k)\,L_1L_2L_3 \end{Bmatrix} \tag{4.57}$$

Here the term $L_1L_2L_3$ is added to permit constant curvature states.

The computation of stiffness and load matrices can again follow the standard patterns, and integration of expressions (4.30) and (4.31) can be done exactly using the general integrals given in Fig. 4.9. However, numerical quadrature is generally used and proves equally efficient (see Chapter 9 of Volume 1). The stiffness matrix requires computation of second derivatives of shape functions and these may be conveniently obtained from

$$\begin{bmatrix} \dfrac{\partial^2 N_i}{\partial x^2} & \dfrac{\partial^2 N_i}{\partial x \partial y} \\ \dfrac{\partial^2 N_i}{\partial y \partial x} & \dfrac{\partial^2 N_i}{\partial y^2} \end{bmatrix} = \frac{1}{4\Delta^2} \begin{bmatrix} b_1 & b_2 & b_3 \\ c_1 & c_2 & c_3 \end{bmatrix} \begin{bmatrix} \dfrac{\partial^2 N_i}{\partial L_1^2} & \dfrac{\partial^2 N_i}{\partial L_1 \partial L_2} & \dfrac{\partial^2 N_i}{\partial L_1 \partial L_3} \\ \dfrac{\partial^2 N_i}{\partial L_2 \partial L_1} & \dfrac{\partial^2 N_i}{\partial L_2^2} & \dfrac{\partial^2 N_i}{\partial L_2 \partial L_3} \\ \dfrac{\partial^2 N_i}{\partial L_3 \partial L_1} & \dfrac{\partial^2 N_i}{\partial L_3 \partial L_2} & \dfrac{\partial^2 N_i}{\partial L_3^2} \end{bmatrix} \begin{bmatrix} b_1 & c_1 \\ b_2 & c_2 \\ b_3 & c_3 \end{bmatrix} \tag{4.58}$$

in which N_i denotes any of the shape functions given in Eq. (4.57).

The element just derived is one first developed in reference 11. Although it satisfies the constant strain criterion (being able to produce constant curvature states) it unfortunately does not pass the patch test for arbitrary mesh configurations. Indeed, this was pointed out in the original reference (which also was the one in which the patch test was mentioned for the first time). However, the patch test is fully satisfied with this element for meshes of triangles created by three sets of equally spaced straight lines. In general, the performance of the element, despite this shortcoming, made the element quite popular in practical applications.[38]

It is possible to amend the element shape functions so that the resulting element passes the patch test in all configurations. An early approach was presented by Kikuchi and Ando[51] by replacing boundary integral terms in the virtual work statement of Eq. (4.26) by

$$\begin{aligned} \delta \Pi_{\text{ext}} = & \int_\Omega \delta w q \, \mathrm{d}\Omega + \sum_e \delta \left[\int_{\Gamma_e} \left(\frac{\partial w}{\partial n} - \frac{\partial \breve{w}}{\partial n} \right) M_n(w) \, \mathrm{d}\Gamma \right] \\ & + \int_{\Gamma_s} \left[\delta w \bar{S}_n + \frac{\partial \delta w}{\partial s} \bar{M}_{ns} \right] \mathrm{d}\Gamma + \int_{\Gamma_n} \frac{\partial \delta \breve{w}}{\partial n} \bar{M}_n \, \mathrm{d}\Gamma \end{aligned} \tag{4.59}$$

in which, Γ_e is the boundary of each element e, $M_n(w)$ is the normal moment computed from second derivatives of the w interpolation, and s is the tangent direction along the element boundaries. The interpolations given by Eq. (4.57) are C_0 conforming and have slopes which match those of adjacent elements at nodes. To correct the slope incompatibility between nodes, a simple interpolation is introduced along each element boundary segment as

$$\frac{\partial \breve{w}}{\partial n} = (1 - s') \left[\frac{\partial w}{\partial x}\bigg|_j m_i + \frac{\partial w}{\partial y}\bigg|_j n_i \right] + s' \left[\frac{\partial w}{\partial x}\bigg|_k m_i + \frac{\partial w}{\partial y}\bigg|_k n_i \right] \tag{4.60}$$

where s' is 0 at node j and 1 at node k, and m_i and n_i are direction cosines with respect to the x and y axes, respectively. The above modification requires boundary integrals in addition to the usual area integrals; however, the final result is one which passes the patch test.

Bergen[44,46,47] and Samuelsson[48] also show a way of producing elements which pass the patch test, but a successful modification useful for general application with elastic and inelastic material behaviour is one derived by Specht.[49] This modification uses three fourth-order terms in place of the three cubic terms of the equation preceding

Eq. (4.55). The particular form of these is so designed that the patch test criterion which we shall discuss in detail later in Sec. 4.7 is identically satisfied. We consider now the nine polynomial functions given by

$$\begin{aligned}\mathbf{P} = [&L_1, L_2, L_3, L_1L_2, L_2L_3, L_3L_1,\\ &L_1^2L_2 + \tfrac{1}{2}L_1L_2L_3\{3(1-\mu_3)L_1 - (1+3\mu_3)L_2 + (1+3\mu_3)L_3\},\\ &L_2^2L_3 + \tfrac{1}{2}L_1L_2L_3\{3(1-\mu_1)L_2 - (1+3\mu_1)L_3 + (1+3\mu_1)L_1\},\\ &L_3^2L_1 + \tfrac{1}{2}L_1L_2L_3\{3(1-\mu_2)L_3 - (1+3\mu_2)L_1 + (1+3\mu_2)L_2\}]\end{aligned} \tag{4.61}$$

where

$$\mu_i = \frac{l_k^2 - l_j^2}{l_i^2} \tag{4.62}$$

and l_i is the length of the triangle side opposite node i.*

The modified interpolation for w is taken as

$$w = \mathbf{P}\boldsymbol{\alpha} \tag{4.63}$$

and, on identification of nodal values and inversion, the shape functions can be written explicitly in terms of the components of the vector $\mathbf{P}$ defined by Eq. (4.61) as

$$\mathbf{N}_i^{\mathrm{T}} = \left\{\begin{array}{c} P_i - P_{i+3} + P_{k+3} + 2(P_{i+6} - P_{k+6}) \\ -b_j(P_{k+6} - P_{k+3}) - b_k P_{i+6} \\ -c_j(P_{k+6} - P_{k+3}) - c_k P_{i+6} \end{array}\right\} \tag{4.64}$$

where i, j, k are the cyclic permutations of 1, 2, 3.

Once again, stiffness and load matrices can be determined either explicitly or using numerical quadrature. The element derived above passes all the patch tests and performs excellently.[41] Indeed, if the quadrature is carried out in a 'reduced' manner using three quadrature points (see Volume 1, Table 9.2 of Sec. 9.11) then the element is one of the best triangles with 9 degrees of freedom that is currently available, as we shall show in the section dealing with numerical comparisons.

4.6 Triangular element of the simplest form (6 degrees of freedom)

If conformity at nodes (C_1 continuity) is to be abandoned, it is possible to introduce even simpler elements than those already described by reducing the element interconnections. A very simple element of this type was first proposed by Morley.[30] In this element, illustrated in Fig. 4.11, the interconnections require continuity of the displacement w at the triangle vertices and of normal slopes at the element mid-sides.

* The constants μ_i are geometric parameters occurring in the expression for normal derivatives. Thus on side l_i the normal derivative is given by

$$\frac{\partial}{\partial n} = \frac{l_i}{4\Delta}\left[\frac{\partial}{\partial L_j} + \frac{\partial}{\partial L_k} - 2\frac{\partial}{\partial L_i} + \mu_i\left(\frac{\partial}{\partial L_k} - \frac{\partial}{\partial L_j}\right)\right]$$

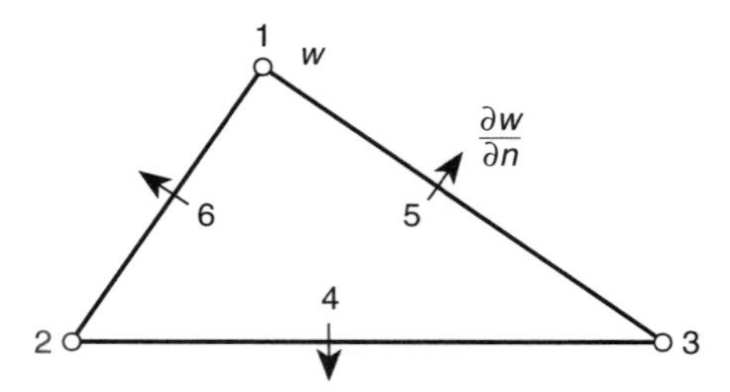

Fig. 4.11 The simplest non-conforming triangle, from Morley,[30] with 6 degrees of freedom.

With 6 degrees of freedom the expansion can be limited to quadratic terms alone, which one can write as

$$w = [\, L_1, \quad L_2, \quad L_3, \quad L_1L_2, \quad L_2L_3, \quad L_3L_1 \,]\boldsymbol{\alpha} \tag{4.65}$$

Identification of nodal variables and inversion leads to the following shape functions: for corner nodes

$$N_i = L_i - L_i(1 - L_i) - \frac{b_i b_k - c_i c_k}{b_j^2 + c_j^2} L_j(1 - L_j) - \frac{b_i b_j - c_i c_j}{b_k^2 + c_k^2} L_k(1 - L_3) \tag{4.66}$$

and for 'normal gradient' nodes

$$N_{i+3} = \frac{2\Delta}{\sqrt{b_i^2 + c_i^2}} L_i(1 - L_i) \tag{4.67}$$

where the symbols are identical to those used in Eq. (4.56) and i, j, k are a cyclic permutation of $1, 2, 3$.

Establishment of stiffness and load matrices follows the standard pattern and we find that once again the element passes fully all the patch tests required. This simple element performs reasonably, as we shall show later, though its accuracy is, of course, less than that of the preceding ones.

It is of interest to remark that the moment field described by the element satisfies exactly interelement equilibrium conditions on the normal moment M_n, as the reader can verify. Indeed, originally this element was derived as an equilibrating one using the complementary energy principle,[52] and for this reason it always gives an upper bound on the strain energy of flexure. This is the simplest possible element as it simply represents the minimum requirements of a constant moment field. An explicit form of stiffness routines for this element is given by Wood.[31]

4.7 The patch test – an analytical requirement

The patch test in its different forms (discussed fully in Chapters 10 and 11 of Volume 1) is generally applied numerically to test the final form of an element. However, the basic requirements for its satisfaction by shape functions that violate compatibility can be forecast accurately if certain conditions are satisfied in the choice of such functions. These conditions follow from the requirement that for constant strain states the virtual work done by internal forces acting at the discontinuity must be zero. Thus if the tractions acting on an element interface of a plate are (see Fig. 4.4)

$$M_n, \qquad M_{ns} \qquad \text{and} \qquad S_n \tag{4.68}$$

and if the corresponding mismatch of virtual displacements are

$$\Delta\theta_n \equiv \Delta\left(\frac{\partial w}{\partial n}\right), \qquad \Delta\theta_s \equiv \Delta\left(\frac{\partial w}{\partial s}\right) \qquad \text{and} \qquad \Delta w \tag{4.69}$$

then ideally we would like the integral given below to be zero, as indicated, at least for the constant stress states:

$$\int_{\Gamma_e} M_n \Delta\theta_n \, \mathrm{d}\Gamma + \int_{\Gamma_e} M_{ns} \Delta\theta_s \, \mathrm{d}\Gamma + \int_{\Gamma_e} S_n \Delta w \, \mathrm{d}\Gamma = 0 \tag{4.70}$$

The last term will always be zero identically for constant M_x, M_y, M_{xy} fields as then $S_x = S_y = 0$ [in the absence of applied couples, see Eq. (4.18)] and we can ensure the satisfaction of the remaining conditions if

$$\int_{\Gamma_e} \Delta\theta_n \, \mathrm{d}\Gamma = 0 \quad \text{and} \quad \int_{\Gamma_e} \Delta\theta_s \, \mathrm{d}\Gamma = 0 \tag{4.71}$$

is satisfied for each straight side Γ_e of the element.

For elements joining at vertices where $\partial w/\partial n$ is prescribed, these integrals will be identically zero only if anti-symmetric cubic terms arise in the departure from linearity and a quadratic variation of normal gradients is absent, as shown in Fig. 4.12(a). This is the motivation for the rather special form of shape function basis chosen to describe the incompatible triangle in Eq. (4.61), and here the first condition of Eq. (4.71) is automatically satisfied. The satisfaction of the second condition of Eq. (4.71) is always ensured if the function w and its first derivatives are prescribed at the corner nodes.

For the purely quadratic triangle of Sec. 4.6 the situation is even simpler. Here the gradients can only be linear, and if their value is prescribed at the element mid-side as shown in Fig. 4.11(b) the integral is identically zero.

The same arguments apparently fail when the rectangular element with the function basis given in Eq. (4.42) is examined. However, the reader can verify by direct

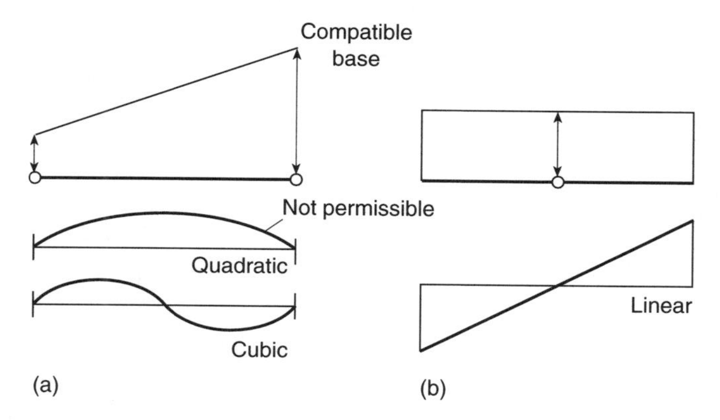

Fig. 4.12 Continuity condition for satisfaction of patch test [$\int(\partial w/\partial n)\,\mathrm{d}s = 0$]; variation of $\partial w/\partial n$ along side. (a) Definition by corner nodes (linear component compatible); (b) definition by one central node (constant component compatible).

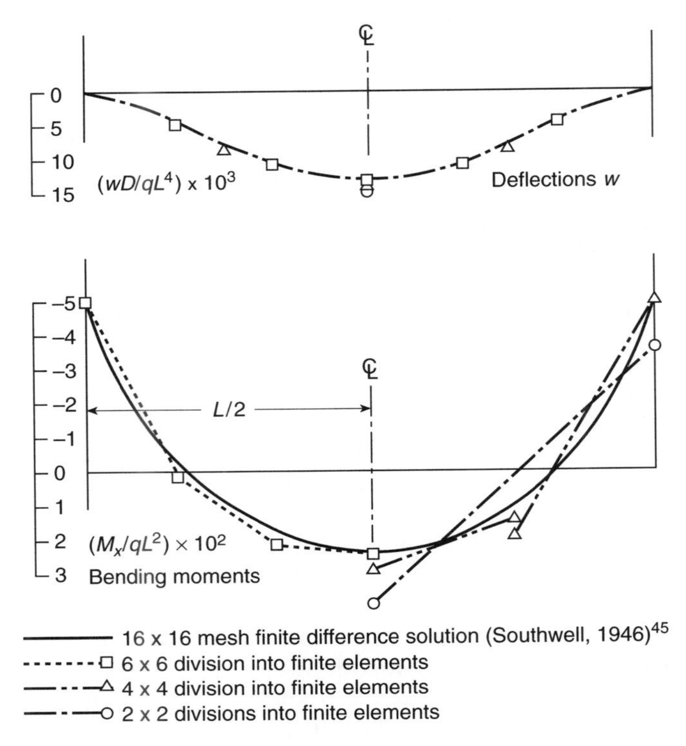

Fig. 4.13 A square plate with clamped edges; uniform load q; square elements.

Table 4.1 Computed central deflection of a square plate for several meshes (rectangular elements)[39]

Mesh	Total number of nodes	Simply supported plate		Clamped plate	
		α^*	$\beta^\dagger$	α^*	$\beta^\dagger$
2×2	9	0.003446	0.013784	0.001480	0.005919
4×4	25	0.003939	0.012327	0.001403	0.006134
8×8	81	0.004033	0.011829	0.001304	0.005803
16×16	169	0.004050	0.011715	0.001283	0.005710
Series (Timoshenko)		0.004062	0.01160	0.00126	0.00560

* $w_{max} = \alpha q L^4/D$ for uniformly distributed load q. † $w_{max} = \beta P L^2/D$ for central concentrated load P. Note: Subdivision of whole plate given for mesh.

Table 4.2 Corner supported square plate

Method	Mesh	Point 1		Point 2	
		w	M_x	w	M_x
Finite element	2×2	0.0126	0.139	0.0176	0.095
	4×4	0.0165	0.149	0.0232	0.108
	6×6	0.0173	0.150	0.0244	0.109
Marcus[53]		0.0180	0.154	0.0281	0.110
Ballesteros and Lee[54]		0.0170	0.140	0.0265	0.109
Multiplier		qL^4/D	qL^2	qL^4/D	qL^2

Note: point 1, centre of side; point 2, centre of plate.

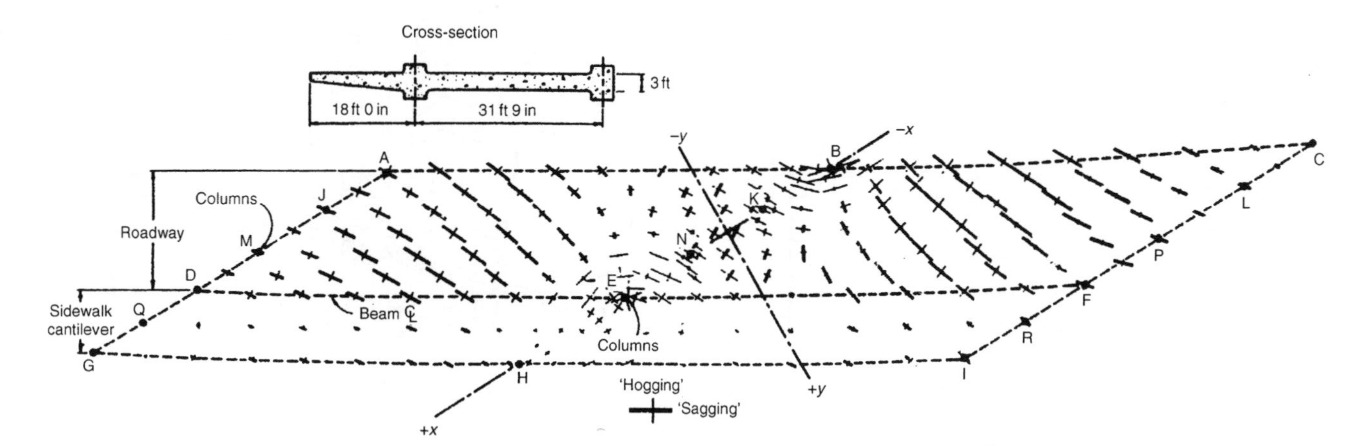

Fig. 4.14 A skew, curved, bridge with beams and non-uniform thickness; plot of principal moments under dead load.

algebra that the integrals of Eqs (4.71) are identically satisfied. Thus, for instance,

$$\int_{-a}^{a} \frac{\partial w}{\partial y}\, \mathrm{d}x = 0 \qquad \text{when} \quad y = \pm b$$

and $\partial w/\partial y$ is taken as zero at the two nodes (i.e. departure from prescribed linear variations only is considered).

The remarks of this section are verified in numerical tests and lead to an intelligent, *a priori*, determination of conditions which make shape functions convergent for incompatible elements.

4.8 Numerical examples

The various plate bending elements already derived – and those to be derived in subsequent sections – have been used to solve some classical plate bending problems. We first give two specific illustrations and then follow these with a general convergence study of elements discussed.

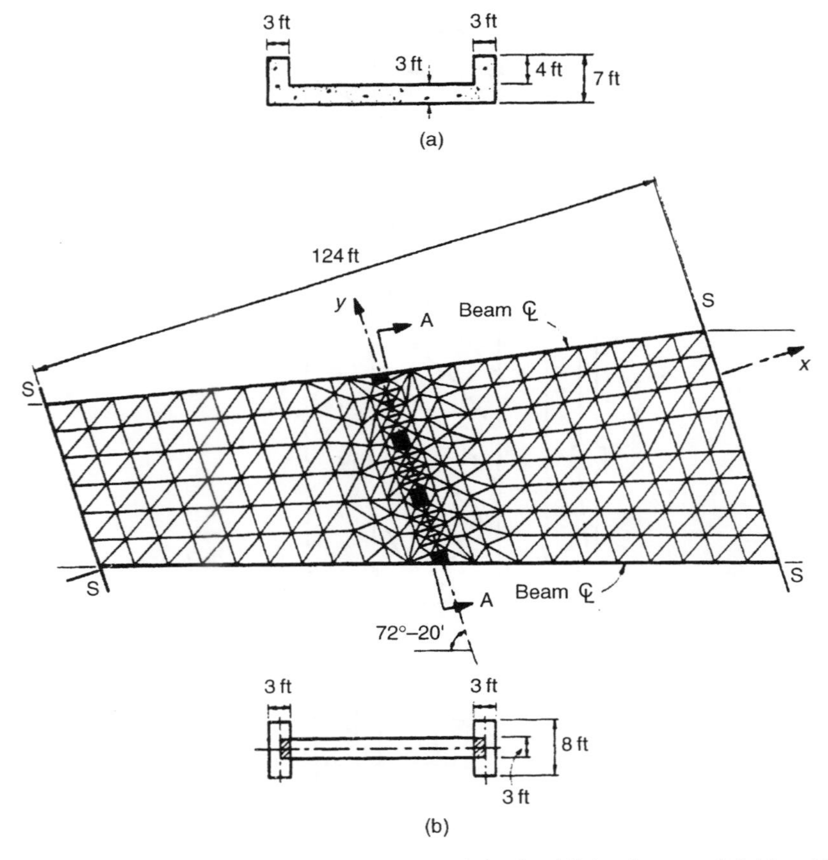

Fig. 4.15 Castleton railway bridge: general geometry and details of finite element subdivision. (a) Typical actual section; (b) idealization and meshing.

Figure 4.13 shows the deflections and moments in a square plate clamped along its edges and solved by the use of the rectangular element derived in Sec. 4.3 and a uniform mesh.[26] Table 4.1 gives numerical results for a set of similar examples solved with the same element,[39] and Table 4.2 presents another square plate with more complex boundary conditions. Exact results are available here and comparisons are made.[53,54]

Figures 4.14 and 4.15 show practical engineering applications to more complex shapes of slab bridges. In both examples the requirements of geometry necessitate the use of a triangular element – with that of reference 11 being used here. Further, in both examples, beams reinforce the slab edges and these are simply incorporated in the analysis on the assumption of concentric behaviour.

Finally in Fig. 4.16(a)–(d) we show the results of a convergence study of the square plate with simply supported and clamped edge conditions for various triangular and

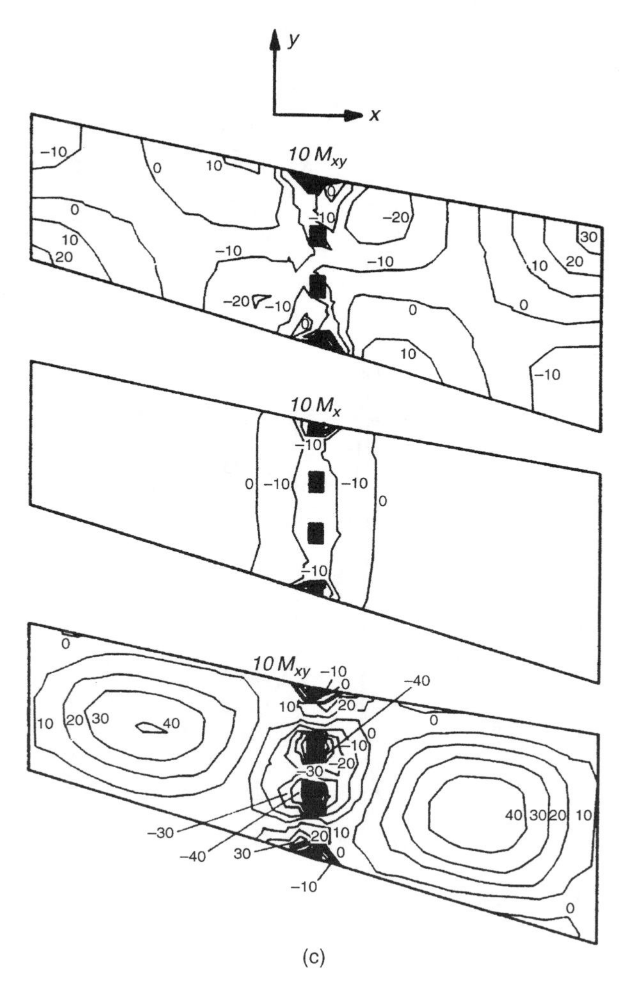

Fig. 4.15 (Continued) Castleton railway bridge: general geometry and details of finite element subdivision. (c) moment components (ton ft ft^{-1}) under uniform load of 150 lb ft^{-2} with computer plot of contours.

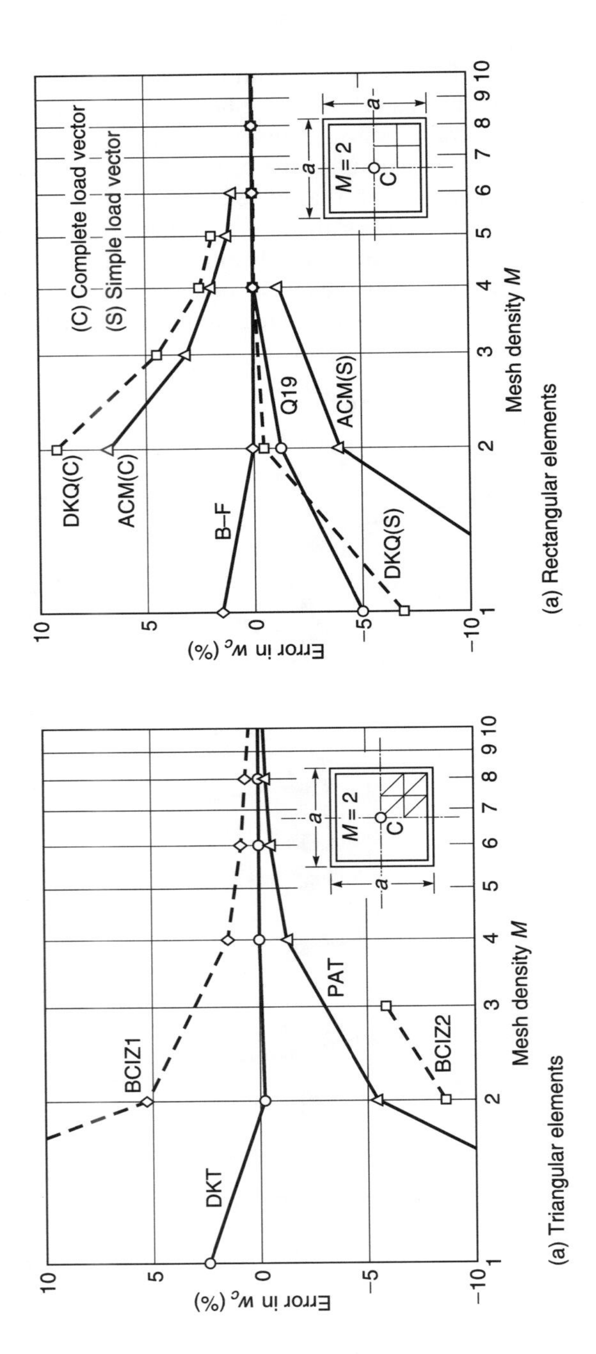

(a) Triangular elements

(a) Rectangular elements

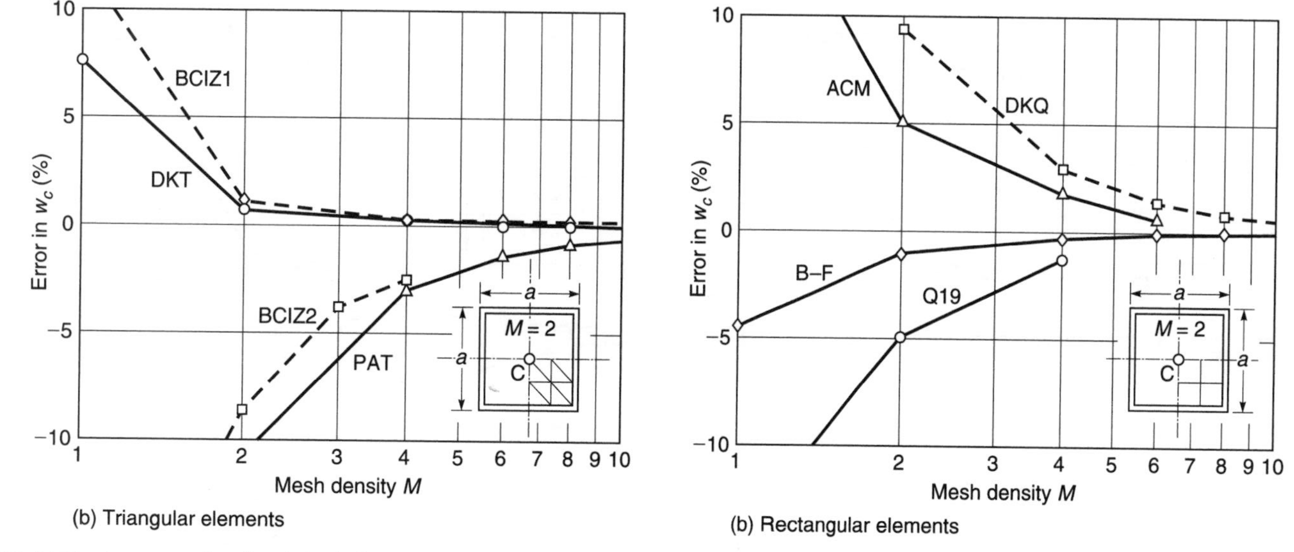

Fig. 4.16 (a) Simply supported uniformly loaded square plate; (b) simply supported square plate with concentrated central load.

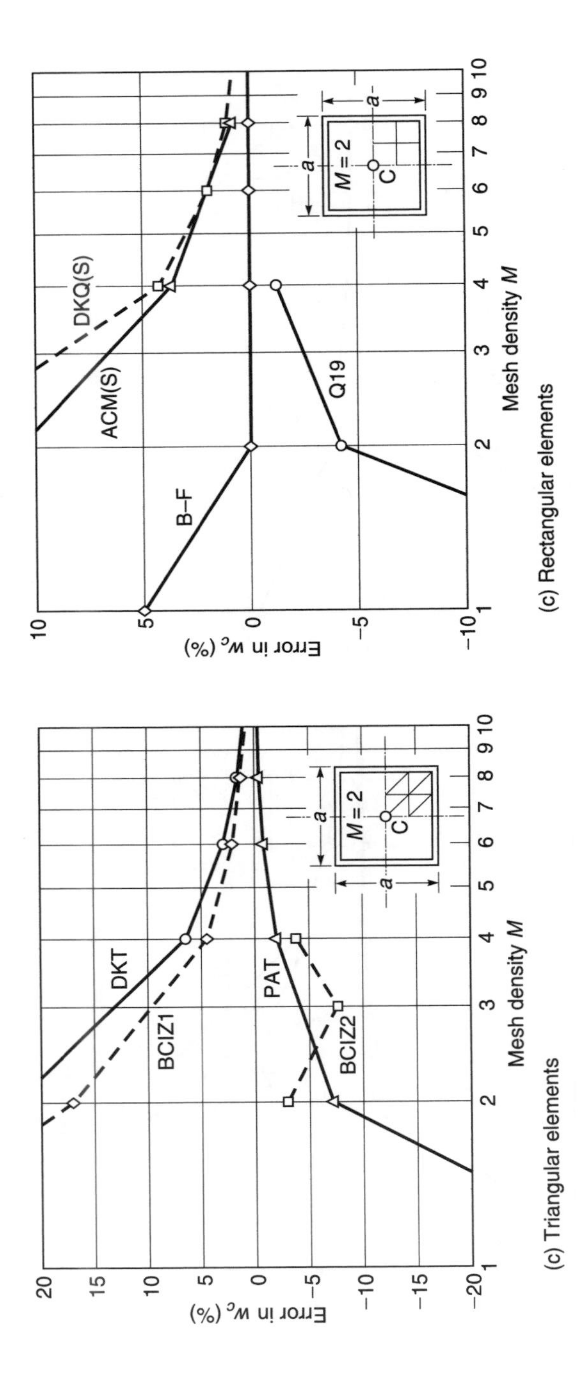

(c) Triangular elements

(c) Rectangular elements

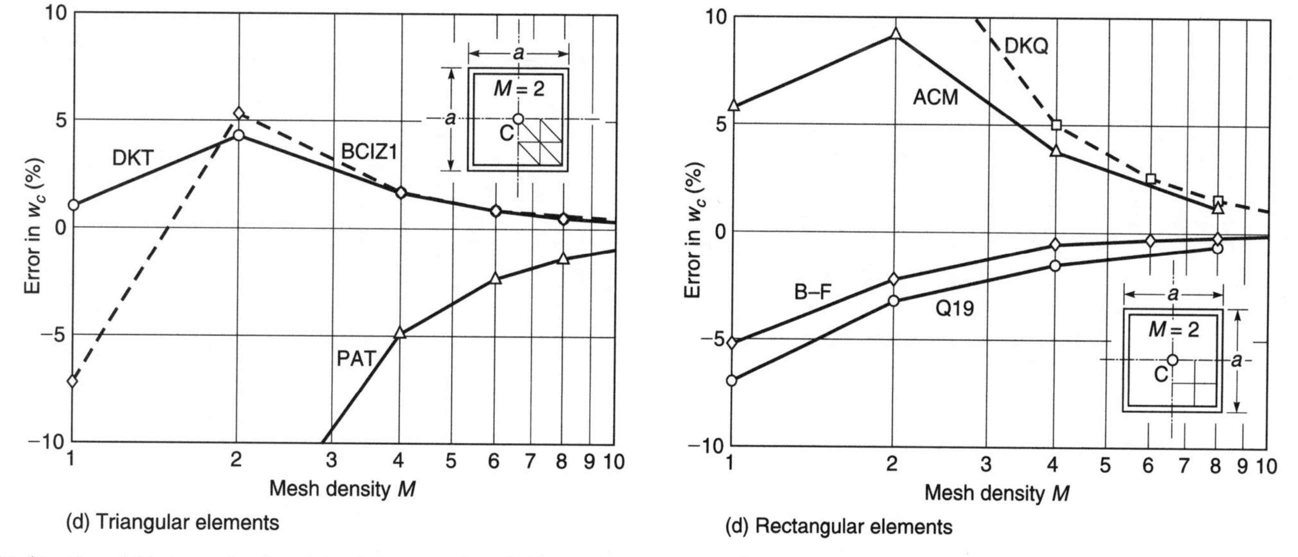

Fig. 4.16 (Continued) (c) clamped uniformly loaded square plate; (d) clamped square plate with concentrated central load. Percentage error in central displacement (see Table 4.3 for key).

Table 4.3 List of elements for comparison of performance in Fig. 4.16: (a) 9 degree-of-freedom triangles; (b) 12 degree-of-freedom rectangles; (c) 16 degree-of-freedom rectangle

Code	Reference	Symbol	Description and comment
(a)			
BCIZ 1	Bazeley *et al.*[11]	◇	Displacement, non-conforming (fails patch test)
PAT	Specht[49]	△	Displacement, non-conforming
BCIZ 2	Bazeley *et al.*[11]	□	Displacement, conforming
(HCT)	Clough and Tocher[10]		
DKT	Stricklin *et al.*[59] and Dhatt[60]	○	Discrete Kirchhoff
(b)			
ACM	Zienkiewicz and Cheung[26]	△	Displacement, non-conforming
Q19	Clough and Felippa[15]	○	Displacement, conforming
DKQ	Batoz and Ben Tohar[61]	□	Displacement, conforming
(c)			
BF	Bogner *et al.*[17]	◇	Displacement conforming

rectangular elements and two load types. This type of diagram is conventionally used for assessing the behaviour of various elements, and we show on it the performance of the elements already described as well as others to which we shall refer to later. Table 4.3 gives the key to the various element 'codes' which include elements yet to be described.[55–58]

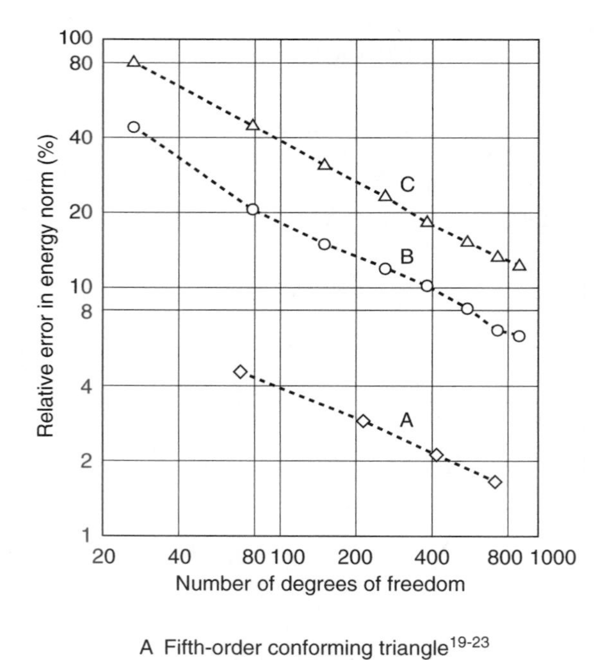

Fig. 4.17 Rate of convergence in energy norm versus degree of freedom for three elements: the problem of a slightly skewed, simply supported plate (80°) with uniform mesh subdivision.[7]

The comparison singles out only one displacement and each plot uses the number of mesh divisions in a quarter of the plate as abscissa. It is therefore difficult to deduce the convergence rate and the performance of elements with multiple nodes. A more convenient plot gives the energy norm $||\mathbf{u}||$, versus the number of degrees of freedom N on a logarithmic scale. We show such a comparison for some elements in Fig. 4.17 for a problem of a slightly skewed, simply supported plate.[7] It is of interest to observe that, owing to the singularity, both high- and low-order elements converge at almost identical rates (though, of course, the former give better overall accuracy). Different rates of convergence would, of course, be obtained if no singularity existed (see Chapter 14 of Volume 1).

Conforming shape functions with nodal singularities

4.9 General remarks

It has already been demonstrated in Sec. 4.3 that it is impossible to devise a simple polynomial function with only three nodal degrees of freedom that will be able to satisfy slope continuity requirements at all locations along element boundaries. The alternative of imposing curvature parameters at nodes has the disadvantage, however, of imposing excessive conditions of continuity (although we will investigate some of the elements that have been proposed from this class). Furthermore, it is desirable from many points of view to limit the nodal variables to three quantities only. These, with simple physical interpretation, allow the generalization of plate elements to shells to be easily interpreted also.

It is, however, possible to achieve C_1 continuity by provision of additional shape functions for which, in general, *second-order derivatives have non-unique values at nodes*. Providing the patch test conditions are satisfied, convergence is again assured.

Such shape functions will be discussed now in the context of triangular and quadrilateral elements. The simple rectangular shape will be omitted as it is a special case of the quadrilateral.

4.10 Singular shape functions for the simple triangular element

Consider for instance either of the following sets of functions:

$$\varepsilon_{jk} = \frac{L_i L_j^2 L_k^2 (L_k - L_j)}{(L_i + L_j)(L_j + L_k)} \tag{4.72}$$

or

$$\varepsilon_{jk} = \frac{L_i L_j^2 L_k^2 (1 + L_i)}{(L_i + L_j)(L_j + L_k)} \tag{4.73}$$

in which once again i, j, k are a cyclic permutation of $1, 2, 3$. Both have the property that along two sides (i–j and i–k) of a triangle (Fig. 4.18) their values and the values

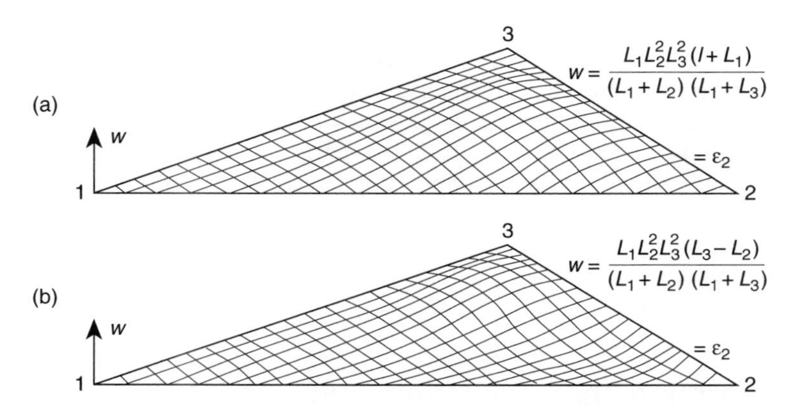

Fig. 4.18 Some singular area coordinate functions.

of their normal slope are zero. On the third side $(j{-}k)$ the function is zero but a normal slope exists. In both, its variation is parabolic. Now, all the functions used to define the non-conforming triangle [see Eq. (4.55)] were cubic and hence permit also a parabolic variation of the normal slope which is not uniquely defined by the two end nodal values (and hence resulted in non-conformity). However, if we specify as an additional variable the *normal slope of* w at a mid-point of each side then, by combining the new functions ε_{jk} with the other functions previously given, a *unique parabolic variation of normal slope* along interelement faces is achieved and a compatible element results.

Apparently, this can be achieved by adding three such additional degrees of freedom to expression (4.55) and proceeding as described above. This will result in an element shown in Fig. 4.19(a), which has six nodes, three corner ones as before and three additional ones at which only normal slope is specified. Such an element requires the definition of a node (or an alternative) to define the normal slope and also involves assembly of nodes with differing numbers of degrees of freedom. It is necessary to define a unique normal slope for the parameter associated with the mid-point of adjacent elements. One simple solution is to use the direction of increasing node number of the adjacent vertices to define a unique normal.

Another alternative, which avoids the above difficulties, is to constrain the mid-side node degree of freedom. For instance, we can assume that the normal slope at the centre-point of a line is given as the average of the two slopes at the ends. This, after suitable transformation, results in a compatible element with exactly the same degrees of freedom as that described in previous sections [see Fig. 4.19(b)].

The algebra involved in the generation of suitable shape functions along the lines described here is quite extensive and will not be given fully.

First, the normal slopes at the mid-sides are calculated from the basic element shape functions [Eq. (4.57)] as

$$\left[\left(\frac{\partial w}{\partial n}\right)_4 \quad \left(\frac{\partial w}{\partial n}\right)_5 \quad \left(\frac{\partial w}{\partial n}\right)_6 \right]^{\mathrm{T}} = \mathbf{Z}\mathbf{a}^{\mathrm{e}} \tag{4.74}$$

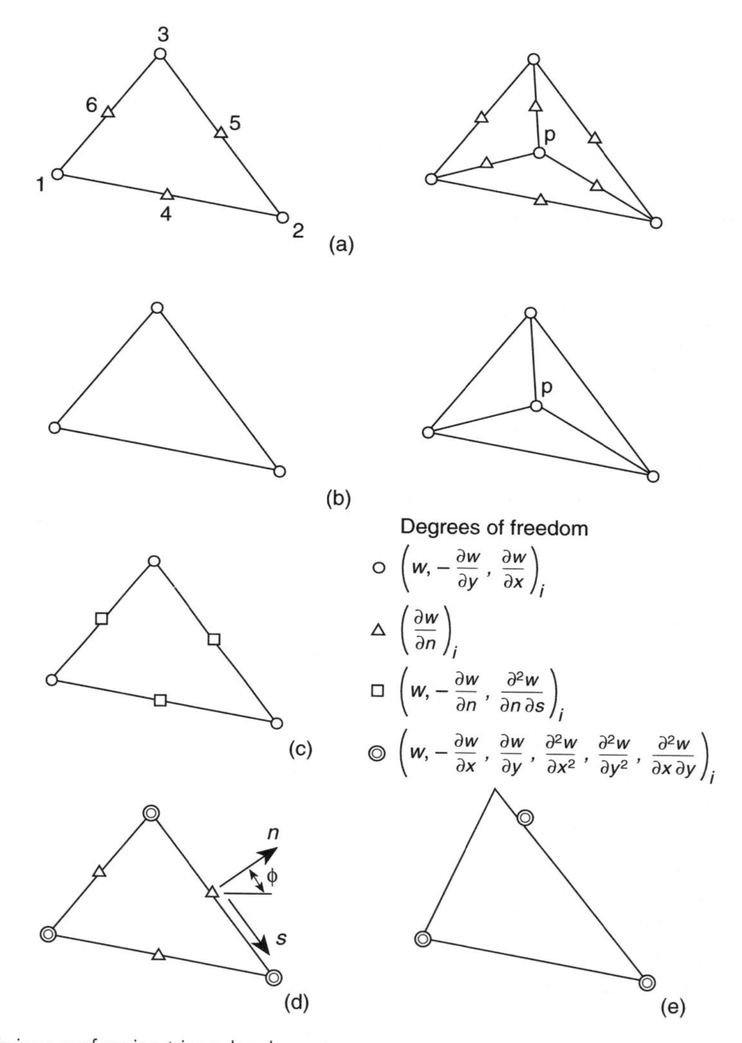

Fig. 4.19 Various conforming triangular elements.

Similarly, the average values of the nodal slopes in directions normal to the sides are calculated from these functions:

$$\left[\left(\frac{\partial w}{\partial n}\right)_4^{\mathrm{a}} \quad \left(\frac{\partial w}{\partial n}\right)_5^{\mathrm{a}} \quad \left(\frac{\partial w}{\partial n}\right)_6^{\mathrm{a}}\right]^{\mathrm{T}} = \bar{\mathbf{Z}}\,\mathbf{a}^{\mathrm{e}} \tag{4.75}$$

The contribution of the ε functions to these slopes is added in proportions of $\varepsilon_{jk} - \gamma_i$ and is simply (as these give unit normal slope)

$$\boldsymbol{\gamma} = [\gamma_1 \quad \gamma_2 \quad \gamma_3]^{\mathrm{T}} \tag{4.76}$$

On combining Eq. (4.57) and the last three relations we have

$$\bar{\mathbf{Z}}\,\mathbf{a}^{\mathrm{e}} = \mathbf{Z}\,\mathbf{a}^{\mathrm{e}} + \boldsymbol{\gamma} \tag{4.77}$$

from which it immediately follows on finding γ that

$$w = \mathbf{N}^0 \mathbf{a}^e + [\varepsilon_{23}, \quad \varepsilon_{31}, \quad \varepsilon_{12}](\bar{\mathbf{Z}} - \mathbf{Z})\mathbf{a}^e \tag{4.78}$$

in which $\mathbf{N}^0$ are the non-conforming shape functions defined in Eq. (4.57). Thus, new shape functions are now available from Eq. (4.78).

An alternative way of generating compatible triangles was developed by Clough and Tocher.[10] As shown in Fig. 4.19(a) each element triangle is first divided into three parts based on an internal point p. For each ijp triangle a complete cubic expansion is written involving 10 terms which may be expressed in terms of the displacement and slopes at each vertex and the mid-side slope along the ij edge.

Matching the values at the vertices for the three sub-triangles produces an element with 15 degrees of freedom: 12 conventional degrees of freedom at nodes 1, 2, 3 and p; and three normal slopes at nodes 4, 5, 6. Full C_1 continuity in the interior of the element is achieved by *constraining* the three parameters at the p node to satisfy continuous normal slope at each internal mid-side. Thus, we achieve an element with 12 degrees of freedom similar to the one previously outlined using the singular shape functions. Constraining the normal slopes on the exterior mid-sides leads to an element with 9 degrees of freedom [see Fig. 4.19(b)].

These elements are achieved at the expense of providing non-unique values of second derivatives at the corners. We note, however, that strains are in general also non-unique in elements surrounding a node (e.g. constant strain triangles in elasticity have different strains in each element surrounding each node). In the previously developed shape functions ε_{jk} an infinite number of values to the second derivatives are obtained at each node depending on the direction the corner is approached. Indeed, the derivation of the Clough and Tocher triangle can be obtained by defining an alternative set of ε functions, as has been shown in reference 11.

As both types of elements lead to almost identical numerical results the preferable one is that leading to simplified computation. If numerical integration is used (as indeed is always strongly recommended for such elements) the form of functions continuously defined over the whole triangle as given by Eqs (4.57) and (4.78) is advantageous, although a fairly high order of numerical integration is necessary because of the singular nature of the functions.

4.11 An 18 degree-of-freedom triangular element with conforming shape functions

An element that presents a considerable improvement over the type illustrated in Fig. 4.19(a) is shown in Fig. 4.19(c). Here, the 12 degrees of freedom are increased to 18 by considering both the values of w and its cross derivative $\partial^2 w/\partial s \partial n$, in addition to the normal slope $\partial w/\partial n$, at element mid-sides.*

Thus an equal number of degrees of freedom is presented at each node. Imposition of the continuity of cross derivatives at *mid-sides* does not involve any additional constraint as this indeed must be continuous in physical situations.

* This is, in fact, identical to specifying both $\partial w/\partial n$ and $\partial w/\partial s$ at the mid-side.

The derivation of this element is given by Irons[14] and it will suffice here to say that in addition to the modes already discussed, fourth-order terms of the type illustrated in Fig. 4.10(d) and 'twist' functions of Fig. 4.18(b) are used. Indeed, it can be simply verified that the element contains *all* 15 terms of the quartic expansion in addition to the 'singularity' functions.

4.12 Compatible quadrilateral elements

Any of the previous triangles can be combined to produce 'composite' compatible quadrilateral elements with or without internal degrees of freedom. Three such quadrilaterals are illustrated in Fig. 4.20 and, in all, no mid-side nodes exist on the external boundaries. This avoids the difficulties of defining a unique parameter and of assembly already mentioned.

In the first, no internal degrees of freedom are present and indeed no improvement on the comparable triangles is expected. In the following two, 3 and 7 internal degrees of freedom exist, respectively. Here, normal slope continuity imposed in the last one does not interfere with the assembly, as internal degrees of freedom are in all cases eliminated by static condensation.[62] Much improved accuracy with these elements has been demonstrated by Clough and Felippa.[15]

An alternative direct derivation of a quadrilateral element was proposed by Sander[12] and Fraeijs de Veubeke.[13,16] This is along the following lines. Within a quadrilateral of Fig. 4.21(a) a complete cubic with 10 constants is taken, giving the first component of the displacement which is defined by three functions. Thus,

$$
\begin{aligned}
w &= w^{a} + w^{b} + w^{c} \\
w^{a} &= \alpha_1 + \alpha_2 x + \cdots + \alpha_{10} y^3
\end{aligned}
\tag{4.79}
$$

The second function w^{b} is defined in a piecewise manner. In the lower triangle of Fig. 4.21(b) it is taken as zero; in the upper triangle a cubic expression with three constants merges with slope discontinuity into the field of the lower triangle. Thus, in *jkm*,

$$
w^{b} = \alpha_{11} y'^2 + \alpha_{12} y'^3 + \alpha_{13} x' y'^2 \tag{4.80}
$$

in terms of the locally specified coordinates x' and y'. Similarly, for the third function, Fig. 4.21(c), $w^{c} = 0$ in the lower triangle, and in *imj* we define

$$
w^{c} = \alpha_{14} y''^2 + \alpha_{15} y''^3 + \alpha_{16} x'' y''^2 \tag{4.81}
$$

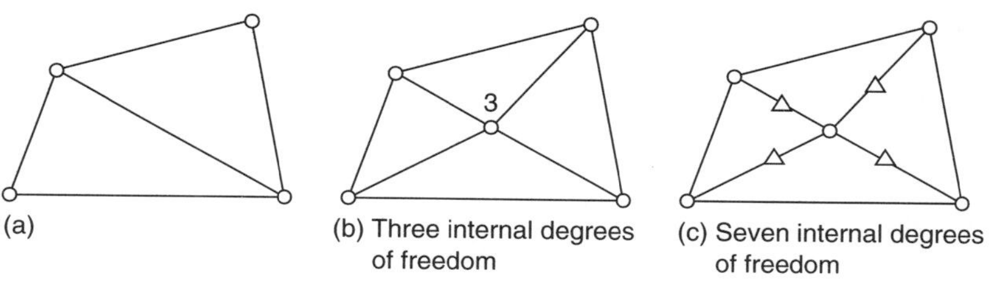

Fig. 4.20 Some composite quadrilateral elements.

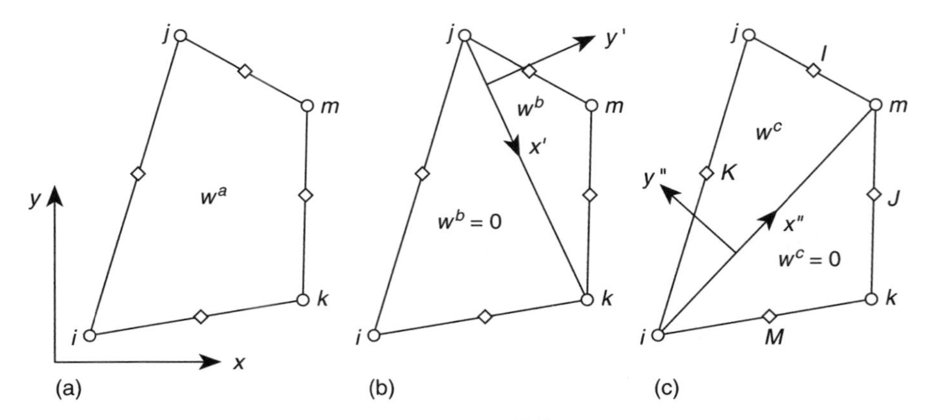

Fig. 4.21 The compatible functions of Fraeijs de Veubeke.[13,16]

The 16 external degrees of freedom are provided by 12 usual corner variables and four normal mid-side slopes and allow the 16 constants α_1 to α_{16} to be found by inversion. Compatibility is assured and once again non-unique second derivatives arise at corners.

Again it is possible to constrain the mid-side nodes if desired and thus obtain a 12 degree-of-freedom element. The expansion can be found explicitly, as shown by Fraeijs de Veubeke, and a useful element generated.[16]

The element described above cannot be formulated if a corner of the quadrilateral is re-entrant. This is not a serious limitation but needs to be considered on occasion if such an element degenerates to a near triangular shape.

4.13 Quasi-conforming elements

The performance of some of the conforming elements discussed in Secs 4.10–4.12 is shown in the comparison graphs of Fig. 4.16. It should be noted that although monotonic convergence in energy norm is now guaranteed, by subdividing each mesh to obtain the next one, the conforming triangular elements of references 10 and 11 perform almost identically but are considerably stiffer and hence less accurate than many of the non-conforming elements previously cited.

To overcome this inaccuracy a *quasi-conforming* or *smoothed* element was derived by Razzaque and Irons.[33,34] For the derivation of this element *substitute shape functions are used.*

The substitute functions are cubic functions (in area coordinates) so designed as to approximate in a least-square sense the singular functions ε and their derivatives used to enforce continuity [see Eqs (4.72)–(4.78)], as shown in Fig. 4.22.

The algebra involved is complex but a full subprogram for stiffness computations is available in reference 33. It is noted that this element performs very similarly to the simper, non-conforming element previously derived for the triangle. It is interesting

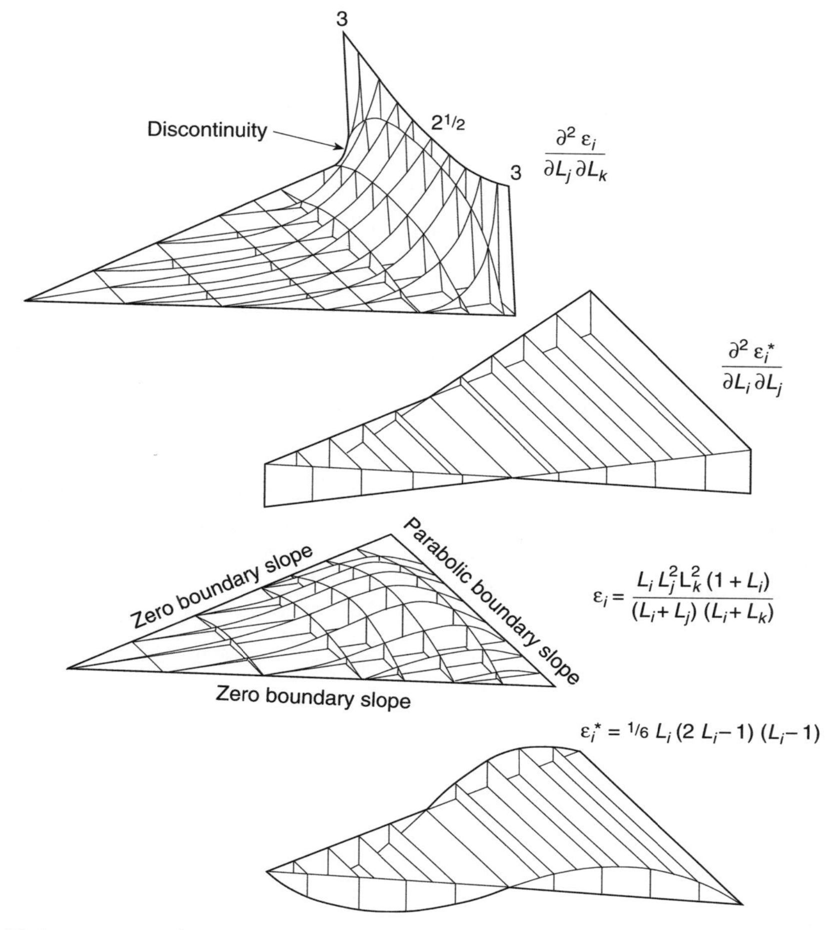

Fig. 4.22 Least-square substitute cubic shape function ε^* in place of rational function ε for plate bending triangles.

to observe that here the non-conforming element is developed by choice and not to avoid difficulties. Its validity, however, is established by patch tests.

Conforming shape functions with additional degrees of freedom

4.14 Hermitian rectangle shape function

With the rectangular element of Fig 4.7 the specification of $\partial^2 w/\partial x \partial y$ as a nodal parameter is always permissible as it does not involve 'excessive continuity'. It is easy to show that for such an element polynomial shape functions giving compatibility can be easily determined.

A polynomial expansion involving 16 constants [equal to the number of nodal parameters w_i, $(\partial w/\partial x)_i$, $(\partial w/\partial y)_i$ and $(\partial^2 w/\partial x \partial y)_i$] could, for instance, be written retaining terms that do not produce a higher-order variation of w or its normal slope along the sides. Many alternatives will be present here and some may not produce invertible **C** matrices [see Eq. (4.41)].

An alternative derivation uses Hermitian polynomials which permit the writing down of suitable functions directly. An Hermitian polynomial

$$H^n_{mi}(x) \tag{4.82}$$

is a polynomial of order $2n+1$ which gives, for $m = 0$ to n,

$$\frac{\mathrm{d}^k H^n_{mi}}{\mathrm{d}x^k} = \begin{cases} 1, & \text{when} \quad k = m \quad \text{and} \quad x = x_i \\ 0, & \text{when} \quad k \neq m \quad \text{or when} \quad x = x_j \end{cases}$$

A set of first-order Hermitian polynomials is thus a set of cubic terms giving shape functions for a line element ij at the ends of which slopes and values of the function are used as variables. Figure 4.23 shows such a set of cubics, and it is easy to verify that the shape functions are given by

$$H^1_{01}(x) = 1 - 3\frac{x^2}{L^2} + 2\frac{x^3}{L^3}$$

$$H^1_{11}(x) = x - 2\frac{x^2}{L} + \frac{x^3}{L^2}$$

$$H^1_{02}(x) = 3\frac{x^2}{L^2} - 2\frac{x^3}{L^3}$$

$$H^1_{12}(x) = -\frac{x^2}{L} + \frac{x^3}{L^2}$$

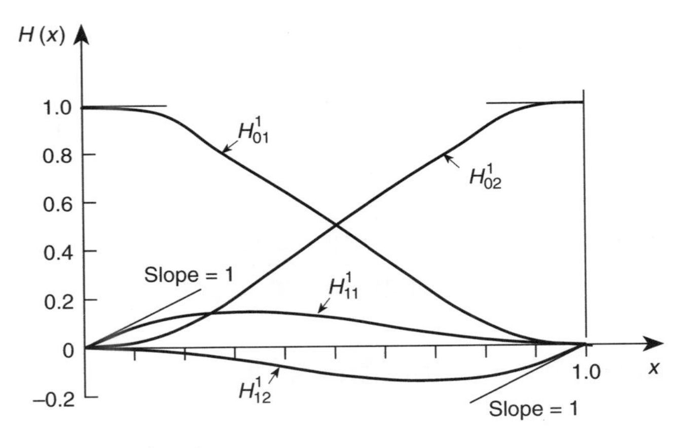

Fig. 4.23 First-order Hermitian functions.

where L is the length of the side. These are precisely the 'beam' functions used in Chapter 2 of Volume 1.

It is easy to verify that the following shape functions

$$\mathbf{N}_i = \left[H^1_{0i}(x)H^1_{0i}(y), \quad H^1_{1i}(x)H^1_{0i}(y), \quad H^1_{0i}(x)H^1_{1i}(y), \quad H^1_{1i}(x)H^1_{1i}(y) \right] \tag{4.83}$$

correspond to the values of

$$w, \quad \frac{\partial w}{\partial x}, \quad \frac{\partial w}{\partial y}, \quad \frac{\partial^2 w}{\partial x \partial y},$$

specified at the corner nodes, taking successively unit values at node i and zero at other nodes.

An element based on these shape functions has been developed by Bogner *et al.*[17] and used with success. Indeed it is the most accurate rectangular element available as indicated by results in Fig. 4.16. A development of this type of element to include continuity of higher derivatives is simple and outlined in reference 18. In their undistorted form the above elements are, as for all rectangles, of very limited applicability.

4.15 The 21 and 18 degree-of-freedom triangle

If continuity of higher derivatives than first is accepted at nodes (thus imposing a certain constraint on non-homogeneous material and discontinuous thickness situations as explained in Sec. 4.2.4), the generation of slope and deflection compatible elements presents less difficulty.

Considering as nodal degrees of freedom

$$w, \quad \frac{\partial w}{\partial x}, \quad \frac{\partial w}{\partial y}, \quad \frac{\partial^2 w}{\partial x^2}, \quad \frac{\partial^2 w}{\partial x \partial y}, \quad \frac{\partial^2 w}{\partial y^2},$$

a triangular element will involve at least 18 degrees of freedom. However, a complete fifth-order polynomial contains 21 terms. If, therefore, we add three normal slopes at the mid-side as additional degrees of freedom a sufficient number of equations appear to exist for which the shape functions can be found with a complete quintic polynomial.

Along any edge we have six quantities determining the variation of w (displacement, slopes, and curvature at corner nodes), that is, specifying a fifth-order variation. Thus, this is uniquely defined and therefore w is continuous between elements. Similarly, $\partial w/\partial n$ is prescribed by five quantities and varies as a fourth-order polynomial. Again this is as required by the slope continuity between elements.

If we write the complete quintic polynomial as*

$$w = \alpha_1 + \alpha_2 x + \cdots + \alpha_{21} y^5 \tag{4.84}$$

* For this derivation use of simple cartesian coordinates is recommended in preference to area coordinates. Symmetry is assured as the polynomial is complete.

using the ordering in the Pascal triangle [see Fig. 8.12 of Volume 1] we can proceed along the lines of the argument used to develop the rectangle in Sec. 4.3 and write

$$w_i = \alpha_1 + \alpha_2 x_i + \cdots + \alpha_{21} y_i^5$$

$$\left(\frac{\partial w}{\partial x}\right)_i = \alpha_2 + \cdots + \alpha_{20} y_i^4$$

$$\left(\frac{\partial w}{\partial y}\right)_i = \alpha_3 + \cdots + 5\alpha_{21} y_i^4$$

$$\left(\frac{\partial^2 w}{\partial x^2}\right)_i = 2\alpha_4 + \cdots + 2\alpha_{19} y_i^3$$

and so on, and finally obtain an expression

$$\mathbf{a}^e = \mathbf{C}\boldsymbol{\alpha} \tag{4.85}$$

in which $\mathbf{C}$ is a 21×21 matrix.

The only apparent difficulty in the process that the reader may experience in forming this is that of the definition of the normal slopes at the mid-side nodes. However, if one notes that

$$\frac{\partial w}{\partial n} = \cos\phi \frac{\partial w}{\partial x} + \sin\phi \frac{\partial w}{\partial y} \tag{4.86}$$

in which ϕ is the angle of a particular side to the x axis, the manner of formulation becomes simple. It is not easy to determine an explicit inverse of $\mathbf{C}$, and the stiffness expressions, etc., are evaluated as in Eqs (4.30)–(4.33) by a numerical inversion.

The existence of the mid-side nodes with their single degree of freedom is an inconvenience. It is possible, however, to constrain these by allowing only a cubic variation of the normal slope along each triangle side. Now, explicitly, the matrix $\mathbf{C}$ and the degrees of freedom can be reduced to 18, giving an element illustrated in Fig. 4.19(e) with three corner nodes and 6 degrees of freedom at each node.

Both of these elements were described in several independently derived publications appearing during 1968 and 1969. The 21 degree-of-freedom element was described independently by Argyris *et al.*,[23] Bell,[19] Bosshard,[22] and Visser,[24] listing the authors alphabetically. The reduced 18 degree-of-freedom version was developed by Argyris *et al.*,[23] Bell,[19] Cowper *et al.*,[21] and Irons.[14] An essentially similar, but more complicated, formulation has been developed by Butlin and Ford,[20] and mention of the element shape functions was made earlier by Withum[63] and Felippa.[64]

It is clear that many more elements of this type could be developed and indeed some are suggested in the above references. A very inclusive study is found in the work of Zenisek,[65] Peano,[66] and others.[67–69] However, it should always be borne in mind that all the elements discussed in this section involve an inconsistency when discontinuous variation of material properties occurs. Further, the existence of higher-order derivatives makes it more difficult to impose boundary conditions and indeed the simple interpretation of energy conjugates as 'nodal forces' is more complex. Thus, the engineer may still feel a justified preference for the more intuitive formulation involving displacements and slopes only, despite the fact that very good accuracy is demonstrated in the references cited for the quartic and quintic elements.

Avoidance of continuity difficulties – mixed and constrained elements

4.16 Mixed formulations – general remarks

Equations (4.13)–(4.18) of this chapter provide for many possibilities to approximate both thick and thin plates by using mixed (i.e. reducible) forms. In these, more than one set of variables is approximated directly, and generally continuity requirements for such approximations can be of either C_1 or C_0 type. The procedures used in mixed formulations generally have been described in Chapters 11–13 of Volume 1, and the reader is referred to these for the general principles involved. The options open are large and indeed so is the number of publications proposing various alternatives. We shall therefore limit the discussion to those that appear most useful.

To avoid constant reference to the beginning of this chapter, the four governing equations (4.13)–(4.18) are rewritten below in their abbreviated form with dependent variable sets $\mathbf{M}$, $\boldsymbol{\theta}$, $\mathbf{S}$, and w:

$$\mathbf{M} - \mathbf{DL}\boldsymbol{\theta} = \mathbf{0} \tag{4.87}$$

$$\frac{1}{\alpha}\mathbf{S} + \boldsymbol{\theta} - \boldsymbol{\nabla} w = \mathbf{0} \tag{4.88}$$

$$\mathbf{L}^{\mathrm{T}}\mathbf{M} + \mathbf{S} = \mathbf{0} \tag{4.89}$$

$$\boldsymbol{\nabla}^{\mathrm{T}}\mathbf{S} + q = 0 \tag{4.90}$$

in which $\alpha = \kappa\,\mathrm{Gt}$. To these, of course, the appropriate boundary conditions can be added. For details of the operators, etc., the fuller forms previously quoted need to be consulted.

Mixed forms that utilize direct approximations to all the four variables are not common. The most obvious set arises from elimination of the moments $\mathbf{M}$, that is

$$\mathbf{L}^{\mathrm{T}}\mathbf{DL}\boldsymbol{\theta} + \mathbf{S} = \mathbf{0} \tag{4.91}$$

$$\frac{1}{\alpha}\mathbf{S} + \boldsymbol{\theta} - \boldsymbol{\nabla} w = \mathbf{0} \tag{4.92}$$

$$\boldsymbol{\nabla}^{\mathrm{T}}\mathbf{S} + q = 0 \tag{4.93}$$

and is the basis of a formulation directly related to the three-dimensional elasticity consideration. This is so important that we shall devote Chapter 5 entirely to it, and, of course, there it can be used for both thick and thin plates. We shall, however, return to one of its derivations in Sec. 4.18.

One of the earliest mixed approaches leaves the variables $\mathbf{M}$ and w to be approximated and eliminates $\mathbf{S}$ and $\boldsymbol{\theta}$. The form given is restricted to thin plates and thus $\alpha = \infty$ is taken.

We now can write for Eqs (4.87) and (4.88),

$$\mathbf{D}^{-1}\mathbf{M} - \mathbf{L}\boldsymbol{\nabla} w = \mathbf{0} \tag{4.94}$$

and for Eqs (4.89) and (4.90),

$$\boldsymbol{\nabla}^{\mathrm{T}}\mathbf{L}^{\mathrm{T}}\mathbf{M} - q = 0 \tag{4.95}$$

The approximation can now be made directly putting

$$\mathbf{M} = \mathbf{N}_M \tilde{\mathbf{M}} \qquad \text{and} \qquad w = \mathbf{N}_w \tilde{\mathbf{w}} \tag{4.96}$$

where $\tilde{\mathbf{M}}$ and $\tilde{\mathbf{w}}$ list the nodal (or other) parameters of the expansions, and $\mathbf{N}_M$ and $\mathbf{N}_w$ are appropriate shape functions.

The approximation equations can, as is well known (see Chapter 3 of Volume 1), be made either via a suitable variational principle or directly in a weighted residual, Galerkin form, both leading to identical results. We choose here the latter, although the first presentations of this approximation by Herrmann[70] and others[71–78] all use the Hellinger–Reissner principle.

A weak form from which the plate approximation may be deduced is given by

$$\delta\Pi = \int_\Omega \delta\mathbf{M}\left(-\mathbf{D}^{-1}\mathbf{M} + \mathbf{L}\boldsymbol{\nabla} w\right) \mathrm{d}\Omega + \int_\Omega \delta w\left(\boldsymbol{\nabla}^{\mathrm{T}}\mathbf{L}^{\mathrm{T}}\mathbf{M} - q\right) \mathrm{d}\Omega + \delta\Pi_{\mathrm{bt}} = 0 \tag{4.97}$$

where $\delta\Pi_{\mathrm{bt}}$ describes appropriate boundary condition terms. Using the Galerkin weighting approximations

$$\delta\mathbf{M} = \mathbf{N}_M \delta\tilde{\mathbf{M}} \qquad \text{and} \qquad \delta w = \mathbf{N}_w \delta\tilde{\mathbf{w}} \tag{4.98}$$

gives on integration by parts the following equation set

$$\begin{bmatrix} \mathbf{A} & \mathbf{C} \\ \mathbf{C}^{\mathrm{T}} & \mathbf{0} \end{bmatrix} \begin{Bmatrix} \tilde{\mathbf{M}} \\ \tilde{\mathbf{w}} \end{Bmatrix} = \begin{Bmatrix} \mathbf{f}_1 \\ \mathbf{f}_2 \end{Bmatrix} \tag{4.99}$$

where

$$\begin{aligned} \mathbf{A} &= -\int_\Omega \mathbf{N}_M^{\mathrm{T}}\mathbf{D}^{-1}\mathbf{N}_M \,\mathrm{d}\Omega & \mathbf{f}_1 &= \int_{\Gamma_t} (\boldsymbol{\nabla}\mathbf{N}_w)^{\mathrm{T}} \begin{Bmatrix} \bar{M}_n \\ \bar{M}_{ns} \end{Bmatrix} \mathrm{d}\Gamma \\ \mathbf{C} &= -\int_\Omega (\mathbf{L}\mathbf{N}_M)^{\mathrm{T}}\boldsymbol{\nabla}\mathbf{N}_w \,\mathrm{d}\Omega & \mathbf{f}_2 &= \int_\Omega \mathbf{N}_w^{\mathrm{T}} q \,\mathrm{d}\Omega + \int_{\Gamma_t} \mathbf{N}_w^{\mathrm{T}} S_n \,\mathrm{d}\Gamma \end{aligned} \tag{4.100}$$

where $\bar{M}_n$ and $\bar{M}_{ns}$ are the prescribed boundary moments, and S_n is the prescribed boundary shear force.

Immediately, it is evident that only C_0 continuity is required for both $\mathbf{M}$ and w interpolation,* and many forms of elements are therefore applicable. Of course, appropriate patch tests for the mixed formulation must be enforced[43] and this requires a necessary condition that

$$n_m \geqslant n_w \tag{4.101}$$

where n_m stands for the number of parameters describing the moment field and n_w the number in the displacement field.

Many excellent elements have been developed by using this type of approximation, though their application is limited because of the difficulty of interconnection with

* It should be observed that, if C_0 continuity to the whole $\mathbf{M}$ field is taken, excessive continuity will arise and it is usual to ensure the continuity of M_n and M_{ns} at interfaces only.

other structures as well as the fact that the coefficient matrix in Eq. (4.99) is indefinite with many zero diagonal terms.

Indeed, a similar fate is encountered in numerous 'equilibrium element' forms in which the moment (stress) field is chosen *a priori* in a manner satisfying Eq. (4.95). Here the research of Fraeijs de Veubeke[77] and others[12,30] has to be noted. It must, however, be observed that the second of these elements[30] is in fact identical to the mixed element developed by Herrmann[71] and Hellan[79] (see also reference 52).

4.17 Hybrid plate elements

Hybrid elements are essentially mixed elements in which the field inside the element is defined by one set of parameters and the one on the element frame by another, as shown in Fig. 4.24. The latter are generally chosen to be of a type identical to other displacement models and thus can be readily incorporated in a general program and indeed used in conjunction with the standard displacement types we have already discussed. The internal parameters can be readily eliminated (being confined to a single element) and thus the difference from displacement forms are confined to the element subprogram. The original concept is attributable to Pian[80,81] who pioneered this approach, and today many variants of the procedures exist in the context of thin plate theory.[82–91]

In the majority of approximations, an equilibrating stress field is assumed to be given by a number of suitable shape functions and unknown parameters. In others, a mixed stress field is taken in the interior. A more refined procedure, introduced by Jirousek,[57,91] assumes in the interior a series solution exactly satisfying all the differential equations involved for a homogeneous field.

All procedures use a suitable linking of the interior parameters with those defined on the boundary by the 'frame parameters'. The procedures for doing this are fully

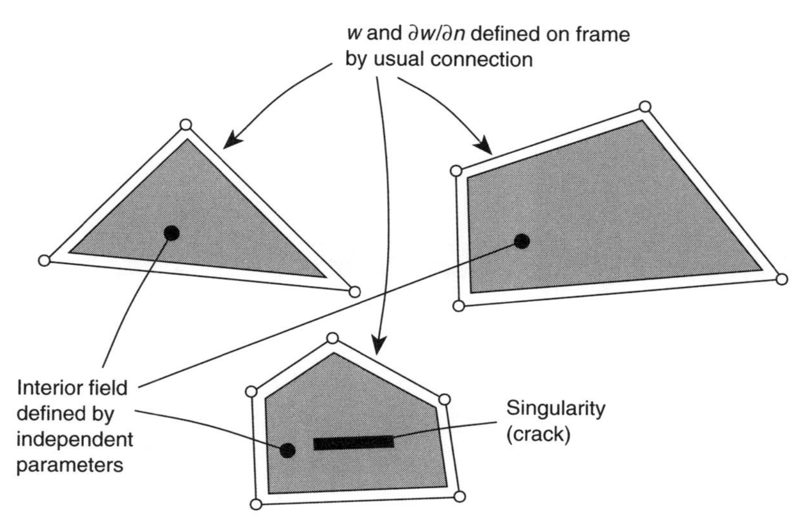

Fig. 4.24 Hybrid elements.

described in Chapter 13 of Volume 1 in the context of elasticity equations, and only a small change of variables is needed to adapt these to the present case. We leave this extension to the reader who can also consult appropriate references for details.

Some remarks need to be made in the context of hybrid elements.

Remark 1. The first is that the number of internal parameters, n_{I}, must be at least as large as the number of frame parameters, n_{F}, which describe the displacements, less the number of rigid-body modes if singularity of the final (stiffness) matrix is to be avoided. Thus, we require that

$$n_{\mathrm{I}} \geqslant n_{\mathrm{F}} - 3 \tag{4.102}$$

for plates.

Remark 2. The second remark is a simple statement that it is possible, but counter-productive, to introduce an excessive number of internal parameters that simply give a more exact solution to a 'wrong' problem in which the frame is constraining the interior of an element. Thus additional accuracy is not achieved overall.

Remark 3. Most of the formulations are available for non-homogeneous plates (and hence non-linear problems). However, this is not true for the Trefftz-hybrid elements[57,91] where an exact solution to the differential equation needs to be available for the element interior. Such solutions are not known for arbitrary non-homogeneous interiors and hence the procedure fails. However, for homogeneous problems the elements can be made much more accurate than any of the others and indeed allow a general polygonal element with singularities and/or internal boundaries to be developed by the use of special functions (see Fig. 4.24). Obviously, this advantage needs to be borne in mind.

A number of elements matching (or duplicating) the displacement method have been developed and the performance of some of the simpler ones is shown in Fig. 4.16. Indeed, it can be shown that many hybrid-type elements duplicate precisely the various incompatible elements that pass the convergence requirement. Thus, it is interesting to note that the triangle of Allman[90] gives precisely the same results as the 'smoothed' Razzaque element of references 33 and 34 or, indeed, the element of Sec. 4.5.

4.18 Discrete Kirchhoff constraints

Another procedure for achieving excellent element performance is achieved as a constrained (mixed) element. Here it is convenient (though by no means essential) to use a variational principle to describe Eqs (4.91) and (4.93). This can be written simply as the minimization of the functional

$$\Pi = \frac{1}{2}\int_{\Omega} (\mathbf{L}\boldsymbol{\theta})^{\mathrm{T}}\mathbf{D}(\mathbf{L}\boldsymbol{\theta})\,\mathrm{d}\Omega + \frac{1}{2}\int_{\Omega} \mathbf{S}^{\mathrm{T}}\frac{1}{\alpha}\mathbf{S}\,\mathrm{d}\Omega - \int_{\Omega} wq\,\mathrm{d}\Omega + \Pi_{\mathrm{bt}} = \text{minimum} \tag{4.103}$$

subject to the constraint that Eq. (4.92) be satisfied, that is, that

$$\frac{1}{\alpha}\mathbf{S} + \boldsymbol{\theta} - \boldsymbol{\nabla} w = \mathbf{0} \tag{4.104}$$

We shall use this form for general thick plates in Chapter 5, but in the case of thin plates with which this chapter is concerned, we can specialize by putting $\alpha = \infty$ and rewrite the above as

$$\Pi = \frac{1}{2}\int_{\Omega}(\mathbf{L}\boldsymbol{\theta})^{\mathrm{T}}\mathbf{D}(\mathbf{L}\boldsymbol{\theta})\,\mathrm{d}\Omega - \int_{\Omega} w q\,\mathrm{d}\Omega + \Pi_{\mathrm{bt}} = \text{minimum} \tag{4.105}$$

subject to

$$\boldsymbol{\theta} - \boldsymbol{\nabla} w = \mathbf{0} \tag{4.106}$$

and we note that the explicit mention of shear forces $\mathbf{S}$ is no longer necessary.

To solve the problem posed by Eqs (4.105) and (4.106) we can

1. approximate w and $\boldsymbol{\theta}$ by independent interpolations of C_0 continuity as

$$w = \mathbf{N}_w \tilde{\mathbf{w}} \qquad \text{and} \qquad \boldsymbol{\theta} = \mathbf{N}_\theta \tilde{\boldsymbol{\theta}} \tag{4.107}$$

2. impose a discrete approximation to the constraint of Eq. (4.106) and solve the minimization problem resulting from substitution of Eq. (4.107) into Eq. (4.105) by either discrete elimination, use of suitable lagrangian multipliers, or penalty procedures.

In the application of the so-called *discrete Kirchhoff constraints*, Eq. (4.106) is approximated by point (or subdomain) *collocation* and direct elimination is used to *reduce the number of nodal parameters*. Of course, the other means of imposing the constraints could be used with *identical* effect and we shall return to these in the next chapter. However, direct elimination is advantageous in reducing the final total number of variables and can be used effectively.

4.18.1 One-dimensional beam example

We illustrate the process to impose discrete constraints on a simple, one-dimensional, example of a beam (or cylindrical bending of a plate) shown in Fig. 4.25. In this, initially the displacements and rotations are taken as determined by a quadratic interpolation of an identical kind and we write in place of Eq. (4.107),

$$\begin{Bmatrix} w \\ \theta \end{Bmatrix} = \sum_{i=1}^{3} N_i \begin{Bmatrix} \tilde{w}_i \\ \tilde{\theta}_i \end{Bmatrix} \tag{4.108}$$

where i are the three element nodes.

The constraint is now applied by point collocation at coordinates x_α and x_β of the beam; that is, we require that at these points

$$\theta - \frac{\partial w}{\partial x} = 0 \tag{4.109}$$

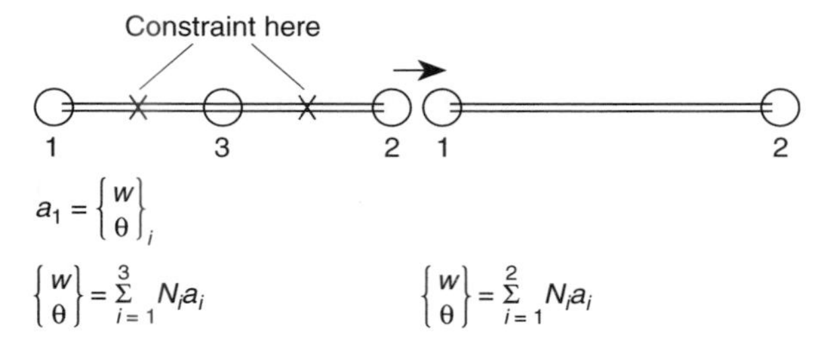

Fig. 4.25 A beam element with independent, Lagrangian, interpolation of w and θ with constraint $\partial w/\partial x - \theta = 0$ applied at points $\times$.

This can be written by using the interpolation of Eq. (4.108) as two simultaneous equations

$$\begin{aligned} \sum_{i=1}^{3} N_i(x_\alpha)\tilde{\theta}_i - \sum_{i=1}^{3} N_i'(x_\alpha)\tilde{w}_i = 0 \\ \sum_{i=1}^{3} N_i(x_\beta)\tilde{\theta}_i - \sum_{i=1}^{3} N_i'(x_\beta)\tilde{w}_i = 0 \end{aligned} \tag{4.110}$$

where

$$N_i(x_\alpha) = N_i(x)|_{x=x_\alpha} \qquad \text{and} \qquad N_i'(x_\alpha) = \left(\frac{\mathrm{d}N_i}{\mathrm{d}x}\right)_{x=x_\alpha}$$

Equations (4.110) can be used to eliminate $\tilde{w}_3$ and $\tilde{\theta}_3$. Writing Eqs (4.110) explicitly we have

$$\mathbf{A}_3 \begin{Bmatrix} \tilde{w}_3 \\ \tilde{\theta}_3 \end{Bmatrix} = \mathbf{A}_1 \begin{Bmatrix} \tilde{w}_1 \\ \tilde{\theta}_1 \end{Bmatrix} + \mathbf{A}_2 \begin{Bmatrix} \tilde{w}_2 \\ \tilde{\theta}_2 \end{Bmatrix} \tag{4.111}$$

where

$$\mathbf{A}_i = \begin{bmatrix} N_i(x_\alpha) - N_i'(x_\alpha) \\ N_i(x_\beta) - N_i'(x_\beta) \end{bmatrix}$$

Substitution of the above into Eq. (4.108) results directly in shape functions from which the centre node has been eliminated, that is,

$$\begin{Bmatrix} w \\ \theta \end{Bmatrix} = \sum_{i=1}^{2} \bar{\mathbf{N}}_i \begin{Bmatrix} \tilde{w}_i \\ \tilde{\theta}_i \end{Bmatrix} \tag{4.112}$$

with

$$\bar{\mathbf{N}}_i = N_i \mathbf{I} + \mathbf{A}_3^{-1} \mathbf{A}_i$$

where $\mathbf{I}$ is a 2×2 identity matrix.

If these functions are used for a beam, we arrive at an element that is convergent. Indeed, in the particular case where x_α and x_β are chosen to coincide with the two Gauss quadrature points the element stiffness coincides with that given by a displacement formulation involving a cubic w interpolation. In fact, the agreement is *exact* for a uniform beam.

For two-dimensional plate elements the situation is a little more complex, but if we imagine x to coincide with the direction tangent to an element side, precisely identical elimination enforces *complete compatibility* along an element side when both gradients of w are specified at the ends. However, with discrete imposition of the constraints it is

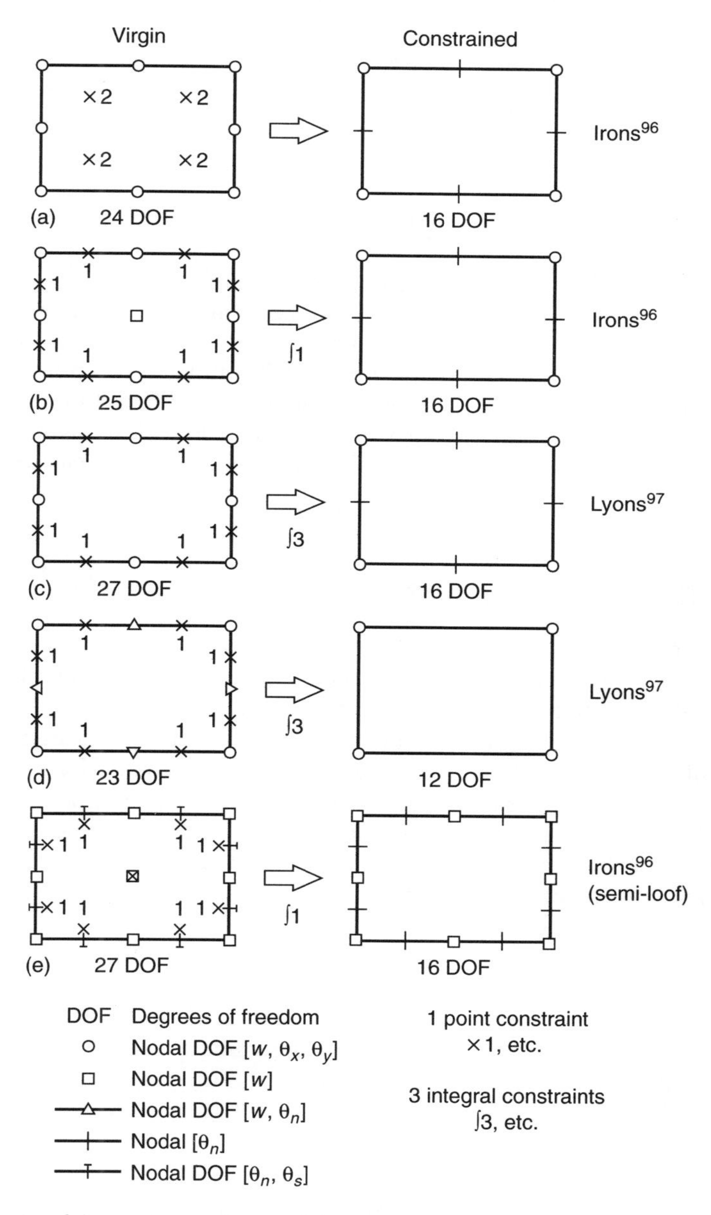

Fig. 4.26 A series of discrete Kirchhoff theory (DKT)-type elements of quadrilateral type.

not clear *a priori* that convergence will always occur – though, of course, one can argue heuristically that collocation applied in numerous directions should result in an acceptable element. Indeed, patch tests turn out to be satisfied by most elements in which the w interpolation (and hence the $\partial w/\partial s$ interpolation) have C_0 continuity.

The constraints frequently applied in practice involve the use of line or subdomain collocation to increase their number (which must, of course, always be less than the number of remaining variables) and such additional constraint equations as

$$
\begin{aligned}
I_\Gamma &\equiv \int_{\Gamma_e} \left(\frac{\partial w}{\partial s} - \theta_s \right) \mathrm{d}s = 0 \\
I_{\Omega x} &\equiv \int_{\Omega_e} \left(\frac{\partial w}{\partial x} - \theta_x \right) \mathrm{d}\Omega = 0 \\
I_{\Omega y} &\equiv \int_{\Omega_e} \left(\frac{\partial w}{\partial y} - \theta_y \right) \mathrm{d}\Omega = 0
\end{aligned}
\tag{4.113}
$$

are frequently used. The algebra involved in the elimination is not always easy and the reader is referred to original references for details pertaining to each particular element.

The concept of discrete Kirchhoff constraints was first introduced by Wempner *et al.*,[92] Stricklin *et al.*,[59] and Dhatt[60] in 1968–69, but it has been applied extensively since.[93–103] In particular, the 9 degree-of-freedom triangle[93,94] and the complex semi-loof element of Irons[96] are elements which have been successfully used.

Figure 4.26 illustrates some of the possible types of quadrilateral elements achieved in these references.

4.19 Rotation-free elements

It is possible to construct elements for thin plates in terms of transverse displacement parameters alone. Nay and Utku used quadratic displacement approximation and minimum potential energy to construct a least-square fit for an element configuration shown in Fig. 4.27(a).[104] The element is non-conforming but passes the patch test and therefore is an admissible form. An alternative, mixed field, construction is given by Oñate and Zárate for a composite element constructed from linear interpolation on each triangle.[105] In this work a mixed variational principle is used together with a special approximation for the curvature. We summarize here the steps in the better approach.

A three-field mixed variational form for a thin plate problem based on the Hu–Washizu functional may be written as

$$
\Pi = \frac{1}{2} \int_A \boldsymbol{\kappa}^{\mathrm{T}} \mathbf{D} \boldsymbol{\kappa} \, \mathrm{d}A - \int_A \mathbf{M}^{\mathrm{T}} [(\mathbf{L}\boldsymbol{\nabla}) w - \boldsymbol{\kappa}] \, \mathrm{d}A - \int_A w q \, \mathrm{d}A + \Pi_{\mathrm{bt}} \tag{4.114}
$$

where now $\boldsymbol{\kappa}$ and $\mathbf{M}$ are mixed variables to be approximated, $(\mathbf{L}\boldsymbol{\nabla})w$ are again second derivatives of displacement w given in Eq. (4.20) and integration is over the area of the plate middle surface. Variation of Eq. (4.114) with respect to $\boldsymbol{\kappa}$ gives the discrete constitutive equation

$$
\int_{A_e} \delta \boldsymbol{\kappa}^{\mathrm{T}} [\mathbf{D} \boldsymbol{\kappa} - \mathbf{M}] \, \mathrm{d}A = 0 \tag{4.115}
$$

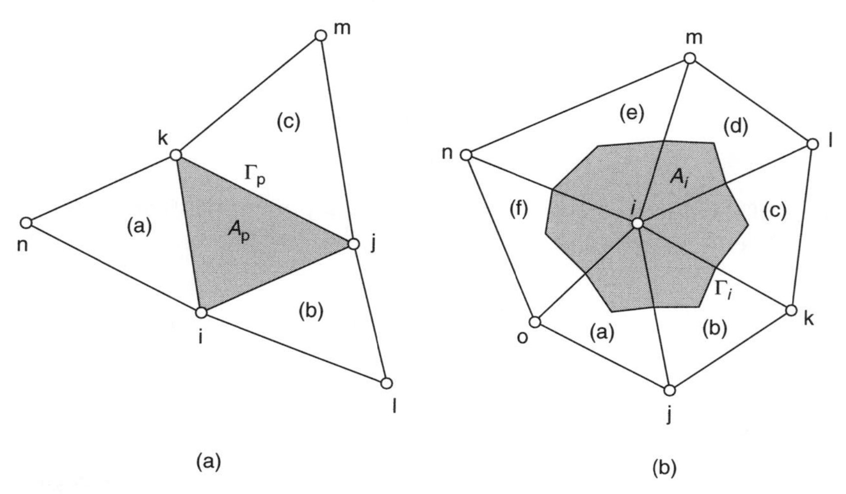

Fig. 4.27 Elements for rotation-free thin plates: (a) patch for Nay and Utku procedure[104] BPT triangle; and (b) patch for BPN triangle.[105]

where A_e is the domain of the patch for the element. Two alternatives for A_e are considered in reference 105 and named BPT and BPN as shown in Figs 4.27(a) and 4.27(b), respectively. For the BPT form the integration is taken over the area of the element 'ijk' with area A_p and boundary Γ_p. For the type BPN integration is over the more complex area A_i with boundary Γ_i. Each, however, are simple to construct. Similarly, variation of Eq. (4.114) with respect to moment gives the discrete curvature relation

$$\int_{A_e} \delta\mathbf{M}^{\mathrm{T}}[(\mathbf{L}\boldsymbol{\nabla})w - \boldsymbol{\kappa}]\,\mathrm{d}A = 0 \tag{4.116}$$

Finally, the equilibrium equations are obtained from the variation with respect to the displacement, and are expressed as

$$\int_{A_e} [(\mathbf{L}\boldsymbol{\nabla})\delta w]^{\mathrm{T}}\mathbf{M}\,\mathrm{d}A - \int_{A_e} \delta w\, q\,\mathrm{dif}\,\mathrm{dA} + \delta_{\mathrm{bt}} = 0 \tag{4.117}$$

A finite element approximation may be constructed in the standard manner by writing

$$\mathbf{M} = N_i^m \tilde{\mathbf{M}}_i, \quad \boldsymbol{\kappa} = N_i^{\kappa}\tilde{\boldsymbol{\kappa}}_i \quad \text{and} \quad w = N_i^w \tilde{\mathbf{w}}_i \tag{4.118}$$

The simplest approximations are for $N_i^m = N_i^{\kappa} = 1$ and linear interpolation over each triangle for N_i^w. Equation (4.115) is easily evaluated; however, the other two integrals have apparent difficulty since a linear interpolation yields zero derivatives within each triangle. Indeed the curvature is now concentrated in the 'kinks' which occur between contiguous triangles. To obtain discrete approximations to the curvature changes an integration by parts is used (see Green's theorem, Appendix G of Volume 1) to rewrite Eq. (4.116) as

$$\int_{A_e} \delta\mathbf{M}^{\mathrm{T}}\boldsymbol{\kappa}\,\mathrm{d}A + \int_{\Gamma_e} \delta\mathbf{M}^{\mathrm{T}}\mathbf{g}\boldsymbol{\nabla}w\,\mathrm{d}\Gamma = 0 \tag{4.119}$$

where

$$\mathbf{g} = \begin{bmatrix} n_x & 0 \\ 0 & n_y \\ n_y & n_x \end{bmatrix} \tag{4.120}$$

is a matrix of the direction cosines for an outward pointing normal vector $\mathbf{n}$ to the boundary Γ_e and

$$\nabla w = \begin{Bmatrix} w_{,x} \\ w_{,y} \end{Bmatrix} \tag{4.121}$$

In these expressions Γ_e is the part of the boundary within the area of integration A_e. Thus, for the element type BPT it is just the contour Γ_p as shown in Fig. 4.27(a). For the element type BPN no slope discontinuity occurs on the boundary Γ_i shown in Fig. 4.27(b); however, it is necessary to integrate along the half sides of each triangle within the patch bounded by Γ_i. The remainder of the derivation is now straightforward and the reader is referred to reference 105 for additional details and results. In this paper results are also presented for thin shells.

We note that the type of element discussed in this section is quite different from those presented previously in that nodes exist outside the boundary of the element. Thus, the definition of an element and the assembly process are somewhat different. In addition, boundary conditions need some special treatments to include in a general manner.[105] Because of these differences we do not consider additional members in this family. We do note, however, that for explicit dynamic programs some advantages occur since no rotation parameters need be integrated. Results for thin shells subjected to impulsive loading are particularly noteworthy.[105]

4.20 Inelastic material behaviour

The preceding discussion has assumed the plate to be a linear elastic material. In many situations it is necessary to consider a more general constitutive behaviour in order to represent the physical problem correctly. For thin plates, only the bending and twisting moment are associated with deformations and are related to the local stresses through

$$\mathbf{M} = \begin{Bmatrix} M_x \\ M_y \\ M_{xy} \end{Bmatrix} = -\int_{-t/2}^{t/2} \begin{Bmatrix} \sigma_x \\ \sigma_y \\ \tau_{xy} \end{Bmatrix} z \, \mathrm{d}z \tag{4.122}$$

Any of the material models discussed in Chapter 3 *which have symmetric stress behaviour with respect to strains* may be used in plate analysis provided an appropriate plane stress form is available. The symmetry is necessary to avoid the generation of in-plane force resultants – which are assumed to decouple from the bending behaviour. If such conditions do not exist it is necessary to use a shell formulation as described in Chapters 6–9.

In practice two approaches are considered – one dealing with the individual lamina using local stress components σ_x, σ_y and τ_{xy} and the other using plate resultant forces M_x, M_y and M_{xy} directly.

4.20.1 Numerical integration through thickness

The most direct approach is to use a *plane stress* form of the stress–strain relation and perform the through-thickness integration numerically. In order to capture the maximum stresses at the top and bottom of the plate it is best to use Gauss–Lobatto-type quadrature formulae[106] where integrals are approximated by

$$\int_{-1}^{1} f(\xi)\,\mathrm{d}\xi \approx f(-1)\,W_1 + \sum_{n=2}^{N-1} f(\xi_n)\,W_n + f(1)\,W_N \tag{4.123}$$

These formula differ from the typical gaussian quadrature considered previously and use the end points on the interval directly. This allows computation of first yield to be more accurate. The values of the parameters ξ_n and W_n are given in Table 4.4 up to the six-point formula. Parameters for higher-order formulae may be found in reference 106.

Noting that the strain components in plates [see Eq. (4.10)] are asymmetric with respect to the middle surface of the plate and that the z coordinate is also asymmetric we can compute the plate resultants by evaluating only half the integral. Accordingly, we may use

$$\mathbf{M} = \begin{Bmatrix} M_x \\ M_y \\ M_{xy} \end{Bmatrix} = -2\int_0^{t/2} \begin{Bmatrix} \sigma_x \\ \sigma_y \\ \tau_{xy} \end{Bmatrix} z\,\mathrm{d}z \tag{4.124}$$

and here a six-point formula or less will generally be sufficient to compute integrals.

Equation (4.29) is replaced by the non-linear equation given as

$$\boldsymbol{\Psi}(\tilde{\mathbf{w}}) = \mathbf{f} - \int_\Omega \mathbf{B}^\mathrm{T}\mathbf{M}\,\mathrm{d}\Omega = \mathbf{0} \tag{4.125}$$

Table 4.4 Gauss–Lobbato quadrature points and weights

N	$\pm\xi_n$	W_n
3	1.0	1/3
	0.0	4/3
4	1.0	1/6
	$\sqrt{0.2}$	5/6
5	1.0	0.1
	$\sqrt{3/7}$	49/90
	0.0	64/90
6	1.0	1/15
	$\sqrt{t_1}$	$0.6/[t_1(1-t_0)^2]$
	$\sqrt{t_2}$	$0.6/[t_2(1-t_0)^2]$
	$t_0 = \sqrt{7}$	$t_1 = (7+2t_0)/21$
		$t_2 = (7-2t_0)/21$

The solution process (for a static case) may now proceed by using, for instance, a Newton–Raphson scheme in which the *tangent moduli* for the plate are obtained by using the tangent moduli for the stress components as

$$d\mathbf{M} = \begin{Bmatrix} dM_x \\ dM_y \\ dM_{xy} \end{Bmatrix} = \left[2 \int_0^{t/2} \mathbf{D}_{\mathrm{T}}^{(\mathrm{ps})} z^2 \, \mathrm{d}z \right] \mathbf{L}\boldsymbol{\nabla} \, dw = \mathbf{D}_{\mathrm{T}} (\mathbf{L}\boldsymbol{\nabla}) \, dw \tag{4.126}$$

where $\mathbf{D}_{\mathrm{T}}^{(\mathrm{ps})}(z)$ is the tangent modulus matrix of a plane stress material model at each lamina level z, and $\mathbf{D}_{\mathrm{T}}$ is the resulting bending tangent stiffness matrix of the plate.

The Newton–Raphson iteration for the displacement increment is computed as

$$\mathbf{K}_{\mathrm{T}}^{(k)} \, d\tilde{\mathbf{w}}^{(k)} = \boldsymbol{\Psi}^{(k)} \tag{4.127}$$

with iterative updates

$$\tilde{\mathbf{w}}^{(k+1)} = \tilde{\mathbf{w}}^{(k)} + d\tilde{\mathbf{w}}^{(k)} \tag{4.128}$$

until a suitable convergence criterion is satisfied. This follows precisely methods previously defined for solids.

4.20.2 Resultant constitutive models

A resultant yield function for plates with Huber–von Mises-type material is given by[107]

$$F(\mathbf{M}) = \left(M_x^2 + M_y^2 - M_x M_y + 3 M_{xy}^2 \right) - M_{\mathrm{u}}^2(\kappa) \leqslant 0 \tag{4.129}$$

where κ is an 'isotropic' hardening parameter and M_{u} denotes a uniaxial yield moment and which for homogeneous plates is generally given by

$$M_{\mathrm{u}} = \tfrac{1}{4} t^2 \sigma_y(\kappa) \tag{4.130}$$

in which σ_y is the material uniaxial yield stress in tension (and compression). We observe that, in the absence of hardening, M_{u} is the moment that exists when the entire cross section is at a yield stress.

4.21 Concluding remarks – which elements?

The extensive bibliography of this chapter outlining the numerous approaches capable of solving the problems of thin, Kirchhoff, plate flexure shows both the importance of the subject in structural engineering – particularly as a preliminary to shell analysis – and the wide variety of possible approaches. Indeed, only part of the story is outlined here, as the next chapter, dealing with thick plate formulation, presents many practical alternatives of dealing with the same problem.

We hope that the presentation, in addition to providing a guide to a particular problem area, is useful in its direct extension to other fields where governing equations lead to C_1 continuity requirements.

Users of practical computer programs will be faced with a problem of 'which element' is to be used to satisfy their needs. We have listed in Table 4.3 some of the more widely known simple elements and compared their performance in Fig. 4.16. The choice is not always unique, and much more will depend on preferences and indeed extensions desired. As will be seen in Chapter 6 for general shell problems, triangular elements are an optimal choice for many applications and configurations. Further, such elements are most easily incorporated if adaptive mesh generation is to be used for achieving errors of predetermined magnitude.

References

1. G. Kirchhoff. Über das Gleichgewicht und die Bewegung einer elastichen Scheibe. *J. Reine und Angewandte Mathematik*, **40**, 51–88, 1850.
2. S.P. Timoshenko and S. Woinowski-Krieger. *Theory of Plates and Shells*, 2nd edition, McGraw-Hill, New York, 1959.
3. L. Bucciarelly and N. Dworsky. *Sophie Germain, An Essay on the History of Elasticity*, Reidel, New York, 1980.
4. E. Reissner. Reflections on the theory of elastic plates. *Appl. Mech. Rev.*, **38**, 1453–64, 1985.
5. E. Reissner. The effect of transverse shear deformation on the bending of elastic plates. *J. Appl. Mech.*, **12**, 69–76, 1945.
6. R.D. Mindlin. Influence of rotatory inertia and shear in flexural motions of isotropic elastic plates. *J. Appl. Mech.*, **18**, 31–8, 1951.
7. I. Babuška and T. Scapolla. Benchmark computation and performance evaluation for a rhombic plate bending problem. *Int. J. Num. Meth. Eng.*, **28**, 155–80, 1989.
8. I. Babuška. The stability of domains and the question of formulation of plate problems. *Appl. Math.*, 463–7, 1962.
9. B.M. Irons and J.K. Draper. Inadequacy of nodal connections in a stiffness solution for plate bending. *Journal of AIAA*, **3**, 5, 1965.
10. R.W. Clough and J.L. Tocher. Finite element stiffness matrices for analysis of plate bending. In J.S. Przemienicki, R.M. Bader, W.F. Bozich, J.R. Johnson and W.J. Mykytow (eds), *Proc. 1st Conf. Matrix Methods in Structural Mechanics*, Volume AFFDL-TR-66-80, pp. 515–45, Air Force Flight Dynamics Laboratory, Wright Patterson Air Force Base, OH, October 1966.
11. G.P. Bazeley, Y.K. Cheung, B.M. Irons and O.C. Zienkiewicz. Triangular elements in bending – conforming and non-conforming solutions. In J.S. Przemienicki, R.M. Bader, W.F. Bozich, J.R. Johnson and W.J. Mykytow (eds), *Proc. 1st Conf. Matrix Methods in Structural Mechanics*, Volume AFFDL-TR-66-80, pp. 547–76, Air Force Flight Dynamics Laboratory, Wright Patterson Air Force Base, OH, October 1966.
12. G. Sander. Bournes supérieures et inérieures dans l'analyse matricielle des plates en flexion–torsion. *Bull. Soc. Royale des Sci. de Liège*, **33**, 456–94, 1964.
13. B. Fraeijs de Veubeke. Bending and stretching of plates. Special models for upper and lower bounds. In J.S. Przemienicki, R.M. Bader, W.F. Bozich, J.R. Johnson and W.J. - Mykytow (eds), *Proc. 1st Conf. Matrix Methods in Structural Mechanics*, Volume AFFDL-TR-66-80, pp. 863–86, Air Force Flight Dynamics Laboratory, Wright Patterson Air Force Base, OH, October 1966.
14. B.M. Irons. A conforming quartic triangular element for plate bending. *Int. J. Num. Meth. Eng.*, **1**, 29–46, 1969.

15. R.W. Clough and C.A. Felippa. A refined quadrilateral element for analysis of plate bending. In L. Berke, R.M. Bader, W.J. Mykytow, J.S. Przemienicki and M.H. Shirk (eds), *Proc. 2nd Conf. Matrix Methods in Structural Mechanics*, Volume AFFDL-TR-68-150, pp. 399–440, Air Force Flight Dynamics Laboratory, Wright Patterson Air Force Base, OH, October 1968.
16. B. Fraeijs de Veubeke. A conforming finite element for plate bending. *International Journal of Solids and Structures*, **4**, 95–108, 1968.
17. F.K. Bogner, R.L. Fox and L.A. Schmit. The generation of interelement-compatible stiffness and mass matrices by the use of interpolation formulae. In J.S. Przemienicki, R.M. Bader, W.F. Bozich, J.R. Johnson and W.J. Mykytow (eds), *Proc. 1st Conf. Matrix Methods in Structural Mechanics*, Volume AFFDITR-66-80, pp. 397–443, Air Force Flight Dynamics Laboratory, Wright Patterson Air Force Base, OH, October 1966.
18. I.M. Smith and W. Duncan. The effectiveness of nodal continuities in finite element analysis of thin rectangular and skew plates in bending. *Int. J. Num. Meth. Eng.*, **2**, 253–8, 1970.
19. K. Bell. A refined triangular plate bending element. *Int. J. Num. Meth. Eng.*, **1**, 101–22, 1969.
20. G.A. Butlin and R. Ford. A compatible plate bending element. Technical Report 68-15, Engineering Department, University of Leicester, Leicester, 1968.
21. G.R. Cowper, E. Kosko, G.M. Lindberg and M.D. Olson. Formulation of a new triangular plate bending element. *Trans. Canad. Aero-Space Inst.*, **1**, 86–90, 1968; see also NRC Aero Report LR514, 1968.
22. W. Bosshard. Ein neues vollverträgliches endliches Element für Plattenbiegung. *Mt. Ass. Bridge Struct. Eng. Bull.*, **28**, 27–40, 1968.
23. J.H. Argyris, I. Fried and D.W. Scharpf. The TUBA family of plate elements for the matrix displacement method. *The Aeronaut. J., Roy. Aeronaut. Soc.*, **72**, 701–9, 1968.
24. W. Visser. *The Finite Element Method in Deformation and Heat Conduction Problems*, PhD dissertation, Techische Hochschule, Delft, 1968.
25. B.M. Irons. Comments on 'complete polynomial displacement fields for finite element method' (by P.C. Dunn). *The Aeronaut. J., Roy. Aeronaut. Soc.*, **72**, 709, 1968.
26. O.C. Zienkiewicz and Y.K. Cheung. The finite element method for analysis of elastic isotropic and orthotropic slabs. *Proc. Inst. Civ. Eng.*, **28**, 471–88, 1964.
27. R.W. Clough. The finite element method in structural mechanics. In O.C. Zienkiewicz and G.S. Holister (eds), *Stress Analysis*, pp. 85–119 John Wiley, Chichester, Sussex, 1965.
28. D.J. Dawe. Parallelogram element in the solution of rhombic cantilever plate problems. *J. Strain Anal*, **1**, 223–30, 1966.
29. J.H. Argyris. Continua and discontinue. In J.S. Przemienicki, R.M. Bader, W.F. Bozich, J.R. Johnson and W.J. Mykytow (eds), *Proc. 1st Conf. Matrix Methods in Structural Mechanics*, Volume AFFDL-TR-66-80, pp. 11–189, Air Force Flight Dynamics Laboratory, Wright Patterson Air Force Base, OH, October 1966.
30. L.S.D. Morley. On the constant moment plate bending element. *J. Strain Anal.*, **6**, 20–4, 1971.
31. R.D. Wood. A shape function routine for the constant moment triangular plate bending element. *Engineering Computations*, **1**, 189–98, 1984.
32. R. Narayanaswami. New triangular plate bending element with transverse shear flexibility. *Journal of AIAA*, **12**, 1761–63, 1974.
33. A. Razzaque. Program for triangular bending element with derivative smoothing. *Int. J. Num. Meth. Eng.*, **5**, 588–9, 1973.
34. B.M. Irons and A. Razzaque. Shape function formulation for elements other than displacement models. In C.A. Brebbia and H. Tottenham (eds), *Proc. of the International*

Conference on Variational Methods in Engineering, Volume II, pp. 4/59–4/72, Southampton University Press, Southampton, 1973.

35. J.E. Walz, R.E. Fulton and N.J. Cyrus. Accuracy and convergence of finite element approximations. In L. Berke, R.M. Bader, W.J. Mykytow, J.S. Przemienicki and M.H. Shirk (eds), *Proc. 2nd Conf. Matrix Methods in Structural Mechanics*, Volume AFFDL-TR68-150, pp. 995–1027, Air Force Flight Dynamics Laboratory, Wright Patterson Air Force Base, OH, October 1968.
36. R.J. Melosh. Structural analysis of solids. *ASCE Structural Journal*, **4**, 205–23, August 1963.
37. A. Adini and R.W. Clough. Analysis of plate bending by the finite element method. Technical Report G-7337, Report to National Science Foundation, USA, 1961.
38. Y.K. Cheung, I.P. King and O.C. Zienkiewicz. Slab bridges with arbitrary shape and support conditions. *Proc. Inst. Civ. Eng.*, **40**, 9–36, 1968.
39. J.L. Tocher and K.K. Kapur. Comment on basis of derivation of matrices for direct stillness method (by R. Melosh). *Journal of AIAA*, **3**, 1215–16, 1965.
40. R.D. Henshell, D. Walters and G.B. Warburton. A new family of curvilinear plate bending elements for vibration and stability. *J. Sound Vibr.*, **20**, 327–343, 1972.
41. R.L. Taylor, O.C. Zienkiewicz, J.C. Simo and A.H.C. Chan. The patch test – a condition for assessing FEM convergence. *Int. J. Num. Meth. Eng.*, **22**, 39–62, 1986.
42. O.C. Zienkiewicz, S. Qu, R.L. Taylor and S. Nakazawa. The patch test for mixed formulations. *Int. J. Num. Meth. Eng.*, **23**, 1873–83, 1986.
43. O.C. Zienkiewicz and D. Lefebvre. Three field mixed approximation and the plate bending problem. *Comm. Appl. Num. Meth.*, **3**, 301–9, 1987.
44. P.G. Bergen and L. Hanssen. A new approach for deriving 'good' element stiffness matrices. In J.R. Whiteman (ed.), *The Mathematics of Finite Elements and Applications*, pp. 483–97. Academic Press, London, 1977.
45. R.V. Southwell. *Relaxation Methods in Theoretical Physics*, 1st edition, Clarendon Press, Oxford, 1946.
46. P.G. Bergan and M.K. Nygard. Finite elements with increased freedom in choosing shape functions. *Int. J. Num. Meth. Eng.*, **20**, 643–63, 1984.
47. C.A. Felippa and P.C. Bergan. A triangular plate bending element based on energy orthogonal free formulation. *Comp. Meth. Appl. Mech. Eng.*, **61**, 129–60, 1987.
48. A. Samuelsson. The global constant strain condition and the patch test. In R. Glowinski, E.Y. Rodin and O.C. Zienkiewicz (eds), *Energy Methods in Finite Element Analysis*, pp. 49–68. John Wiley, Chichester, Sussex, 1979.
49. B. Specht. Modified shape functions for the three node plate bending element passing the patch test. *Int. J. Num. Meth. Eng.*, **26**, 705–15, 1988.
50. T.J.R. Hughes. *The Finite Element Method: Linear Static and Dynamic Analysis*, Prentice-Hall, Englewood Cliffs, NJ, 1987.
51. F. Kikuchi and Y. Ando. A new variational functional for the finite element method and its application to plate and shell problems. *Nuclear Engineering and Design*, **21**(1), 95–113, 1972.
52. L.S.D. Morley. The triangular equilibrium element in the solution of plate bending problems. *Aero. Q.*, **19**, 149–69, 1968.
53. H. Marcus. *Die Theorie elastisher Geweve und ihre Anwendung auf die Berechnung biegsamer Platten*, Springer, Berlin, 1932.
54. P. Ballesteros and S.L. Lee. Uniformly loaded rectangular plate supported at the corners. *Int. J. Mech. Sci.*, **2**, 206–11, 1960.
55. J.L. Batoz, K.-J. Bathe and L.W. Ho. A study of three-node triangular plate bending elements. *Int. J. Num. Meth. Eng.*, **15**, 1771–812, 1980.
56. M.M. Hrabok and T.M. Hrudey. A review and catalogue of plate bending finite elements. *Computers and Structures*, **19**, 479–95, 1984.

57. J. Jirousek and L. Guex. The hybrid-Trefftz finite element model and its application to plate bending. *Int. J. Num. Meth. Eng.*, **23**, 651–93, 1986.
58. A. Razzaque. *Finite Element Analysis of Plates and Shells*, PhD thesis, Civil Engineering Department, University of Wales, Swansea, 1972.
59. J.A. Stricklin, W. Haisler, P. Tisdale and K. Gunderson. A rapidly converging triangle plate element. *Journal of AIAA*, **7**, 180–1, 1969.
60. G.S. Dhatt. Numerical analysis of thin shells by curved triangular elements based on discrete Kirchholf hypotheses. In W.R. Rowan and R.M. Hackett (eds), *Proc. Symp. on Applications of FEM in Civil Engineering*, Vandervilt University, Nashville, TN, 1969.
61. J.L. Batoz and M.B. Tahar. Evaluation of a new quadrilateral thin plate bending element. *Int. J. Num. Meth. Eng.*, **18**, 1655–77, 1982.
62. E.L. Wilson. The static condensation algorithm. *Int. J. Num. Meth. Eng.*, **8**, 199–203, 1974.
63. D. Withum. Berechnung von Platten nach dem Ritzchen Verfahren mit Hilfe dreieckförmiger Meshnetze. Technical Report, Mittl. Inst. Statik Tech. Hochschule, Hanover, 1966.
64. C.A. Felippa. *Refined Finite Element Analysis of Linear and Non-linear Two-dimensional Structures*. PhD dissertation, Department of Civil Engineering, SEMM, University of California, Berkeley, CA, 1966.
65. A. Zanisek. Interpolation polynomials on the triangle. *Int. J. Num. Meth. Eng.*, **10**, 283–96, 1976.
66. A. Peano. Conforming approximation for Kirchhoff plates and shells. *Int. J. Num. Meth. Eng.*, **14**, 1273–91, 1979.
67. J.J. Göel. Construction of basic functions for numerical utilization of Ritz's method. *Numerische Math.*, **12**, 435–47, 1968.
68. G. Birkhoff and L. Mansfield. Compatible triangular finite elements. *J. Math. Anal. Appl.*, **47**, 531–53, 1974.
69. C.L. Lawson. C^1-compatible interpolation over a triangle. Technical Report RM 33–770, NASA Jet Propulsion Laboratory, Pasadena, CA, 1976.
70. L.R. Herrmann. Finite element bending analysis of plates. In J.S. Przemienicki, R.M. Bader, W.F. Bozich, J.R. Johnson and W.J. Mykytow (eds), *Proc. 1st Conf. Matrix Methods in Structural Mechanics*, Volume AFFDITR-66-80, pp. 577–602, Air Force Flight Dynamics Laboratory, Wright-Patterson Air Force Base, OH, 1965.
71. L.R. Herrmann. Finite element bending analysis of plates. *Proc. Am. Soc. Civ. Eng.*, **94**(EM5), 13–25, 1968.
72. W. Visser. A refined mixed type plate bending element. *Journal of AIAA*, **7**, 1801–3, 1969.
73. J.C. Boot. On a problem arising from the derivation of finite element matrices using Reissner's principle. *Int. J. Num. Meth. Eng.*, **12**, 1879–82, 1978.
74. A. Chaterjee and A.V. Setlur. A mixed finite element formulation for plate problems. *Int. J. Num. Meth. Eng.*, **4**, 67–84, 1972.
75. J.W. Harvey and S. Kelsey. Triangular plate bending elements with enforced compatibility. *Journal of AIAA*, **9**, 1023–6, 1971.
76. B. Fraeijs de Veubeke and O.C. Zienkiewicz. Strain energy bounds in finite element analysis by slab analogy. *J. Strain Anal*, **2**, 265–71, 1967.
77. B. Fraeijs de Veubeke. An equilibrium model for plate bending. *International Journal of Solids and Structures*, **4**, 447–68, 1968.
78. J. Bron and G. Dhatt. Mixed quadrilateral elements for plate bending. *Journal of AIAA*, **10**, 1359–61, 1972.
79. K. Hellan. Analysis of elastic plates in flexure by a simplified finite element method. Technical Report C146, Acta Polytechnica Scandinavica, Trondheim, 1967.

80. T.H.H. Pian. Derivation of element stiffness matrices by assumed stress distribution. *Journal of AIAA*, **2**, 1332–6, 1964.
81. T.H.H. Pian and R Tong. Basis of finite element methods for solid continua. *Int. J. Num. Meth. Eng.*, **1**, 3–28, 1969.
82. R.J. Allwood and G.M.M. Cornes. A polygonal finite element for plate bending problems using the assumed stress approach. *Int. J. Num. Meth. Eng.*, **1**, 135–60, 1969.
83. B.E. Greene, R.E. Jones, R.M. McLay and D.R. Strome. Generalized variational principles in the finite element method. *Journal of AIAA*, **7**, 1254–60, 1969.
84. P. Tong. New displacement hybrid models for solid continua. *Int. J. Num. Meth. Eng.*, **2**, 73–83, 1970.
85. B.K. Neale, R.D. Henshell and G. Edwards. Hybrid plate bending elements. *J. Sound Vibr.*, **22**, 101–12, 1972.
86. R.D. Cook. Two hybrid elements for analysis of thick, thin and sandwich plates. *Int. J. Num. Meth. Eng.*, **5**, 277–99, 1972.
87. C. Johnson. On the convergence of a mixed finite element method for plate bending problems. *Num. Math.*, **21**, 43–62, 1973.
88. R.D. Cook and S.G. Ladkany. Observations regarding assumed-stress hybrid plate elements. *Int. J. Num. Meth. Eng.*, **8**, 513–20, 1974.
89. I. Torbe and K. Church. A general quadrilateral plate element. *Int. J. Num. Meth. Eng.*, **9**, 855–68, 1975.
90. D.J. Allman. A simple cubic displacement model for plate bending. *Int. J. Num. Meth. Eng.*, **10**, 263–81, 1976.
91. J. Jirousek. Improvement of computational efficiency of the 9 degrees of freedom triangular hybrid-Trefftz plate bending element (letter to editor), *Int. J. Num. Meth. Eng.*, **23**, 2167–8, 1986.
92. G.A. Wempner, J.T. Oden and D.K. Cross. Finite element analysis of thin shells. *Proc. Am. Soc. Civ. Eng.*, **94**(EM6), 1273–94, 1968.
93. G. Dhatt. An efficient triangular shell element. *Journal of AIAA*, **8**, 2100–2, 1970.
94. J.L. Batoz and G. Dhatt. Development of two simple shell elements. *Journal of AIAA*, **10**, 237–8, 1972.
95. J.T. Baldwin, A. Razzaque and B.M. Irons. Shape function subroutine for an isoparametric thin plate element. *Int. J. Num. Meth. Eng.*, **7**, 431–40, 1973.
96. B.M. Irons. The semi-Loof shell element. In D.G. Ashwell and R.H. Gallagher (eds), *Finite Elements for Thin Shells and Curved Members*, pp. 197–222. John Wiley, Chichester, Sussex, 1976.
97. L.P.R. Lyons. *A General Finite Element System with Special Analysis of Cellular Structures*. PhD thesis, Imperial College of Science and Technology, London, 1977.
98. R.A.F. Martins and D.R.J. Owen. Thin plate semi-Loof element for structural analysis including stabililty and structural vibration. *Int. J. Num. Meth. Eng.*, **12**, 1667–76, 1978.
99. J.L. Batoz. An explicit formulation for an efficient triangular plate bending element. *Int. J. Num. Meth. Eng.*, **18**, 1077–89, 1982.
100. M.A. Crisfield. A new model thin plate bending element using shear constraints: a modified version of Lyons' element. *Comp. Meth. Appl. Mech. Eng.*, **38**, 93–120, 1983.
101. M.A. Crisfield. A qualitative Mindlin element using shear constraints. *Computers and Structures*, **18**, 833–52, 1984.
102. M.A. Crisfield. *Finite Elements and Solution Procedures for Structural Analysis, Volume 1, Linear Analysis*, Pineridge Press, Swansea, 1986.
103. G. Dhatt, L. Marcotte and Y. Matte. A new triangular discrete Kirchholl plate-shell element. *Int. J. Num. Meth. Eng.*, **23**, 453–70, 1986.
104. R.A. Nay and S. Utku. An alternative for the finite element method. *Variational Methods in Engineering*, 1, 1972.

105. E. Oñate and F. Zárate. Rotation-free triangular plate and shell elements. *Int. J. Num. Meth. Eng.*, **47**, 557–603, 2000.
106. M. Abramowitz and I.A. Stegun (eds), *Handbook of Mathematical Functions*, Dover Publications, New York, 1965.
107. G.S. Shapiro. On yield surfaces for ideal plastic shells. In *Problems of Continuum Mechanics*, pp. 414–8, SIAM, Philadelphia, PA, 1961.

5

'Thick' Reissner–Mindlin plates – irreducible and mixed formulations

5.1 Introduction

We have already introduced in Chapter 4 the full theory of thick plates from which the thin plate, Kirchhoff, theory arises as the limiting case. In this chapter we shall show how the numerical solution of thick plates can easily be achieved and how, in the limit, an alternative procedure for solving all problems of Chapter 4 appears.

To ensure continuity we repeat below the governing equations [Eqs (4.13)–(4.18), or Eqs (4.87)–(4.90)]. Referring to Fig. 4.3 of Chapter 4 and the text for definitions, we remark that all the equations could equally well be derived from full three-dimensional analysis of a flat and relatively thin portion of an elastic continuum illustrated in Fig. 5.1. All that it is now necessary to do is to assume that whatever form of the approximating shape functions in the xy plane those in the z direction are only linear. Further, it is assumed that σ_z stress is zero, thus eliminating the effect of vertical strain.* The first approximations of this type were introduced quite early[1,2] and the elements then derived are exactly of the Reissner–Mindlin type discussed in Chapter 4.

The equations from which we shall start and on which we shall base all subsequent discussion are thus

$$\mathbf{M} - \mathbf{D}\mathbf{L}\boldsymbol{\theta} = \mathbf{0} \tag{5.1}$$

[see Eqs (4.13) and (4.87)],

$$\mathbf{L}^{\mathrm{T}}\mathbf{M} + \mathbf{S} = \mathbf{0} \tag{5.2}$$

[see Eqs (4.18) and (4.89)],

$$\frac{1}{\alpha}\mathbf{S} + \boldsymbol{\theta} - \boldsymbol{\nabla} w = \mathbf{0} \tag{5.3}$$

where $\alpha = \kappa G t$ is the shear rigidity [see Eqs (4.15) and (4.88)] and

$$\boldsymbol{\nabla}^{\mathrm{T}}\mathbf{S} + q = 0 \tag{5.4}$$

* Reissner includes the effect of σ_z in bending but, for simplicity, this is disregarded here.

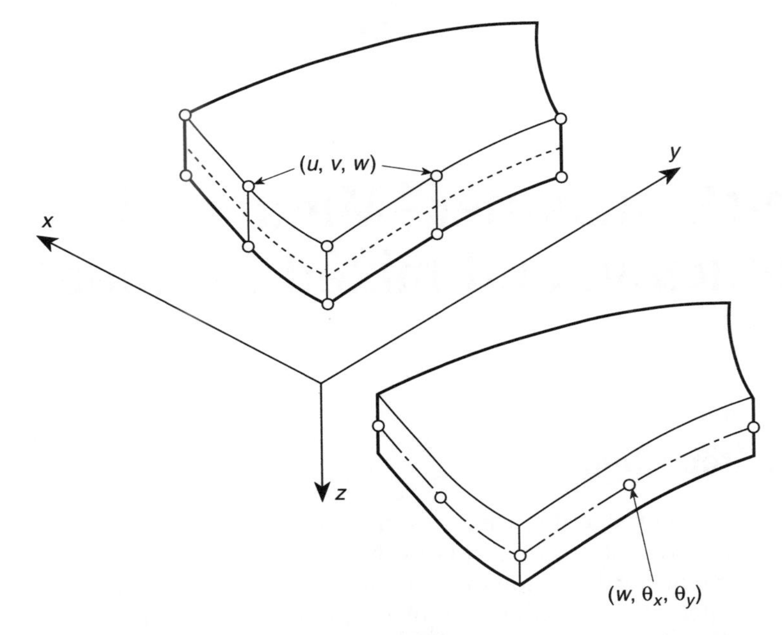

Fig. 5.1 An isoparametric three-dimensional element with linear interpolation in the transverse (thickness) direction and the 'thick' plate element.

[see Eqs (4.17) and (4.90)]. In the above the moments **M**, the transverse shear forces **S**, and the elastic matrices **D** are as defined in Chapter 4, and

$$\mathbf{L} = \begin{bmatrix} \dfrac{\partial}{\partial x} & 0 \\ 0 & \dfrac{\partial}{\partial y} \\ \dfrac{\partial}{\partial y} & \dfrac{\partial}{\partial x} \end{bmatrix} \tag{5.5}$$

defines the strain-displacement operator on rotations $\boldsymbol{\theta}$, and its transpose the equilibrium operator on moments, **M**. Boundary conditions are of course imposed on w and $\boldsymbol{\theta}$ or the corresponding plate forces S_n, M_n, M_{ns} in the manner discussed in Sec. 4.2.2.

It is convenient to eliminate **M** from Eqs (5.1)–(5.4) and write the system of three equations [Eqs (4.91)–(4.93)] as

$$\mathbf{L}^{\mathrm{T}}\mathbf{D}\mathbf{L}\boldsymbol{\theta} + \mathbf{S} = \mathbf{0} \tag{5.6}$$

$$\frac{1}{\alpha}\mathbf{S} + \boldsymbol{\theta} - \boldsymbol{\nabla} w = \mathbf{0} \tag{5.7}$$

$$\boldsymbol{\nabla}^{\mathrm{T}}\mathbf{S} + q = 0 \tag{5.8}$$

This equation system can serve as the basis on which a mixed discretization is built – or alternatively can be reduced further to yield an irreducible form. In Chapter 4 we have dealt with the irreducible form which is given by a fourth-order equation in terms of w alone and which could only serve for solution of *thin plate* problems, that is, when $\alpha = \infty$ [Eq. (4.21)]. On the other hand, it is easy to derive an alternative irreducible form which is valid only if $\alpha \neq \infty$. Now the shear forces can be eliminated yielding two equations;

$$\mathbf{L}^{\mathrm{T}}\mathbf{D}\mathbf{L}\boldsymbol{\theta} + \alpha(\boldsymbol{\nabla} w - \boldsymbol{\theta}) = \mathbf{0} \tag{5.9}$$

$$\boldsymbol{\nabla}^{\mathrm{T}}[\alpha(\boldsymbol{\nabla} w - \boldsymbol{\theta})] + q = 0 \tag{5.10}$$

This is an *irreducible* system corresponding to minimization of the total potential energy

$$\Pi = \frac{1}{2}\int_{\Omega}(\mathbf{L}\boldsymbol{\theta})^{\mathrm{T}}\mathbf{D}\mathbf{L}\boldsymbol{\theta}\,\mathrm{d}\Omega + \frac{1}{2}\int_{\Omega}(\boldsymbol{\nabla} w - \boldsymbol{\theta})^{\mathrm{T}}\alpha(\boldsymbol{\nabla} w - \boldsymbol{\theta})\,\mathrm{d}\Omega$$
$$- \int_{\Omega} wq\,\mathrm{d}\Omega + \Pi_{\mathrm{bt}} = \text{minimum} \tag{5.11}$$

as can easily be verified. In the above the first term is simply the bending energy and the second the shear distortion energy [see Eq. (4.103)].

Clearly, this irreducible system is only possible when $\alpha \neq \infty$, but it can, obviously, be interpreted as a solution of the potential energy given by Eq. (4.103) for 'thin' plates with the constraint of Eq. (4.104) being imposed in a *penalty manner* with α being now a penalty parameter. Thus, as indeed is physically evident, the thin plate formulation is simply a limiting case of such analysis.

We shall see that the penalty form can yield a satisfactory solution only when discretization of the corresponding mixed formulation satisfies the necessary convergence criteria.

The thick plate form now permits independent specification of three conditions at each point of the boundary. The options which exist are:

$$\begin{array}{ccc} w & \text{or} & S_n \\ \theta_n & \text{or} & M_n \\ \theta_s & \text{or} & M_{ns} \end{array}$$

in which the subscript n refers to a normal direction to the boundary and s a tangential direction. Clearly, now there are many combinations of possible boundary conditions.

A 'fixed' or 'clamped' situation exists when all three conditions are given by displacement components, which are generally zero, as

$$w = \theta_n = \theta_s = 0$$

and a free boundary when all conditions are the 'resultant' components

$$S_n = M_n = M_{ns} = 0$$

When we discuss the so-called simply supported conditions (see Sec. 4.2.2), we shall usually refer to the specification

$$w = 0 \qquad \text{and} \qquad M_n = M_{ns} = 0$$

as a 'soft' support (and indeed the most realistic support) and to

$$w = 0 \qquad M_n = 0 \qquad \text{and} \qquad \theta_s = 0$$

as a 'hard' support. The latter in fact replicates the thin plate assumptions and, incidentally, leads to some of the difficulties associated with it.

Finally, there is an important difference between thin and thick plates when 'point' loads are involved. In the thin plate case the displacement w remains finite at locations where a point load is applied; however, for thick plates the presence of shearing deformation leads to an infinite displacement (as indeed three-dimensional elasticity theory also predicts). In finite element approximations one always predicts a finite displacement at point locations with the magnitude increasing without limit as a mesh is refined near the loads. Thus, it is meaningless to compare the deflections at point load locations for different element formulations and we will not do so in this chapter. It is, however, possible to compare the total strain energy for such situations and here we immediately observe that for cases in which a single point load is involved the displacement provides a direct measure for this quantity.

5.2 The irreducible formulation – reduced integration

The procedures for discretizing Eqs (5.9) and (5.10) are straightforward. First, the two displacement variables are approximated by appropriate shape functions and parameters as

$$\boldsymbol{\theta} = \mathbf{N}_\theta \tilde{\boldsymbol{\theta}} \qquad \text{and} \qquad w = \mathbf{N}_w \tilde{\mathbf{w}} \tag{5.12}$$

We recall that the rotation parameters $\boldsymbol{\theta}$ may be transformed into physical rotations about the coordinate axes, $\hat{\boldsymbol{\theta}}$, [see Fig. 4.7], using

$$\hat{\boldsymbol{\theta}} = \mathbf{T}\boldsymbol{\theta} \quad \text{where} \quad \mathbf{T} = \begin{bmatrix} 0 & 1 \\ -1 & 0 \end{bmatrix} \tag{5.13}$$

These are often more convenient for calculations and are essential in shell developments. The approximation equations now are obtained directly by the use of the total potential energy principle [Eq. (5.11)], the Galerkin process on the weak form, or by the use of virtual work expressions. Here we note that the appropriate generalized strain components, corresponding to the moments $\mathbf{M}$ and shear forces $\mathbf{S}$, are

$$\boldsymbol{\varepsilon}_m = \mathbf{L}\boldsymbol{\theta} = (\mathbf{L}\mathbf{N}_\theta)\tilde{\boldsymbol{\theta}} \tag{5.14}$$

and

$$\boldsymbol{\varepsilon}_s = \boldsymbol{\nabla} w - \boldsymbol{\theta} = \boldsymbol{\nabla}\mathbf{N}_w \tilde{\mathbf{w}} - \mathbf{N}_\theta \tilde{\boldsymbol{\theta}} \tag{5.15}$$

We thus obtain the discretized problem

$$\left(\int_\Omega (\mathbf{L}\mathbf{N}_\theta)^{\mathrm{T}} \mathbf{D} \mathbf{L}\mathbf{N}_\theta \,\mathrm{d}\Omega + \int_\Omega \mathbf{N}_\theta^{\mathrm{T}} \alpha \mathbf{N}_\theta \,\mathrm{d}\Omega\right)\tilde{\boldsymbol{\theta}} - \left(\int_\Omega \mathbf{N}_\theta^{\mathrm{T}} \alpha \boldsymbol{\nabla}\mathbf{N}_w \,\mathrm{d}\Omega\right)\tilde{\mathbf{w}} = \mathbf{f}_\theta \tag{5.16}$$

and

$$-\left(\int_\Omega (\boldsymbol{\nabla}\mathbf{N}_w)^{\mathrm{T}}\alpha\mathbf{N}_\theta \,\mathrm{d}\Omega\right)\tilde{\boldsymbol{\theta}} + \left(\int_\Omega (\boldsymbol{\nabla}\mathbf{N}_w)^{\mathrm{T}}\alpha\,\boldsymbol{\nabla}\mathbf{N}_w \,\mathrm{d}\Omega\right)\tilde{\mathbf{w}} = \mathbf{f}_w \tag{5.17}$$

or simply

$$\begin{bmatrix} \mathbf{K}_{ww} & \mathbf{K}_{w\theta} \\ \mathbf{K}_{\theta w} & \mathbf{K}_{\theta\theta} \end{bmatrix} \begin{Bmatrix} \tilde{\mathbf{w}} \\ \tilde{\boldsymbol{\theta}} \end{Bmatrix} = \mathbf{K}\mathbf{a} = (\mathbf{K}_b + \mathbf{K}_s)\mathbf{a} = \begin{Bmatrix} \mathbf{f}_w \\ \mathbf{f}_\theta \end{Bmatrix} = \mathbf{f} \tag{5.18}$$

with

$$\mathbf{a}^{\mathrm{T}} = [\tilde{\mathbf{w}},\ \tilde{\boldsymbol{\theta}}] \qquad \tilde{\boldsymbol{\theta}}^{\mathrm{T}} = [\tilde{\boldsymbol{\theta}}_x,\ \tilde{\boldsymbol{\theta}}_y,]$$

$$\mathbf{K}_b = \begin{bmatrix} \mathbf{0} & \mathbf{0} \\ \mathbf{0} & \mathbf{K}^b_{\theta\theta} \end{bmatrix}$$

$$\mathbf{K}_s = \begin{bmatrix} \mathbf{K}^s_{ww} & \mathbf{K}^s_{w\theta} \\ \mathbf{K}^s_{\theta w} & \mathbf{K}^s_{\theta\theta} \end{bmatrix}$$

where the arrays are defined by

$$\begin{aligned} \mathbf{K}^s_{ww} &= \int_\Omega (\boldsymbol{\nabla}\mathbf{N}_w)^{\mathrm{T}}\alpha\,\boldsymbol{\nabla}\mathbf{N}_w \,\mathrm{d}\Omega \\ \mathbf{K}^s_{\theta w} &= -\int_\Omega \mathbf{N}^{\mathrm{T}}_\theta \alpha\,\boldsymbol{\nabla}\mathbf{N}_w \,\mathrm{d}\Omega = (\mathbf{K}^s_{w\theta})^{\mathrm{T}} \\ \mathbf{K}^s_{\theta\theta} &= \int_\Omega \mathbf{N}^{\mathrm{T}}_\theta \alpha \mathbf{N}_\theta \,\mathrm{d}\Omega \\ \mathbf{K}^b_{\theta\theta} &= \int_\Omega (\mathbf{L}\mathbf{N}_\theta)^{\mathrm{T}}\mathbf{D}\,\mathbf{L}\mathbf{N}_\theta \,\mathrm{d}\Omega \end{aligned} \tag{5.19}$$

and forces are given by

$$\begin{aligned} \mathbf{f}_w &= \int_\Omega \mathbf{N}^{\mathrm{T}}_w q \,\mathrm{d}\Omega + \int_{\Gamma_s} \mathbf{N}^{\mathrm{T}}_w \bar{S}_n \,\mathrm{d}\Gamma \\ \mathbf{f}_\theta &= \int_{\Gamma_m} \mathbf{N}^{\mathrm{T}}_\theta \bar{\mathbf{M}} \,\mathrm{d}\Gamma \end{aligned} \tag{5.20}$$

where $\bar{S}_n$ is the prescribed shear on boundary Γ_s, and $\bar{\mathbf{M}}$ is the prescribed moment on boundary Γ_m.

The formulation is straightforward and there is little to be said about it *a priori*. Since the form contains only first derivatives apparently any C_0 shape functions of a two-dimensional kind can be used to interpolate the two rotations and the lateral displacement. Figure 5.2 shows some rectangular (or with isoparametric distortion, quadrilateral) elements used in the early work.[1–3] All should, in principle, be convergent as C_0 continuity exists and constant strain states are available. In Fig. 5.3 we show what in fact happens with a fairly fine subdivision of quadratic serendipity and lagrangian rectangles as the ratio of thickness to span, t/L, varies.

We note that the magnitude of the coefficient α is best measured by the ratio of the bending to shear rigidities and we could assess its value in a non-dimensional form.

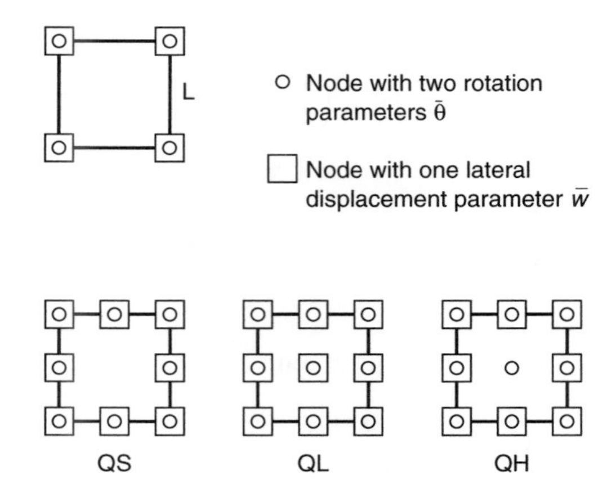

Fig. 5.2 Some early thick plate elements.

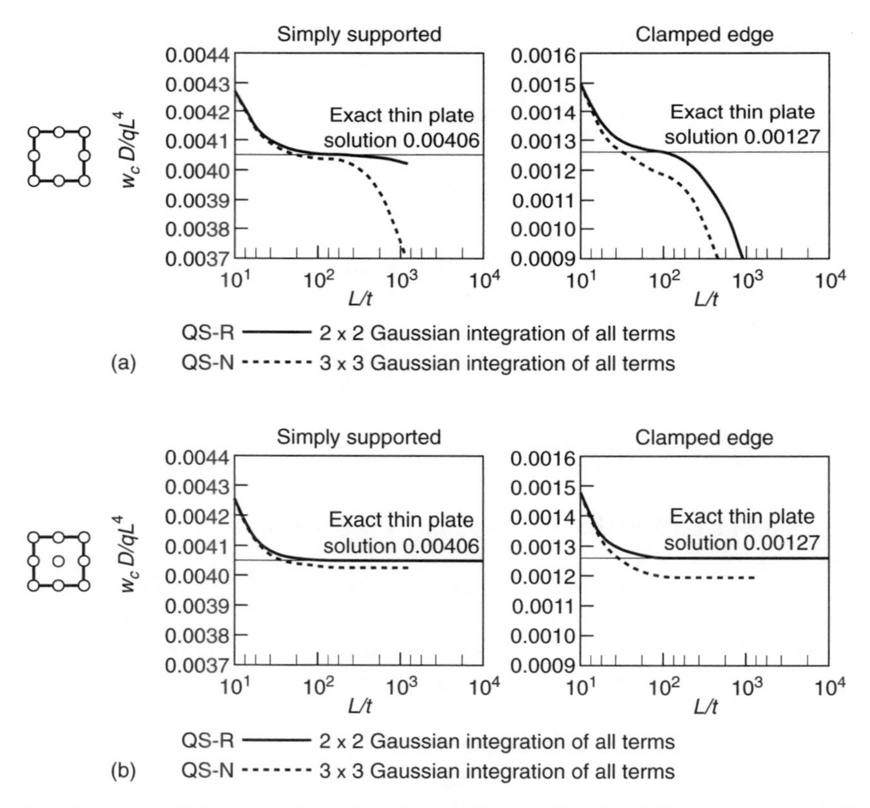

Fig. 5.3 Performance of (a) quadratic serendipity (QS) and (b) Lagrangian (QL) elements with varying span-to-thickness L/t, ratios, uniform load on a square plate with 4×4 normal subdivisions in a quarter. R is reduced 2×2 quadrature and N is normal 3×3 quadrature.

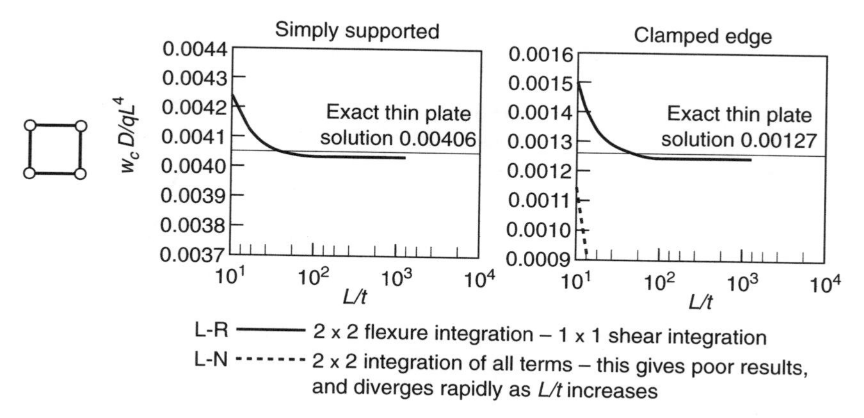

Fig. 5.4 Performance of bilinear elements with varying span-to-thickness, L/t, values.

Thus, for an isotropic material with $\alpha = Gt$ this ratio becomes

$$\frac{12(1-\nu^2)GtL^2}{Et^3} \propto \left(\frac{L}{t}\right)^2 \tag{5.21}$$

Obviously, 'thick' and 'thin' behaviour therefore depends on the L/t ratio.

It is immediately evident from Fig. 5.3 that, while the answers are quite good for smaller L/t ratios, the serendipity quadratic fully integrated elements (QS) rapidly depart from the thin plate solution, and in fact tend to zero results (locking) when this ratio becomes large. For lagrangian quadratics (QL) the answers are better, but again as the plate tends to be thin they are on the small side.

The reason for this 'locking' performance is similar to the one we considered for the nearly incompressible problem in Chapters 11 and 12 of Volume 1. In the case of plates the shear constraint implied by Eq. (5.7), and used to eliminate the shear resultant, is too strong if the terms in which this is involved are fully integrated. Indeed, we see that the effect is more pronounced in the serendipity element than in the lagrangian one. In early work the problem was thus mitigated by using a *reduced* quadrature, either on all terms, which we label R in the figure,[4,5] or only on the offending shear terms selectively[6,7] (labelled S). The dramatic improvement in results is immediately noted.

The same improvement in results is observed for linear quadrilaterals in which the full (exact) integration gives results that are totally unacceptable (as shown in Fig. 5.4), but where a reduced integration on the shear terms (single point) gives excellent performance,[8] although a carefull assessment of the element stiffness shows it to be rank deficient in an 'hourglass' mode in transverse displacements. (Reduced integration on all terms gives additional matrix singularity.)

A remedy thus has been suggested; however, it is not universal. We note in Fig. 5.3 that even without reduction of integration order, lagrangian elements perform better in the quadratic expansion. In cubic elements (Fig. 5.5), however, we note that (a) almost no change occurs when integration is 'reduced' and (b), again, lagrangian-type elements perform very much better.

Many heuristic arguments have been advanced for devising better elements,[9–12] all making use of reduced integration concepts. Some of these perform quite well, for

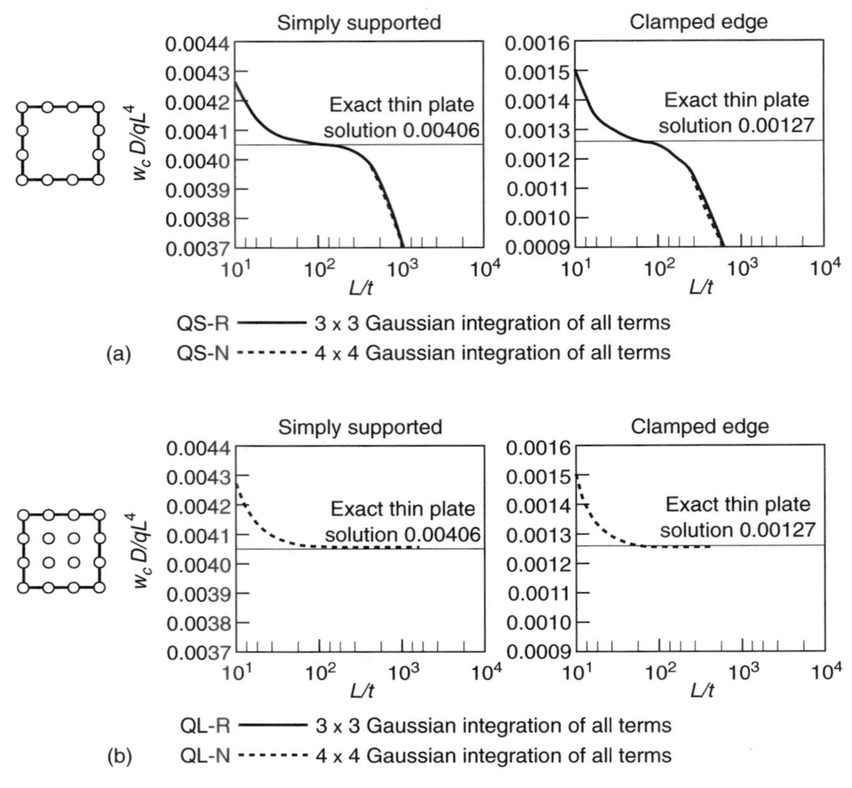

Fig. 5.5 Performance of cubic quadrilaterals: (a) serendipity (QS) and (b) lagrangian (QL) with varying span-to-thickness, L/t, values.

example the so-called 'heterosis' element of Hughes and Cohen[9] illustrated in Fig. 5.3 (in which the serendipity type interpolation is used on w and a lagrangian one on $\boldsymbol{\theta}$), but all of the elements suggested in that era fail on some occasions, either locking or exhibiting singular behaviour. Thus such elements are not 'robust' and should not be used universally.

A better explanation of their failure is needed and hence an understanding of how such elements could be designed. In the next section we shall address this problem by considering a mixed formulation. The reader will recognize here arguments used in Volume 1 which led to a better understanding of the failure of some straightforward elasticity elements as incompressible behaviour was approached. The situation is completely parallel here.

5.3 Mixed formulation for thick plates

5.3.1 The approximation

The problem of thick plates can, of course, be solved as a mixed one starting from Eqs (5.6)–(5.8) and approximating directly each of the variables $\boldsymbol{\theta}$, $\mathbf{S}$ and w

independently. Using Eqs (5.6)–(5.8), we construct a weak form as

$$\begin{aligned}
\int_\Omega \delta w \left[\boldsymbol{\nabla}^{\mathrm{T}}\mathbf{S} + q\right] \mathrm{d}\Omega = 0 \\
\int_\Omega \delta\boldsymbol{\theta}^{\mathrm{T}} \left[\mathbf{L}^{\mathrm{T}}\mathbf{D}\mathbf{L}\boldsymbol{\theta} + \mathbf{S}\right] \mathrm{d}\Omega = 0 \\
\int_\Omega \delta\mathbf{S}^{\mathrm{T}} \left[\frac{1}{\alpha}\mathbf{S} + \boldsymbol{\theta} - \boldsymbol{\nabla} w\right] \mathrm{d}\Omega = 0
\end{aligned} \tag{5.22}$$

We now write the independent approximations, using the standard Galerkin procedure, as

$$\begin{aligned}
\mathbf{w} = \mathbf{N}_w \tilde{\mathbf{w}} \qquad & \boldsymbol{\theta} = \mathbf{N}_\theta \tilde{\boldsymbol{\theta}} \qquad \text{and} \qquad \mathbf{S} = \mathbf{N}_s \tilde{\mathbf{S}} \\
\delta\mathbf{w} = \mathbf{N}_w \delta\tilde{\mathbf{w}} \qquad & \delta\boldsymbol{\theta} = \mathbf{N}_\theta \delta\tilde{\boldsymbol{\theta}} \qquad \text{and} \qquad \delta\mathbf{S} = \mathbf{N}_s \delta\tilde{\mathbf{S}}
\end{aligned} \tag{5.23}$$

though, of course, other interpolation forms can be used, as we shall note later.

After appropriate integrations by parts of Eq. (5.22), we obtain the discrete symmetric equation system (changing some signs to obtain symmetry)

$$\begin{bmatrix} \mathbf{0} & \mathbf{0} & \mathbf{E} \\ \mathbf{0} & \mathbf{K}_b & \mathbf{C} \\ \mathbf{E}^{\mathrm{T}} & \mathbf{C}^{\mathrm{T}} & \mathbf{H} \end{bmatrix} \begin{Bmatrix} \tilde{\mathbf{w}} \\ \tilde{\boldsymbol{\theta}} \\ \tilde{\mathbf{S}} \end{Bmatrix} = \begin{Bmatrix} \mathbf{f}_w \\ \mathbf{f}_\theta \\ \mathbf{0} \end{Bmatrix} \tag{5.24}$$

where

$$\begin{aligned}
\mathbf{K}_b &= \int_\Omega (\mathbf{L}\mathbf{N}_\theta)^{\mathrm{T}} \mathbf{D} (\mathbf{L}\mathbf{N}_\theta)\, \mathrm{d}\Omega \\
\mathbf{E} &= \int_\Omega (\boldsymbol{\nabla}\mathbf{N}_w)^{\mathrm{T}} \mathbf{N}_s \, \mathrm{d}\Omega \\
\mathbf{C} &= -\int_\Omega \mathbf{N}_\theta^{\mathrm{T}} \mathbf{N}_s \, \mathrm{d}\Omega \\
\mathbf{H} &= -\int_\Omega \mathbf{N}_s^{\mathrm{T}} \frac{1}{\alpha} \mathbf{N}_s \, \mathrm{d}\Omega
\end{aligned} \tag{5.25}$$

and where $\mathbf{f}_w$ and $\mathbf{f}_\theta$ are as defined in Eq. (5.20).

The above represents a typical three-field mixed problem of the type discussed in Sec. 11.5.1 of Volume 1, which has to satisfy certain criteria for stability of approximation as the thin plate limit (which can now be solved exactly) is approached. For this limit we have

$$\alpha = \infty \qquad \text{and} \qquad \mathbf{H} = \mathbf{0} \tag{5.26}$$

In this limiting case it can readily be shown that *necessary* criteria of solution stability for any element assembly and boundary conditions are that

$$n_\theta + n_w \geqslant n_s \qquad \text{or} \qquad \alpha_{\mathrm{P}} \equiv \frac{n_\theta + n_w}{n_s} \geqslant 1 \tag{5.27}$$

and

$$n_s \geqslant n_w \qquad \text{or} \qquad \beta_{\mathrm{P}} \equiv \frac{n_s}{n_w} \geqslant 1 \tag{5.28}$$

where n_θ, n_s and n_w are the number of $\tilde{\boldsymbol{\theta}}$, $\tilde{\mathbf{S}}$ and $\tilde{\mathbf{w}}$ parameters in Eqs (5.23).

When the necessary count condition is not satisfied then the *equation system will be singular*. Equations (5.27) and (5.28) must be satisfied for the whole system but, in addition, they need to be satisfied for element patches if local instabilities and oscillations are to be avoided.[13–15]

The above criteria will, as we shall see later, help us to design suitable thick plate elements which show convergence to correct thin plate solutions.

5.3.2 Continuity requirements

The approximation of the form given in Eqs (5.24) and (5.25) implies certain continuities. It is immediately evident that C_0 continuity is needed for rotation shape functions $\mathbf{N}_\theta$ (as products of first derivatives are present in the approximation), but that either $\mathbf{N}_w$ or $\mathbf{N}_s$ can be discontinuous. In the form given in Eq. (5.25) a C_0 approximation for w is implied; however, after integration by parts a form for C_0 approximation of $\mathbf{S}$ results. Of course, physically only the component of $\mathbf{S}$ normal to boundaries should be continuous, as we noted also previously for moments in the mixed form discussed in Sec. 4.16.

In all the early approximations discussed in the previous section, C_0 continuity was assumed for both $\boldsymbol{\theta}$ and w variables, this being very easy to impose. We note that such continuity cannot be described as *excessive* (as no physical conditions are violated), but we shall show later that very successful elements also can be generated with discontinuous w interpolation (which is indeed not motivated by physical considerations).

For $\mathbf{S}$ it is obviously more convenient to use a completely discontinuous interpolation as then the shear can be eliminated at the element level and the final stiffness matrices written simply in standard $\tilde{\boldsymbol{\theta}}$, $\tilde{\mathbf{w}}$ terms for element boundary nodes. We shall show later that some formulations permit a limit case where α^{-1} is identically zero while others require it to be non-zero.

The continuous interpolation of the normal component of $\mathbf{S}$ is, as stated above, physically correct in the absence of line or point loads. However, with such interpolation, elimination of $\tilde{\mathbf{S}}$ is not possible and the retention of such additional system variables is usually too costly to be used in practice and has so far not been adopted. However, we should note that an iterative solution process applicable to mixed forms described in Sec. 11.6 of Volume 1 can reduce substantially the cost of such additional variables.[16]

5.3.3 Equivalence of mixed forms with discontinuous S interpolation and reduced (selective) integration

The equivalence of penalized mixed forms with discontinuous interpolation of the constraint variable and of the corresponding irreducible forms with the same penalty variable was demonstrated in Sec. 12.5 of Volume 1 following work of Malkus and Hughes for incompressible problems.[17] Indeed, an exactly analogous proof can be used for the present case, and we leave the details of this to the reader; however, below we summarize some equivalencies that result.

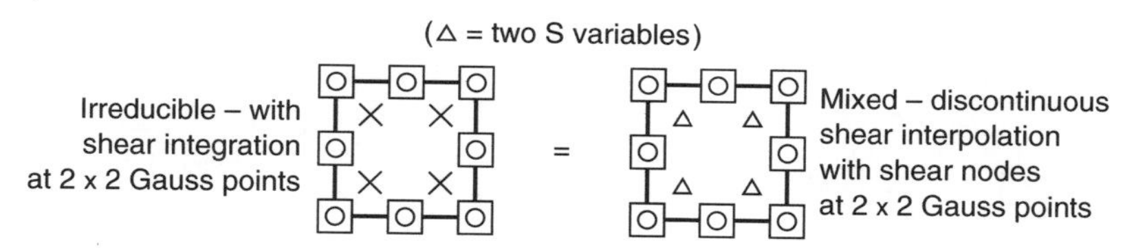

Fig. 5.6 Equivalence of mixed form and reduced shear integration in quadratic serendipity rectangle.

Thus, for instance, we consider a serendipity quadrilateral, shown in Fig. 5.6(a), in which integration of shear terms (involving α) is made at four Gauss points (i.e. 2×2 reduced quadrature) in an irreducible formulation [see Eqs (5.16)–(5.20)], we find that the answers are *identical* to a mixed form in which the **S** variables are given by a bilinear interpolation from nodes placed at the same Gauss points.

This result can also be argued from the limitation principle first given by Fraeijs de Veubeke.[18] This states that if the mixed form in which the stress is independently interpolated is precisely capable of reproducing the stress variation which is given in a corresponding irreducible form then the analysis results will be identical. It is clear that the four Gauss points at which the shear stress is sampled can only define a bilinear variation and thus the identity applies here.

The equivalence of reduced integration with the mixed discontinuous interpolation of **S** will be useful in our discussion to point out reasons why many elements mentioned in the previous section failed. However, in practice, it will be found equally convenient (and often more effective) to use the mixed interpolation explicitly and eliminate the **S** variables by element-level condensation rather than to use special integration rules. Moreover, in more general cases where the material properties lead to coupling between bending and shear response (e.g. elastic–plastic behaviour) use of selective reduced integration is not convenient. It must also be pointed out that the equivalence fails if α varies within an element or indeed if the isoparametric mapping implies different interpolations. In such cases the mixed procedures are generally more accurate.

5.4 The patch test for plate bending elements

5.4.1 Why elements fail

The nature and application of the patch test have changed considerably since its early introduction. As shown in references 13–15 and 19–23 (and indeed as discussed in Chapters 10–12 of Volume 1 in detail), this test can prove, in addition to *consistency* requirements (which were initially the only item tested), the *stability* of the approximation by requiring that for a patch consisting of an assembly of one or more elements the stiffness matrices are non-singular whatever the boundary conditions imposed.

To be absolutely sure of such non-singularity the test must, at the final stage, be performed numerically. However, we find that the 'count' conditions given in Eqs (5.27) and (5.28) are *necessary* for avoiding such non-singularity. Frequently, they also prove *sufficient* and make the numerical test only a final confirmation.[14,15]

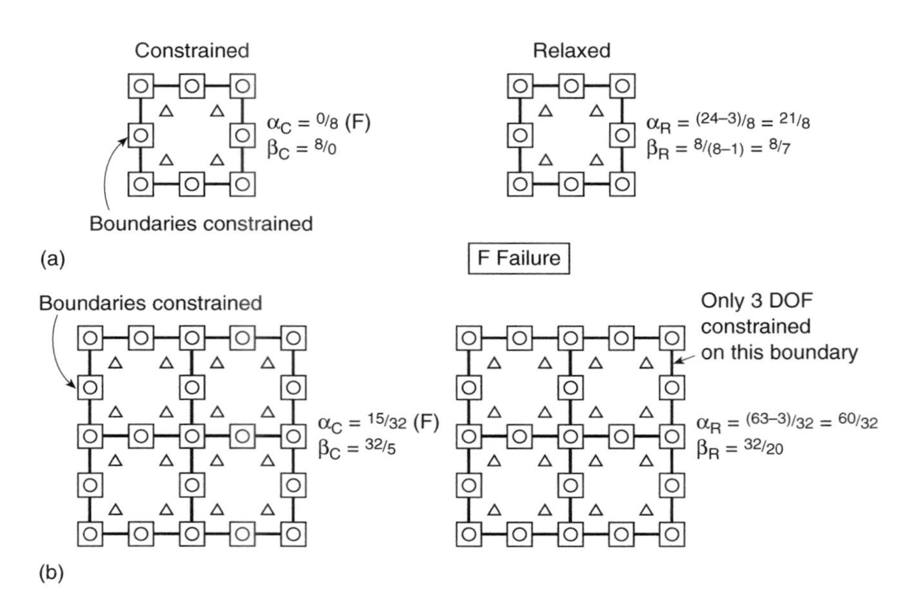

Fig. 5.7 'Constrained' and 'relaxed' patch test/count for serendipity (quadrilateral). (In the *C* test all boundary displacements are fixed. In the *R* test only three boundary displacements are fixed, eliminating rigid body modes.) (a) Single-element test; (b) four-element test.

We shall demonstrate how the simple application of such counts immediately indicates *which elements fail* and which have a chance of success. Indeed, it is easy to show why the original quadratic serendipity element with reduced integration (QS-R) is not robust.

In Fig. 5.7 we consider this element in a single-element and four-element patch subjected to so-called *constrained* boundary conditions, in which all displacements on the external boundary of the patch are prescribed and a *relaxed* boundary condition in which only three displacements (conveniently two θ's and one w) eliminate the rigid body modes. To ease the presentation of this figure, as well as in subsequent tests, we shall simply quote the values of α_P and β_P parameters as defined in Eqs (5.27) and (5.28) with subscript replaced by C or R to denote the constrained or relaxed tests, respectively. The symbol F will be given to any failure to satisfy the *necessary* condition. In the tests of Fig. 5.7 we note that both patch tests fail with the parameter α_C being less than 1, and hence the elements will lock under certain circumstances (or show a singularity in the evaluation of $\mathbf{S}$). A failure in the relaxed tests generally predicts a singularity in the final stiffness matrix of the assembly, and this is also where frequently computational failures have been observed.

As the mixed and reduced integration elements are identical in this case we see immediately why the element fails in the problem of Fig. 5.3 (more severely under clamped conditions). Indeed, it is clear why in general the performance of lagrangian-type elements is better as it adds further degrees of freedom to increase n_θ (and also n_w unless heterosis-type interpolation is used).[9]

In Table 5.1 we show a list of the α_P and β_P values for single and four element patches of various rectangles, and again we note that none of these satisfies completely the *necessary* requirements, and therefore none can be considered

Table 5.1 Quadrilateral mixed elements: patch count

Element	Reference	Single element patch				Four element patch			
		α_C	β_C	α_R	β_R	α_C	β_C	α_R	β_R
I									
	Q8S4[4]	$\frac{0}{8}$ (F)	$\frac{8}{0}$	$\frac{21}{8}$	$\frac{8}{7}$	$\frac{15}{32}$ (F)	$\frac{32}{5}$	$\frac{60}{32}$	$\frac{32}{20}$
	Q9S4[6,7]	$\frac{3}{8}$ (F)	$\frac{8}{1}$	$\frac{16}{8}$	$\frac{8}{8}$	$\frac{27}{32}$ (F)	$\frac{32}{9}$	$\frac{72}{32}$	$\frac{32}{20}$
	Q9HS4[9]	$\frac{2}{8}$ (F)	$\frac{8}{0}$	$\frac{15}{8}$	$\frac{8}{7}$	$\frac{23}{32}$ (F)	$\frac{32}{5}$	$\frac{68}{32}$	$\frac{32}{20}$
	Q12S9[4]	$\frac{0}{18}$ (F)	$\frac{18}{0}$	$\frac{23}{18}$	$\frac{18}{11}$	$\frac{27}{72}$ (F)	$\frac{72}{9}$	$\frac{96}{72}$	$\frac{72}{32}$
	Q16S9[25]	$\frac{12}{18}$ (F)	$\frac{18}{4}$	$\frac{45}{18}$	$\frac{18}{15}$	$\frac{75}{72}$	$\frac{72}{25}$	$\frac{150}{72}$	$\frac{72}{50}$
	Q4S1[4,7,8]	$\frac{0}{2}$ (F)	$\frac{2}{0}$	$\frac{9}{2}$	$\frac{2}{3}$ (F)	$\frac{3}{8}$ (F)	$\frac{8}{1}$	$\frac{24}{8}$	$\frac{8}{8}$
II									
	Q4S1B1[26] Q4S1B1L[26,27]	$\frac{2}{2}$	$\frac{2}{0}$	$\frac{11}{2}$	$\frac{2}{3}$ (F)	$\frac{11}{8}$	$\frac{8}{1}$	$\frac{32}{8}$	$\frac{8}{8}$
	Q4S2B2L[28]	$\frac{4}{4}$	$\frac{4}{0}$	$\frac{13}{4}$	$\frac{4}{4}$	$\frac{19}{16}$	$\frac{16}{1}$	$\frac{40}{16}$	$\frac{16}{8}$

[F] Failure to satisfy necessary conditions.

robust. However, it is interesting to note that the elements closest to satisfaction of the count perform best, and this explains why the heterosis elements[24] are quite successful and indeed why the lagrangian cubic is nearly robust and often is used with success.[25]

Of course, similar approximation and counts can be made for various triangular elements. We list some typical and obvious ones, together with patch test counts, in the first part of Table 5.2. Again, none perform adequately and all will result in locking and spurious modes in finite element applications.

We should note again that the failure of the patch test (with regard to stability) means that under some circumstances the element will fail. However, in many problems a reasonable performance can still be obtained and non-singularity observed in its performance, providing consistency is, of course, also satisfied.

Table 5.2 Triangular mixed elements: patch count

Element	Reference	Single element patch				Six element patch			
		α_C	β_C	α_R	β_R	α_C	β_C	α_R	β_R
I									
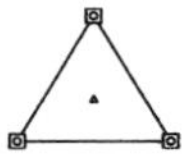	T3S1	$\frac{0}{2}$ (F)	$\frac{2}{0}$	$\frac{6}{2}$	$\frac{2}{2}$	$\frac{3}{12}$ (F)	$\frac{12}{1}$	$\frac{18}{12}$	$\frac{12}{6}$
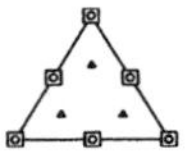	T6S3	$\frac{0}{6}$ (F)	$\frac{6}{0}$	$\frac{15}{6}$	$\frac{6}{5}$	$\frac{21}{36}$ (F)	$\frac{36}{7}$	$\frac{54}{36}$	$\frac{36}{18}$
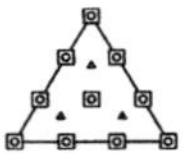	T10S3	$\frac{3}{6}$ (F)	$\frac{6}{6}$	$\frac{27}{6}$	$\frac{6}{9}$ (F)	$\frac{57}{36}$	$\frac{36}{19}$	$\frac{108}{36}$	$\frac{36}{36}$
II									
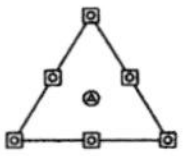	T6S1B1	$\frac{2}{2}$	$\frac{2}{0}$	$\frac{17}{2}$	$\frac{2}{5}$ (F)	$\frac{33}{12}$	$\frac{12}{7}$	$\frac{66}{12}$	$\frac{12}{18}$ (F)
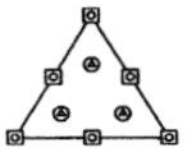	T6S3B3[15]	$\frac{6}{6}$	$\frac{6}{0}$	$\frac{21}{6}$	$\frac{6}{5}$	$\frac{75}{36}$	$\frac{36}{7}$	$\frac{108}{36}$	$\frac{36}{18}$
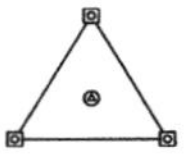	T3S1B1L[29]	$\frac{2}{2}$	$\frac{2}{0}$	$\frac{8}{2}$	$\frac{2}{2}$	$\frac{20}{12}$	$\frac{12}{6}$	$\frac{35}{12}$	$\frac{12}{11}$
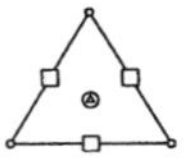	T3S1B1A[30]	$\frac{2}{2}$	$\frac{2}{0}$	$\frac{8}{2}$	$\frac{2}{2}$	$\frac{15}{12}$	$\frac{12}{1}$	$\frac{30}{12}$	$\frac{12}{6}$

F Failure to satisfy necessary conditions.

Numerical patch test

While the 'count' condition of Eqs (5.27) and (5.28) is a necessary one for stability of patches, on occasion singularity (and hence instability) can still arise even with its satisfaction. For this reason numerical tests should always be conducted ascertaining the rank sufficiency of the stiffness matrices and also testing the consistency condition.

In Chapter 10 of Volume 1 we discussed in detail the consistency test for irreducible forms in which a single variable set $\mathbf{u}$ occurred. It was found that with a second-order operator the discrete equations should satisfy *at least* the solution corresponding to a linear field $\mathbf{u}$ exactly, thus giving constant strains (or first derivatives) and stresses. For the mixed equation set [Eqs (5.6)–(5.8)] again the lowest-order exact solution that has to be satisfied corresponds to:

1. constant values of moments or $\mathbf{L}\boldsymbol{\theta}$ and hence a linear $\boldsymbol{\theta}$ field;
2. linear w field;
3. constant $\mathbf{S}$ field.

The exact solutions for which plate elements commonly are tested and where full satisfaction of nodal equations is required consist of:

1. arbitrary constant **M** fields and arbitrary linear $\boldsymbol{\theta}$ fields with zero shear forces ($\mathbf{S} = \mathbf{0}$); here a quadratic w form is assumed still yielding an exact finite element solution;
2. constant **S** and linear w fields yielding a constant $\boldsymbol{\theta}$ field. The solution requires a distributed couple on the right-hand side of Eq. (5.6) and this was not included in the original formulation. A simple procedure is to disregard the satisfaction of the moment equilibrium in this test. This may be done simply by inserting a very large value of the bending rigidity **D**.

5.4.2 Design of some useful elements

The simple patch count test indicates how elements could be designed to pass it, and thus avoid the singularity (instability). Equation (5.28) is always trivial to satisfy for elements in which **S** is interpolated independently in each element. In a single-element test it will be necessary to restrain at least one $\tilde{\mathbf{w}}$ degree-of-freedom to prevent rigid body translations. Thus, the *minimum* number of terms which can be included in **S** for each element is always one less than the number of $\tilde{\mathbf{w}}$ parameters in each element. As patches with more than one element are constructed the number of w parameters will increase proportionally with the number of nodes and the number of shear constraints increase by the number of elements. For both quadrilateral and triangular elements the requirement that $n_s \geqslant n_w - 1$ *for no boundary restraints* ensures that Eq. (5.28) is satisfied on all patches for both constrained and relaxed boundary conditions. Failure to satisfy this simple requirement explains clearly why certain of the elements in Tables 5.1 and 5.2 failed the single-element patch test for the relaxed boundary condition case.

Thus, a successful satisfaction of the count condition requires now only the consideration of Eq. (5.27). In the remainder of this chapter we will discuss two approaches which can successfully satisfy Eq. (5.27). The first is the use of discrete collocation constraints in which Eq. (5.7) is enforced at preselected points on the boundary and occasionally in the interior of elements. Boundary constraints are often 'shared' between two elements and thus reduce the rate at which n_s increases. The other approach is to introduce bubble or enhanced modes for the rotation parameters in the interior of elements. Here, for convenience, we refer to both as a 'bubble mode' approach. The inclusion of at least as many bubble modes as shear modes will automatically satisfy Eq. (5.27). This latter approach is similar to that used in Sec. 12.7 of Volume 1 to stabilize elements for solving the (nearly) incompressible problem and is a clear violation of 'intuition' since for the thin plate problem the rotations appear as derivatives of w. Its use in this case is justified by patch counts and performance.

5.5 Elements with discrete collocation constraints

5.5.1 General possibilities of discrete collocation constraints – quadrilaterals

The possibility of using conventional interpolation to achieve satisfactory performance of mixed-type elements is limited, as is apparent from the preceding discussion.

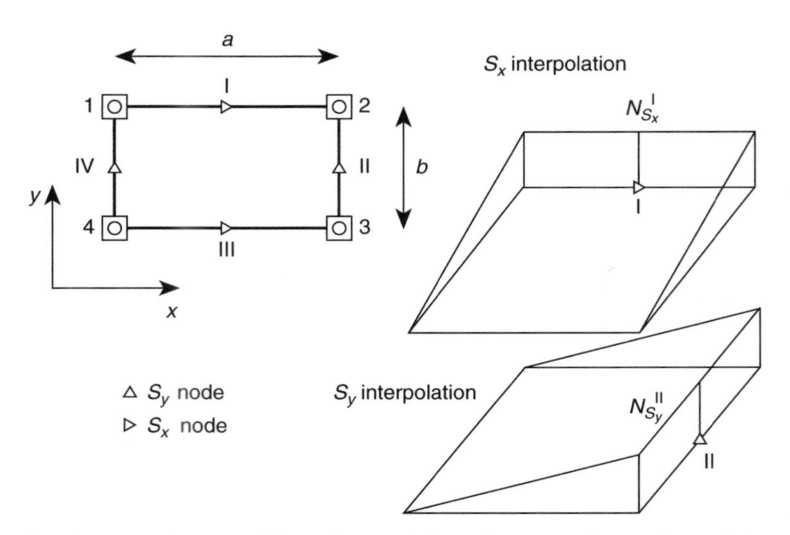

Fig. 5.8 Collocation constraints on a bilinear element: independent interpolation of S_x and S_y.

One feasible alternative is that of increasing the element order, and we have already observed that the cubic lagrangian interpolation nearly satisfies the stability requirement and often performs well.[2,7,25] However, the complexity of the formulation is formidable and this direction is not often used.

A different approach uses collocation constraints for the shear approximation [see Eq. (5.7)] on the element boundaries, thus limiting the number of **S** parameters and making the patch count more easily satisfied. This direction is indicated in the work of Hughes and Tezduyar,[31] Bathe and co-workers,[32,33] and Hinton and Huang,[34,35] as well as in generalizations by Zienkiewicz *et al.*,[36] and others.[37–44] The procedure bears a close relationship to the so-called DKT (discrete Kirchhoff theory) developed in Chapter 4 (see Sec. 4.18) and indeed explains why these, essentially thin plate, approximations are successful.

The key to the discrete formulation is evident if we consider Fig. 5.8, where a simple bilinear element is illustrated. We observe that with a C_0 interpolation of $\boldsymbol{\theta}$ and w, the shear strain

$$\gamma_x = \frac{\partial w}{\partial x} - \theta_x \tag{5.29}$$

is uniquely determined at any point of the side 1–2 (such as point I, for instance) and that hence [by Eq. (5.3)]

$$S_x = \alpha \gamma_x \tag{5.30}$$

is also uniquely determined there.

Thus, if a node specifying the shear resultant distribution were placed at that point and if the constraints [or satisfaction of Eq. (5.3)] were only imposed there, then

1. the nodal value of S_x would be shared by adjacent elements (assuming continuity of α);

2. the nodal values of S_x would be prescribed if the $\tilde{\boldsymbol{\theta}}$ and $\tilde{\mathbf{w}}$ values were constrained as they are in the constrained patch test.

Indeed if α, the shear rigidity, were to vary between adjacent elements the values of S_x would only differ by a multiplying constant and arguments remain essentially the same.

The prescription of the shear field in terms of such boundary values is simple. In the case illustrated in Fig. 5.8 we interpolate independently

$$S_x = \mathbf{N}_{sx} \tilde{\mathbf{S}}_x \qquad \text{and} \qquad S_y = \mathbf{N}_{sy} \tilde{\mathbf{S}}_y \tag{5.31}$$

using the shape functions

$$\begin{aligned} \mathbf{N}_{sx} &= \frac{1}{y_I - y_{III}} [y - y_{III}, \quad y_I - y] \\ \mathbf{N}_{sy} &= \frac{1}{x_{II} - x_{IV}} [x - x_{IV}, \quad x_{II} - x] \end{aligned} \tag{5.32}$$

as illustrated. Such an interpolation, of course, defines $\mathbf{N}_s$ of Eq. (5.23).

The introduction of the discrete constraint into the analysis is a little more involved. We can proceed by using different (Petrov–Galerkin) weighting functions, and in particular applying a Dirac delta weighting or point collocation to Eq. (5.3) in the approximate form. However, it is advantageous here to return to the constrained variational principle [see Eq. (4.103)] and seek stationarity of

$$\Pi = \frac{1}{2} \int_\Omega (\mathbf{L}\boldsymbol{\theta})^{\mathrm{T}} \mathbf{D} \mathbf{L} \boldsymbol{\theta} \, \mathrm{d}\Omega + \frac{1}{2} \int_\Omega \mathbf{S}^{\mathrm{T}} \frac{1}{\alpha} \mathbf{S} \, \mathrm{d}\Omega - \int_\Omega w q \, \mathrm{d}\Omega + \Pi_{\mathrm{bt}} = \text{stationary} \tag{5.33}$$

where the first term on the right-hand side denotes the *bending* and the second the *transverse shear* energy. In the above we again use the approximations

$$\begin{aligned} \boldsymbol{\theta} &= \mathbf{N}_\theta \tilde{\boldsymbol{\theta}} \qquad & w &= \mathbf{N}_w \tilde{\mathbf{w}} \\ \mathbf{S} &= \mathbf{N}_s \tilde{\mathbf{S}} \qquad & \mathbf{N}_s &= [\mathbf{N}_{sx}, \mathbf{N}_{sy}] \end{aligned} \tag{5.34}$$

subject to the constraint Eq. (5.3):

$$\mathbf{S} = \alpha(\boldsymbol{\nabla} w - \boldsymbol{\theta}) \tag{5.35}$$

being applied directly in a discrete manner, that is, by collocation at such points as I to IV in Fig. 5.8 and appropriate direction selection. We shall eliminate $\mathbf{S}$ from the computation but before proceeding with any details of the algebra it is interesting to observe the relation of the element of Fig. 5.8 to the patch test, noting that we still have a mixed problem requiring the count conditions to be satisfied. (This indeed is the element of references 32 and 33.) We show the counts on Fig. 5.9 and observe that although they fail in the four-element assembly the margin is small here (and for larger patches, counts are satisfactory).* The results given by this element are quite good, as will be shown in Sec. 5.9.

The discrete constraints and the boundary-type interpolation can of course be used in other forms. In Fig. 5.10 we illustrate the quadratic element of Huang and Hinton.[34,35] Here two points on each side of the quadrilateral define the shears S_x

* Reference 33 reports a mathematical study of stability for this element.

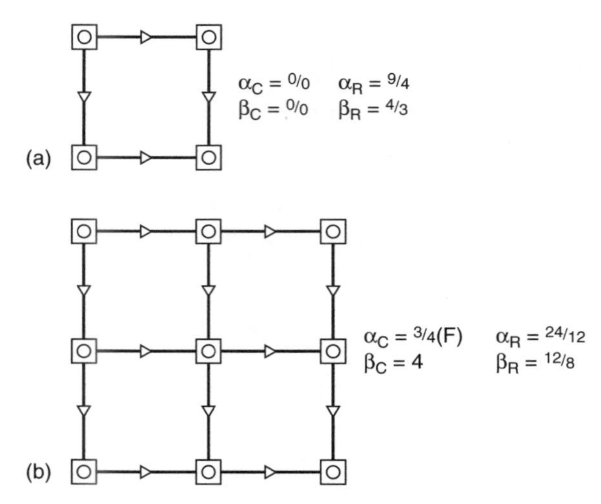

Fig. 5.9 Patch test on (a) one and (b) four elements of the type given in Fig 5.8. (Observe that in a constrained test boundary values of **S** are prescribed.)

and S_y but in addition four internal parameters are introduced as shown. Now both the boundary and internal 'nodes' are again used as collocation points for imposing the constraints.

The count for single-element and four-element patches is given in Table 5.3. This element only fails in a single-element patch under constrained conditions, and again numerical verification shows generally good performance. Details of numerical examples will be given later.

It is clear that with discrete constraints many more alternatives for design of satisfactory elements that pass the patch test are present. In Table 5.3 several

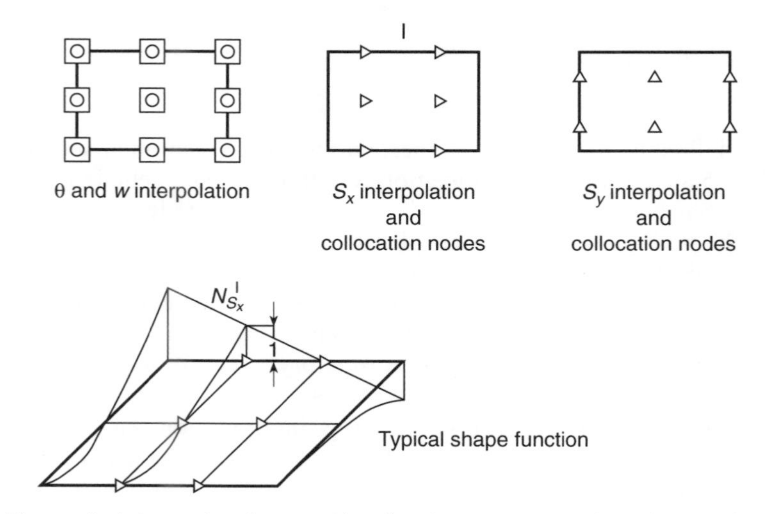

Fig. 5.10 The quadratic lagrangian element with collocation constraints on boundaries and in the internal domain.[34,35]

Table 5.3 Elements with collocation constraints: patch count. Degrees of freedom: □, $w-1$; ○, $\theta-2$; △, $S-1$; ∩, θ_n-1

Element	Reference	Single element patch				Four element patch			
		α_C	β_C	α_R	β_R	α_C	β_C	α_R	β_R
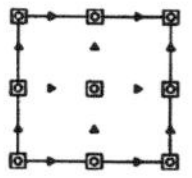	Q9D12[34,35]	$\frac{3}{4}$ (F)	$\frac{4}{1}$	$\frac{24}{12}$	$\frac{12}{8}$	$\frac{27}{24}$	$\frac{24}{9}$	$\frac{72}{40}$	$\frac{40}{24}$
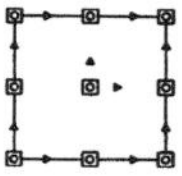	Q9D10	$\frac{3}{2}$	$\frac{2}{1}$	$\frac{24}{10}$	$\frac{10}{8}$	$\frac{27}{16}$	$\frac{16}{9}$	$\frac{72}{32}$	$\frac{32}{24}$
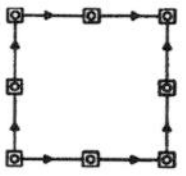	Q8D8	$\frac{0}{0}$	$\frac{0}{0}$	$\frac{21}{8}$	$\frac{8}{7}$	$\frac{15}{8}$	$\frac{8}{5}$	$\frac{60}{24}$	$\frac{24}{21}$
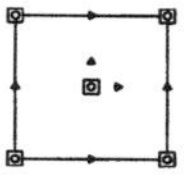	Q5D6	$\frac{3}{2}$	$\frac{2}{1}$	$\frac{12}{6}$	$\frac{6}{4}$	$\frac{15}{12}$	$\frac{12}{5}$	$\frac{36}{20}$	$\frac{20}{12}$
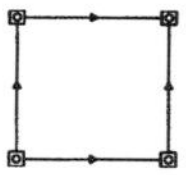	Q4D4[32,33]	$\frac{0}{0}$	$\frac{0}{0}$	$\frac{8}{4}$	$\frac{4}{3}$	$\frac{3}{4}$ (F)	$\frac{4}{1}$	$\frac{24}{12}$	$\frac{12}{8}$

Element	Reference	Single element patch				Six element patch			
		α_C	β_C	α_R	β_R	α_C	β_C	α_R	β_R
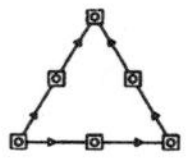	T6D6[36]	$\frac{0}{0}$	$\frac{0}{0}$	$\frac{15}{6}$	$\frac{6}{5}$	$\frac{21}{12}$	$\frac{12}{7}$	$\frac{43}{24}$	$\frac{24}{23}$
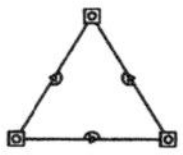	T3D3[36,44]	$\frac{0}{0}$	$\frac{0}{0}$	$\frac{9}{3}$	$\frac{3}{2}$	$\frac{9}{6}$	$\frac{6}{1}$	$\frac{45}{12}$	$\frac{12}{6}$

[F] Failure to satisfy necessary conditions.

quadrilaterals and triangles that satisfy the count conditions are illustrated. In the first a modification of the Hinton–Huang element with reduced internal shear constraints is shown (second element). Here biquadratic 'bubble functions' are used in the interior shear component interpolation, as shown in Fig. 5.11. Similar improvements in the count can be achieved by using a serendipity-type interpolation, but now, of course, the distorted performance of the element may be impaired (for reasons we discussed in Volume 1, Sec. 9.7). Addition of bubble functions on all the w and $\boldsymbol{\theta}$ parameters can, as shown, make the Bathe–Dvorkin fully satisfy the count condition. We shall pursue this further in Sec. 5.6.

All quadrilateral elements can, of course, be mapped isoparametrically, remembering of course that components of Shear S_ξ and S_η parallel to the ξ, η coordinates have to be used to ensure the preservation of the desirable constrained properties previously discussed. Such 'directional' shear interpolation is also essential when considering triangular elements, to which the next section is devoted. Before, doing this, however, we shall complete the algebraic derivation of element properties.

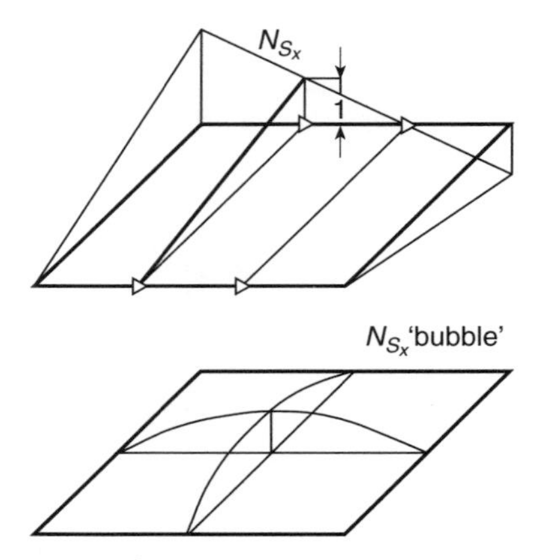

Fig. 5.11 A biquadratic hierarchical bubble for S_x.

5.5.2 Element matrices for discrete collocation constraints

The starting point here will be to use the variational principle given by Eq. (5.33) with the shear variables eliminated directly.

The application of the discrete constraints of Eq. (5.35) allows the 'nodal' parameters $\tilde{\mathbf{S}}$ defining the shear force distribution to be determined explicitly in terms of the $\tilde{\mathbf{w}}$ and $\tilde{\boldsymbol{\theta}}$ parameters. This gives in general terms

$$\tilde{\mathbf{S}} = \alpha\left[\mathbf{Q}_w\tilde{\mathbf{w}} + \mathbf{Q}_\theta\tilde{\boldsymbol{\theta}}\right] \tag{5.36}$$

in each element. For instance, for the rectangular element of Fig. 5.8 we can write

$$\begin{aligned} S_x^I &= \alpha\left[\frac{\tilde{w}_2 - \tilde{w}_1}{a} - \frac{\tilde{\theta}_{x1} + \tilde{\theta}_{x2}}{2}\right] \\ S_y^{II} &= \alpha\left[\frac{\tilde{w}_2 - \tilde{w}_3}{b} - \frac{\tilde{\theta}_{y2} + \tilde{\theta}_{y3}}{2}\right] \\ S_x^{III} &= \alpha\left[\frac{\tilde{w}_3 - \tilde{w}_4}{a} - \frac{\tilde{\theta}_{x3} + \tilde{\theta}_{x4}}{2}\right] \\ S_y^{IV} &= \alpha\left[\frac{\tilde{w}_1 - \tilde{w}_4}{b} - \frac{\tilde{\theta}_{y1} + \tilde{\theta}_{y4}}{2}\right] \end{aligned} \tag{5.37}$$

which can readily be rearranged into the form of Eq. (5.36) as

$$\mathbf{Q}_w = \frac{1}{ab}\begin{bmatrix} -b & b & 0 & 0 \\ 0 & a & -a & 0 \\ 0 & 0 & b & -b \\ a & 0 & 0 & -a \end{bmatrix} \quad \text{and} \quad \mathbf{Q}_\theta = -\frac{1}{2}\begin{bmatrix} 1 & 0 & 1 & 0 & 0 & 0 & 0 & 0 \\ 0 & 0 & 0 & 1 & 0 & 1 & 0 & 0 \\ 0 & 0 & 0 & 0 & 1 & 0 & 1 & 0 \\ 0 & 1 & 0 & 0 & 0 & 0 & 0 & 1 \end{bmatrix}$$

where

$$\tilde{\mathbf{S}} = [\, S_x^I \quad S_x^{II} \quad S_x^{III} \quad S_y^{IV} \,]^{\mathrm{T}}$$
$$\tilde{\mathbf{w}} = [\, \tilde{w}_1 \quad \tilde{w}_2 \quad \tilde{w}_3 \quad \tilde{w}_4 \,]^{\mathrm{T}}$$
$$\tilde{\boldsymbol{\theta}} = [\, \tilde{\boldsymbol{\theta}}_1 \quad \tilde{\boldsymbol{\theta}}_2 \quad \tilde{\boldsymbol{\theta}}_3 \quad \tilde{\boldsymbol{\theta}}_4 \,]^{\mathrm{T}}$$
$$\tilde{\boldsymbol{\theta}}_i = [\, \tilde{\theta}_{xi} \quad \tilde{\theta}_{yi} \,]$$

Including the above discrete constraint conditions in the variational principle of Eq. (5.33) we obtain

$$\Pi = \frac{1}{2}\int_\Omega \left(\mathbf{L}\mathbf{N}_\theta\tilde{\boldsymbol{\theta}}\right)^{\mathrm{T}}\mathbf{D}\,\mathbf{L}\mathbf{N}_\theta\tilde{\boldsymbol{\theta}}\,\mathrm{d}\Omega + \frac{1}{2}\int_\Omega [\mathbf{N}_s(\mathbf{Q}_\theta\tilde{\boldsymbol{\theta}} + \mathbf{Q}_w\tilde{\mathbf{w}})]^{\mathrm{T}}\alpha\,[\mathbf{N}_s(\mathbf{Q}_\theta\tilde{\boldsymbol{\theta}} + \mathbf{Q}_w\tilde{\mathbf{w}})]\,\mathrm{d}\Omega$$
$$- \int_\Omega wq\,\mathrm{d}\Omega + \Pi_{\mathrm{bt}} = \text{minimum} \tag{5.38}$$

This is a constrained potential energy principle from which on minimization we obtain the system of equations

$$\begin{bmatrix} \mathbf{K}_{ww} & \mathbf{K}_{w\theta} \\ \mathbf{K}_{\theta w} & \mathbf{K}_{\theta\theta} \end{bmatrix} \begin{Bmatrix} \tilde{\mathbf{w}} \\ \tilde{\boldsymbol{\theta}} \end{Bmatrix} = \begin{Bmatrix} \mathbf{f}_w \\ \mathbf{f}_\theta \end{Bmatrix} \tag{5.39}$$

The element contributions are

$$\begin{aligned}
\mathbf{K}_{ww} &= \mathbf{Q}_w^{\mathrm{T}}\mathbf{K}_{ss}\mathbf{Q}_w \\
\mathbf{K}_{w\theta} &= \mathbf{Q}_w^{\mathrm{T}}\mathbf{K}_{ss}\mathbf{Q}_\theta = \mathbf{K}_{\theta w}^{\mathrm{T}} \\
\mathbf{K}_{\theta\theta} &= \int_\Omega (\mathbf{L}\mathbf{N}_\theta)^{\mathrm{T}}\mathbf{D}(\mathbf{L}\mathbf{N}_\theta)\,\mathrm{d}\Omega + \mathbf{Q}_\theta^{\mathrm{T}}\mathbf{K}_{ss}\mathbf{Q}_\theta \\
\mathbf{K}_{ss} &= \int_\Omega \mathbf{N}_s^{\mathrm{T}}\alpha\mathbf{N}_s\,\mathrm{d}\Omega
\end{aligned} \tag{5.40}$$

with the force terms identical to those defined in Eq. (5.20).

These general expressions derived above can be used for any form of discrete constraint elements described and present no computational difficulties.

In the preceding we have imposed the constraints by point collocation of nodes placed on external boundaries or indeed the interior of the element. Other integrals could be used without introducing any difficulties in the final construction of the stiffness matrix. One could, for instance, require integrals such as

$$\int_\Gamma W\left[S_s - \alpha\left(\frac{\partial w}{\partial s} - \theta_s\right)\right]\mathrm{d}\Gamma = 0$$

on segments of the boundary, or

$$\int_\Omega W\left[S_s - \alpha\left(\frac{\partial w}{\partial s} - \theta_s\right)\right]\mathrm{d}\Omega = 0$$

in the interior of an element. All would achieve the same objective, providing elimination of the S_s parameters is still possible.

The use of discrete constraints can easily be shown to be equivalent to use of *substitute shear strain matrices* in the irreducible formulation of Eq. (5.18). This makes introduction of such forms easy in standard computer programs. Details of such an approach are given by Oñate *et al.*[42,43]

5.5.3 Relation to the discrete Kirchhoff formulation

In Chapter 4, Sec. 4.18, we have discussed the so-called discrete Kirchhoff theory (DKT) formulation in which the Kirchhoff constraints [i.e. Eq. (5.35) with $\alpha = \infty$] were applied in a discrete manner. The reason for the success of such discrete constraints was not obvious previously, but we believe that the formulation presented here in terms of the mixed form fully explains its basis. It is well known that the study of mixed forms frequently reveals the robustness or otherwise of irreducible approaches.

In Chapter 12 of Volume 1 we explained why certain elements of irreducible form perform well as the limit of incompressibility is approached and why others fail. Here an analogous situation is illustrated.

It is clear that every one of the elements so far discussed has its analogue in the DKT form. Indeed, the thick plate approach we have adopted here with $\alpha \neq \infty$ is simply a penalty approach to the DKT constraints in which direct elimination of variables was used. Many opportunities for development of interesting and perhaps effective plate elements are thus available for both the thick and thin range.

We shall show in the next section some triangular elements and their DKT counterparts. Perhaps the whole range of the present elements should be termed 'QnDc' and 'TnDc' (discrete Reissner–Mindlin) elements in order to ease the classification. Here 'n' is number of displacement nodes and 'c' number of shear constraints.

5.5.4 Collocation constraints for triangular elements

Figure 5.12 illustrates a triangle in which a straightforward quadratic interpolation of $\boldsymbol{\theta}$ and w is used. In this we shall take the shear forces to be given as a complete linear field defined by six shear force values on the element boundaries in directions parallel to these. The shear 'nodes' are located at Gauss points along each edge and the constraint collocation is made at the same position.

Writing the interpolation in standard area coordinates [see Chapter 8 of Volume 1] we have

$$\mathbf{S} = \sum_{i=1}^{3} L_i \mathbf{a}_i \tag{5.41}$$

where $\mathbf{a}_i$ are six parameters which can be determined by writing the tangential shear at the six constraint nodes. Introducing the tangent vector to each edge of the element as

$$\mathbf{e}_j = \begin{Bmatrix} e_{jx} \\ e_{jy} \end{Bmatrix} \tag{5.42}$$

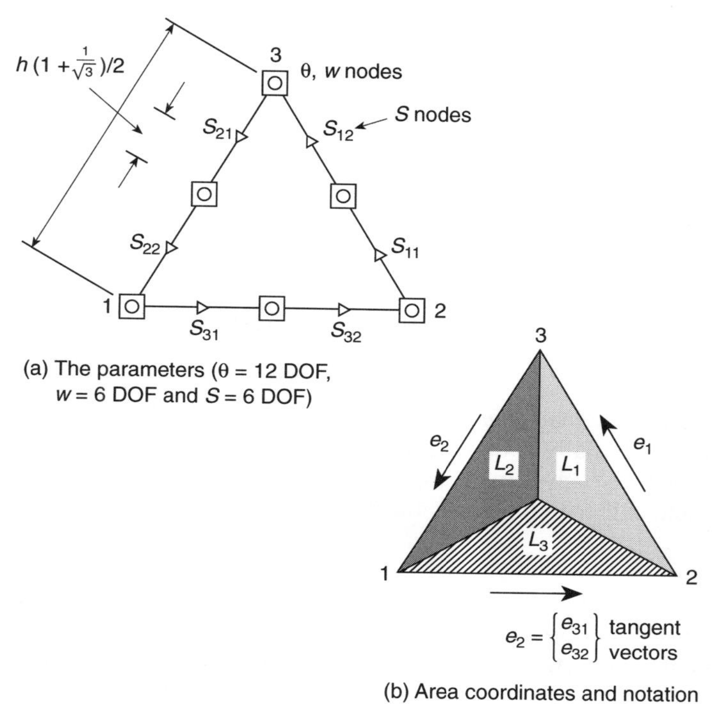

Fig. 5.12 The T6D6 triangular plate element.

a tangential component of shear on the j-edge (for which $L_j = 0$) is obtained from

$$S_j = \mathbf{S} \cdot \mathbf{e}_j \tag{5.43}$$

Evaluation of Eq. (5.43) at the two Gauss points (defined on interval 0 to 1)

$$g_1 = \frac{1}{2\sqrt{3}}(\sqrt{3} - 1) \quad \text{and} \quad g_2 = \frac{1}{2\sqrt{3}}(\sqrt{3} + 1) \tag{5.44}$$

yields a set of six equations which can be used to determine the six parameters $\mathbf{a}$ in terms of the tangential edge shears $\tilde{S}_{j1}$ and $\tilde{S}_{j2}$. The final solution for the shear interpolation then becomes

$$\mathbf{S} = \sum_{i=1}^{3} \frac{L_i}{\Delta_i} \begin{bmatrix} e_{ky}, & -e_{jy} \\ -e_{kx}, & e_{jx} \end{bmatrix} \begin{Bmatrix} g_1 \tilde{S}_{j1} + g_2 \tilde{S}_{j2} \\ g_1 \tilde{S}_{k1} + g_2 \tilde{S}_{k2} \end{Bmatrix} \tag{5.45}$$

in which i, j, k is a cyclic permutation of $1, 2, 3$. This defines uniquely the shape functions of Eq. (5.23) and, on application of constraints, expresses finally the shear field of nodal displacements $\tilde{\mathbf{w}}$ and rotations $\tilde{\boldsymbol{\theta}}$ in the manner of Eq. (5.36). The full derivation of the above expression is given in reference 36, and the final derivation of element matrices follows the procedures of Eqs (5.38)–(5.40).

The element derived satisfies fully the patch test count conditions as shown in Table 5.3 as the T6D6 element. This element yields results which are generally 'too flexible'. An alternative triangular element which shows considerable improvement

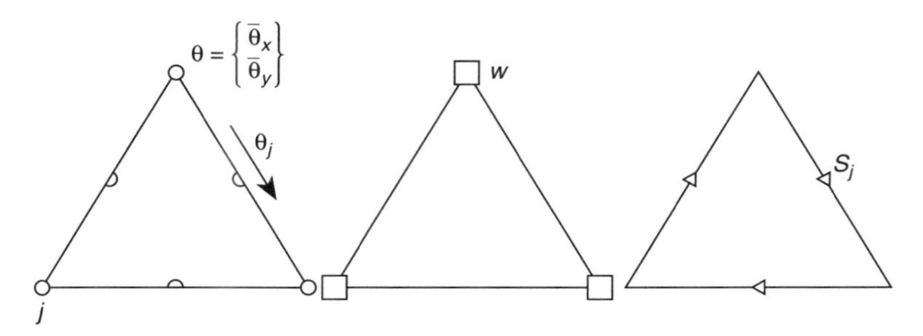

Fig. 5.13 The T3D3 (discrete Reissner–Mindlin) triangle of reference 44 with w linear, $\boldsymbol{\theta}$ constrained quadratic, and S_j constant and parallel to sides.

in performance is indicated in Fig. 5.13. Here the w displacement is interpolated linearly and $\boldsymbol{\theta}$ is initially a quadratic but is constrained to have linear behaviour along the tangent to each element edge.[44] Only a single shear constraint is introduced on each element side with the shear interpolation obtained from Eq. (5.45) by setting

$$\tilde{S}_{j1} = \tilde{S}_{j2} = \tilde{S}_j \tag{5.46}$$

The 'count' conditions are again fully satisfied for single and multiple element patches as shown in Table 5.3.

This element is of particular interest as it turns out to be the exact equivalent of the DKT triangle with 9 degrees of freedom which gave a satisfactory solution for thin plates.[45–47] Indeed, in the limit the two elements have an identical performance, though of course the T3D3 element is applicable also to plates with shear deformation. We note that the original DKT element can be modified in a different manner to achieve shear deformability[48] and obtain similar results. However, this element as introduced in reference 48 is not fully convergent.

5.6 Elements with rotational bubble or enhanced modes

As a starting point for this class of elements we may consider a standard functional of Reissner type given by

$$\Pi = \frac{1}{2}\int_\Omega (\mathbf{L}\boldsymbol{\theta})^{\mathrm{T}}\mathbf{D}\mathbf{L}\boldsymbol{\theta}\,\mathrm{d}\Omega - \frac{1}{2}\int_\Omega \mathbf{S}^{\mathrm{T}}\alpha^{-1}\mathbf{S}\,\mathrm{d}\Omega + \int_\Omega \mathbf{S}^{\mathrm{T}}(\boldsymbol{\nabla}w - \boldsymbol{\theta})\,\mathrm{d}\Omega - \int_\Omega wq\,\mathrm{d}\Omega + \Pi_{\mathrm{bt}}$$
$$= \text{stationary} \tag{5.47}$$

in which approximations for w, $\boldsymbol{\theta}$ and $\mathbf{S}$ are required.

Three triangular elements designed by introducing 'bubble modes' for rotation parameters are found to be robust, and at the same time excellent performers. None of these elements is 'obvious', and they all use an interpolation of rotations that is of higher or equal order than that of w. Figure 5.14 shows the degree-of-freedom assignments for these triangular elements and the second part of Table 5.2 shows again their performance in patches.

The quadratic element (T6S3B3) was devised by Zienkiewicz and Lefebvre[15] starting from a quadratic interpolation for w and $\boldsymbol{\theta}$. The shear $\mathbf{S}$ is interpolated by

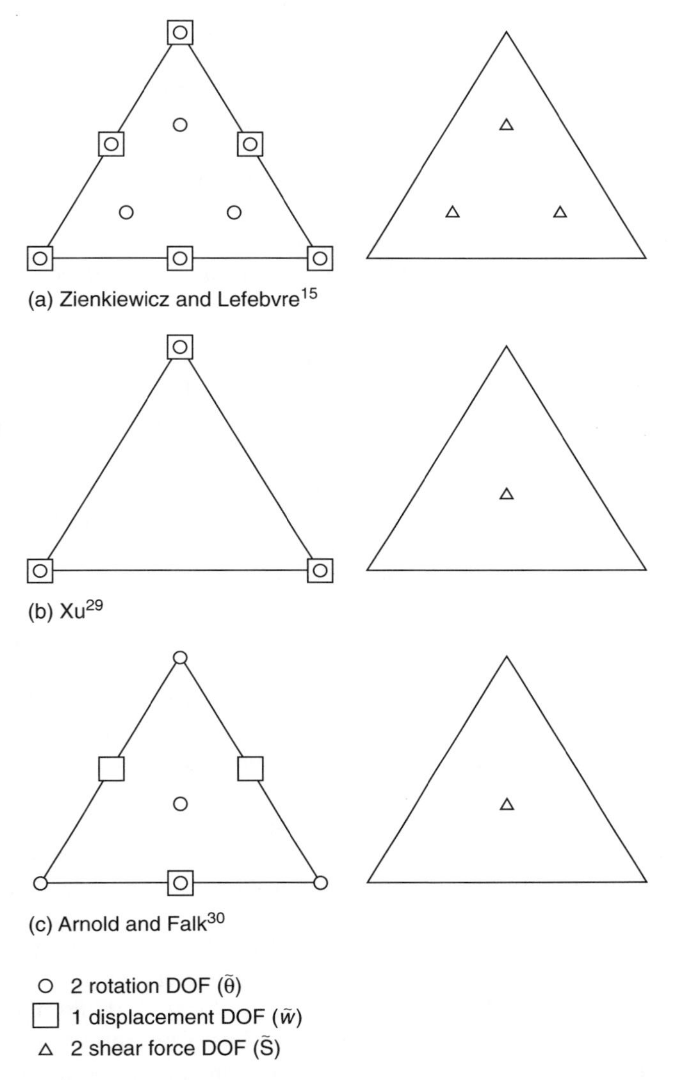

Fig. 5.14 Three robust triangular elements: (a) the T6S3B3 element of Zienkiewicz and Lefebvre;[15] (b) the T6S1B1 element of Xu;[29] (c) the T3S1B1A element of Arnold and Falk.[30]

a complete linear polynomial in each element, giving here six parameters, $\tilde{\mathbf{S}}$. Three hierarchical quartic bubbles are added to the rotations giving the interpolation

$$\boldsymbol{\theta} = \sum_{i=1}^{6} N_i(L_k)\tilde{\boldsymbol{\theta}}_i + \sum_{j=1}^{3} \Delta N_j(L_k)\Delta\tilde{\boldsymbol{\theta}}_j$$

where $N_i(L_k)$ are conventional quadratic interpolations on the triangle (see Sec. 8.8.2 of Volume 1), and shape functions for the quartic bubble modes are given as

$$\Delta N_j(L_k) = L_j\,(L_1\,L_2\,L_3)$$

Thus, we have introduced six additional rotation parameters but have left the number of w parameters unaltered from those given by a quadratic interpolation. This element has very desirable properties and excellent performance when the integral for $\mathbf{K}_b$ in Eq. (5.25) is computed by using seven points (see Table 9.2 in Chapter 9 of Volume 1) and the other integrals are computed by using four points. Slightly improved answers can be obtained by using a mixed formulation for the bending terms. In the mixed form the bending moments are approximated as discontinuous quadratic interpolations for each component and a $\mathbf{u} - \boldsymbol{\sigma}$ form as described in Sec. 11.4.2 of Volume 1 employed. Using this approach we replace the first integral in Eq. (5.47) as follows

$$\frac{1}{2}\int_{\Omega}(\mathbf{L}\boldsymbol{\theta})^{\mathrm{T}}\mathbf{D}\mathbf{L}\boldsymbol{\theta}\,\mathrm{d}\Omega \rightarrow \int_{\Omega}(\mathbf{L}\boldsymbol{\theta})^{\mathrm{T}}\mathbf{M}\,\mathrm{d}\Omega - \frac{1}{2}\int_{\Omega}\mathbf{M}^{\mathrm{T}}\mathbf{D}^{-1}\mathbf{M}\,\mathrm{d}\Omega \tag{5.48}$$

All other terms in Eq. (5.47) remain the same. The mixed element computed in this way is denoted as T6S3B3M in subsequent results.

Since the T6S3B3 type elements use a complete quadratic to describe the displacement and rotation field, an isoparametric mapping may be used to produce curved-sided elements and, indeed, curved-shell elements. Furthermore, by design this type passes the count test and by numerical testing is proved to be fully robust when used to analyse both thick and thin plate problems.[15,49] Since the w displacement interpolation is a standard quadratic interpolation the element may be joined compatibly to tetrahedral or prism solid elements which have six-noded faces.

A linear triangular element [T3S1B1 – Fig. 5.14(b)] with a total of 9 nodal degrees of freedom adds a single cubic bubble to the linear rotational interpolation and uses linear interpolation for w with constant discontinuous shear. This element satisfies all count conditions for solution (see Table 5.2); however, without further enhancements it locks as the thin plate limit is approached.[29] As we have stated previously the count condition is necessary but not sufficient to define successful elements and numerical testing is always needed. In a later section we discuss a 'linked interpolation' modification which makes this element robust.

A third element employing bubble modes [T3S1B1A – Fig. 5.14(c)] was devised by Arnold and Falk.[30] It is of interest to note that this element uses a discontinuous (non-conforming) w interpolation with parameters located at the mid-side of each triangle. The rotation interpolation is a standard linear interpolation with an added cubic bubble. The shear interpolation is constant on each element. This element is a direct opposite of the triangular element of Morley discussed in Chapter 4 in that location of the displacement and rotation parameters is reversed. The location of the displacement parameters, however, precludes its use in combination with standard solid elements. Thus, this element is of little general interest.

The introduction of successful bubble modes in quadrilaterals is more difficult. The first condition examined was the linear quadrilateral with a single bubble mode (Q4S1B1). For this element the patch count test fails when only a single element is considered but for assemblies above four elements it is passed and much hope was placed on this condition.[27] Unfortunately, a singular mode with a single zero eigenvalue persists in all assemblies when the completely relaxed support conditions are considered. Despite this singularity the element does not lock and usually gives an excellent performance.[27]

To avoid, however, any singularity it is necessary to have at least three shear stress components and a similar number of rotation components of bubble form. No simple way of achieving a three-term interpolation exist but a successful four-component form was obtained by Auricchio and Taylor.[28] This four-term interpolation for shear is given by

$$\mathbf{S} = \begin{bmatrix} J_{11}^0 & J_{21}^0 & J_{11}^0\eta & J_{21}^0\xi \\ J_{12}^0 & J_{22}^0 & J_{12}^0\eta & J_{22}^0\xi \end{bmatrix} \begin{Bmatrix} \tilde{S}_1 \\ \tilde{S}_2 \\ \tilde{S}_3 \\ \tilde{S}_4 \end{Bmatrix} \tag{5.49}$$

The Jacobian transformation J_{ij}^0 is identical to that introduced when describing the Pian–Sumihara element in Sec. 11.4.4 of Volume 1 and is computed as

$$\begin{aligned} J_{11}^0 &= x_{,\xi}\big|_{\xi=\eta=0}, \qquad & J_{12}^0 &= x_{,\eta}\big|_{\xi=\eta=0} \\ J_{21}^0 &= y_{,\xi}\big|_{\xi=\eta=0}, \qquad & J_{22}^0 &= y_{,\eta}\big|_{\xi=\eta=0} \end{aligned} \tag{5.50}$$

To satisfy Eq. (5.27) it is necessary to construct a set of four bubble modes. An appropriate form is found to be

$$\Delta\mathbf{N}_b = \frac{1}{j} N_b(\xi, \eta) \begin{bmatrix} J_{22}^0 & -J_{12}^0 & J_{22}^0\eta & -J_{12}^0\xi \\ -J_{21}^0 & J_{11}^0 & -J_{21}^0\eta & J_{11}^0\xi \end{bmatrix} \tag{5.51}$$

in which j is the determinant of the Jacobian transformation $\mathbf{J}$ (i.e. not the determinant of $\mathbf{J}^0$) and $N_b = (1-\xi^2)(1-\eta^2)$ is a bubble mode. Thus, the rotation parameters are interpolated by using

$$\boldsymbol{\theta} = N_i \tilde{\boldsymbol{\theta}} + \Delta\mathbf{N}_b\, \Delta\tilde{\boldsymbol{\theta}}_b \tag{5.52}$$

where N_i are the standard bilinear interpolations for the four-noded quadrilateral.

The element so achieved (Q4S2B2) is fully stable.

5.7 Linked interpolation – an improvement of accuracy

In the previous section we outlined various procedures which are effective in ensuring the necessary count conditions and which are, therefore, essential to make the elements 'work' without locking or singularity. In this section we shall try to improve the interpolation used to increase the accuracy without involving additional parameters.

The reader will here observe that in the primary interpolation we have used equal-order polynomials to interpolate both the displacement (w) and the rotations ($\boldsymbol{\theta}$). Clearly, if we consider the limit of thin plate theory

$$\boldsymbol{\theta} = \begin{Bmatrix} \theta_x \\ \theta_y \end{Bmatrix} = \boldsymbol{\nabla} w \tag{5.53}$$

and hence one order lower interpolation for $\boldsymbol{\theta}$ appears necessary. To achieve this we introduce here the concept of *linked interpolation* in which the primary expression is written as

$$\boldsymbol{\theta} = \mathbf{N}_{\theta}\tilde{\boldsymbol{\theta}} \tag{5.54}$$

and

$$w = \mathbf{N}_{w}\tilde{\mathbf{w}} + \mathbf{N}_{w\theta}\tilde{\boldsymbol{\theta}} \tag{5.55}$$

Such an expression ensures now that one order higher polynomials can be introduced for the representation of w without adding additional element parameters. This procedure can, of course, be applied to any of the elements we have listed in Tables 5.1 and 5.2 to improve the accuracy attainable. We shall here develop such linking for two types of elements in which the essential interpolations are linear on each edge.

We thus improve the triangular T3S1B1 and by linking L to its formulation we arrive at T3S1B1L. The same procedure can, of course, be applied to the quadrilaterals Q4S1B1 and Q4S2B2 of which only the second is unconditionally stable and add the letter L.

Similar interpolations have also been used by Tessler and Hughes[50,51] and termed 'anisoparametric'. The earliest appearance of linked interpolation appears in the context of beams by Fraeijs de Veubeke.[18] Additional studies in the context of beams have been performed leading to general families of elements.[52,58] In the context of thick plates linked interpolation on a three-noded triangle was introduced by Lynn and co-workers[54,55] and first extended to permit also thin plate analysis by Xu.[29,56] Additional presentations dealing with the simple triangular element with nine degrees of freedom in its reduced form [see Eq. (5.72)] have been given by Taylor and co-workers.[44,57,58]

Quadrilateral elements employing linked interpolation have been developed by Crisfield,[59] Zienkiewicz and co-workers,[26,27] and Auricchio and Taylor.[28]

5.7.1 Derivation of the linking function

The derivation of the linking function $\mathbf{N}_{w\theta}$ is somewhat more complex than that of the basic shape function. For the linear triangle T3 and the linear quadrilaterals Q4 we require that these functions be:

1. uniquely defined by the two nodal rotations $\boldsymbol{\theta}_i$ at the ends of each side to ensure C_0 continuity;
2. must introduce quadratic terms in the w interpolation.

We shall write the interpolation as

$$w = N_i\tilde{w}_i + \frac{1}{8}\sum_{k=1}^{n_{el}} (N_{w\theta})_k\, l_{ij}\,(\theta_{si} - \theta_{sj}) \tag{5.56}$$

where n_{el} is the number of vertex nodes on the element (i.e. 3 or 4), l_{ij} is the length of the i–j side, θ_{si} is the rotation at node i in the tangent direction of the kth side, and $(N_{w\theta})_k$ are shape functions defining the quadratic w along the side but still maintaining zero w at corner nodes. For the triangular element these are the shape functions are identical to those arising in the plane six-noded element at mid-side nodes and are given by (see Chapter 8 of Volume 1)

$$\mathbf{N}_{w\theta} = 4\,[\,L_1 L_2 \quad L_2 L_3 \quad L_3 L_1\,] \tag{5.57}$$

and for the quadrilateral element these are the shape functions for the eight-noded serendipity functions given by

$$\mathbf{N}_{w\theta} = \tfrac{1}{2}\,[\,(1-\xi^2)(1-\eta) \quad (1+\xi)(1-\eta^2) \quad (1-\xi^2)(1+\eta) \quad (1-\xi)(1-\eta^2)\,] \tag{5.58}$$

The development of one shape function for the three-noded triangular element with a total of 9 degrees of freedom is here fully developed using the linked interpolation concept. The process to develop a linked interpolation for the transverse displacement, w, starts with a full quadratic expansion written in hierarchical form. Thus, for a triangle we have the interpolation in area coordinates

$$w = L_1 w_1 + L_2 w_2 + L_3 w_3 + 4L_1 L_2 \alpha_1 + 4L_2 L_3 \alpha_2 + 4L_3 L_1 \alpha_3 \tag{5.59}$$

where w_i are the nodal displacements and α_i are hierarchical parameters. The hierarchical parameters are then expressed in terms of rotational parameters. Along any edge, say the 1–2 edge (where $L_3 = 0$), the displacement is given by

$$w = L_1 w_1 + L_2 w_2 + 4L_1 L_2 \alpha_1 \tag{5.60}$$

The expression used to eliminate α_1 is deduced by constraining the transverse edge shear to be a constant. Along the edge the transverse shear is given by

$$\gamma_{12} = \frac{\partial w}{\partial s} - \theta_s \tag{5.61}$$

where s is the coordinate tangential to the edge and θ_s is the component of the rotation along the edge. The derivative of Eq. (5.60) along the edge is given by

$$\frac{\partial w}{\partial s} = \frac{w_2 - w_1}{l_{12}} + 4\,\frac{L_1 - L_2}{l_{12}}\,\alpha_1 \tag{5.62}$$

where l_{12} is the length of the 1–2 side. Assuming a linear interpolation for θ_s along the edge we have

$$\theta_s = L_1 \theta_{s1} + L_2 \theta_{s2} \tag{5.63}$$

which may also be expressed as

$$\theta_s = \tfrac{1}{2}(\theta_{s1} + \theta_{s2}) + \tfrac{1}{2}(\theta_{s1} - \theta_{s2})(L_1 - L_2) \tag{5.64}$$

after noting that $L_1 + L_2 = 1$ along the edge. The transverse shear may now be given as

$$\gamma_{12} = \frac{w_2 - w_1}{l_{12}} - \frac{1}{2}(\theta_{s1} + \theta_{s2}) + \left[\frac{4}{l_{12}}\alpha_1 - \frac{1}{2}(\theta_{s1} - \theta_{s2})\right](L_1 - L_2) \tag{5.65}$$

Constraining the strain to be constant gives

$$\alpha_1 = \frac{l_{12}}{8}(\theta_{s1} - \theta_{s2}) \tag{5.66}$$

yielding the 'linked' edge interpolation

$$w = L_1 w_1 + L_2 w_2 + \tfrac{1}{2} l_{12} L_1 L_2 (\theta_{s1} - \theta_{s2}) \tag{5.67}$$

The normal rotations may now be expressed in terms of the nodal cartesian components by using

$$\theta_s = \cos\phi_{12}\theta_x + \sin\phi_{12}\theta_y \tag{5.68}$$

where ϕ_{12} is the angle that the normal to the edge makes with the x-axis. Repeating this process for the other two edges gives the final interpolation for the transverse displacement.

A similar process can be followed to develop the linked interpolations for the quadrilateral element. The reader can verify that the use of the constant 1/8 ensures that constant shear strain on the element side occurs. Further, a rigid body rotation with $\theta_i^k = \theta_j^k$ in the element causes no straining. Finally, with rotation θ_i^k being the same for adjacent elements C_0 continuity is ensured.

We have not considered here elements with quadratic or higher basic interpolation. The linking obviously proceeds on similar lines and some elements with excellent performance can thus be achieved.

5.8 Discrete 'exact' thin plate limit

Discretization of Eq. (5.47) using interpolations of the form*

$$w = \mathbf{N}_w\tilde{\mathbf{w}} + \mathbf{N}_{w\theta}\tilde{\boldsymbol{\theta}}, \qquad \boldsymbol{\theta} = \mathbf{N}_\theta\tilde{\boldsymbol{\theta}}, + \Delta\mathbf{N}_b\Delta\tilde{\boldsymbol{\theta}}_b, \qquad \mathbf{S} = \mathbf{N}_s\tilde{\mathbf{S}} \tag{5.69}$$

leads to the algebraic system of equations

$$\begin{bmatrix} \mathbf{0} & \mathbf{0} & \mathbf{0} & \mathbf{K}_{sw}^{\mathrm{T}} \\ \mathbf{0} & \mathbf{K}_{\theta\theta} & \mathbf{K}_{b\theta}^{\mathrm{T}} & \mathbf{K}_{s\theta}^{\mathrm{T}} \\ \mathbf{0} & \mathbf{K}_{b\theta} & \mathbf{K}_{bb} & \mathbf{K}_{sb}^{\mathrm{T}} \\ \mathbf{K}_{sw} & \mathbf{K}_{s\theta} & \mathbf{K}_{sb} & \mathbf{K}_{ss} \end{bmatrix} \begin{Bmatrix} \tilde{\mathbf{w}} \\ \tilde{\boldsymbol{\theta}} \\ \Delta\tilde{\boldsymbol{\theta}}_b \\ \tilde{\mathbf{S}} \end{Bmatrix} = \begin{Bmatrix} \mathbf{f}_w \\ \mathbf{f}_\theta \\ \mathbf{0} \\ \mathbf{0} \end{Bmatrix} \tag{5.70}$$

where, for simplicity, only the forces $\mathbf{f}_w$ and $\mathbf{f}_\theta$ due to transverse load q and boundary conditions are included [see Eq. (5.20)]. The arrays appearing in Eq. (5.70) are given by

$$\begin{aligned}
\mathbf{K}_{\theta\theta} &= \int_\Omega (\mathbf{L}\mathbf{N}_\theta)^{\mathrm{T}}\mathbf{D}(\mathbf{L}\mathbf{N}_\theta)\,\mathrm{d}\Omega, & \mathbf{K}_{ss} &= -\int_\Omega \mathbf{N}_s\boldsymbol{\alpha}^{-1}\mathbf{N}_s\,\mathrm{d}\Omega, \\
\mathbf{K}_{b\theta} &= \int_\Omega (\mathbf{L}\Delta\mathbf{N}_b)^{\mathrm{T}}\mathbf{D}(\mathbf{L}\mathbf{N}_\theta)\,\mathrm{d}\Omega, & \mathbf{K}_{s\theta} &= \int_\Omega \mathbf{N}_s^{\mathrm{T}}[\boldsymbol{\nabla}\Delta\mathbf{N}_{w\theta} - \mathbf{N}_\theta]\,\mathrm{d}\Omega, \\
\mathbf{K}_{bb} &= \int_\Omega (\mathbf{L}\Delta\mathbf{N}_b)^{\mathrm{T}}\mathbf{D}(\mathbf{L}\Delta\mathbf{N}_b)\,\mathrm{d}\Omega, & \mathbf{K}_{sb} &= -\int_\Omega \mathbf{N}_s^{\mathrm{T}}\Delta\mathbf{N}_b\,\mathrm{d}\Omega, \\
\mathbf{K}_{sw} &= \int_\Omega \mathbf{N}_s^{\mathrm{T}}\boldsymbol{\nabla}\mathbf{N}_w\,\mathrm{d}\Omega
\end{aligned} \tag{5.71}$$

* The term $\mathbf{N}_{w\theta}$ will be exploited in the next part of this section and thus is included for completeness.

Adopting a static condensation process at the element level[60] in which the internal rotational parameters are eliminated first, followed by the shear parameters, yields the element stiffness matrix in terms of the element $\tilde{\mathbf{w}}$ and $\tilde{\boldsymbol{\theta}}$ parameters given by

$$\begin{bmatrix} \mathbf{K}_{sw}^{\mathrm{T}}\mathbf{A}_{ss}^{-1}\mathbf{K}_{sw} & -\mathbf{K}_{sw}^{\mathrm{T}}\mathbf{A}_{ss}^{-1}\mathbf{A}_{s\theta} \\ -\mathbf{A}_{s\theta}^{\mathrm{T}}\mathbf{A}_{ss}^{-1}\mathbf{K}_{sw} & -\mathbf{A}_{\theta\theta} + \mathbf{A}_{s\theta}^{\mathrm{T}}\mathbf{A}_{ss}^{-1}\mathbf{A}_{s\theta} \end{bmatrix} \begin{Bmatrix} \tilde{\mathbf{w}} \\ \tilde{\boldsymbol{\theta}} \end{Bmatrix} = \begin{Bmatrix} \mathbf{f}_w \\ \mathbf{f}_\theta \end{Bmatrix} \tag{5.72}$$

in which

$$\begin{aligned} \mathbf{A}_{ss} &= \mathbf{K}_{sb}\mathbf{K}_{bb}^{-1}\mathbf{K}_{sb}^{\mathrm{T}} - \mathbf{K}_{ss}, \qquad \mathbf{A}_{s\theta} = \mathbf{K}_{sb}\mathbf{K}_{bb}^{-1}\mathbf{K}_{b\theta} - \mathbf{K}_{s\theta} \\ \mathbf{A}_{\theta\theta} &= \mathbf{K}_{b\theta}^{\mathrm{T}}\mathbf{K}_{bb}^{-1}\mathbf{K}_{b\theta} - \mathbf{K}_{\theta\theta} \end{aligned} \tag{5.73}$$

This solution strategy requires the inverse of $\mathbf{K}_{bb}$ and $\mathbf{A}_{ss}$ only. In particular, we note that the inverse of $\mathbf{A}_{ss}$ can be performed even if $\mathbf{K}_{ss}$ is zero (provided the other term is non-singular). The vanishing of $\mathbf{K}_{ss}$ defines the *thin plate limit*. Thus, the above strategy leads to a solution process in which the thin plate limit is defined without recourse to a penalty method. Indeed, all terms in the process generally are not subject to ill-conditioning due to differences in large and small numbers. In the context of thick and thin plate analysis this solution strategy has been exploited with success in references 28 and 58.

5.9 Performance of various 'thick' plate elements – limitations of thin plate theory

The performance of both 'thick' and 'thin' elements is frequently compared for the examples of clamped and simply supported square plates, though, of course, more stringent tests can and should be devised. Figure 5.15(a)–5.15(d) illustrates the behaviour of various elements we have discussed in the case of a span-to-thickness ratio (L/t) of 100, which generally is considered within the scope of thin plate theory. The results are indeed directly comparable to those of Fig. 4.16 of Chapter 4, and it is evident that here the thick plate elements perform as well as the best of the thin plate forms.

It is of interest to note that in Fig. 5.15 we have included some elements that do not fully pass the patch test and hence are not robust. Many such elements are still used as their failure occurs only occasionally – although new developments should always strive to use an element which is robust.

All 'robust' elements of the thick plate kind can be easily mapped isoparametrically and their performance remains excellent and convergent. Figure 5.16 shows isoparametric mapping used on a curved-sided mesh in the solution of a circular plate for two elements previously discussed. Obviously, such a lack of sensitivity to distortion will be of considerable advantage when shells are considered, as we shall show in Chapter 8.

Of course, when the span-to-thickness ratio decreases and thus shear deformation importance increases, the thick plate elements are capable of yielding results not obtainable with thin plate theory. In Table 5.4 we show some results for a simply supported, uniformly loaded plate for two L/t ratios and in Table 5.5 we show results for the clamped uniformly loaded plate for the same ratios. In this example we show

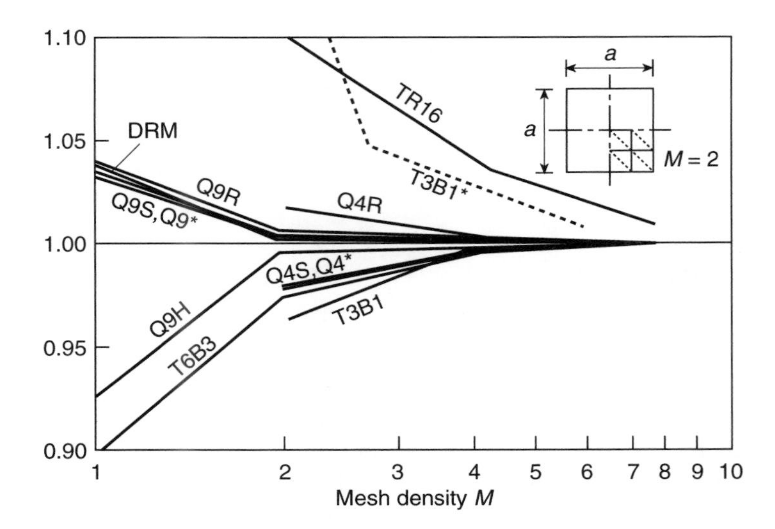

(a) Centre displacement normalized with respect to thin plate theory for simply supported, uniformly loaded square plate

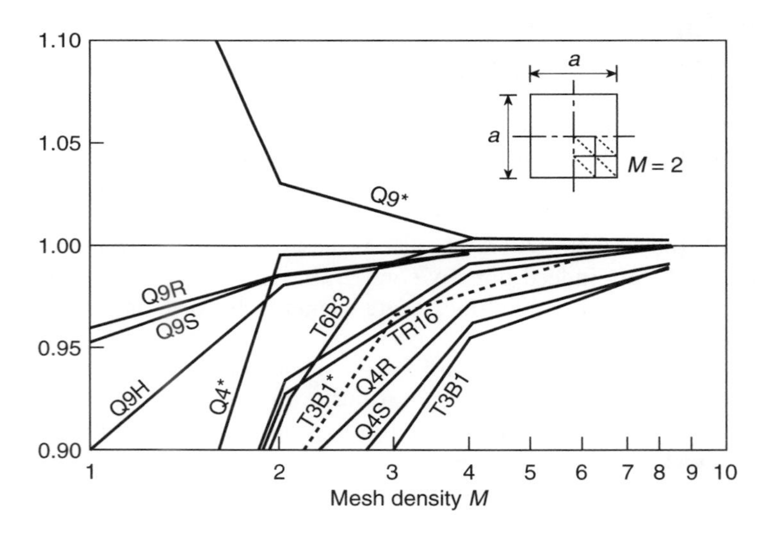

(b) Moment at Gauss point nearest centre (or central point) normalized by centre moment of thin plate theory for simply supported, uniformly loaded square plate

Fig. 5.15 Convergence study for relatively thin plate ($L/t = 100$): (a) centre displacement (simply supported, uniform load, square plate); (b) moment at Gauss point nearest centre (simply supported, uniform load, square plate). Tables 5.1 and 5.2 give keys to elements used.

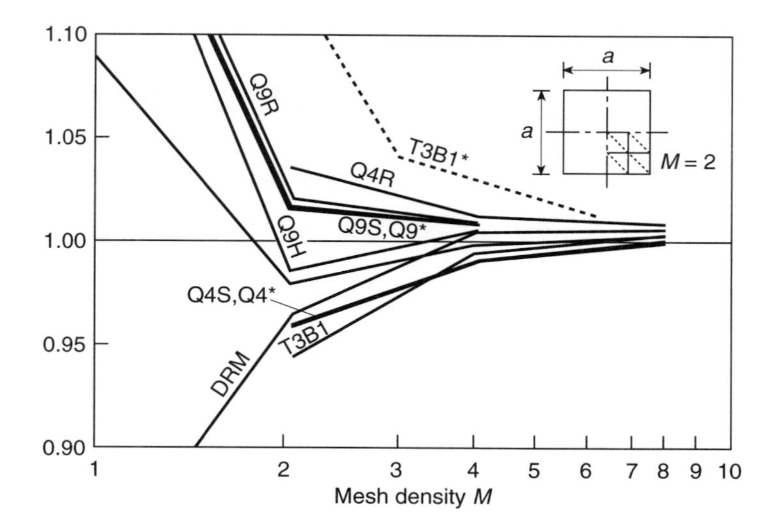

(c) Centre displacement normalized with respect to thin plate theory for clamped, uniformly loaded square plate

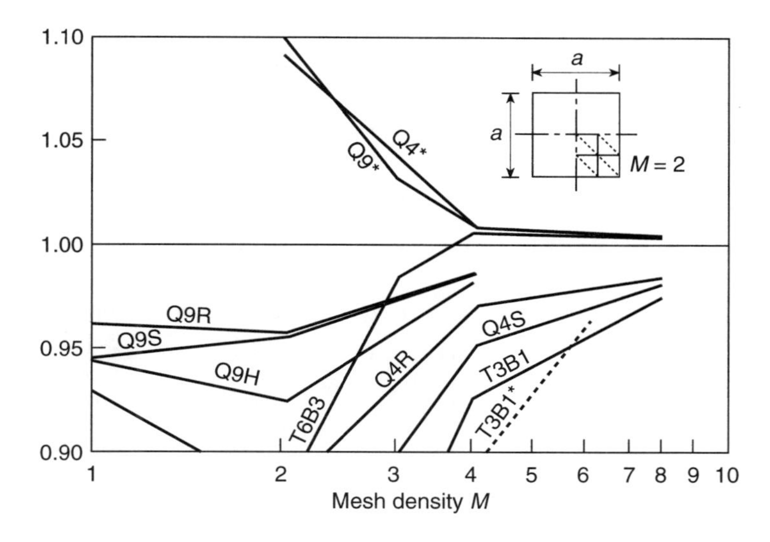

(d) Moment at Gauss point nearest centre (or central point) normalized by centre moments of thin plate theory for clamped, uniformly loaded square plate

Fig. 5.15 Convergence study for relatively thin plate ($L/t = 100$): (c) centre displacement (clamped, uniform load, square plate); (d) moment at Gauss point nearest centre (clamped, uniform load, square plate). Tables 5.1 and 5.2 give keys to elements used.

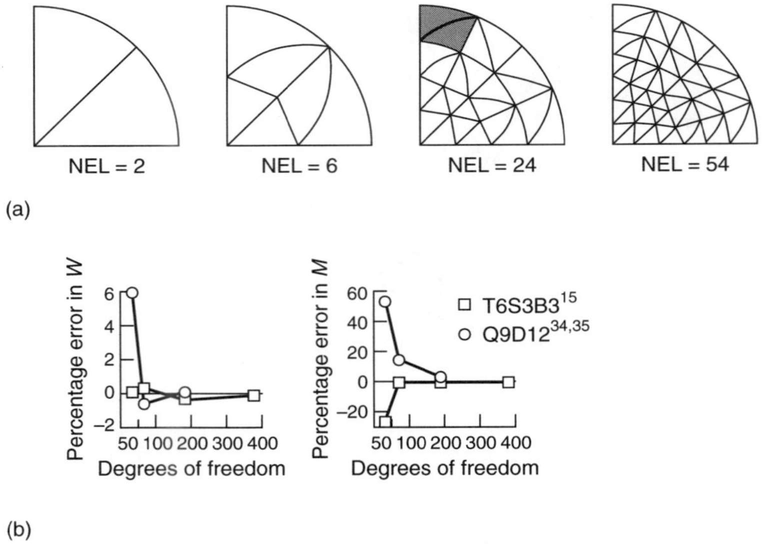

NEL = 2
NEL = 6
NEL = 24
NEL = 54
(a)
Percentage error in W
6
4
2
0
−2
50 100 200 300 400
Degrees of freedom
Percentage error in M
60
40
20
0
−20
50 100 200 300 400
Degrees of freedom
T6S3B3[15]
Q9D12[34,35]
(b)

Fig. 5.16 Mapped curvilinear elements in solution of a circular clamped plate under uniform load: (a) meshes used; (b) percentage error in centre displacement and moment.

also the effect of the *hard* and *soft* simple support conditions. In the hard support we assume just as in thin plates that the rotation along the support (θ_s) is zero. In the soft support case we take, more rationally, a zero twisting moment along the support (see Chapter 4, Sec. 4.2.2).

Table 5.4 Centre displacement of a simply supported plate under uniform load for two L/t ratios; $E = 10.92$, $\nu = 0.3$, $L = 10$, $q = 1$

Mesh, M	$L/t = 10$; $w \times 10^{-1}$		$L/t = 1000$; $w \times 10^{-7}$	
	hard support	soft support	hard support	soft support
2	4.2626	4.6085	4.0389	4.2397
4	4.2720	4.5629	4.0607	4.1297
8	4.2727	4.5883	4.0637	4.0928
16	4.2728	4.6077	4.0643	4.0773
32	4.2728	4.6144	4.0644	4.0700
Series	4.2728		4.0624	

Table 5.5 Centre displacement of a clamped plate under uniform load for two L/t ratios; $E = 10.92$, $\nu = 0.3$, $L = 10$, $q = 1$

Mesh, M	$L/t = 10$; $w \times 10^{-1}$	$L/t = 1000$; $w \times 10^{-7}$
2	1.4211	1.1469
4	1.4858	1.2362
8	1.4997	1.2583
16	1.5034	1.2637
32	1.5043	1.2646
Series	1.499[26]	1.2653

It is immediately evident that:

1. the thick plate ($L/t = 10$) shows deflections converging to very different values depending on the support conditions, both being considerably larger than those given by thin plate theory;
2. for the thin plate ($L/t = 1000$) the deflections converge uniformly to the thin (Kirchhoff) plate results for *hard* support conditions, but for *soft* support conditions give answers about 0.2 per cent higher in center deflection.

It is perhaps an insignificant difference that occurs in this example between the support conditions but this can be more pronounced in different plate configurations.

In Fig. 5.17 we show the results of a study of a simply supported rhombic plate with $L/t = 100$ and 1000. For this problem an accurate Kirchhoff plate theory solution is available,[61] but as will be noticed the thick plate results converge uniformly to a displacement nearly 4 per cent in excess of the thin plate solution for all the $L/t = 100$ cases.

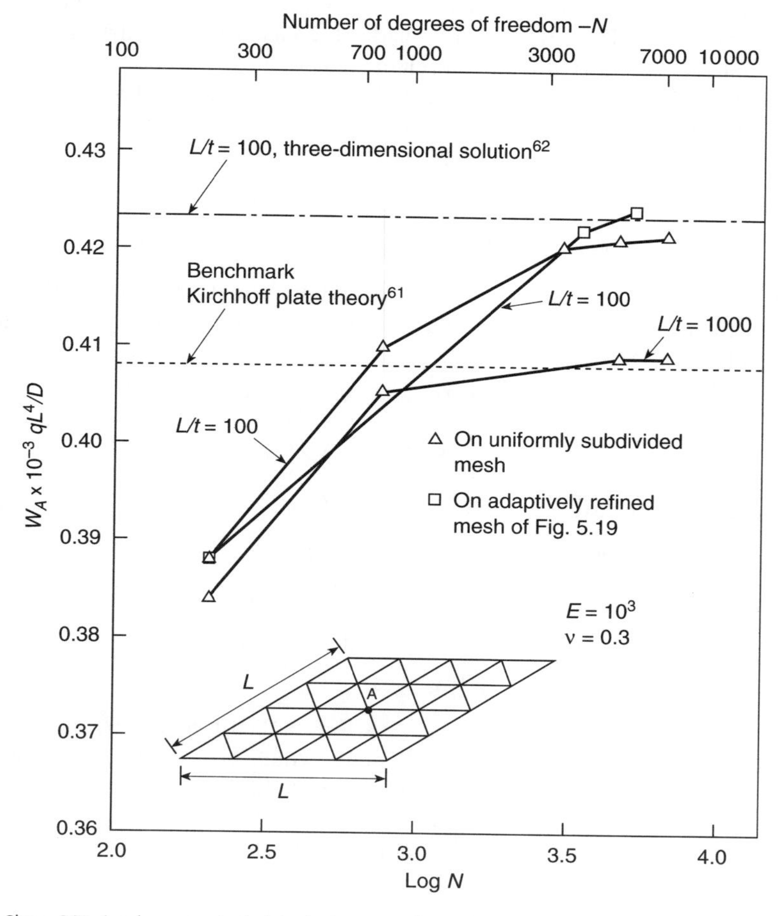

Fig. 5.17 Skew 30° simply supported plate (soft support); maximum deflection at A, the centre for various degrees of freedom N. The triangular element of reference 15 is used.

This problem is illustrative of the substantial difference that can on occasion arise in situations that fall well within the limits assumed by conventional thin plate theory ($L/t = 100$), and for this reason the problem has been thoroughly investigated by Babuška and Scapolla,[62] who solve it as a fully three-dimensional elasticity problem using support conditions of the 'soft' type which appear to be the closest to physical reality. Their three-dimensional result is very close to the thick plate solution, and confirms its validity and, indeed, superiority over the thin plate forms. However, we note that for very thin plates, even with soft support, convergence to the thin plate results occurs.

5.10 Forms without rotation parameters

It is possible to formulate the thick plate theory without direct use of rotation parameters. Such an approach has advantages for problems with large rotations where use of rotation parameters leads to introduction of trigonometric functions (e.g. see Chapter 11). Here we again consider the case of a cylindrical bending of a plate (or a straight beam) where each element is defined by coordinates at the two ends. Starting from a four-noded rectangular element in which the origin of a local cartesian coordinate system passes through the mid-surface (centroid) of the element we may write interpolations as (Fig. 5.18)

$$\begin{aligned} x &= N_i(\xi,\eta)\tilde{x}_i + N_j(\xi,\eta)\tilde{x}_j + N_k(\xi,\eta)\tilde{x}_k + N_l(\xi,\eta)\tilde{x}_l \\ y &= N_i(\xi,\eta)\tilde{y}_i + N_j(\xi,\eta)\tilde{y}_j + N_k(\xi,\eta)\tilde{y}_k + N_l(\xi,\eta)\tilde{y}_l \end{aligned} \tag{5.74}$$

in which N_i, etc. are the usual 4-noded bilinear shape functions. Noting the rectangular form of the element, these interpolations may be rewritten in terms of alternative parameters [Fig. 5.18(b)] as

$$\begin{aligned} x &= N_1(\xi)\tilde{x}_1 + N_2(\xi)\tilde{x}_2 \\ y &= \frac{\eta}{2}\left[N_1(\xi)\tilde{t}_1 + N_2(\xi)\tilde{t}_2\right] \end{aligned} \tag{5.75}$$

where shape functions are

$$N_1(\xi) = \tfrac{1}{2}(1-\xi), \qquad N_2(\xi) = \tfrac{1}{2}(1+\xi) \tag{5.76}$$

and new nodal parameters are related to the original ones through

$$\begin{aligned} \tilde{x}_1 &= \tfrac{1}{2}(\tilde{x}_i + \tilde{x}_l) \quad \text{and} \quad \tilde{t}_1 = \tilde{y}_l - \tilde{y}_k \\ \tilde{x}_2 &= \tfrac{1}{2}(\tilde{x}_j + \tilde{x}_k) \quad \text{and} \quad \tilde{t}_2 = \tilde{y}_k - \tilde{y}_j \end{aligned} \tag{5.77}$$

Since the element is rectangular $\tilde{t}_1 = \tilde{t}_2 = \tilde{t}$ [Fig. 5.18(b)]; however, the above interpolations can be generalized easily to elements which are tapered. We can now use isoparametric concepts to write the displacement field for the element as

$$\begin{aligned} u &= N_1(\xi)\left[\tilde{u}_1 + \frac{\eta\tilde{t}}{2\tilde{t}}\Delta\tilde{u}_1\right] + N_2(\xi)\left[\tilde{u}_2 + \frac{\eta\tilde{t}}{2\tilde{t}}\Delta\tilde{u}_2\right] \\ v &= N_1(\xi)\left[\tilde{v}_1 + \frac{\eta\tilde{t}}{2\tilde{t}}\Delta\tilde{v}_1\right] + N_2(\xi)\left[\tilde{v}_2 + \frac{\eta\tilde{t}}{2\tilde{t}}\Delta\tilde{v}_2\right] \end{aligned} \tag{5.78}$$

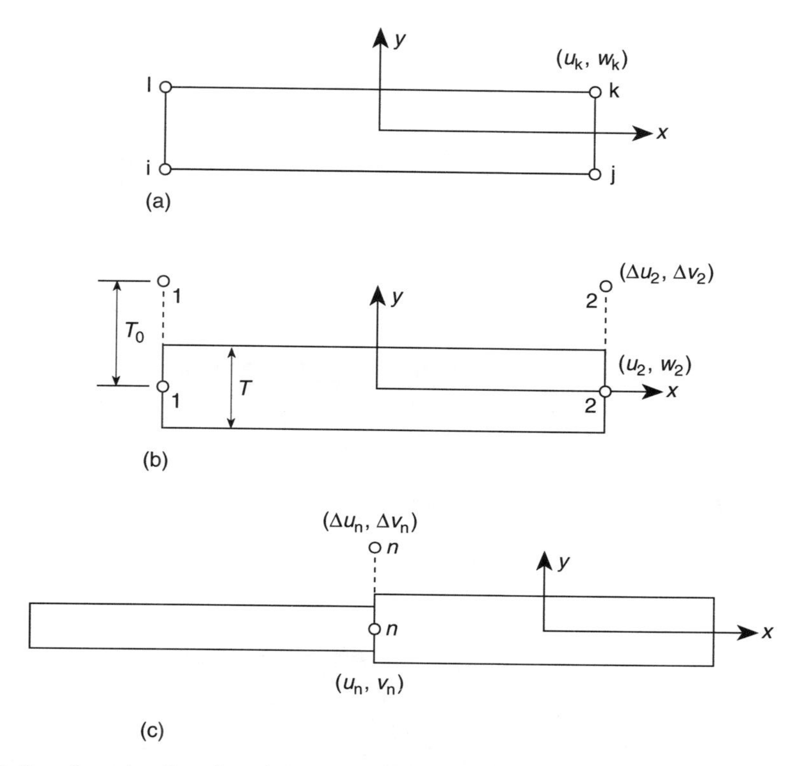

Fig. 5.18 One-dimensional bending of plates (or beams): (a) geometry, Q4 element; (b) geometry, no rotation parameters; (c) joining elements with different thickness.

in which $\bar{t}$ is a 'thickness' parameter chosen to permit elements of different thickness to be joined at a common node n [Fig. 5.18(c)].

It is evident that the above interpolations are identical to those originally written for the rectangular quadrilateral element in Chapter 8 of Volume 1. Only the parameters are different. Based on results obtained in Volume 1 we also know the element will not perform well in bending situations because of 'shear locking', especially when the aspect ratio of the element thickness to length becomes very small. In order to improve the behaviour we introduce a three-field approximation by using the enhanced strain concept described in Sec. 11.5 of Volume 1. Accordingly, the mixed strain approximation will be taken as

$$\begin{Bmatrix} \varepsilon_x \\ \varepsilon_y \\ \gamma_{xy} \end{Bmatrix} = \begin{bmatrix} N_{i,x} & \dfrac{\eta \tilde{t}}{2\bar{t}} N_{i,x} & 0 & 0 \\ 0 & 0 & 0 & \dfrac{1}{\bar{t}} N_i \\ 0 & \dfrac{1}{\bar{t}} N_i & N_{i,x} & \dfrac{\eta \tilde{t}}{2\bar{t}} N_{i,x} \end{bmatrix} \begin{Bmatrix} \tilde{u}_i \\ \Delta\tilde{u}_i \\ \tilde{v}_i \\ \Delta\tilde{v}_i \end{Bmatrix} + \begin{bmatrix} \xi & 0 & 0 & 0 \\ 0 & \eta & 0 & 0 \\ 0 & 0 & \xi & \eta \end{bmatrix} \begin{Bmatrix} \beta_1 \\ \beta_2 \\ \beta_3 \\ \beta_4 \end{Bmatrix} \tag{5.79}$$

where β_i are parameters of the enhanced strains (see Sec. 11.5.4, Volume 1). The remainder of the development is straightforward using the form given in Sec. 11.5

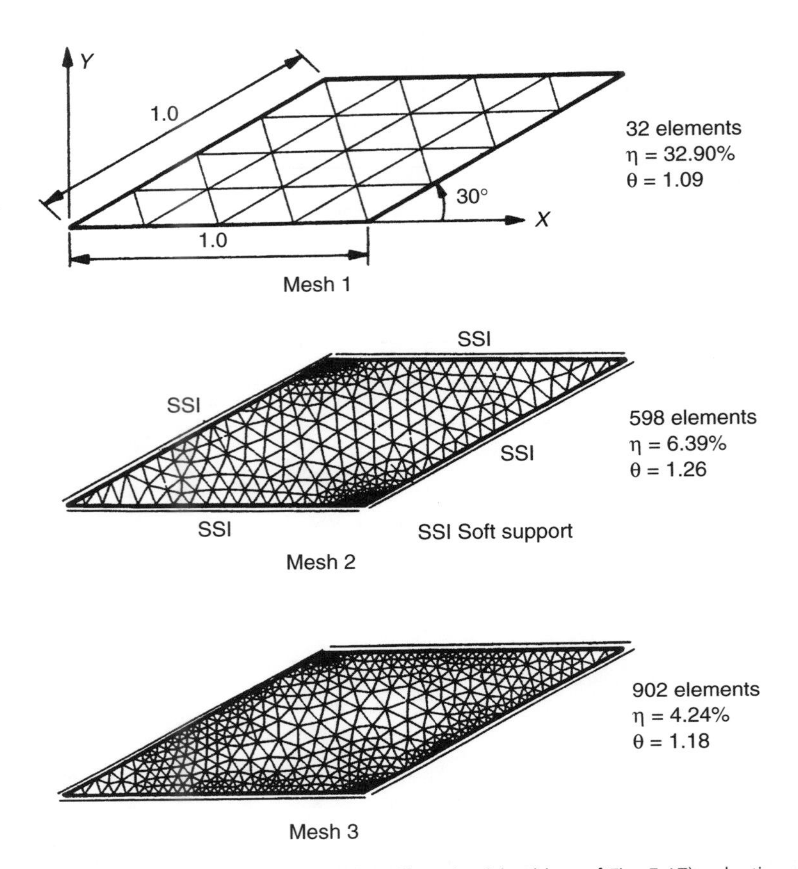

Fig. 5.19 Simply supported 30° skew plate with uniform load (problem of Fig. 5.17); adaptive analysis to achieve 5 per cent accuracy; $L/t = 100$, $\nu = 0.3$, six-node element T6S3B3;[15] θ = effectivity index, η = percentage error in energy norm of estimator.

of Volume 1 and is left as an exercise for the reader. We do note that here it is not necessary to use a constitutive equation which has been reduced to give zero stress in the through-thickness (y) direction. By including additional enhanced terms in the thickness direction one may use the three-dimensional constitutive equations directly. Such developments have been pursued for plate and shell applications.[63–66] We note that while the above form can be used for flat surfaces and easily extended for smoothly curved surfaces it has difficulties when 'kinks' or multiple branches are encountered as then there is no unique 'thickness' direction. Thus, considerable additional work remains to be done to make this a generally viable approach.

5.11 Inelastic material behaviour

We have discussed in some detail the problem of inelastic behaviour in Sec. 4.19 of Chapter 4. The procedures of dealing with the same situation when using the

Reissner–Mindlin theory are nearly identical and here we will simply refer the reader to the literature on the subject[67,68] and to the previous chapter.

5.12 Concluding remarks – adaptive refinement

The simplicity of deriving and using elements in which independent interpolation of rotations and displacements is postulated and shear deformations are included assures popularity of the approach. The final degrees of freedom used are exactly the same type as those used in the direct approach to thin plate forms in Chapter 4, and at no additional complexity shear deformability is included in the analysis.

If care is used to ensure robustness, elements of the type discussed in this chapter are generally applicable and indeed could be used with similar restrictions to other finite element approximations requiring C_1 continuity in the limit.

The ease of element distortion will make elements of the type discussed here the first choice for curved element solutions and they can easily be adapted to non-linear material behaviour. Extension to geometric non-linearity is also possible; however, in this case the effects of in-plane forces must be included and this renders the problem identical to shell theory. We shall discuss this more in Chapter 11.

In Chapters 14 and 15 of Volume 1 we discussed the need for an adaptive approach in which error estimation is used in conjunction with mesh generation to obtain answers of specified accuracy. Such adaptive procedures are easily used in plate bending problems with an almost identical form of error estimation.[69]

In Figs 5.19 and 5.20 we show a sequence of automatically generated meshes for the problem of the skew plate. It is of particular interest to note:

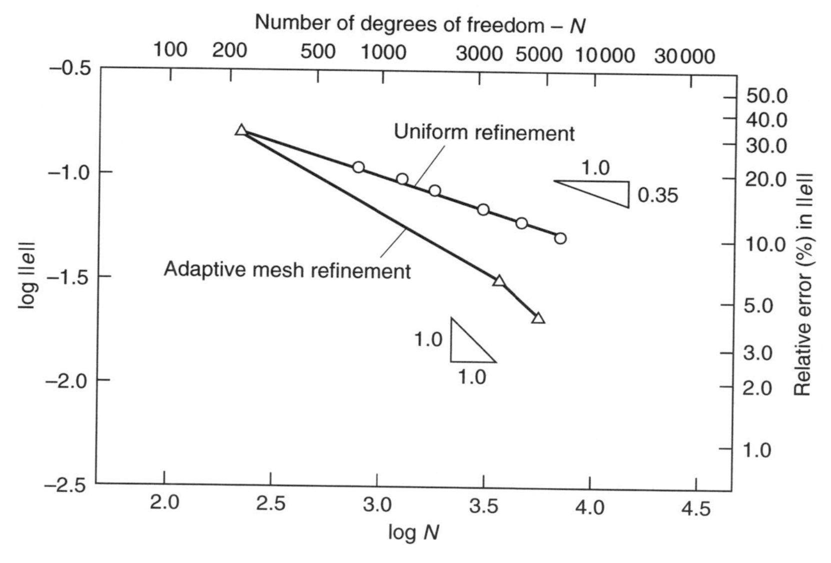

Fig. 5.20 Energy norm rate of convergence for the 30° skew plate of Fig. 5.17 for uniform and adaptive refinement; adaptive analysis to achieve 5 per cent accuracy.

1. the initial refinement in the vicinity of corner singularity;
2. the final refinement near the simply support boundary conditions where the effects of transverse shear will lead to a 'boundary layer'.

Indeed, such boundary layers can occur near all boundaries of shear deformable plates and it is usually found that the shear error represents a very large fraction of the total error when approximations are made.

References

1. S. Ahmad, B.M. Irons and O.C. Zienkiewicz. Curved thick shell and membrane elements with particular reference to axi-symmetric problems. In L. Berke, R.M. Bader, W.J. Mykytow, J.S. Przemienicki and M.H. Shirk (eds), *Proc. 2nd Conf. Matrix Methods in Structural Mechanics*, Volume AFFDL-TR-68-150, pp. 539–72, Air Force Flight Dynamics Laboratory, Wright Patterson Air Force Base, OH, October 1968.
2. S. Ahmad, B.M. Irons and O.C. Zienkiewicz. Analysis of thick and thin shell structures by curved finite elements. *Int. J. Num. Meth. Eng.*, **2**, 419–51, 1970.
3. S. Ahmad. *Curved Finite Elements in the Analysis of Solids, Shells and Plate Structures*, PhD thesis, University of Wales, Swansea, 1969.
4. O.C. Zienkiewicz, J. Too and R.L. Taylor. Reduced integration technique in general analysis of plates and shells. *Int. J. Num. Meth. Eng.*, **3**, 275–90, 1971.
5. S.F. Pawsey and R.W. Clough. Improved numerical integration of thick slab finite elements. *Int. J. Num. Meth. Eng.*, **3**, 575–86, 1971.
6. O.C. Zienkiewicz and E. Hinton. Reduced integration, function smoothing and non-conformity in finite element analysis. *J. Franklin Inst.*, **302**, 443–61, 1976.
7. E.D.L. Pugh, E. Hinton and O.C. Zienkiewicz. A study of quadrilateral plate bending elements with reduced integration. *Int. J. Num. Meth. Eng.*, **12**, 1059–79, 1978.
8. T.J.R. Hughes, R.L. Taylor and W. Kanoknukulchai. A simple and efficient finite element for plate bending. *Int. J. Num. Meth. Eng.*, **11**, 1529–43, 1977.
9. T.J.R. Hughes and M. Cohen. The 'heterosis' finite element for plate bending. *Computers and Structures*, **9**, 445–50, 1978.
10. T.J.R. Hughes, M. Cohen and M. Harou. Reduced and selective integration techniques in the finite element analysis of plates. *Nuclear Engineering and Design*, **46**, 203–22, 1978.
11. E. Hinton and N. Biĉaniĉ. A comparison of Lagrangian and serendipity Mindlin plate elements for free vibration analysis. *Computers and Structures*, **10**, 483–93, 1979.
12. R.D. Cook. *Concepts and Applications of Finite Element Analysis*, John Wiley, Chichester, Sussex, 1982.
13. R.L. Taylor, O.C. Zienkiewicz, J.C. Simo and A.H.C. Chan. The patch test – a condition for assessing FEM convergence. *Int. J. Num. Meth. Eng.*, **22**, 39–62, 1986.
14. O.C. Zienkiewicz, S. Qu, R.L. Taylor and S. Nakazawa. The patch test for mixed formulations. *Int. J. Num. Meth. Eng.*, **23**, 1873–83, 1986.
15. O.C. Zienkiewicz and D. Lefebvre. A robust triangular plate bending element of the Reissner–Mindlin type. *Int. J. Num. Meth. Eng.*, **26**, 1169–84, 1988.
16. O.C. Zienkiewicz, J.P. Vilotte, S. Toyoshima and S. Nakazawa. Iterative method for constrained and mixed approximation: an inexpensive improvement of FEM performance. *Comp. Meth. Appl. Mech. Eng.*, **51**, 3–29, 1985.
17. D.S. Malkus and T.J.R. Hughes. Mixed finite element methods in reduced and selective integration techniques: a unification of concepts. *Comp. Meth. Appl. Mech. Eng.*, **15**, 63–81, 1978.

18. B. Fraeijs de Veubeke. Displacement and equilibrium models in finite element method. In O.C. Zienkiewicz and G.S. Holister (eds), *Stress Analysis*, pp. 145–97. John Wiley, Chichester, Sussex, 1965.
19. K.-J. Bathe. *Finite Element Procedures*, Prentice-Hall, Englewood Cliffs, NJ, 1996.
20. W.X. Zhong. FEM patch test and its convergence. Technical Report 97-3001, Research Institute Engineering Mechanics, Dalian University of Technology, 1997, [in Chinese].
21. W.X. Zhong. Convergence of FEM and the conditions of patch test. Technical Report 97-3002, Research Institute Engineering Mechanics, Dalian University of Technology, 1997, [in Chinese].
22. I. Babuška and R. Narasimhan. The Babuška–Brezzi condition and the patch test: an example. *Comp. Meth. Appl. Mech. Eng.*, **140**, 183–99, 1997.
23. O.C. Zienkiewicz and R.L. Taylor. The finite element patch test revisited: a computer test for convergence, validation and error estimates. *Comp. Meth. Appl. Mech. Eng.*, **149**, 523–44, 1997.
24. T.J.R. Hughes and W.K. Liu. Implicit–explicit finite elements in transient analyses: part I and part II. *J. Appl. Mech.*, **45**, 371–8, 1978.
25. K.-J. Bathe and L.W. Ho. Some results in the analysis of thin shell structures. In W. Sunderlich *et al.* (eds), *Nonlinear Finite Element Analysis in Structural Mechanics*, pp. 122–56, Springer-Verlag, Berlin, 1981.
26. O.C. Zienkiewicz, Z. Xu, L.F. Zeng, A. Samuelsson and N.-E. Wiberg. Linked interpolation for Reissner–Mindlin plate elements: part I – a simple quadrilateral element. *Int. J. Num. Meth. Eng.*, **36**, 3043–56, 1993.
27. Z. Xu, O.C. Zienkiewicz and L.F. Zeng. Linked interpolation for Reissner–Mindlin plate elements: part III – an alternative quadrilateral. *Int. J. Num. Meth. Eng.*, **37**, 1437–43, 1994.
28. F. Auricchio and R.L. Taylor. A shear deformable plate element with an exact thin limit. *Comp. Meth. Appl. Mech. Eng.*, **118**, 393–412, 1994.
29. Z. Xu. A simple and efficient triangular finite element for plate bending. *Acta Mechanica Sinica*, **2**, 185–92, 1986.
30. D.N. Arnold and R.S. Falk. A uniformly accurate finite element method for Mindlin–Reissner plate. Technical Report IMA Preprint Series No. 307, Institute for Mathematics and its Application, University of Maryland, College Park, MD, 1987.
31. T.J.R. Hughes and T. Tezduyar. Finite elements based upon Mindlin plate theory with particular reference to the four node bilinear isoparametric element. *J. Appl. Mech.*, **46**, 587–96, 1981.
32. E.N. Dvorkin and K.-J. Bathe. A continuum mechanics based four node shell element for general non-linear analysis. *Engineering Computations*, **1**, 77–88, 1984.
33. K.-J. Bathe and A.B. Chaudhary. A solution method for planar and axisymmetric contact problems. *Int. J. Num. Meth. Eng.*, **21**, 65–88, 1985.
34. H.C. Huang and E. Hinton. A nine node Lagrangian Mindlin element with enhanced shear interpolation. *Engineering Computations*, **1**, 369–80, 1984.
35. E. Hinton and H.C. Huang. A family of quadrilateral Mindlin plate elements with substitute shear strain fields. *Computers and Structures*, **23**, 409–31, 1986.
36. O.C. Zienkiewicz, R.L. Taylor, P. Papadopoulos and E. Oñate. Plate bending elements with discrete constraints; new triangular elements. *Computers and Structures*, **35**, 505–22, 1990.
37. K.-J. Bathe and F. Brezzi. On the convergence of a four node plate bending element based on Mindlin–Reissner plate theory and a mixed interpolation. In J.R. Whiteman (ed.), *The Mathematics of Finite Elements and Applications*, Volume V, pp. 491–503, Academic Press, London, 1985.
38. H.K. Stolarski and M.Y.M. Chiang. On a definition of the assumed shear strains in formulation of the C^0 plate elements. *European Journal of Mechanics, A/Solids*, **8**, 53–72, 1989.

39. H.K. Stolarski and M.Y.M. Chiang. Thin-plate elements with relaxed Kirchhoff constraints. *Int. J. Num. Meth. Eng.*, **26**, 913–33, 1988.
40. R.S. Rao and H.K. Stolarski. Finite element analysis of composite plates using a weak form of the Kirchhoff constraints. *Finite Elements in Analysis and Design*, **13**, 191–208, 1993.
41. K.-J. Bathe, M.L. Bucalem and F. Brezzi. Displacement and stress convergence of four MITC plate bending elements. *Engineering Computations*, **7**, 291–302, 1990.
42. E. Oñate, R.L. Taylor and O.C. Zienkiewicz. Consistent formulation of shear constrained Reissner–Mindlin plate elements. In C. Kuhn and H. Mang (eds), *Discretization Methods in Structural Mechanics*, pp. 169–80. Springer-Verlag, Berlin, 1990.
43. E. Oñate, O.C. Zienkiewicz, B. Suárez and R.L. Taylor. A general methodology for deriving shear constrained Reissner–Mindlin plate elements. *Int. J. Num. Meth. Eng.*, **33**, 345–67, 1992.
44. P. Papadopoulos and R.L. Taylor. A triangular element based on Reissner–Mindlin plate theory. *Int. J. Num. Meth. Eng.*, **30**, 1029–49, 1990.
45. G.S. Dhatt. Numerical analysis of thin shells by curved triangular elements based on discrete Kirchhoff hypotheses. In W.R. Rowan and R.M. Hackett (eds), *Proc. Symp. on Applications of FEM in Civil Engineering*, Vandervilt University, Nashville, TN, 1969.
46. J.L. Batoz, K.-J. Bathe and L.W. Ho. A study of three-node triangular plate bending elements. *Int. J. Num. Meth. Eng.*, **15**, 1771–812, 1980.
47. J.L. Batoz. An explicit formulation for an efficient triangular plate bending element. *Int. J. Num. Meth. Eng.*, **18**, 1077–89, 1982.
48. J.L. Batoz and P. Lardeur. A discrete shear triangular nine d.o.f. element for the analysis of thick to very thin plates. *Int. J. Num. Meth. Eng.*, **28**, 533–60, 1989.
49. T. Tu. *Performance of Reissner–Mindlin elements*, PhD thesis, Department of Mathematics, Rutgers University, New Brunswick, NJ, 1998.
50. A. Tessler and T.J.R. Hughes. A three node Mindlin plate element with improved transverse shear. *Comp. Meth. Appl. Mech. Eng.*, **50**, 71–101, 1985.
51. A. Tessler. A C_0 anisoparametric three node shallow shell element. *Comp. Meth. Appl. Mech. Eng.*, **78**, 89–103.
52. A. Tessler and S.B. Dong. On a hierarchy of conforming Timoshenko beam elements. *Computers and Structures*, **14**, 335–44, 1981.
53. M.A. Crisfield. *Finite Elements and Solution Procedures for Structural Analysis, Volume 1, Linear Analysis*, Pineridge Press, Swansea, 1986.
54. L.F. Greimann and P.P. Lynn. Finite element analysis of plate bending with transverse shear deformation. *Nuclear Engineering and Design*, **14**, 223–30, 1970.
55. P.P. Lynn and B.S. Dhillon. Triangular thick plate bending elements. In *First Int. Conf. Struct. Mech. in Reactor Tech.*, p. M 6/5, Berlin, 1971.
56. Z. Xu. A thick–thin triangular plate element. *Int. J. Num. Meth. Eng.*, **33**, 963–73, 1992.
57. R.L. Taylor and F. Auricchio. Linked interpolation for Reissner–Mindlin plate elements: part II – a simple triangle. *Int. J. Num. Meth. Eng.*, **36**, 3057–66, 1993.
58. F. Auricchio and R.L. Taylor. A triangular thick plate finite element with an exact thin limit. *Finite Elements in Analysis and Design*, **19**, 57–68, 1995.
59. M.A. Crisfield. *Non-linear Finite Element Analysis of Solids and Structures*, Volume 1, John Wiley, Chichester, Sussex, 1991.
60. E.L. Wilson. The static condensation algorithm. *Int. J. Num. Meth. Eng.*, **8**, 199–203, 1974.
61. L.S.D. Morley. *Skew Plates and Structures*, International Series of Monographs in Aeronautics and Astronautics, Macmillan, New York, 1963.
62. I. Babuška and T. Scapolla. Benchmark computation and performance evaluation for a rhombic plate bending problem. *Int. J. Num. Meth. Eng.*, **28**, 155–80, 1989.

63. N. Büchter, E. Ramm and D. Roehl. Three-dimensional extension of nonlinear shell formulations based on the enhanced assumed strain concept. *Int. J. Num. Meth. Eng.*, **37**, 2551–68, 1994.
64. M. Braun, M. Bischoff and E. Ramm. Nonlinear shell formulations for complete three-dimensional constitutive laws include composites and laminates. *Computational Mechanics*, **15**, 1–18, 1994.
65. P. Betsch, F. Gruttmann and E. Stein. A 4-node finite shell element for the implementation of general hyperelastic 3d-elasticity at finite strains. *Comp. Meth. Appl. Mech. Eng.*, **130**, 57–79, 1996.
66. M. Bischoff and E. Ramm. Shear deformable shell elements for large strains and rotations. *Int. J. Num. Meth. Eng.*, **40**, 4427–49, 1997.
67. P. Papadopoulos and R.L. Taylor. Elasto-plastic analysis of Reissner–Mindlin plates. *Appl. Mech. Rev.*, **43**(5), S40–S50, 1990.
68. P. Papadopoulos and R.L. Taylor. An analysis of inelastic Reissner–Mindlin plates. *Finite Elements in Analysis and Design*, **10**, 221–33, 1991.
69. O.C. Zienkiewicz and J.Z. Zhu. Error estimation and adaptive refinement for plate bending problems. *Int. J. Num. Meth. Eng.*, **28**, 2839–53, 1989.

6

Shells as an assembly of flat elements

6.1 Introduction

A shell is, in essence, a structure that can be derived from a plate by initially forming the middle surface as a singly (or doubly) curved surface. The same assumptions as used in thin plates regarding the transverse distribution of strains and stresses are again valid. However, the way in which the shell supports external loads is quite different from that of a flat plate. The stress resultants acting on the middle surface of the shell now have both tangential and normal components which carry a major part of the load, a fact that explains the economy of shells as load-carrying structures and their well-deserved popularity.

The derivation of detailed governing equations for a curved shell problem presents many difficulties and, in fact, leads to many alternative formulations, each depending on the approximations introduced. For details of classical shell treatment the reader is referred to standard texts on the subject, for example, the well-known treatise by Flügge[1] or the classical book by Timoshenko and Woinowski-Krieger.[2]

In the finite element treatment of shell problems to be described in this chapter the difficulties referred to above are eliminated, at the expense of introducing a further approximation. This approximation is of a physical, rather than mathematical, nature. In this it is assumed that the behaviour of a continuously curved surface can be adequately represented by the behaviour of a surface built up of small flat elements. Intuitively, as the size of the subdivision decreases it would seem that convergence must occur and indeed experience indicates such a convergence.

It will be stated by many shell experts that when we compare the *exact* solution of a shell approximated by flat facets to the exact solution of a truly curved shell, considerable differences in the distribution of bending moments, etc., occur. It is arguable if this is true, but for simple elements the discretization error is approximately of the same order and excellent results can be obtained with flat shell element approximation. The mathematics of this problem is discussed in detail by Ciarlet.[3]

In a shell, the element generally will be subject both to bending and to 'in-plane' force resultants. For a flat element these cause independent deformations, provided the local deformations are small, and therefore the ingredients for obtaining the necessary stiffness matrices are available in the material already covered in the preceding chapters and Volume 1.

In the division of an arbitrary shell into flat elements only triangular elements can be used for doubly curved surfaces. Although the concept of the use of such elements in the analysis was suggested as early as 1961 by Greene *et al.*,[4] the success of such analysis was hampered by the lack of a good stiffness matrix for triangular plate elements in bending.[5-8] The developments described in Chapters 4 and 5 open the way to adequate models for representing the behaviour of shells with such a division.

Some shells, for example those with general cylindrical shapes (can be well represented by flat elements of rectangular or quadrilateral shape provided the mesh subdivision does not lead to 'warped' elements). With good stiffness matrices available for such elements the progress here has been more satisfactory. Practical problems of arch dam design and others for cylindrical shape roofs have been solved quite early with such subdivisions.[9,10]

Clearly, the possibilities of analysis of shell structures by the finite element method are enormous. Problems presented by openings, variation of thickness, or anisotropy are no longer of consequence.

A special case is presented by axisymmetrical shells. Although it is obviously possible to deal with these in the way described in this chapter, a simpler approach can be used. This will be presented in Chapters 7–9.

As an alternative to the type of analysis described here, curved shell elements could be used. Here curvilinear coordinates are essential and general procedures in Chapter 9 of volume 1 can be extended to define these. The physical approximation involved in flat elements is now avoided at the expense of reintroducing an arbitrariness of various shell theories. Several approaches using a direct displacement approximation are given in references 11–31, and the use of 'mixed variational principles are given in references 32–35.

A very simple and effective way of deriving curved shell elements is to use the so-called 'shallow' shell theory approach.[13,14,36,37] Here the variables u, v, w define the *tangential* and *normal* components of displacement to the curved surface. If all the elements are assumed to be tangential to each other, no need arises to transfer these from local to global values. The element is assumed to be 'shallow' with respect to a local coordinate system representing its projection on a plane defined by nodal points, and its strain energy is defined by appropriate equations that include derivatives with respect to *coordinates in the plane of projection*. Thus, precisely the same shape functions can be used as in flat elements discussed in this chapter and all integrations are in fact carried out in the 'plane' as before.

Such shallow shell elements, by coupling the effects of membrane and bending strain in the energy expression, are slightly more efficient than flat ones where such coupling occurs on the interelement boundary only. For simple, small elements the gains are marginal, but with few higher order large elements advantages appear. A good discussion of such a formulation is given in reference 22.

For many practical purposes the flat element approximation gives very adequate answers and also permits an easy coupling with edge beam and rib members, a facility sometimes not present in a curved element formulation. Indeed, in many practical problems the structure is in fact composed of flat surfaces, at least in part, and these can be simply reproduced. For these reasons curved general thin shell forms will not be discussed here and instead a general formulation of thick curved shells (based directly on three-dimensional behaviour and avoiding the shell equation ambiguities) will be presented in Chapter 8. The development of curved elements for general shell theories also can be effected in a direct manner; however, additional transformations over those discussed in this chapter are involved. The interested reader is referred to references 38 and 39 for additional discussion on this approach. In many respects the differences in the two approaches are quite similar, as shown by Bischoff and Ramm.[40]

In most arbitrary shaped, curved shell elements the coordinates used are such that complete smoothness of the surface between elements is not guaranteed. The shape discontinuity occurring there, and indeed on any shell where 'branching' occurs, is precisely of the same type as that encountered in this chapter and therefore the methodology of assembly discussed here is perfectly general.

6.2 Stiffness of a plane element in local coordinates

Consider a typical polygonal flat element in a *local* coordinate system $\bar{x}\bar{y}\bar{z}$ subject simultaneously to 'in-plane' and 'bending' actions (Fig. 6.1).

Taking first the *in-plane* (plane stress) action, we know from Chapter 4 of Volume 1 that the state of strain is uniquely described in terms of the $\bar{u}$ and $\bar{v}$ displacement of each typical node i. The minimization of the total potential energy led to the stiffness matrices described there and gives 'nodal' forces due to displacement parameters $\bar{\mathbf{a}}^{\mathrm{p}}$ as

$$(\bar{\mathbf{f}}^e)^{\mathrm{p}} = (\bar{\mathbf{K}}^e)^{\mathrm{p}}\,\bar{\mathbf{a}}^{\mathrm{p}} \qquad \text{with} \qquad \bar{\mathbf{a}}_i^{\mathrm{p}} = \begin{Bmatrix} \bar{u}_i \\ \bar{v}_i \end{Bmatrix} \qquad \bar{\mathbf{f}}_i^{\mathrm{p}} = \begin{Bmatrix} F_{\bar{x}i} \\ F_{\bar{y}i} \end{Bmatrix} \tag{6.1}$$

Similarly, when bending was considered in Chapters 4 and 5, the state of strain was given uniquely by the nodal displacement in the $\bar{z}$ direction ($\bar{w}$) and the two rotations

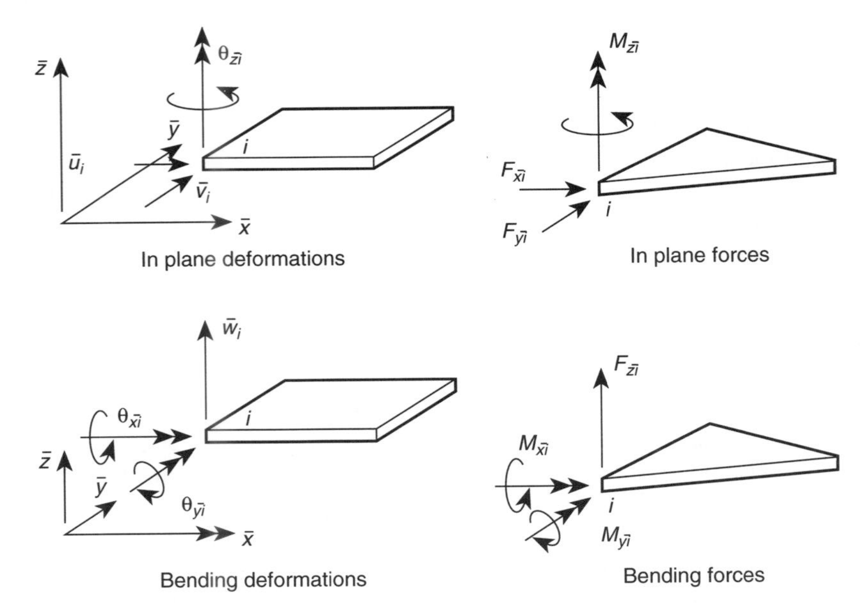

Fig. 6.1 A flat element subject to 'in-plane' and 'bending' actions.

$\theta_{\bar{x}}$ and $\theta_{\bar{y}}$. This resulted in stiffness matrices of the type

$$(\bar{\mathbf{f}}^e)^{\mathrm{b}} = (\bar{\mathbf{K}}^e)^{\mathrm{b}}\bar{\mathbf{a}}^{\mathrm{b}} \qquad \text{with} \qquad \bar{\mathbf{a}}_i^{\mathrm{b}} = \begin{Bmatrix} \bar{w}_i \\ \theta_{\bar{x}i} \\ \theta_{\bar{y}i} \end{Bmatrix} \qquad \bar{\mathbf{f}}_i^{\mathrm{b}} = \begin{Bmatrix} F_{\bar{z}i} \\ M_{\bar{x}i} \\ M_{\bar{y}i} \end{Bmatrix} \tag{6.2}$$

Before combining these stiffnesses it is important to note two facts. The first is that the displacements prescribed for 'in-plane' forces do not affect the bending deformations and vice versa. The second is that the rotation $\theta_{\bar{z}}$ does not enter as a parameter into the definition of deformations in either mode. While one could neglect this entirely at the present stage it is convenient, for reasons which will be apparent later when assembly is considered, to take this rotation into account and associate with it a fictitious couple $M_{\bar{z}}$. The fact that it does not enter into the minimization procedure can be accounted for simply by inserting an appropriate number of zeros into the stiffness matrix.

Redefining the combined nodal displacement as

$$\bar{\mathbf{a}}_i = \begin{bmatrix} \bar{u}_i & \bar{v}_i & \bar{w}_i & \theta_{\bar{x}i} & \theta_{\bar{y}i} & \theta_{\bar{z}i} \end{bmatrix}^{\mathrm{T}} \tag{6.3}$$

and the appropriate nodal 'forces' as

$$\bar{\mathbf{f}}_i^e = \begin{bmatrix} F_{\bar{x}i} & F_{\bar{y}i} & F_{\bar{z}i} & M_{\bar{x}i} & M_{\bar{y}i} & M_{\bar{z}i} \end{bmatrix}^{\mathrm{T}} \tag{6.4}$$

we can write

$$\bar{\mathbf{K}}^e\,\bar{\mathbf{a}} = \bar{\mathbf{f}}^e \tag{6.5}$$

The stiffness matrix is now made up from the following submatrices

$$\bar{\mathbf{K}}_{rs} = \left[\begin{array}{cc:ccc:c} \multicolumn{2}{c:}{\bar{\mathbf{K}}_{rs}^{\mathrm{p}}} & 0 & 0 & 0 & 0 \\ & & 0 & 0 & 0 & 0 \\ \hdashline 0 & 0 & & & & 0 \\ 0 & 0 & & \bar{\mathbf{K}}_{rs}^{\mathrm{b}} & & 0 \\ 0 & 0 & & & & 0 \\ \hdashline 0 & 0 & 0 & 0 & 0 & 0 \end{array}\right] \tag{6.6}$$

if we note that

$$\bar{\mathbf{a}}_i = \begin{bmatrix} \bar{\mathbf{a}}_i^{\mathrm{p}} & \bar{\mathbf{a}}_i^{\mathrm{b}} & \theta_{\bar{z}i} \end{bmatrix}^{\mathrm{T}} \tag{6.7}$$

The above formulation is valid for any shape of polygonal element and, in particular, for the two important types illustrated in Fig. 6.1.

6.3 Transformation to global coordinates and assembly of elements

The stiffness matrix derived in the previous section used a system of local coordinates as the 'reference plane', and forces and bending components also are originally derived for this system.

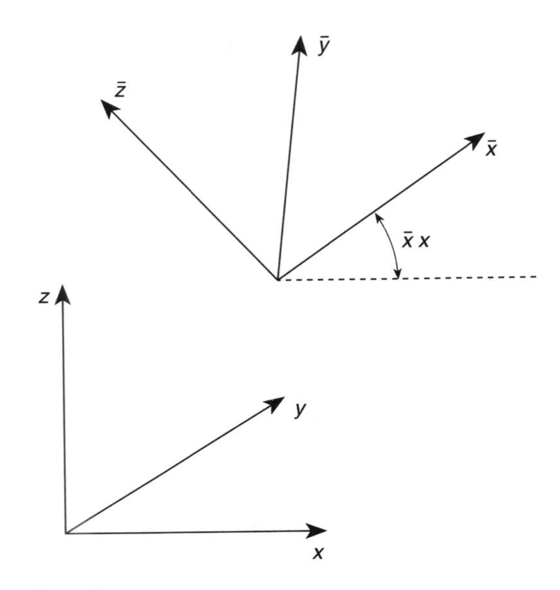

Fig. 6.2 Local and global coordinates.

Transformation of coordinates to a common global system (which will be denoted by xyz with the local system still $\bar{x}\bar{y}\bar{z}$) will be necessary to assemble the elements and to write the appropriate equilibrium equations.

In addition it will be initially more convenient to specify the element nodes by their global coordinates and to establish from these the local coordinates, thus requiring an inverse transformation. All the transformations are accomplished by a simple process.

The two systems of coordinates are shown in Fig 6.2. The forces and displacements of a node transform from the global to the local system by a matrix $\mathbf{T}$ giving

$$\bar{\mathbf{a}}_i = \mathbf{T}\,\mathbf{a}_i \qquad \bar{\mathbf{f}}_i = \mathbf{T}\mathbf{f}_i \tag{6.8}$$

in which

$$\mathbf{T} = \begin{bmatrix} \mathbf{\Lambda} & \mathbf{0} \\ \mathbf{0} & \mathbf{\Lambda} \end{bmatrix} \tag{6.9}$$

with $\mathbf{\Lambda}$ being a 3×3 matrix of direction cosines between the two sets of axes,[41,42] that is,

$$\mathbf{\Lambda} = \begin{bmatrix} \cos(\bar{x}, x) & \cos(\bar{x}, y) & \cos(\bar{x}, z) \\ \cos(\bar{y}, x) & \cos(\bar{y}, y) & \cos(\bar{y}, z) \\ \cos(\bar{z}, x) & \cos(\bar{z}, y) & \cos(\bar{z}, z) \end{bmatrix} = \begin{bmatrix} \Lambda_{\bar{x}x} & \Lambda_{\bar{x}y} & \Lambda_{\bar{x}z} \\ \Lambda_{\bar{y}x} & \Lambda_{\bar{y}y} & \Lambda_{\bar{y}z} \\ \Lambda_{\bar{z}x} & \Lambda_{\bar{z}y} & \Lambda_{\bar{z}z} \end{bmatrix} \tag{6.10}$$

where $\cos(\bar{x}, x)$ is the cosine of the angle between the $\bar{x}$-axis and the x-axis, and so on.

By the rules of orthogonal transformation the inverse of $\mathbf{T}$ is given by its transpose (see Sec. 1.8 of Volume 1); thus we have

$$\mathbf{a}_i = \mathbf{T}^{\mathrm{T}}\,\bar{\mathbf{a}}_i \qquad \mathbf{f}_i = \mathbf{T}^{\mathrm{T}}\bar{\mathbf{f}}_i \tag{6.11}$$

which permits the stiffness matrix of an element in the global coordinates to be computed as

$$\mathbf{K}^e_{rs} = \mathbf{T}^\mathrm{T} \bar{\mathbf{K}}^e_{rs} \mathbf{T} \tag{6.12}$$

in which $\bar{\mathbf{K}}^e_{rs}$ is determined by Eq. (6.6) in the local coordinates.

The determination of the local coordinates follows a similar pattern. The relationship between global and local systems is given by

$$\begin{Bmatrix} \bar{x} \\ \bar{y} \\ \bar{z} \end{Bmatrix} = \boldsymbol{\Lambda} \begin{Bmatrix} x - x_0 \\ y - y_0 \\ z - z_0 \end{Bmatrix} \tag{6.13}$$

where x_0, y_0, z_0 is the distance from the origin of the global coordinates to the origin of the local coordinates. As in the computation of stiffness matrices for flat plane and bending elements the position of the origin is immaterial, this transformation will always suffice for determination of the local coordinates in the plane (or a plane parallel to the element).

Once the stiffness matrices of all the elements have been determined in a common global coordinate system, the assembly of the elements and forces follow the standard solution pattern. The resulting displacements calculated are referred to the global system, and before the stresses can be computed it is necessary to change these to the local system for each element. The usual stress calculations for 'in-plane' and 'bending' components can then be used.

6.4 Local direction cosines

The determination of the direction cosine matrix $\boldsymbol{\Lambda}$ gives rise to some algebraic difficulties and, indeed, is not unique since the direction of one of the local axes is arbitrary, provided it lies in the plane of the element. We shall first deal with the assembly of rectangular elements in which this problem is particularly simple; later we shall consider the case for triangular elements arbitrarily orientated in space.

6.4.1 Rectangular elements

Such elements are limited in use to representing a cylindrical or box type of surface. It is convenient to take one side of each element and the corresponding $\bar{x}$-axis parallel to the global x-axis. For a typical element *ijkm*, illustrated in Fig 6.3, it is now easy to calculate all the relevant direction cosines. Direction cosines of $\bar{x}$ are, obviously,

$$\Lambda_{\bar{x}x} = 1 \qquad \Lambda_{\bar{x}y} = \Lambda_{\bar{x}z} = 0 \tag{6.14}$$

The direction cosines of the $\bar{y}$ axis have to be obtained by consideration of the coordinates of the various nodal points. Thus,

$$\begin{aligned} \Lambda_{\bar{y}x} &= 0 \\ \Lambda_{\bar{y}y} &= \frac{y_m - y_i}{\sqrt{(y_m - y_i)^2 + (z_m - z_i)^2}} \end{aligned} \tag{6.15}$$

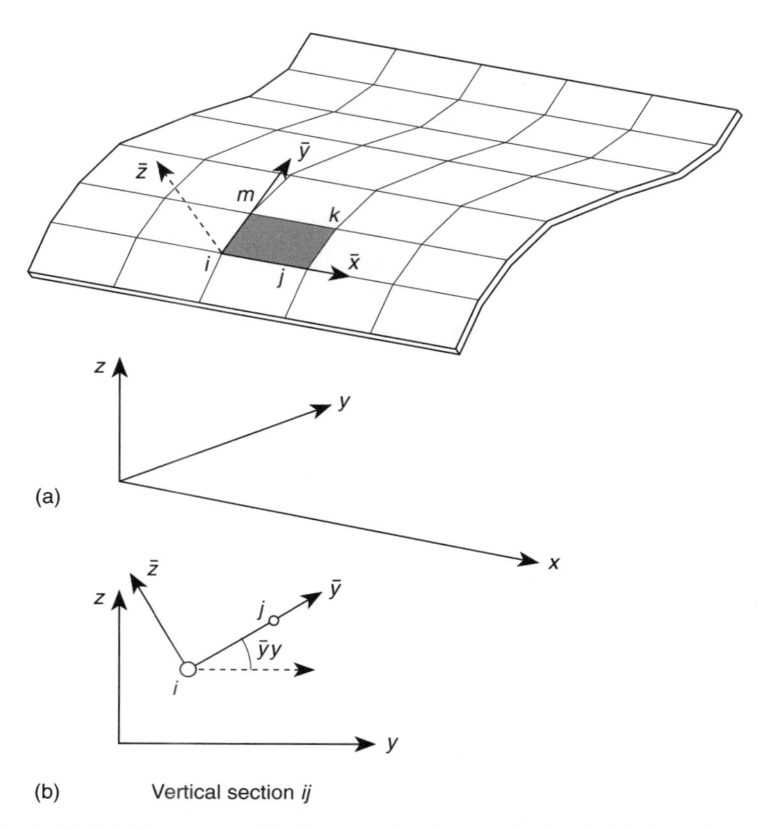

Fig. 6.3 A cylindrical shell as an assembly of rectangular elements: local and global coordinates.

$$\Lambda_{\bar{y}z} = \frac{z_m - z_i}{\sqrt{(y_m - y_i)^2 + (z_m - z_i)^2}} \tag{6.16}$$

are simple geometrical relations which can be obtained by consideration of the sectional plane passing vertically through *im* in the z direction. Similarly, from the same section, we have for the $\bar{z}$ axis

$$\Lambda_{\bar{z}x} = 0$$

$$\Lambda_{\bar{z}y} = \frac{z_i - z_m}{\sqrt{(y_m - y_i)^2 + (z_m - z_i)^2}} = -\Lambda_{\bar{y}z} \tag{6.17}$$

$$\Lambda_{\bar{z}z} = \frac{y_m - y_i}{\sqrt{(y_m - y_i)^2 + (z_m - z_i)^2}} = \Lambda_{\bar{y}y} \tag{6.18}$$

Clearly, the numbering of points in a consistent fashion is important to preserve the correct signs of the expression.

6.4.2 Triangular elements arbitrarily orientated in space

An arbitrary shell divided into triangular elements is shown in Fig. 6.4(a). Each element has an orientation in which the angles with the coordinate planes are arbitrary. The problem of defining local axes and their direction cosines is therefore more complex than in the previous simple example. The most convenient way of dealing with the problem is to use some properties of geometrical vector algebra (see Appendix F, Volume 1).

One arbitrary but convenient choice of local axis direction is given here. We shall specify that the $\bar{x}$ axis is to be directed along the side ij of the triangle, as shown in Fig. 6.4(b).

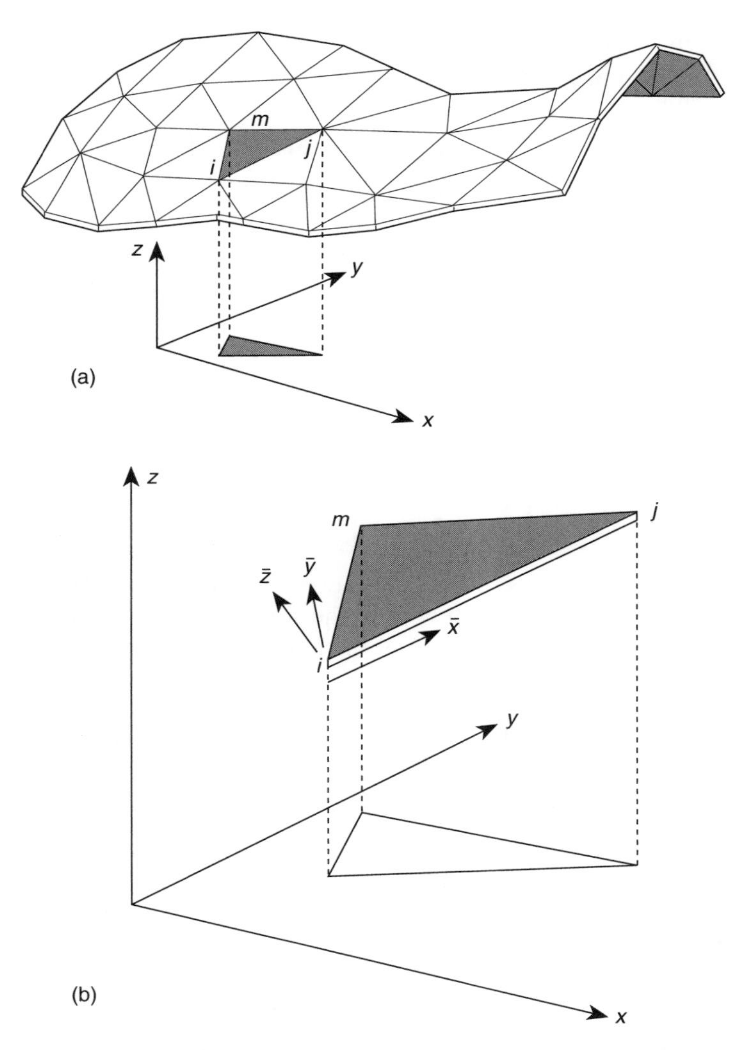

Fig. 6.4 (a) An assemblage of triangular elements representing an arbitrary shell; (b) local and global coordinates for a triangular element.

The vector $\mathbf{V}_{ji}$ defines this side and in terms of global coordinates we have

$$\mathbf{V}_{ji} = \begin{Bmatrix} x_j - x_i \\ y_j - y_i \\ z_j - z_i \end{Bmatrix} = \begin{Bmatrix} x_{ji} \\ y_{ji} \\ z_{ji} \end{Bmatrix} \tag{6.19}$$

The direction cosines are given by dividing the components of this vector by its length, that is, defining a vector of unit length

$$\mathbf{v}_{\bar{x}} = \begin{Bmatrix} \Lambda_{\bar{x}x} \\ \Lambda_{\bar{x}y} \\ \Lambda_{\bar{x}z} \end{Bmatrix} = \frac{1}{l_{ji}} \begin{Bmatrix} x_{ji} \\ y_{ji} \\ z_{ji} \end{Bmatrix} \qquad \text{with} \qquad l_{ji} = \sqrt{x_{ji}^2 + y_{ji}^2 + z_{ji}^2} \tag{6.20}$$

Now, the $\bar{z}$ direction, which must be normal to the plane of the triangle, needs to be established. We can obtain this direction from a 'vector' cross-product of two sides of the triangle. Thus,

$$\mathbf{V}_{\bar{z}} = \mathbf{V}_{ji} \times \mathbf{V}_{mi} = \begin{Bmatrix} y_{ji}z_{mi} - z_{ji}y_{mi} \\ z_{ji}x_{mi} - x_{ji}z_{mi} \\ x_{ji}y_{mi} - y_{ji}x_{mi} \end{Bmatrix} = \begin{Bmatrix} yz_{ijm} \\ zx_{ijm} \\ xy_{ijm} \end{Bmatrix} \tag{6.21}$$

represents a vector normal to the plane of the triangle whose length, by definition (see Appendix F of Volume 1), is equal to twice the area of the triangle. Thus,

$$l_{\bar{z}} = 2\Delta = \sqrt{(yz_{ijm})^2 + (zx_{ijm})^2 + (xy_{ijm})^2}$$

The direction cosines of the $\bar{z}$-axis are available simply as the direction cosines of $\mathbf{V}_{\bar{z}}$, and we have a unit vector

$$\mathbf{v}_{\bar{z}} = \begin{Bmatrix} \Lambda_{\bar{z}x} \\ \Lambda_{\bar{z}y} \\ \Lambda_{\bar{z}z} \end{Bmatrix} = \frac{1}{2\Delta} \begin{Bmatrix} y_{ji}z_{mi} - z_{ji}y_{mi} \\ z_{ji}x_{mi} - x_{ji}z_{mi} \\ x_{ji}y_{mi} - y_{ji}x_{mi} \end{Bmatrix} \tag{6.22}$$

Finally, the direction cosines of the $\bar{y}$-axis are established in a similar manner as the direction cosines of a vector normal both to the $\bar{x}$ direction and to the $\bar{z}$ direction. If vectors of unit length are taken in each of these directions [as in fact defined by Eqs (6.20)–(6.22)] we have simply

$$\mathbf{v}_{\bar{y}} = \begin{Bmatrix} \Lambda_{\bar{y}x} \\ \Lambda_{\bar{y}y} \\ \Lambda_{\bar{y}z} \end{Bmatrix} = \mathbf{v}_{\bar{z}} \times \mathbf{v}_{\bar{x}} = \begin{Bmatrix} \Lambda_{\bar{z}y}\Lambda_{\bar{x}z} - \Lambda_{\bar{z}z}\Lambda_{\bar{x}y} \\ \Lambda_{\bar{z}z}\Lambda_{\bar{x}x} - \Lambda_{\bar{z}x}\Lambda_{\bar{x}z} \\ \Lambda_{\bar{z}x}\Lambda_{\bar{x}y} - \Lambda_{\bar{z}y}\Lambda_{\bar{x}x} \end{Bmatrix} \tag{6.23}$$

without having to divide by the length of the vector, which is now simply unity.

The vector operations involved can be written as a special computer routine in which vector products, normalizing (i.e. division by length), etc., are automatically carried out[43] and there is no need to specify in detail the various operations given above.

In the preceding outline the direction of the $\bar{x}$ axis was taken as lying along one side of the element. A useful alternative is to specify this by the section of the triangle plane with a plane parallel to one of the coordinate planes. Thus, for instance, if we desire to erect the $\bar{x}$ axis along a horizontal contour of the triangle (i.e. a section parallel to the xy plane) we can proceed as follows.

First, the normal direction cosines $\mathbf{v}_{\bar{z}}$ are defined as in Eq. (6.23). Now, the matrix of direction cosines of $\bar{x}$ has to have a zero component in the z direction and thus we have

$$\mathbf{v}_{\bar{x}} = \begin{Bmatrix} \Lambda_{\bar{x}x} \\ \Lambda_{\bar{x}y} \\ 0 \end{Bmatrix} \tag{6.24}$$

As the length of the vector is unity

$$\Lambda_{\bar{x}x}^2 + \Lambda_{\bar{x}y}^2 = 1 \tag{6.25}$$

and as further the *scalar* product of the $\mathbf{v}_{\bar{x}}$ and $\mathbf{v}_{\bar{z}}$ must be zero, we can write

$$\Lambda_{\bar{x}x}\Lambda_{\bar{z}x} + \Lambda_{\bar{x}y}\Lambda_{\bar{z}y} = 0 \tag{6.26}$$

and from these two equations $\mathbf{v}_{\bar{x}}$ can be uniquely determined. Finally, as before

$$\mathbf{v}_{\bar{y}} = \mathbf{v}_{\bar{z}} \times \mathbf{v}_{\bar{x}} \tag{6.27}$$

It should be noted that this transformation will be singular if there is no line in the plane of the element which is parallel to the xy plane, and some other orientation must then be selected. Yet another alternative of a specification of the $\bar{x}$ axis is given in Chapter 8 where we discuss the development of 'shell' elements directly from the three-dimensional equations of solids.

6.5 'Drilling' rotational stiffness – 6 degree-of-freedom assembly

In the formulation described above a difficulty arises if all the elements meeting at a node are co-planar. This situation will happen for flat (folded) shell segments and at straight boundaries of developable surfaces (e.g. cylinders or cones). The difficulty is due to the assignment of a zero stiffness in the $\theta_{\bar{z}i}$ direction of Fig. 6.1 and the fact that classical shell equations do not produce equations associated with this rotational parameter. Inclusion of the third rotation and the associated 'force' $F_{\bar{z}i}$ has obvious benefits for a finite element model in that both rotations and displacements at nodes may be treated in a very simple manner using the transformations just presented.

If the set of assembled equilibrium equations in *local coordinates* is considered at such a point we have six equations of which the last (corresponding to the $\theta_{\bar{z}}$ direction) is simply

$$0\theta_{\bar{z}} = 0 \tag{6.28}$$

As such, an equation of this type presents no special difficulties (solution programs usually detect the problem and issue a warning). However, if the global coordinate directions differ from the local ones and a transformation is accomplished, the six equations mask the fact that the equations are singular. Detection of this singularity is somewhat more difficult and depends on round-off in each computer system.

A number of alternatives have been presented that avoid the presence of this singular behaviour. Two simple ones are:

1. assemble the equations (or just the rotational parts) at points where elements are co-planar in local coordinates (and delete the $0\theta_{\bar{z}} = 0$ equation); and/or
2. insert an arbitrary stiffness coefficient $\bar{k}_{\theta_{\bar{z}}}$ at such points only.

This leads in the local coordinates to replacing Eq. (6.28) by

$$\bar{k}_{\theta_{\bar{z}}} \theta_{\bar{z}_i} = 0 \tag{6.29}$$

which, on transformation, leads to a perfectly well-behaved set of equations from which, by usual processes, all displacements now including $\theta_{\bar{z}i}$, are obtained. As $\theta_{\bar{z}i}$ does not affect the stresses and indeed is uncoupled from all equilibrium equations any value of $\bar{k}_{\theta_{\bar{z}}}$ similar to values already in Eq. (6.6) can be inserted as an external stiffness without affecting the result.

These two approaches lead to programming complexity (as a decision on the co-planar nature is necessary) and an alternative is to modify the formulation so that the rotational parameters arise more naturally and have a real physical significance. This has been a topic of much study[44–56] and the $\theta_{\bar{z}}$ parameter introduced in this way is commonly called a *drilling* degree of freedom, on account of its action to the surface of the shell. An early application considering the rotation as an additional degree of freedom in plane analysis is contained in reference 14. In reference 8 a set of rotational stiffness coefficients was used in a general shell program for all elements whether co-planar or not. These were defined such that in local coordinates overall equilibrium is not disturbed. This may be accomplished by adding to the formulation for each element the term

$$\Pi^* = \Pi + \int_\Omega \alpha_n E t^n \left(\theta_{\bar{z}} - \bar{\theta}_{\bar{z}}\right)^2 \mathrm{d}\Omega \tag{6.30}$$

in which the parameter α_n is a fictitious elastic parameter and $\bar{\theta}_{\bar{z}}$ is a mean rotation of each element which permits the element to satisfy local equilibrium in a weak sense. The above is a generalization of that proposed in reference 8 where the value of n is unity in the scaling value t^n. Since the term will lead to a stiffness that will be in terms of rotation parameters the scaling indicated above permits values proportional to those generated by the bending rotations – namely, proportional to t cubed. In numerical experiments this scaling leads to less sensitivity in the choice of α_n. For a triangular element in which a linear interpolation is used for $\theta_{\bar{z}}$ minimization with respect to $\theta_{\bar{z}i}$ leads to the form

$$\begin{Bmatrix} M_{\bar{z}i} \\ M_{\bar{z}j} \\ M_{\bar{z}m} \end{Bmatrix} = \frac{1}{36} \alpha_n E t^n \Delta \begin{bmatrix} 1 & -0.5 & -0.5 \\ -0.5 & 1 & -0.5 \\ -0.5 & -0.5 & 1 \end{bmatrix} \tag{6.31}$$

where α_n is yet to be specified. This additional stiffness does in fact affect the results where nodes are not co-planar and indeed represents an approximation; however, effects of varying α_n over fairly wide limits are quite small in many applications. For instance in Table 6.1 a set of displacements of an arch dam analysed in reference 8 is given for various values of α_1. For practical purposes extremely small values of α_n are possible, providing a large computer word length is used.[57]

The analysis of the spherical test problem proposed by MacNeal and Harter as a standard test[58] is indicated in Fig. 6.5. For this test problem a constant strain triangular membrane together with the discrete Kirchhoff triangular plate bending element

Table 6.1 Nodal rotation coefficient in dam analysis[8]

α_1	1.00	0.50	0.10	0.03	0.00
Radial displacement (mm)	61.13				

is combined with the rotational treatment. The results for regular meshes are shown in Table 6.2 for several values of α_3 and mesh subdivisions.

The above development, while quite easy to implement, retains the original form of the membrane interpolations. For triangular elements with corner nodes only, the membrane form utilizes linear displacement fields that yield only constant strain terms. Most bending elements discussed in Chapters 4 and 5 have bending strains with higher than constant terms. Consequently, the membrane error terms will dominate the behaviour of many shell problem solutions. In order to improve the situation it is desirable to increase the order of interpolation. Using conventional interpolations this implies the introduction of additional nodes on each element (e.g. see Chapter 8 of Volume 1); however, by utilizing a drill parameter these interpolations can be transformed to a form that permits a 6 degree-of-freedom assembly at each vertex node. Quadratic interpolations along the edge of an element can be expressed as

$$\bar{\mathbf{u}}(\xi) = N_i(\xi)\,\bar{\mathbf{u}}_i + N_j(\xi)\,\bar{\mathbf{u}}_j + N_k(\xi)\,\Delta\bar{\mathbf{u}}_k \tag{6.32}$$

where $\bar{\mathbf{u}}_i$ are nodal displacements $(\bar{u}_i, \bar{v}_i)$ at an end of the edge (vertex), similarly $\bar{\mathbf{u}}_j$ is the other end, and $\Delta\bar{\mathbf{u}}_k$ are hierarchical displacements at the centre of the edge (Fig. 6.6).

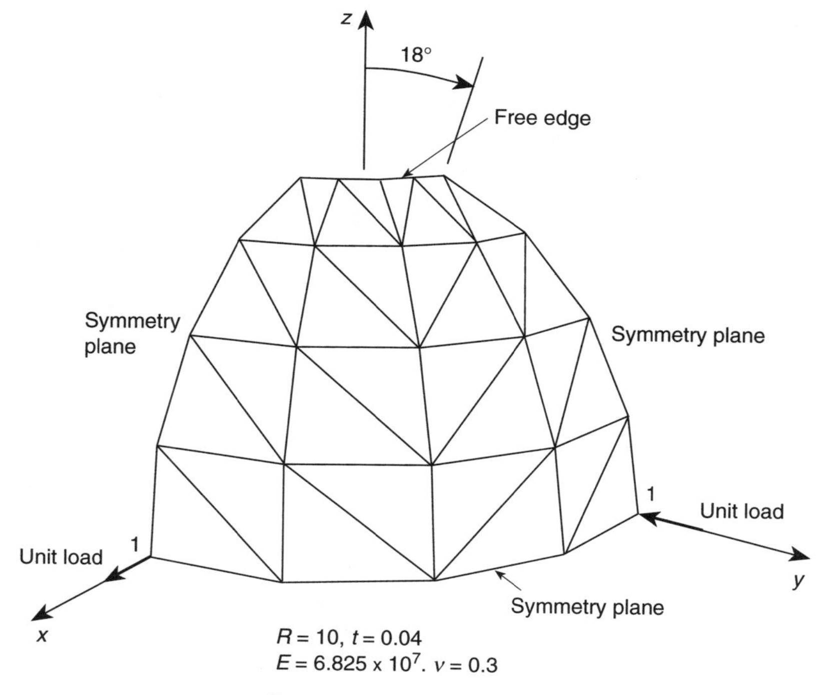

Fig. 6.5 Spherical shell test problem.[58]

Table 6.2 Sphere problem: radial displacement at load

Mesh	α_3 value					
	10.0	1.00	0.100	0.010	0.001	0.000
4 × 4	0.0639	0.0919	0.0972	0.0979	0.0980	0.0980
8 × 8	0.0897	0.0940	0.0945	0.0946	0.0946	0.0946
16 × 16	0.0926	0.0929	0.0929	0.0929	0.0930	0.0930

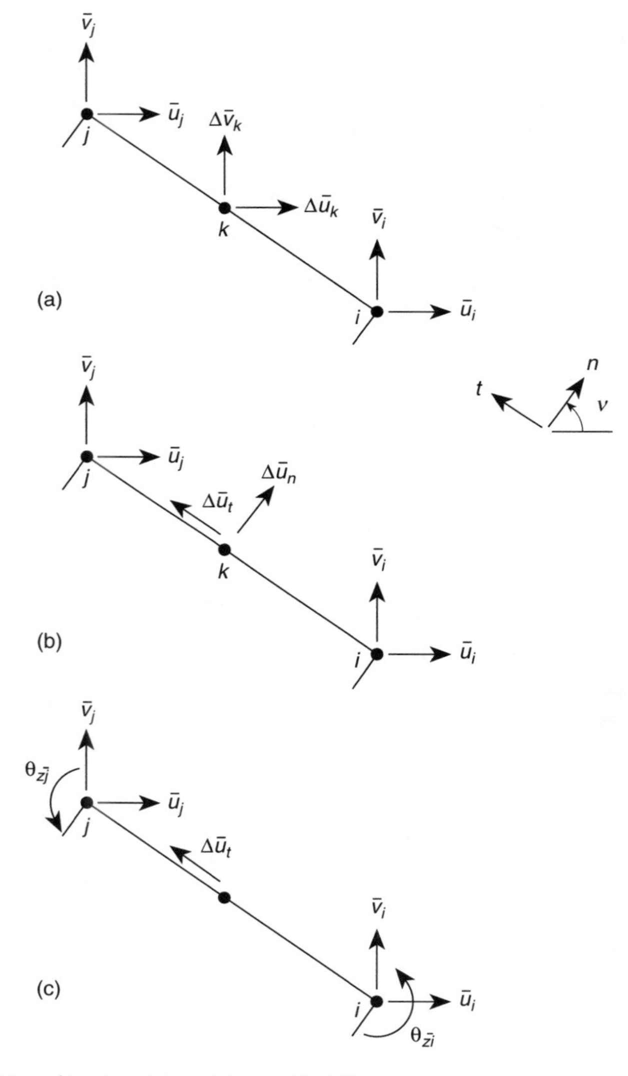

Fig. 6.6 Construction of in-plane interpolations with drilling parameters.

The centre displacement parameters may be expressed in terms of *normal* $(\Delta \bar{u}_n)$ and *tangential* $(\Delta \bar{u}_t)$ components as

$$\Delta \bar{\mathbf{u}}_k = \Delta \bar{u}_n \mathbf{n} + \Delta \bar{u}_t \mathbf{t} \tag{6.33}$$

where **n** is a unit outward normal and **t** is a unit tangential vector to the edge:

$$\mathbf{n} = \begin{Bmatrix} \cos\nu \\ \sin\nu \end{Bmatrix} \quad \text{and} \quad \mathbf{t} = \begin{Bmatrix} -\sin\nu \\ \cos\nu \end{Bmatrix} \tag{6.34}$$

where ν is the angle that the normal makes with the $\bar{x}$ axis. The normal displacement component may be expressed in terms of drilling parameters at each end of the edge (assuming a quadratic expansion).[44,52] Accordingly,

$$\Delta u_n = \tfrac{1}{8} l_{ij} (\theta_{\bar{z}j} - \theta_{\bar{z}i}) \tag{6.35}$$

in which l_{ij} is the length of the ij side. This construction produces an interpolation on each edge given by

$$\bar{\mathbf{u}}(\xi) = N_i(\xi)\bar{\mathbf{u}}_i + N_j(\xi)\bar{\mathbf{u}}_j + N_k(\xi)\left[\tfrac{1}{8} l_{ij} (\theta_{\bar{z}j} - \theta_{\bar{z}i}) + \Delta \bar{u}_t \mathbf{t}\right] \tag{6.36}$$

The reader will undoubtedly observe the similarity here with the process used to develop linked interpolation for the bending element (see Sec. 5.7).

The above interpolation may be further simplified by constraining the $\Delta \bar{u}_t$ parameters to zero. We note, however, that these terms are beneficial in a three-node triangular element. If a common sign convention is used for the hierarchical tangential displacement at each edge, this tangential component maintains compatibility of displacement even in the presence of a kink between adjacent elements. For example, an appropriate sign convention can be accomplished by directing a positive component in the direction in which the end (vertex) node numbers increase. The above structure for the in-plane displacement interpolations may be used for either an irreducible or a mixed element model and generates stiffness coefficients that include terms for the $\theta_{\bar{z}}$ parameters as well as those for $\bar{u}$ and $\bar{v}$. It is apparent, however, that the element generated in this manner must be singular (i.e. has spurious zero-energy modes) since for equal values of the end rotation the interpolation is independent of the $\theta_{\bar{z}}$ parameters. Moreover, when used in non-flat shell applications the element is not free of local equilibrium errors. This later defect may be removed by using the procedure identified above in Eq. (6.30), and results for a quadrilateral element generated according to this scheme are given by Jetteur[53] and Taylor.[54]

A structure of the plane stress problem which includes the effects of a drill rotation field is given by Reissner[59] and is extended to finite element applications by Hughes and Brezzi.[51] A variational formulation for the in-plane problem may be stated as [see Eq. (2.29) in Volume 1]

$$\Pi_d(\bar{\mathbf{u}}, \theta_{\bar{z}}, \tau) = \frac{1}{2}\int_\Omega \boldsymbol{\varepsilon}^{\mathrm{T}} \mathbf{D} \boldsymbol{\varepsilon} \, \mathrm{d}\Omega + \int_\Omega \tau (\omega_{\bar{x}\bar{y}} - \theta_{\bar{z}}) \, \mathrm{d}\Omega \tag{6.37}$$

where τ is a *skew-symmetric* stress component and $\omega_{\bar{x}\bar{y}}$ is the rotational part of the displacement gradient, which for the $\bar{x}\bar{y}$ plane is given by

$$\omega_{\bar{x}\bar{y}} = \frac{\partial \bar{v}}{\partial \bar{x}} - \frac{\partial \bar{u}}{\partial \bar{y}} \tag{6.38}$$

In addition to the terms shown in Eq. (6.37), terms associated with initial stress and strain as well as boundary and body load must be appended for the general shell problem as discussed in Chapters 2 and 4 of Volume 1.

A variation of Eq. (6.37) with respect to τ gives the constraint that the skew-symmetric part of the displacement gradients is the rotation $\theta_{\bar{z}}$. Conversely, variation with respect to $\theta_{\bar{z}}$ gives the result that τ must vanish. Thus, the equations generated from Eq. (6.37) are those of the conventional membrane but include the rotation field. A penalty form of the above equations suitable for finite element applications may be constructed by modifying Eq. (6.37) to

$$\bar{\Pi}_d = \Pi_d - \int_\Omega \frac{1}{\alpha_\tau Et} \tau^2 \, d\Omega \tag{6.39}$$

where α_τ is a penalty number.

It is important to use this mixed representation of the problem with the mixed patch test to construct viable finite element models. Use of constant τ and isoparametric interpolation of $\theta_{\bar{z}}$ in each element together with the interpolations for the displacement approximation given by Eq. (6.36) lead to good triangular and quadrilateral membrane elements. Applications to shell solutions using this form are given by Ibrahimbegovic *et al.*[56] Also the solution for a standard barrel vault problem is contained in Sec. 6.8.

6.6 Elements with mid-side slope connections only

Many of the difficulties encountered with the nodal assembly in global coordinates disappear if the element is so constructed as to require only the continuity of displacements u, v, and w at the corner nodes, with continuity of the normal slope being imposed along the element sides. Clearly, the corner assembly is now simple and the introduction of the sixth nodal variable is unnecessary. As the normal slope rotation along the sides is the same both in local and in global coordinates its transformation there is unnecessary – although again it is necessary to have a unique definition of parameters for the adjacent elements.

Elements of this type arise naturally in hybrid forms (see Chapter 13 of Volume 1) and we have already referred to a plate bending element of a suitable type in Sec. 4.6. This element of the simplest possible kind has been used in shell problems by Dawe[25] with some success. A considerably more sophisticated and complex element of such type is derived by Irons[29] and named ‘semi-loof’. This element is briefly mentioned in Chapter 4 and although its derivation is far from simple it performs well in many situations.

6.7 Choice of element

Numerous membrane and bending element formulations are now available, and, in both, conformity is achievable in flat assemblies. Clearly, if the elements are not co-planar conformity will, in general, be violated and only approached in the limit as smooth shell conditions are reached.

It would appear consistent to use expansions of similar accuracy in both the membrane and bending approximations but much depends on which action is dominant. For thin shells, the simplest triangular element would thus appear to be one with a linear in-plane displacement field and a quadratic bending displacement – thus approximating the stresses as constants in membrane and in bending actions. Such an element is used by Dawe[25] but gives rather marginal (though convergent) results.

In the examples shown we use the following elements which give quite adequate performance.

Element A: this is a mixed rectangular membrane with four corner nodes (Sec. 11.4.4 of Volume 1) combined with the non-conforming bending rectangle with four corner nodes (Sec.4.3). This was first used in references 9 and 10.

Element B: this is a constant strain triangle with three nodes (the basic element of Chapter 4 of Volume 1) combined with the incompatible bending triangle with 9 degrees of freedom (Sec. 4.5). Use of this in the shell context is given in references 8 and 60.

Element C: in this a more consistent linear strain triangle with six nodes is combined with a 12 degree-of-freedom bending triangle using shape function smoothing. This element has been introduced by Razzaque.[61]

Element D: this is a four-node quadrilateral with drilling degrees of freedom [Eq. (6.36) with $\Delta \bar{u}_t$ constrained to zero] combined with a discrete Kirchhoff quadrilateral.[54,62]

6.8 Practical examples

The first example given here is that for the solution of an arch dam shell. The simple geometrical configuration, shown in Fig. 6.7, was taken for this particular problem as results of model experiments and alternative numerical approaches were available.

A division based on rectangular elements (type A) was used as the simple cylindrical shape permitted this, although a rather crude approximation for the fixed foundation had to be used.

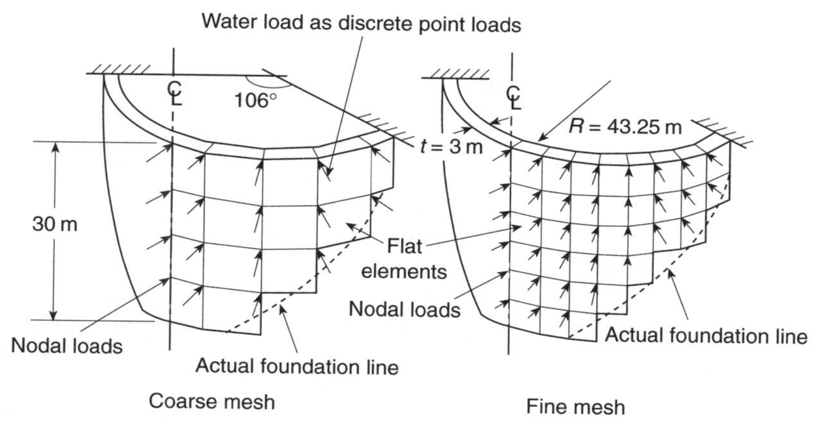

Fig. 6.7 An arch dam as an assembly of rectangular elements.

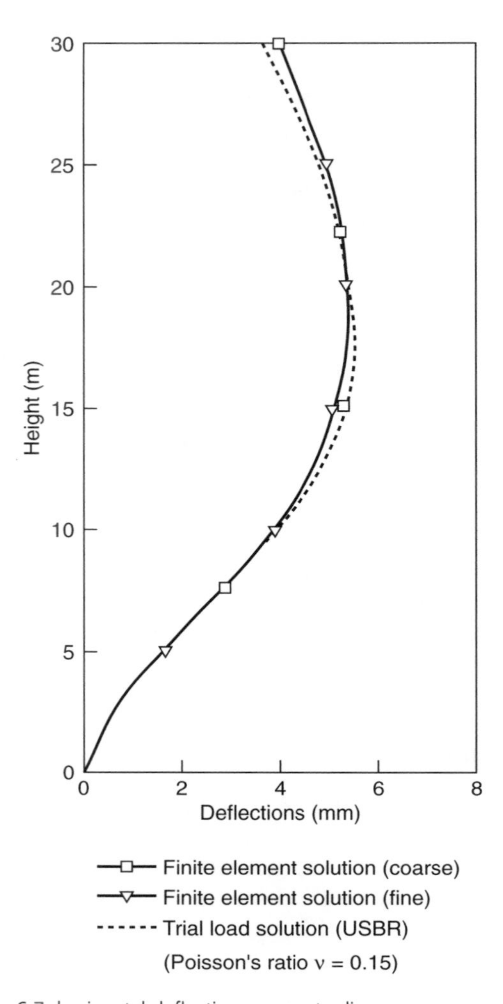

Fig. 6.8 Arch dam of Fig. 6.7: horizontal deflections on centre-line.

Two sizes of division into elements are used, and the results given in Figs 6.8 and 6.9 for deflections and stresses on the centre-line section show that little change occurred by the use of the finer mesh. This indicates that the convergence of both the physical approximation to the true shape by flat elements and of the mathematical approximation involved in the finite element formulation is more than adequate. For comparison, stresses and deflection obtained using the USBR trial load solution (another approximate method) are also shown.

A large number of examples have been computed by Parekh[60] using the triangular, non-conforming element (type B), and indeed show for equal division a general improvement over the conforming triangular version presented by Clough and Johnson.[7] Some examples of such analyses are now shown.

A doubly curved arch dam was similarly analysed using the triangular flat element (type B) representation. The results show an even better approximation.[8]

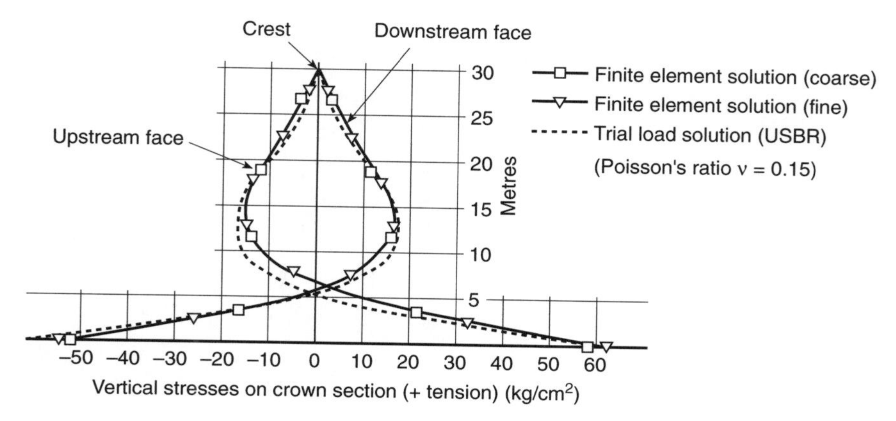

Fig. 6.9 Arch dam of Fig. 6.7: vertical stresses on centre-line.

6.8.1 Cooling tower

This problem of a general axisymmetric shape could be more efficiently dealt with by the axisymmetric formulations to be presented in Chapters 7 and 9. However, here this example is used as a general illustration of the accuracy attainable. The answers against which the numerical solution is compared have been derived by Albasiny and

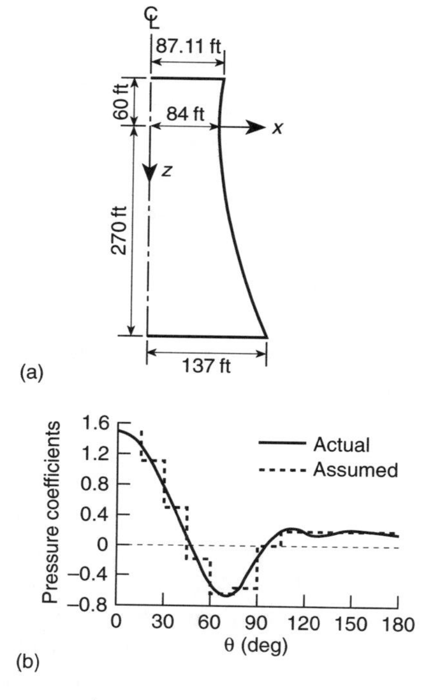

Fig. 6.10 Cooling tower: geometry and pressure load variation about circumference.

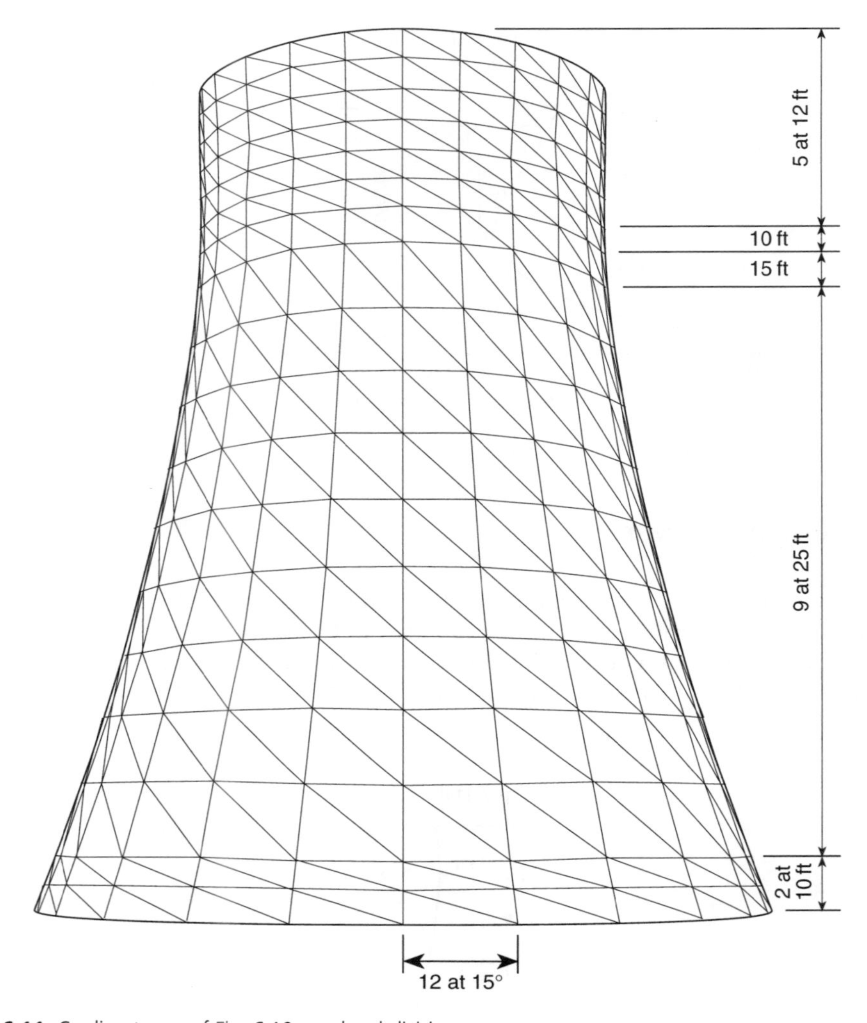

Fig. 6.11 Cooling tower of Fig. 6.10: mesh subdivisions.

Martin.[63] Figures 6.10 to 6.12 show the geometry of the mesh used and some results for a 5 inch and a 7 inch thick shell. Unsymmetric wind loading is used here.

6.8.2 Barrel vault

This typical shell used in many civil engineering applications is solved using analytical methods by Scordelis and Lo[64] and Scordelis.[65] The barrel is supported on rigid diaphragms and is loaded by its own weight. Figures 6.13 and 6.14 show some comparative answers, obtained by elements of type B, C and D. Elements of type C are obviously more accurate, involving more degrees of freedom, and with a mesh of 6×6 elements the results are almost indistinguishable from analytical ones. This problem has become

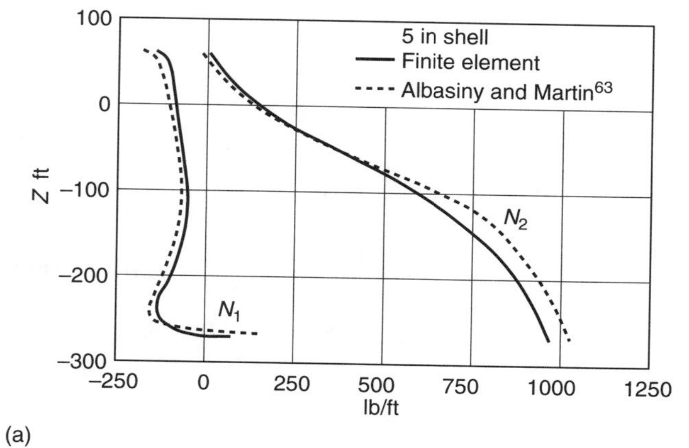

(a)

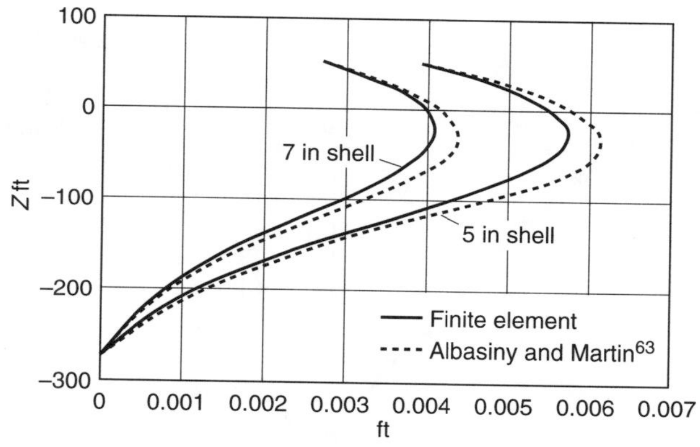

(b)

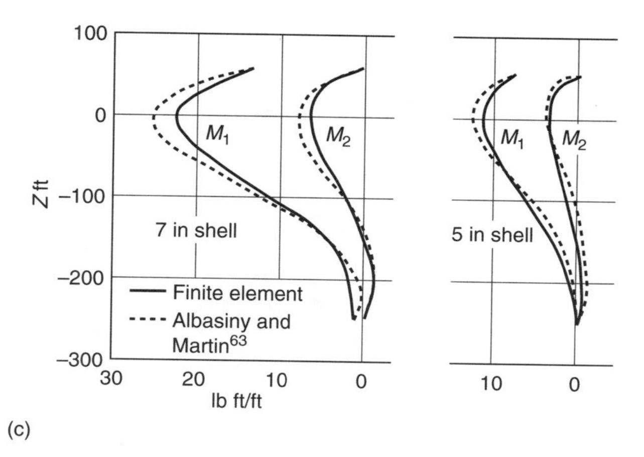

(c)

Fig. 6.12 Cooling tower of Fig. 6.10: (a) membrane forces at $\theta = 0°$; N_1, tangential; N_2, meridional; (b) radial displacements at $\theta = 0°$; (c) moments at $\theta = 0°$; M_1, tangential; M_2, meridional.

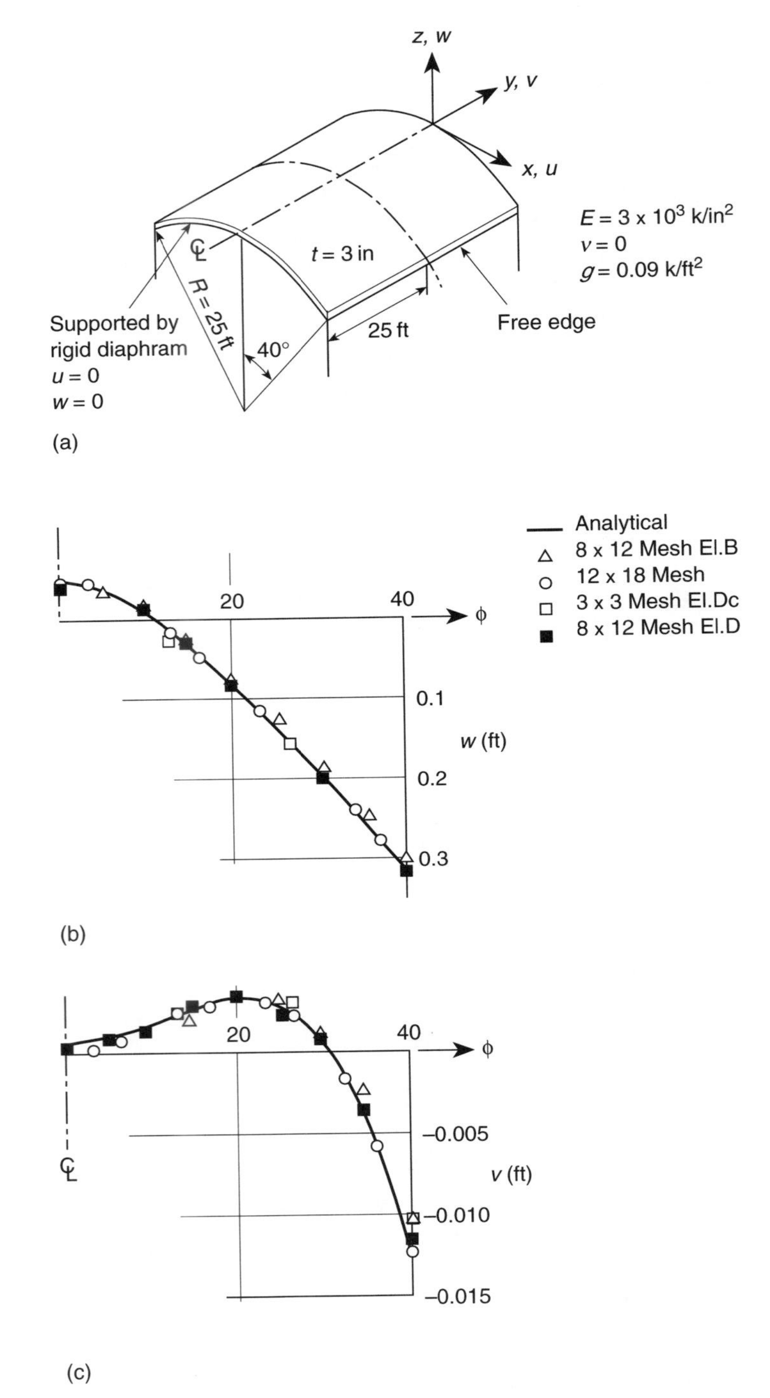

Fig. 6.13 Barrel (cylindrical) vault: flat element model results. (a) Barrel vault geometry and properties; (b) vertical displacement of centre section; (c) longitudinal displacement of support.

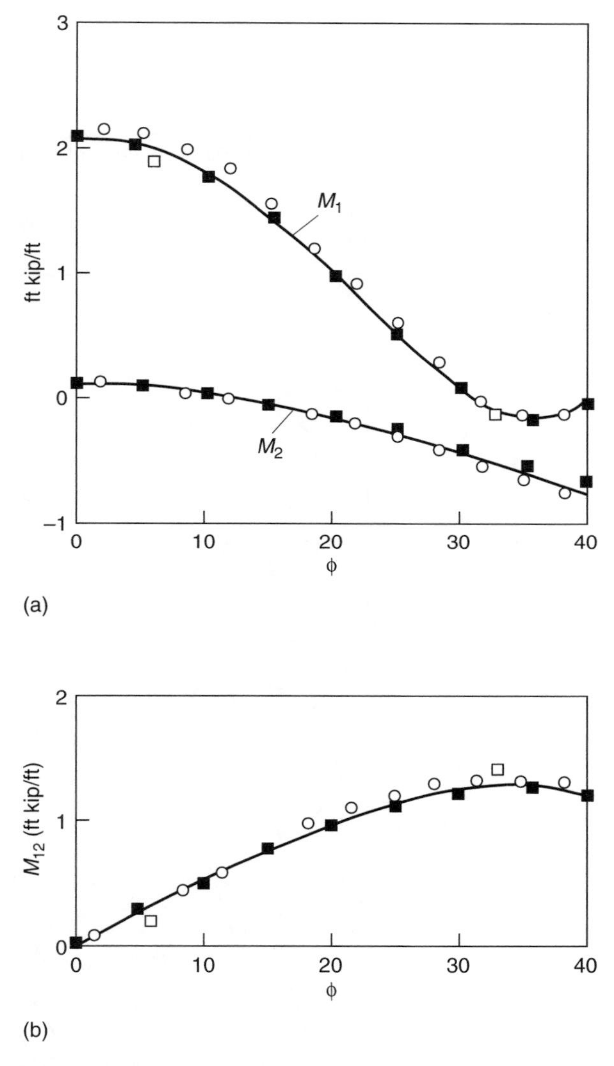

Fig. 6.14 Barrel vault of Fig. 6.13. (a) M_1, transverse; M_2, longitudinal; centre-line moments; (b) M_{12}, twisting moment at support.

a classic on which various shell elements are compared and we shall return to it in Chapter 8. It is worthwhile remarking that only a few, second-order, curved elements give superior results to those presented here with a flat element approximation.

6.8.3 Folded plate structure

As no analytical solution of this problem is known, comparison is made with a set of experimental results obtained by Mark and Riesa.[66]

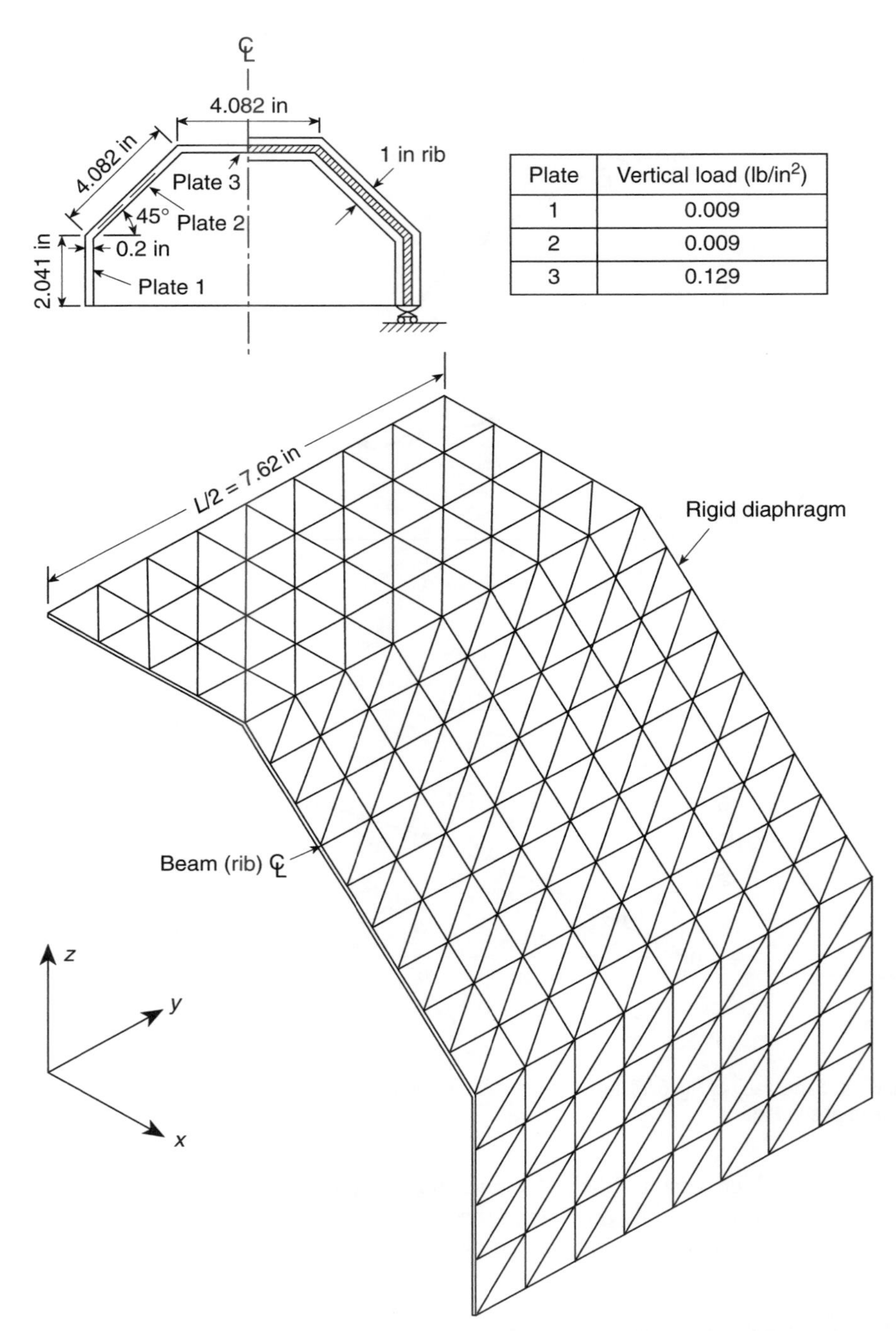

Plate	Vertical load (lb/in^2)
1	0.009
2	0.009
3	0.129

Fig. 6.15 A folded plate structure;[67] model geometry, loading and mesh, $E = 3560\,\text{lb/in}^2$, $\nu = 0.43$.

This example presents a problem in which actual flat finite element representation is physically exact. Also a frame stiffness is included by suitable superposition of beam elements – thus illustrating also the versatility and ease by which different types of elements may be used in a single analysis.

Figures 6.15 and 6.16 show the results using elements of type B. Similar applications are of considerable importance in the analysis of box-type bridge structures, etc.

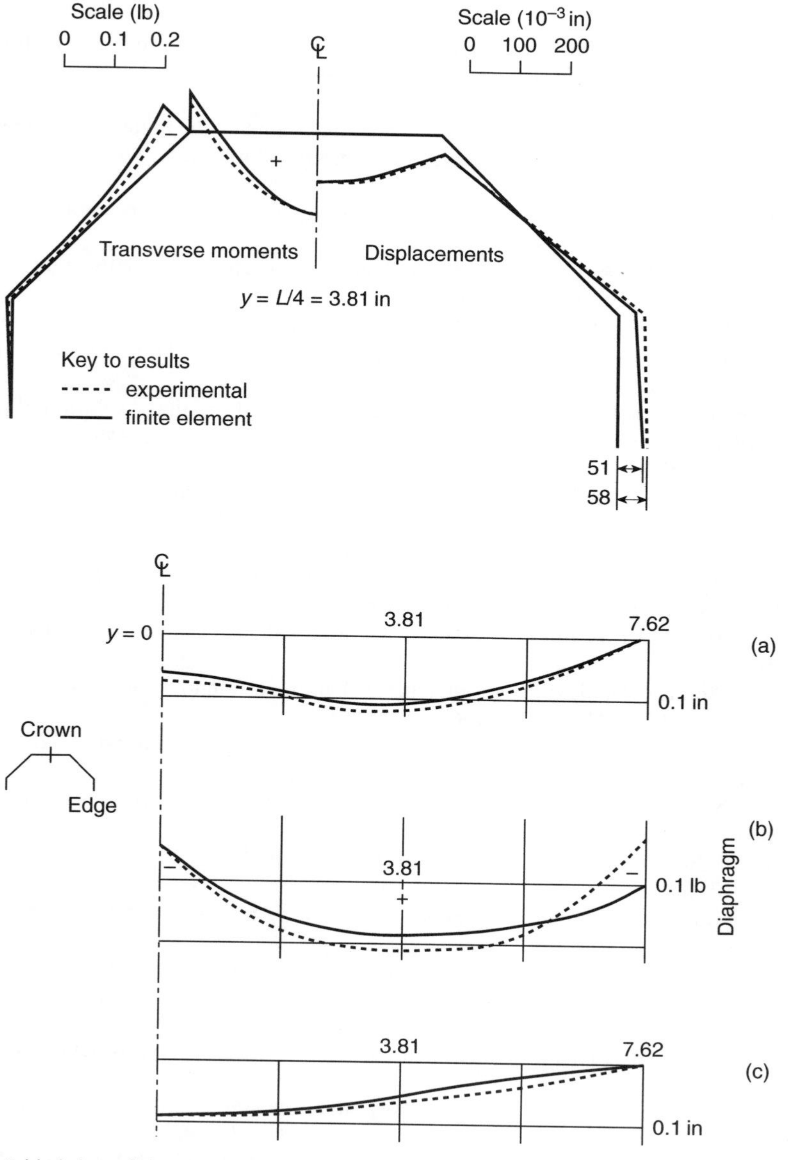

Fig. 6.16 Folded plate of Fig. 6.15; moments and displacements on centre section. (a) Vertical displacements along the crown; (b) longitudinal moments along the crown; (c) horizontal displacements along edge.

References

1. W. Flügge. *Stresses in Shells*, Springer-Verlag, Berlin, 1960.
2. S.P. Timoshenko and S. Woinowski-Krieger. *Theory of Plates and Shells*, 2nd edition, McGraw-Hill, New York, 1959.
3. P.G. Ciarlet. Conforming finite element method for shell problem. In J. Whiteman (ed.), *The Mathematics of Finite Elements and Application*, Volume II, pp. 105–23, Academic Press, London, 1977.
4. B.E. Greene, D.R. Strome and R.C. Weikel. Application of the stiffness method to the analysis of shell structures. In *Proc. Aviation Conf. of American Society of Mechanical Engineers*, ASME, Los Angeles, CA, March 1961.
5. R.W. Clough and J.L. Tocher. Analysis of thin arch dams by the finite element method. In *Proc. Symp. on Theory of Arch Dams*, Southampton University, 1964; Pergamon Press, Oxford, 1965.
6. J.H. Argyris. Matrix displacement analysis of anisotropic shells by triangular elements. *J. Roy. Aero. Soc.*, **69**, 801–5, 1965.
7. R.W. Clough and C.P. Johnson. A finite element approximation for the analysis of thin shells. *J. Solids Struct.*, **4**, 43–60, 1968.
8. O.C. Zienkiewicz, C.J. Parekh and I.P. King. Arch dams analysed by a linear finite element shell solution program. In *Proc. Symp. on Theory of Arch Dams*, Southampton University, 1964; Pergamon Press, Oxford, 1965.
9. O.C. Zienkiewicz and Y.K. Cheung. Finite element procedures in the solution of plate and shell problems. In O.C. Zienkiewicz and G.S. Holister (eds), *Stress Analysis*, pp. 120–41, John Wiley, Chichester, Sussex, 1965.
10. O.C. Zienkiewicz and Y.K. Cheung. Finite element methods of analysis for arch dam shells and comparison with finite difference procedures. In *Proc. Symp. on Theory of Arch Dams*, Southampton University, 1964; Pergamon Press, Oxford, 1965.
11. R.H. Gallagher. Shell elements. In *World Conf. on Finite Element Methods in Structural Mechanics*, Bournemouth, England, October 1975.
12. F.K. Bogner, R.L. Fox and L.A. Schmit. A cylindrical shell element. *Journal of AIAA*, **5**, 745–50, 1966.
13. J. Connor and C. Brebbia. Stiffness matrix for shallow rectangular shell element. *Proc. Am. Soc. Civ. Eng.*, **93**(EM1), 43–65, 1967.
14. A.J. Carr. A refined element analysis of thin shell structures including dynamic loading. Technical Report SEL 67-9, University of California, Berkeley, CA, 1967.
15. S. Utku. Stiffness matrices for thin triangular elements of non-zero Gaussian curvature. *Journal of AIAA*, **5**, 1659–67, 1967.
16. G. Cantin. Strain–displacement relationships for cylindrical shells. *Journal of AIAA*, **6**, 1787–8, 1968.
17. G. Cantin and R.W. Clough. A curved, cylindrical shell finite element. *AIAA Journal*, **6**, 1057–62, 1968.
18. G. Bonnes, G. Dhatt, Y.M. Giroux and L.P.A. Robichaud. Curved triangular elements for analysis of shells. In L. Berke, R.M. Bader, W.J. Mykytow, J.S. Przemienicki and M.H. Shirk (eds), *Proc. 2nd Conf. Matrix Methods in Structural Mechanics*, Volume AFFDL-TR-68-150, pp. 617–39, Air Force Flight Dynamics Laboratory, Wright Patterson Air Force Base, OH, October 1968.
19. G.E. Strickland and W.A. Loden. A doubly curved triangular shell element. In L. Berke, R.M. Bader, W.J. Mykytow, J.S. Przemienicki and M.H. Shirk (eds), *Proc. 2nd Conf. Matrix Methods in Structural Mechanics*, Volume AFFDL-TR-68-150, pp. 641–66, Air Force Flight Dynamics Laboratory, Wright Patterson Air Force Base, OH, October 1968.

20. B.E. Greene, R.E. Jones and R.W. McLay. Dynamic analysis of shells using doubly curved finite elements. In L. Berke, R.M. Bader, W.J. Mykytow, J.S. Przemienicki and M.H. Shirk (eds), *Proc. 2nd Conf. Matrix Methods in Structural Mechanics*, Volume AFFDLTR-68-150, pp. 185–212, Air Force Flight Dynamics Laboratory, Wright Patterson Air Force Base, OH, October 1968.
21. S. Ahmad. *Curved Finite Elements in the Analysis of Solids, Shells and Plate Structures*, PhD thesis, Department of Civil Engineering, University of Wales, Swansea, 1969.
22. G.R. Cowper, G.M. Lindberg, and M.D. Olson. A shallow shell finite element of triangular shape. *International Journal of Solids and Structures*, **6**, 1133–56, 1970.
23. G. Dupuis and J.J. Goël. A curved finite element for thin elastic shells. *International Journal of Solids and Structures*, **6**, 987–96, 1970.
24. S.W. Key and Z.E. Beisinger. The analysis of thin shells by the finite element method. In *High Speed Computing of Elastic Structures*, Volume 1, pp. 209–52, University of Liège Press, Liège, 1971.
25. D.J. Dawe. The analysis of thin shells using a facet element. Technical Report CEGB RD/B/N2038, Berkeley Nuclear Laboratory, England, 1971.
26. D.J. Dawe. Rigid-body motions and strain–displacement equations of curved shell finite elements. *Int. J. Mech. Sci.*, **14**, 569–78, 1972.
27. D.G. Ashwell and A. Sabir. A new cylindrical shell finite element based on simple independent strain functions. *Int. J. Mech. Sci.*, **4**, 37–47, 1973.
28. G.R. Thomas and R.H. Gallagher. A triangular thin shell finite element: linear analysis. Technical Report CR-2582, NASA, 1975.
29. B.M. Irons. The semi-loof shell element. In D.G. Ashwell and R.H. Gallagher (eds), *Finite Elements for Thin Shells and Curved Members*, pp. 197–222, John Wiley, Chichester, Sussex, 1976.
30. D.G. Ashwell. Strain elements with application to arches, rings, and cylindrical shells. In D.G. Ashwell and R.H. Gallagher (eds), *Finite Elements for Thin Shells and Curved Members*, John Wiley, Chichester, Sussex, 1976.
31. N. Carpenter, H. Stolarski and T. Belytschko. A flat triangular shell element with improved membrane interpolation. *Communications in Applied Numerical Method,* **1**, 161–8, 1985.
32. C. Pratt. Shell finite element via Reissner's principle. *International Journal of Solids and Structures*, **5**, 1119–33, 1969.
33. J. Connor and G. Will. A mixed finite element shallow shell formulation. In R.H. Gallagher *et al.* (eds), *Advances in Matrix Methods of Structural Analysis and Design*, pp. 105–37, University of Alabama Press, Tuscaloosa, AL, 1969.
34. L.R. Herrmann and W.E. Mason. Mixed formulations for finite element shell analysis. In *Conf. on Computer-oriented Analysis of Shell Structures*, Paper AFFDL-TR-71-79, June 1971.
35. G. Edwards and J.J. Webster. Hybrid cylindrical shell elements. In D.G. Ashwell and R.H. Gallagher (eds), *Finite Elements for Thin Shells and Curved Members*, pp. 171–95, John Wiley, Chichester, Sussex, 1976.
36. H. Stolarski and T. Belytschko. Membrane locking and reduced integration for curved elements. *J. Appl. Mech.*, **49**, 172–6, 1982.
37. P. Jetteur and F. Frey. A four node Marguerre element for non-linear shell analysis. *Engineering Computations*, **3**, 276–82, 1986.
38. J.C. Simo and D.D. Fox. On a stress resultant geometrically exact shell model. Part I: formulation and optimal parametrization. *Comp. Meth. Appl. Mech. Eng.*, **72**, 267–304, 1989.
39. J.C. Simo, D.D. Fox and M.S. Rifai. On a stress resultant geometrically exact shell model. Part II: the linear theory; computational aspects. *Comp. Meth. Appl. Mech. Eng.*, **73**, 53–92, 1989.

40. M. Bischoff and E. Ramm. Solid-like shell or shell-like solid formulation? A personal view. In W. Wunderlich (ed.), *Proc. Eur. Conf. on Comp. Mech.* (*ECCM'99 CD-ROM*), Munich, September 1999.
41. I.H. Shames and F.A. Cozzarelli. *Elastic and Inelastic Stress Analysis*, revised edition, Taylor & Francis, Washington, DC, 1997.
42. L.E. Malvern. *Introduction to the Mechanics of a Continuous Medium*, Prentice-Hall, Englewood Cliffs, NJ, 1969.
43. S. Ahmad, B.M. Irons and O.C. Zienkiewicz. A simple matrix–vector handling scheme for three-dimensional and shell analysis. *Int. J. Num. Meth. Eng.*, **2**, 509–22, 1970.
44. D.J. Allman. A compatible triangular element including vertex rotations for plane elasticity analysis. *Computers and Structures*, **19**, 1–8, 1984.
45. D.J. Allman. A quadrilateral finite element including vertex rotations for plane elasticity analysis. *Int. J. Num. Meth. Eng.*, **26**, 717–30, 1988.
46. D.J. Allman. Evaluation of the constant strain triangle with drilling rotations. *Int. J. Num. Meth. Eng.*, **26**, 2645–55, 1988.
47. P.G. Bergan and C.A. Felippa. A triangular membrane element with rotational degrees of freedom. *Comp. Meth. Appl. Mech. Eng.*, **50**, 25–69, 1985.
48. P.G. Bergan and C.A. Felippa. Efficient implementation of a triangular membrane element with drilling freedoms. In T.J.R. Hughes and E. Hinton (eds), *Finite Element Methods for Plate and Shell Structures*, Volume 1, pp. 128–52, Pineridge Press, Swansea, 1986.
49. R.D. Cook. On the Allman triangle and a related quadrilateral element. *Computers and Structures*, **2**, 1065–7, 1986.
50. R.D. Cook. A plane hybrid element with rotational d.o.f. and adjustable stiffness. *Int. J. Num. Meth. Eng.*, **24**, 1499–508, 1987.
51. T.J.R. Hughes and F. Brezzi. On drilling degrees-of-freedom. *Comp. Meth. Appl. Mech. Eng.*, **72**, 105–21, 1989.
52. R.L. Taylor and J.C. Simo. Bending and membrane elements for analysis of thick and thin shells. In G.N. Pande and J. Middleton (eds), *Proc. NUMETA 85 Conf.*, Volume 1, pp. 587–91, A.A. Balkema, Rotterdam, 1985.
53. P. Jetteur. Improvement of the quadrilateral JET shell element for a particular class of shell problems. Technical Report IREM 87/1, École Polytechnique Fédérale de Lausanne, Centre-Est, Ecublens, 1015 Lausanne, February 1987.
54. R.L. Taylor. Finite element analysis of linear shell problems. In J.R. Whiteman (ed.), *The Mathematics of Finite Elements and Applications VI*, pp. 191–203, Academic Press, London, 1988.
55. R.H. MacNeal and R.L Harter. A refined four-noded membrane element with rotational degrees of freedom. *Computers and Structures*, **28**, 75–88, 1988.
56. A. Ibrahimbegovic, R.L. Taylor and E.L. Wilson. A robust quadrilateral membrane finite element with drilling degrees of freedom. *Int. J. Num. Meth. Eng.*, **30**, 445–57, 1990.
57. R.W. Clough and E.L. Wilson. Dynamic finite element analysis of arbitrary thin shells. *Computers and Structures*, **1**, 33–56, 1971.
58. R.H. MacNeal and R.L. Harter. A proposed standard set of problems to test finite element accuracy. *Journal of Finite Elements in Analysis and Design*, **1**, 3–20, 1985.
59. E. Reissner. A note on variational theorems in elasticity. *International Journal of Solids and Structures*, **1**, 93–5, 1965.
60. C.J. Parekh. *Finite Element Solution System*, PhD thesis, Department of Civil Engineering, University of Wales, Swansea, 1969.
61. A. Razzaque. *Finite Element Analysis of Plates and Shells*, PhD thesis, Civil Engineering Department, University of Wales, Swansea, 1972.
62. J.L. Batoz and M.B. Tahar. Evaluation of a new quadrilateral thin plate bending element. *Int. J. Num. Meth. Eng.*, **18**, 1655–77, 1982.

63. E.L. Albasiny and D.W. Martin. Bending and membrane equilibrium in cooling towers. *Proc. Am. Soc. Civ. Eng.*, **93**(EM3), 1–17, 1967.
64. A.C. Scordelis and K.S. Lo. Computer analysis of cylindrical shells. *J. Am. Concr. Inst.*, **61**, 539–61, 1964.
65. A.C. Scordelis. Analysis of cylindrical shells and folded plates. In *Concrete Thin Shells*, Report SP 28-N, American Concrete Institute, Farmington, MI, 1971.
66. R. Mark and J.D. Riesa. Photoelastic analysis of folded plate structures. *Proc. Am. Soc. Civ. Eng.*, **93**(EM4), 79–83, 1967.

7

Axisymmetric shells

7.1 Introduction

The problem of axisymmetric shells is of sufficient practical importance to include in this chapter special methods dealing with their solution. While the general method described in the previous chapter is obviously applicable here, it will be found that considerable simplification can be achieved if account is taken of axial symmetry of the structure. In particular, if both the shell and the loading are axisymmetric it will be found that the elements become 'one-dimensional'. This is the simplest type of element, to which little attention was given in earlier chapters.

The first approach to the finite element solution of axisymmetric shells was presented by Grafton and Strome.[1] In this, the elements are simple conical frustra and a direct approach via displacement functions is used. Refinements in the derivation of the element stiffness are presented in Popov *et al.*[2] and in Jones and Strome.[3] An extension to the case of unsymmetrical loads, which was suggested in Grafton and Strome, is elaborated in Percy *et al.*[4] and others.[5,6]

Later, much work was accomplished to extend the process to curved elements and indeed to refine the approximations involved. The literature on the subject is considerable, no doubt promoted by the interest in aerospace structures, and a complete bibliography is here impractical. References 7–15 show how curvilinear coordinates of various kinds can be introduced to the analysis, and references 9 and 14 discuss the use of additional nodeless degrees of freedom in improving accuracy. 'Mixed' formulations (Chapter 11 of Volume 1) have found here some use.[16] Early work on the subject is reviewed comprehensively by Gallagher[17,18] and Stricklin.[19]

In axisymmetric shells, in common with all other shells, both bending and 'in-plane' or 'membrane' forces will occur. These will be specified uniquely in terms of the generalized 'strains', which now involve extensions and changes in curvatures of the middle surface. If the displacement of each point of the middle surface is specified, such 'strains' and the internal stress resultants, or simply 'stresses', can be determined by formulae available in standard texts dealing with shell theory.[20–22]

7.2 Straight element

As a simple example of an axisymmetric shell subjected to axisymmetric loading we consider the case shown in Figs 7.1 and 7.2 in which the displacement of a point on the middle surface of the meridian plane at an angle ϕ measured positive from the x-axis is uniquely determined by two components $\bar{u}$ and $\bar{w}$ in the tangential (s) and normal directions, respectively.

Using the Kirchhoff–Love assumption (which excludes transverse shear deformations) and assuming that the angle ϕ does not vary (i.e. elements are straight), the four strain components are given by[20–22]

$$\boldsymbol{\varepsilon} = \begin{Bmatrix} \varepsilon_s \\ \varepsilon_\theta \\ \chi_s \\ \chi_\theta \end{Bmatrix} = \begin{Bmatrix} \mathrm{d}\bar{u}/\mathrm{d}s \\ [\bar{u}\cos\phi - \bar{w}\sin\phi]/r \\ -\mathrm{d}^2\bar{w}/\mathrm{d}s^2 \\ -(\mathrm{d}\bar{w}/\mathrm{d}s)\cos\phi/r \end{Bmatrix} \tag{7.1}$$

This results in the four internal stress resultants shown in Fig. 7.1 that are related to the strains by an elasticity matrix $\mathbf{D}$:

$$\boldsymbol{\sigma} = \begin{Bmatrix} N_s \\ N_\theta \\ M_s \\ M_\theta \end{Bmatrix} = \mathbf{D}\boldsymbol{\varepsilon} \tag{7.2}$$

For an isotropic shell the elasticity matrix becomes

$$\mathbf{D} = \frac{Et}{1-\nu^2}\begin{bmatrix} 1 & \nu & 0 & 0 \\ \nu & 1 & 0 & 0 \\ 0 & 0 & t^2/12 & \nu t^2/12 \\ 0 & 0 & \nu t^2/12 & t^2/12 \end{bmatrix} \tag{7.3}$$

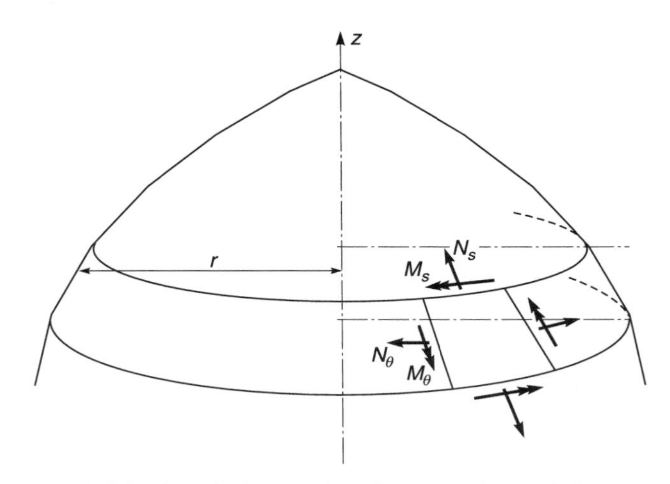

Fig. 7.1 Axisymmetric shell, loading, displacements, and stress resultants; shell represented as a stack of conical frustra.

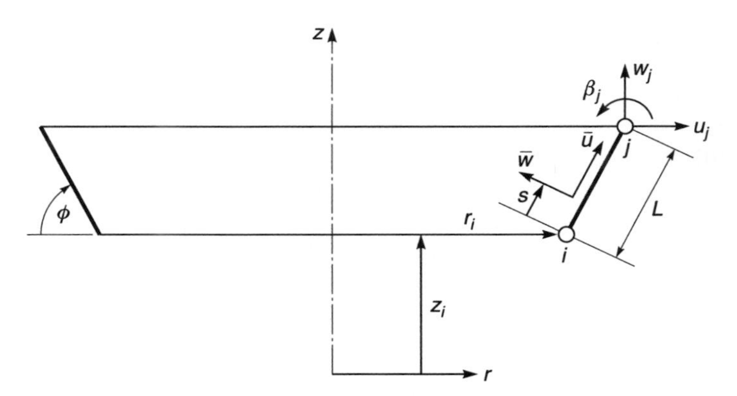

Fig. 7.2 An element of an axisymmetric shell.

the upper part being a plane stress and the lower a bending stiffness matrix with shear terms omitted as 'thin' conditions are assumed.

7.2.1 Element characteristics – axisymmetrical loads

Let the shell be divided by nodal circles into a series of conical frustra, as shown in Fig. 7.2. The nodal displacements at points 1 and 2 for a typical 1–2 element such as i and j will have to define uniquely the deformations of the element via prescribed shape functions.

At each node the radial and axial displacements, u and w, and a rotation, β, will be used as parameters. From virtual work by edge forces we find that all three components are necessary as the shell can carry in-plane forces and bending moments. The displacements of a node i can thus be defined by three components, the first two being in global directions r and z,

$$\mathbf{a}_i = \begin{Bmatrix} u_i \\ w_i \\ \beta_i \end{Bmatrix} \tag{7.4}$$

The simplest elements with two nodes, i and j, thus possess 6 degrees of freedom, determined by the element displacements

$$\mathbf{a}^e = \begin{Bmatrix} \mathbf{a}_1 \\ \mathbf{a}_2 \end{Bmatrix} \tag{7.5}$$

The displacements within the element have to be uniquely determined by the nodal displacements $\mathbf{a}^e$ and the position s (as shown in Fig. 7.2) and maintain slope and displacement continuity.

Thus in local (s) coordinates we have

$$\bar{\mathbf{u}} = \begin{Bmatrix} \bar{u} \\ \bar{w} \end{Bmatrix} = \mathbf{N}(s)\,\mathbf{a}^e \tag{7.6}$$

Based on the strain–displacement relations (7.1) we observe that $\bar{u}$ can be of C_0 type while $\bar{w}$ must be of type C_1. The simplest approximation takes $\bar{u}$ varying linearly with s and $\bar{w}$ as cubic in s. We shall then have six undetermined constants which can be determined from nodal values of u, w, and β.

At the node i,

$$\begin{Bmatrix} \bar{u}_i \\ \bar{w}_i \\ (\mathrm{d}\bar{w}/\mathrm{d}s)_i \end{Bmatrix} = \begin{bmatrix} \cos\phi & \sin\phi & 0 \\ -\sin\phi & \cos\phi & 0 \\ 0 & 0 & 1 \end{bmatrix} \begin{Bmatrix} u_i \\ w_i \\ \beta_i \end{Bmatrix} = \mathbf{T}\mathbf{a}_i \tag{7.7}$$

Introducing the interpolations

$$\bar{u} = N_1^u(\xi)\,\bar{u}_1 + N_2^u(\xi)\,\bar{u}_2 \tag{7.8}$$

$$\bar{w} = N_1^w(\xi)\,\bar{w}_1 + N_2^w(\xi)\,\bar{w}_2 + \frac{L}{2}\left[N_1^\beta(\xi)\left(\frac{\mathrm{d}\bar{w}}{\mathrm{d}s}\right)_1 + N_2^\beta(\xi)\left(\frac{\mathrm{d}\bar{w}}{\mathrm{d}s}\right)_2\right] \tag{7.9}$$

where N_i^u are the usual linear interpolations in ξ $(-1 \leqslant \xi \leqslant 1)$

$$N_1^u = \tfrac{1}{2}(1-\xi) \qquad \text{and} \qquad N_2^u = \tfrac{1}{2}(1+\xi)$$

and N_i^w and N_i^β are the Hermitian interpolations of order 0 and 1 given as (see Chapter 4, Sec. 4.14)

$$N_1^w = \tfrac{1}{4}(2 - 3\xi + \xi^3) \qquad \text{and} \qquad N_2^w = \tfrac{1}{4}(2 + 3\xi - \xi^3)$$

and

$$N_1^\beta = \tfrac{1}{4}(1 - \xi - \xi^2 + \xi^3) \qquad \text{and} \qquad N_2^\beta = \tfrac{1}{4}(-1 - \xi + \xi^2 + \xi^3)$$

in which, placing the origin of the meridian coordinate s at the i node,

$$s = N_2^u(\xi)\,L = \tfrac{1}{2}(1+\xi)\,L$$

The global coordinates for the conical frustrum may also be expressed by using the N_i^u interpolations as

$$\begin{aligned} r &= N_1^u(\xi)\,r_1 + N_2^u(\xi)\,r_2 \\ z &= N_1^u(\xi)\,z_1 + N_2^u(\xi)\,z_2 \end{aligned} \tag{7.10}$$

and used to compute the length L as

$$L = \sqrt{(r_2 - r_1)^2 + (z_2 - z_1)^2}$$

Writing the interpolations as

$$\bar{\mathbf{u}} = \begin{bmatrix} N_i^u & 0 & 0 \\ 0 & N_i^w & N_i^\beta \end{bmatrix} \begin{Bmatrix} \bar{u}_i \\ \bar{w}_i \\ (\mathrm{d}\bar{w}/\mathrm{d}s)_i \end{Bmatrix} = \bar{\mathbf{N}}_i \bar{\mathbf{u}}_i \tag{7.11}$$

we can now write the global interpolation as

$$\mathbf{u} = \begin{Bmatrix} u \\ w \end{Bmatrix} = \begin{bmatrix} \bar{\mathbf{N}}_1\mathbf{T} & \bar{\mathbf{N}}_2\mathbf{T} \end{bmatrix}\mathbf{a}^e = \mathbf{N}\mathbf{a}^e \tag{7.12}$$

From Eq. (7.12) it is a simple matter to obtain the strain matrix **B** by use of the definition Eq. (7.1). This gives

$$\boldsymbol{\varepsilon} \equiv \mathbf{B}\mathbf{a}^e = \begin{bmatrix} \bar{\mathbf{B}}_1\mathbf{T} & \bar{\mathbf{B}}_2\mathbf{T} \end{bmatrix} \mathbf{a}^e \tag{7.13}$$

in which, noting from Eq. (7.7) that $u = \cos\phi\bar{u} - \sin\phi\bar{w}$, we have

$$\bar{\mathbf{B}}_i = \begin{bmatrix} \mathrm{d}N_i^u/\mathrm{d}s & 0 & 0 \\ N_i^u \cos\phi/r & -N_i^w \sin\phi/r & -N_i^\beta \sin\phi/r \\ 0 & -\mathrm{d}^2N_i^w/\mathrm{d}s^2 & -\mathrm{d}^2N_i^\beta/\mathrm{d}s^2 \\ 0 & -(\mathrm{d}N_i^w/\mathrm{d}s)\cos\phi/r & -(\mathrm{d}N_i^\beta/\mathrm{d}s)\cos\phi/r \end{bmatrix} \tag{7.14}$$

Derivatives are evaluated by using

$$\frac{\mathrm{d}N_i}{\mathrm{d}s} = \frac{2}{L}\frac{\mathrm{d}N_i}{\mathrm{d}\xi} \quad \text{and} \quad \frac{\mathrm{d}^2N_i}{\mathrm{d}s^2} = \frac{4}{L^2}\frac{\mathrm{d}^2N_i}{\mathrm{d}\xi^2}$$

Now all the 'ingredients' required for computing the stiffness matrix (or load, stress, and initial stress matrices) by standard formulae are known. The integrations required are carried out over the area, A, of each element, that is, with

$$\mathrm{d}A = 2\pi r\,\mathrm{d}s = \pi r L\,\mathrm{d}\xi \tag{7.15}$$

with ξ varying from -1 to 1.

Thus, the stiffness matrix **K** becomes, in local coordinates,

$$\bar{\mathbf{K}}_{mn} = \pi L \int_{-1}^{1} \bar{\mathbf{B}}_m^{\mathrm{T}} \mathbf{D} \bar{\mathbf{B}}_n r\,\mathrm{d}\xi \tag{7.16}$$

On transformation, the stiffness $\mathbf{K}_{mn}$ of the global matrix is given by

$$\mathbf{K}_{mn} = \mathbf{T}^{\mathrm{T}} \bar{\mathbf{K}}_{mn} \mathbf{T} \tag{7.17}$$

Once again it is convenient to evaluate the integrals numerically and the form above is written for Gaussian quadrature (see Table 9.1, Volume 1). Grafton and Strome[1] give an explicit formula for the stiffness matrix based on a single average value of the integrand (one-point Gaussian quadrature) and using a **D** matrix corresponding to an orthotropic material. Percy *et al.*[4] and Klein[5] used a seven-point numerical integration; however, it is generally recommended to use only two-points to obtain all arrays (especially if inertia forces are added, since one point then would yield a rank deficient mass matrix).

It should be remembered that if any external line loads or moments are present, their full circumferential value must be used in the analysis, just as was the case with axisymmetric solids discussed in Chapter 5 of Volume 1.

7.2.2 Additional enhanced mode

A slight improvement to the above element may be achieved by adding an *enhanced strain* mode to the ε_s component. Here this is achieved by following the procedures outlined in Chapter 12 of Volume 1, and we can observe that the necessary condition not to affect a constant value of N_s is given by

$$2\pi \int_L \varepsilon_s^{(\text{en})} r \, \mathrm{d}s = \pi L \int_{-1}^{1} \varepsilon_s^{(\text{en})} r \, \mathrm{d}\xi = 0 \tag{7.18}$$

where $\varepsilon_s^{(\text{en})}$ denotes the enhanced strain component. A simple mode may thus be defined as

$$\varepsilon_s^{(\text{en})} = \frac{\xi}{r} \alpha_{\text{en}} = B_{\text{en}} \alpha_{\text{en}} \tag{7.19}$$

in which α_{en} is a parameter to be determined. For the linear elastic case considered above the mode may be determined from

$$\begin{bmatrix} K_{\text{en}} & \mathbf{G} \\ \mathbf{G}^{\mathrm{T}} & \mathbf{K} \end{bmatrix} \begin{Bmatrix} \alpha_{\text{en}} \\ \mathbf{a}^e \end{Bmatrix} = \begin{Bmatrix} \mathbf{0} \\ \mathbf{f} \end{Bmatrix} \tag{7.20}$$

where

$$\begin{aligned} K_{\text{en}} &= 2\pi L \int_{-1}^{1} B_{\text{en}} D_{11} B_{\text{en}} r \, \mathrm{d}\xi \\ \mathbf{G}_i &= 2\pi L \int_{-1}^{1} [B_{\text{en}} \quad 0 \quad 0 \quad 0] \mathbf{D} \mathbf{B}_i r \, \mathrm{d}\xi \end{aligned} \tag{7.21}$$

Now a partial solution may be performed by means of static condensation[23] to obtain the stiffness for assembly

$$\bar{\mathbf{K}} = \mathbf{K} - \mathbf{G}^{\mathrm{T}} K_{\text{en}}^{-1} \mathbf{G} \tag{7.22}$$

The effect of the added mode is most apparent in the force resultant N_s where solution oscillations are greatly reduced. This improvement is not needed for the purely elastic case but is more effective when the material properties are inelastic where the oscillations can cause errors in behaviour, such as erratic yielding in elasto-plastic solutions.

7.2.3 Examples and accuracy

In the treatment of axisymmetric shells described here, continuity between the shell elements is satisfied at all times. For an axisymmetric shell of polygonal meridian shape, therefore, convergence will always occur.

The problem of the physical approximation to a curved shell by a polygonal shape is similar to the one discussed in Chapter 6. Intuitively, convergence can be expected, and indeed numerous examples indicate this.

When the loading is such as to cause predominantly membrane stresses, discrepancies in bending moment values exist (even with reasonably fine subdivision). Again, however, these disappear as the size of the subdivisions decreases, particularly if correct sampling is used (see Chapter 14 of Volume 1). This is necessary to eliminate the physical approximation involved in representing the shell as a series of conical frustra.

Figures 7.3 and 7.4 illustrate some typical examples taken from the Grafton and Strome paper which show quite remarkable accuracy. In each problem it should be noted that small elements are needed near free edges to capture the 'boundary layer' nature of shell solutions.

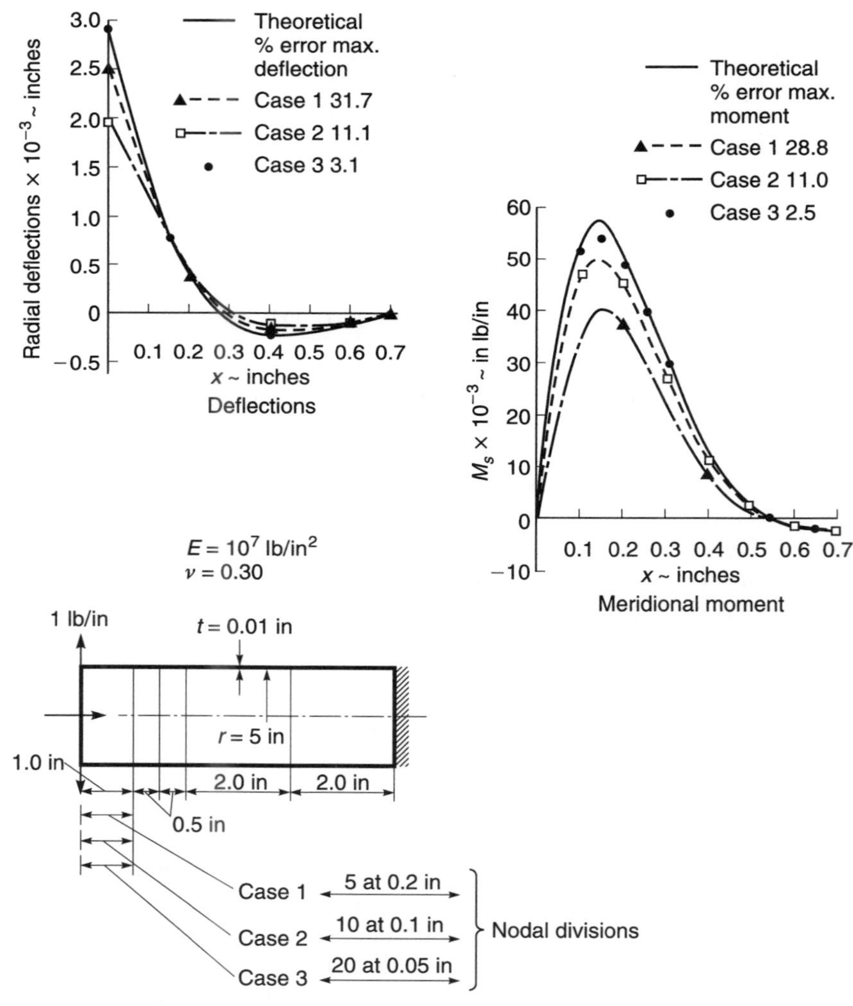

Fig. 7.3 A cylindrical shell solution by finite elements, from Grafton and Strome.[1]

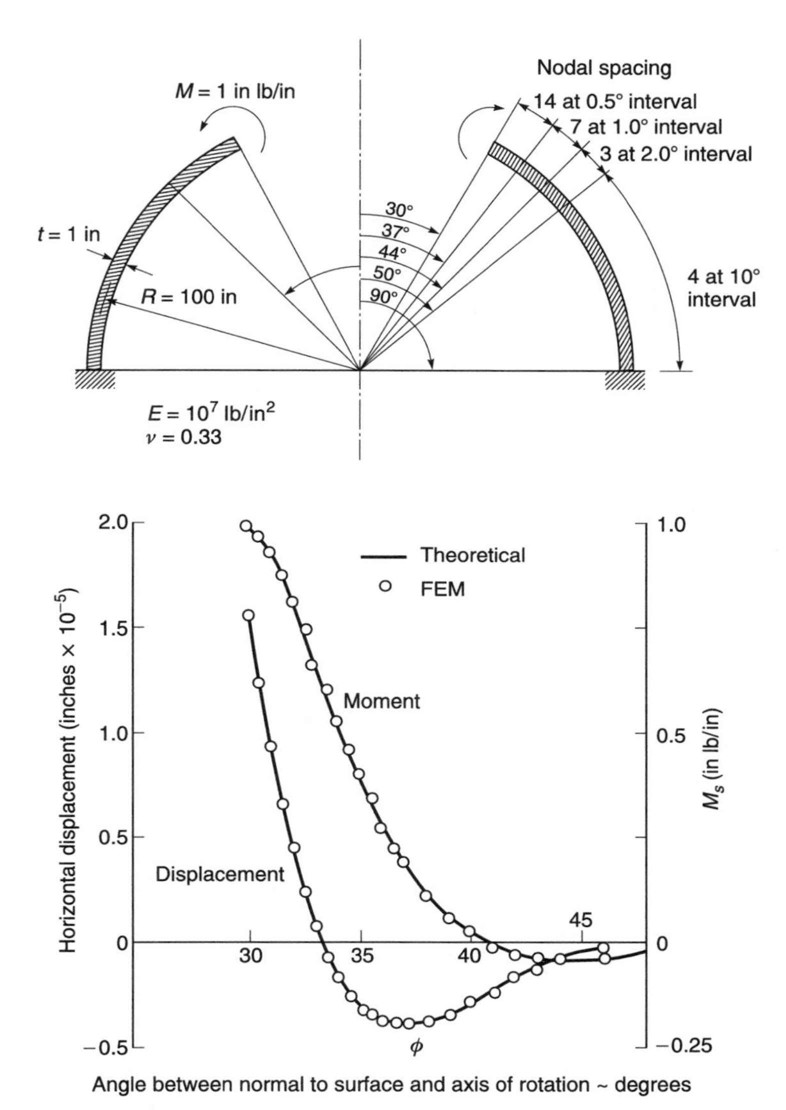

Fig. 7.4 A hemispherical shell solution by finite elements, from Grafton and Strome.[1]

7.3 Curved elements

Use of curved elements has already been described in Chapter 9 of Volume 1, in the context of analyses that involved only first derivatives in the definition of strain. Here second derivatives exist [see Eq. (7.1)] and some of the theorems of Chapter 8 of Volume 1 are no longer applicable.

It was previously mentioned that many possible definitions of curved elements have been proposed and used in the context of axisymmetric shells. The derivation used

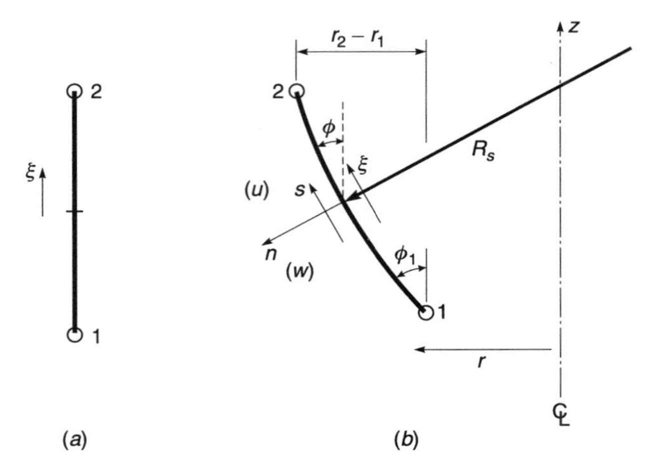

Fig. 7.5 Curved, isoparametric, shell element for axisymmetric problems: (a) parent element; (b) curvilinear coordinates.

here is one due to Delpak[14] and, to use the nomenclature of Chapter 8, Volume 1, is of the subparametric type.

The basis of curved element definition is one that gives a common tangent between adjacent elements (or alternatively, a specified tangent direction). This is physically necessary to avoid 'kinks' in the description of a smooth shell.

If a general curved form of a shell of revolution is considered, as shown in Fig. 7.5, the expressions for strain quoted in Eq. (7.1) have to be modified to take into account the curvature of the shell in the meridian plane.[20,21] These now become

$$\boldsymbol{\varepsilon} = \begin{Bmatrix} \varepsilon_s \\ \varepsilon_\theta \\ \chi_s \\ \chi_\theta \end{Bmatrix} = \begin{Bmatrix} \mathrm{d}\bar{u}/\mathrm{d}s + \bar{w}/R_s \\ [\bar{u}\cos\phi - \bar{w}\sin\phi]/r \\ -\mathrm{d}^2\bar{w}/\mathrm{d}s^2 - \mathrm{d}(\bar{u}/R_s)/\mathrm{d}s \\ -[(\mathrm{d}\bar{w}/\mathrm{d}s + \bar{u}/R_s)]\cos\phi/r \end{Bmatrix} \tag{7.23}$$

In the above the angle ϕ is a function of s, that is,

$$\frac{\mathrm{d}r}{\mathrm{d}s} = \cos\phi \qquad \text{and} \qquad \frac{\mathrm{d}z}{\mathrm{d}s} = \sin\phi$$

R_s is the principal radius in the meridian plane, and the second principal curvature radius R_θ is given by

$$r = R_\theta \sin\phi$$

The reader can verify that for $R_s = \infty$ Eq. (7.23) coincides with Eq. (7.1).

7.3.1 Shape functions for a curved element

We shall now consider the 1–2 element to be curved as shown in Fig. 7.5(b), where the coordinate is in 'parent' form ($-1 \leqslant \xi \leqslant 1$) as shown in Fig. 7.5(a). The coordinates and the unknowns are 'mapped' in the manner of Chapter 9 of Volume 1. As we wish to interpolate a quantity with slope continuity we can write for a typical function ψ

$$\psi = \sum_{i=1}^{2} \left[N_i^w(\xi)\,\psi_i + N_i^\beta \left(\frac{d\psi}{d\xi} \right)_i \right] = \mathbf{N}\tilde{\boldsymbol{\psi}} \tag{7.24}$$

where again the order 1 Hermitian interpolations have been used. We can now simultaneously use these functions to describe variations of the global displacements u and w as*

$$\begin{aligned} u &= \sum_{i=1}^{2} \left[N_i^w(\xi)\,u_i + N_i^\beta \left(\frac{du}{d\xi} \right)_i \right] \\ w &= \sum_{i=1}^{2} \left[N_i^w(\xi)\,w_i + N_i^\beta \left(\frac{dw}{d\xi} \right)_i \right] \end{aligned} \tag{7.25}$$

and of the coordinates r and z which define the shell (mid-surface). Indeed, if the thickness of the element is also variable the same interpolation could be applied to it. Such an element would then be isoparametric (see Chapter 9 of Volume 1). Accordingly, we can define the geometry as

$$\begin{aligned} r &= \sum_{i=1}^{2} \left[N_i^w(\xi)\,r_i + N_i^\beta \left(\frac{dr}{d\xi} \right)_i \right] \\ z &= \sum_{i=1}^{2} \left[N_i^w(\xi)\,z_i + N_i^\beta \left(\frac{dz}{d\xi} \right)_i \right] \end{aligned} \tag{7.26}$$

and, provided the nodal values in the above can be specified, a one-to-one relation between ξ and the position on the curved element surface is defined [Fig. 7.5(b)].

While specification of r_i and z_i is obvious, at the ends only the slope

$$\cot \phi_i = -\left(\frac{dr}{dz} \right)_i \tag{7.27}$$

is defined. The specification to be adopted with regard to the derivatives occurring in Eq. (7.26) depends on the *scaling* of ξ along the tangent length s. Only the ratio

$$\left(\frac{dr}{dz} \right)_i = \frac{(dr/d\xi)_i}{(dz/d\xi)_i} \tag{7.28}$$

is unambiguously specified. Thus $(dr/d\xi)_i$ or $(dz/d\xi)_i$ can be given an arbitrary value. Here, however, practical considerations intervene as with the wrong choice a very uneven relationship between s and ξ will occur. Indeed, with an unsuitable choice the shape of the curve can depart from the smooth one illustrated and loop between the end values.

To achieve a reasonably uniform spacing it suffices for well-behaved surfaces to approximate

$$\frac{dr}{d\xi} \approx \frac{\Delta r}{\Delta \xi} = \frac{r_2 - r_1}{2} \qquad \text{or} \qquad \frac{dz}{d\xi} \approx \frac{\Delta z}{\Delta \xi} = \frac{z_2 - z_1}{2} \tag{7.29}$$

* One immediate difference will be observed from that of the previous formulation. Now both displacement components vary in a cubic manner along an element while previously a linear variation of the tangential displacement was permitted. This additional degree of freedom does not, however, introduce excessive constraints provided the shell thickness is itself continuous.

using whichever is largest and noting that the whole range of ξ is 2 between the nodal points.

7.3.2 Strain expressions and properties of curved elements

The variation of global displacements are specified by Eq. (7.25) while the strains are described in locally directed displacements in Eq. (7.23). Some transformations are therefore necessary before the strains can be determined.

We can express the locally directed displacements $\bar{u}$ and $\bar{w}$ in terms of the global displacements by using Eq. (7.7), that is,

$$\begin{Bmatrix} \bar{u} \\ \bar{w} \end{Bmatrix} = \begin{bmatrix} \cos\phi & \sin\phi \\ -\sin\phi & \cos\phi \end{bmatrix} \begin{Bmatrix} u \\ w \end{Bmatrix} = \bar{\mathbf{T}}\mathbf{u} \tag{7.30}$$

where ϕ is the angle of the tangent to the curve and the r axis (Fig. 7.5). We note that this transformation may be expressed in terms of the ξ coordinate using Eqs (7.27) and (7.28) and the interpolations for r and z. With this transformation the continuity of displacement between adjacent elements is achieved by matching the global nodal displacements u_i and w_i. However, in the development for the conical element we have specified continuity of *rotation* of the cross-section only. Here we shall allow usually the continuity of both s derivatives in displacements. Thus, the parameters

$$\frac{\mathrm{d}u}{\mathrm{d}s} \quad \text{and} \quad \frac{\mathrm{d}w}{\mathrm{d}s}$$

will be given common values at nodes. As

$$\frac{\mathrm{d}u}{\mathrm{d}s}\frac{\mathrm{d}s}{\mathrm{d}\xi} = \frac{\mathrm{d}u}{\mathrm{d}\xi} \quad \text{and} \quad \frac{\mathrm{d}w}{\mathrm{d}s}\frac{\mathrm{d}s}{\mathrm{d}\xi} = \frac{\mathrm{d}w}{\mathrm{d}\xi} \tag{7.31}$$

and

$$\frac{\mathrm{d}s}{\mathrm{d}\xi} = \sqrt{\left(\frac{\mathrm{d}r}{\mathrm{d}\xi}\right)^2 + \left(\frac{\mathrm{d}z}{\mathrm{d}\xi}\right)^2}$$

no difficulty exists in substituting these new variables in Eqs (7.25) and (7.30) which now take the form

$$\bar{\mathbf{u}} = \mathbf{N}(\xi)\,\mathbf{a}^e \qquad \text{with} \qquad \mathbf{a}_i = [u_i \quad w_i \quad (\mathrm{d}u/\mathrm{d}s)_i \quad (\mathrm{d}w/\mathrm{d}s)_i]^{\mathrm{T}} \tag{7.32}$$

The form of the 2×4 shape function submatrices $\mathbf{N}_i$ can now be explicitly determined by using the above transformations in Eq. (7.25).[14] We note that the meridian radius of curvature R_s can be calculated explicitly from the mapped, parametric, form of the element by using

$$R_s = \frac{\left[(\mathrm{d}r/\mathrm{d}\xi)^2 + (\mathrm{d}z/\mathrm{d}\xi)^2\right]^{3/2}}{(\mathrm{d}r/\mathrm{d}\xi)(\mathrm{d}^2z/\mathrm{d}\xi^2) - (\mathrm{d}z/\mathrm{d}\xi)(\mathrm{d}^2r/\mathrm{d}\xi^2)} \tag{7.33}$$

in which all the derivatives are directly determined from expression (7.26).

If shells that branch or in which abrupt thickness changes occur are to be treated, the nodal parameters specified in Eq. (7.32) are not satisfactory. It is better to rewrite these as

$$\mathbf{a}_i = [u_i \quad w_i \quad \beta_i \quad (\mathrm{d}\bar{u}/\mathrm{d}s)_i]^{\mathrm{T}} \tag{7.34}$$

where β_i, equal to $(\mathrm{d}\bar{w}/\mathrm{d}s)_i$, is the nodal rotation, and to connect only the first three parameters. The fourth is now an unconnected element parameter with respect to which, however, the usual treatment is still carried out. Transformations needed in the above are implied in Eq. (7.7).

In the derivation of the **B** matrix expressions which define the strains, both first and second derivatives with respect to s occur, as seen in the definition of Eq. (7.23). If we observe that the derivatives can be obtained by the simple (chain) rules already implied in Eq. (7.31), for any function F we can write

$$\frac{\mathrm{d}F}{\mathrm{d}\xi} = \frac{\mathrm{d}F}{\mathrm{d}s}\frac{\mathrm{d}s}{\mathrm{d}\xi} \qquad \frac{\mathrm{d}^2F}{\mathrm{d}\xi^2} = \frac{\mathrm{d}^2F}{\mathrm{d}s^2}\left(\frac{\mathrm{d}s}{\mathrm{d}\xi}\right)^2 + \frac{\mathrm{d}F}{\mathrm{d}s}\left(\frac{\mathrm{d}^2s}{\mathrm{d}\xi^2}\right) \tag{7.35}$$

and all the expressions of **B** can be found.

Finally, the stiffness matrix is obtained in a similar way as in Eq. (7.16), changing the variable

$$\mathrm{d}s = \frac{\mathrm{d}s}{\mathrm{d}\xi}\,\mathrm{d}\xi \tag{7.36}$$

and integrating ξ within the limits -1 and $+1$. Once again the quantities contained in the integral expressions prohibit exact integration, and numerical quadrature must be used. As this is carried out in one coordinate only it is not very time-consuming and an adequate number of Gauss points can be used to determine the stiffness (generally three points suffice). Initial stress and other load matrices are similarly obtained.

The particular isoparametric formulation presented in summary form here differs somewhat from the alternatives of references 7, 8, 13 and 15 and has the advantage that, because of its *isoparametric* form, rigid body displacement modes and indeed the states of constant first derivatives are available. Proof of this is similar to that contained in Sec. 9.5 of Volume 1. The fact that the forms given in the alternative formulations have strain under rigid body nodal displacements may not be serious in some applications, as discussed by Haisler and Stricklin.[24] However, in some modes of non-axisymmetric loads (see Chapter 9) this incompleteness may be a serious drawback and may indeed lead to very wrong results.

Constant states of curvature cannot be obtained for a *finite* element of any kind described here and indeed are not physically possible. When the size of the element decreases it will be found that such arbitrary constant curvature states are available in the limit (see Sec. 10.10 in Volume 1).

7.3.3 Additional nodeless variables

As in the straight frustrum element, addition of nodeless (enhanced) variables in the analysis of axisymmetric shells is particularly valuable when large curved elements are capable of reproducing with good accuracy the geometric shapes. Thus an addition of a set of internal, hierarchical, element variables

$$\sum_{j=1}^{n} \hat{N}_j \Delta a_j \tag{7.37}$$

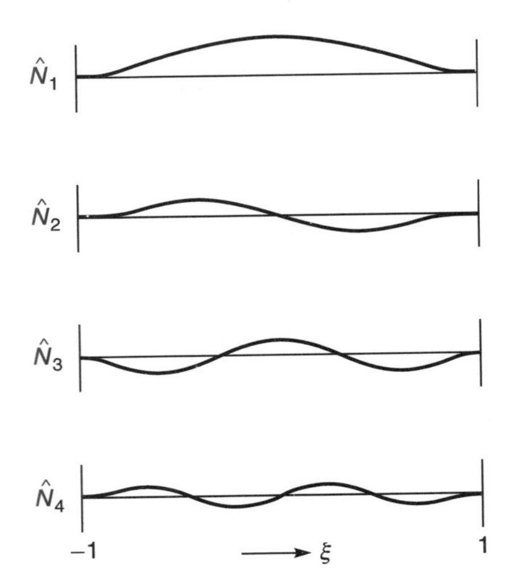

Fig. 7.6 Internal shape functions for a linear element.

to the definition of the normal displacement defined in Eq. (7.6) or Eq. (7.25), in which Δa_j is a set of internal parameters and $\hat{N}_j$ is a set of functions having zero values and zero first derivatives at the nodal points, allows considerable improvement in representation of the displacements to be achieved without violating any of the convergence requirements (see Chapter 2 of Volume 1). For tangential displacements the requirement of zero first derivatives at nodes could be omitted. Webster also uses such additional functions in the context of straight elements.[9] In transient situations where these modes affect the mass matrix one can also use these functions as a basis for developing *enhanced strain modes* (see Sec. 7.2.3 and Chapters 11 and 12 of Volume 1) since these by definition do not influence the assumed displacement field and, hence, the mass and surface loading terms.

Whether the element is in fact straight or curved does not matter and indeed we can supplement the definitions of displacements contained in Eq. (7.25) by Eq. (7.37) for each of the components. If this is done only in the displacement definition and *not* in the coordinate definition [Eq. (7.26)] the element becomes now of the category of subparametric.* As proved in Chapter 9 of Volume 1, the same advantages are retained as in isoparametric forms.

The question as to the expression to be used for additional, internal shape functions is of some importance though the choice is wide. While it is no longer necessary to use polynomial representation, Delpak does so and uses a special form of Lègendre polynomial (hierarchical functions). The general shapes are shown in Fig. 7.6.

* While it would obviously be possible to include the new shape function in the element coordinate definition, little practical advantage would be gained as a cubic represents realistic shapes adequately. Further, the development would then require 'fitting' the a_j for coordinates to the shape, complicating even further the development of derivatives.

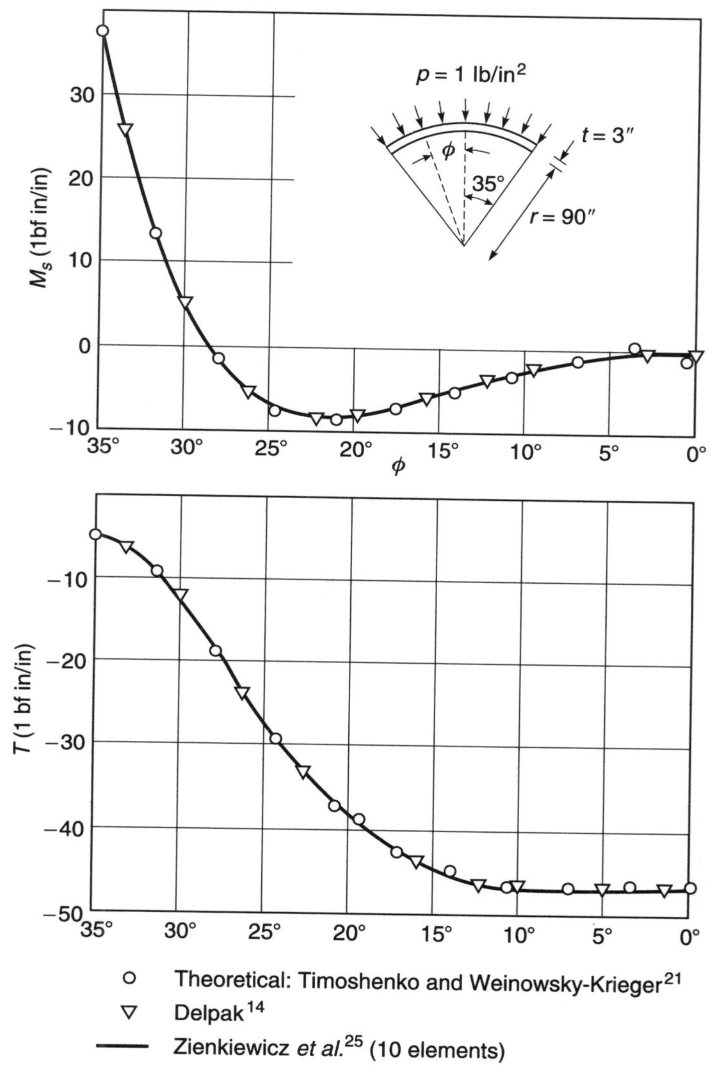

Fig. 7.7 Spherical dome under uniform pressure.

A series of examples shown in Figs 7.7–7.9 illustrate the applications of the isoparametric curvilinear element of the previous section with additional internal parameters.

In Fig. 7.7 a spherical dome with clamped edges is analysed and compared with analytical results of reference 21. Figures 7.8 and 7.9 show, respectively, more complex examples. In the first a torus analysis is made and compared with alternative finite element and analytical results.[12,13,25–7] The second case is one where branching occurs, and here alternative analytical results are given by Kraus.[28]

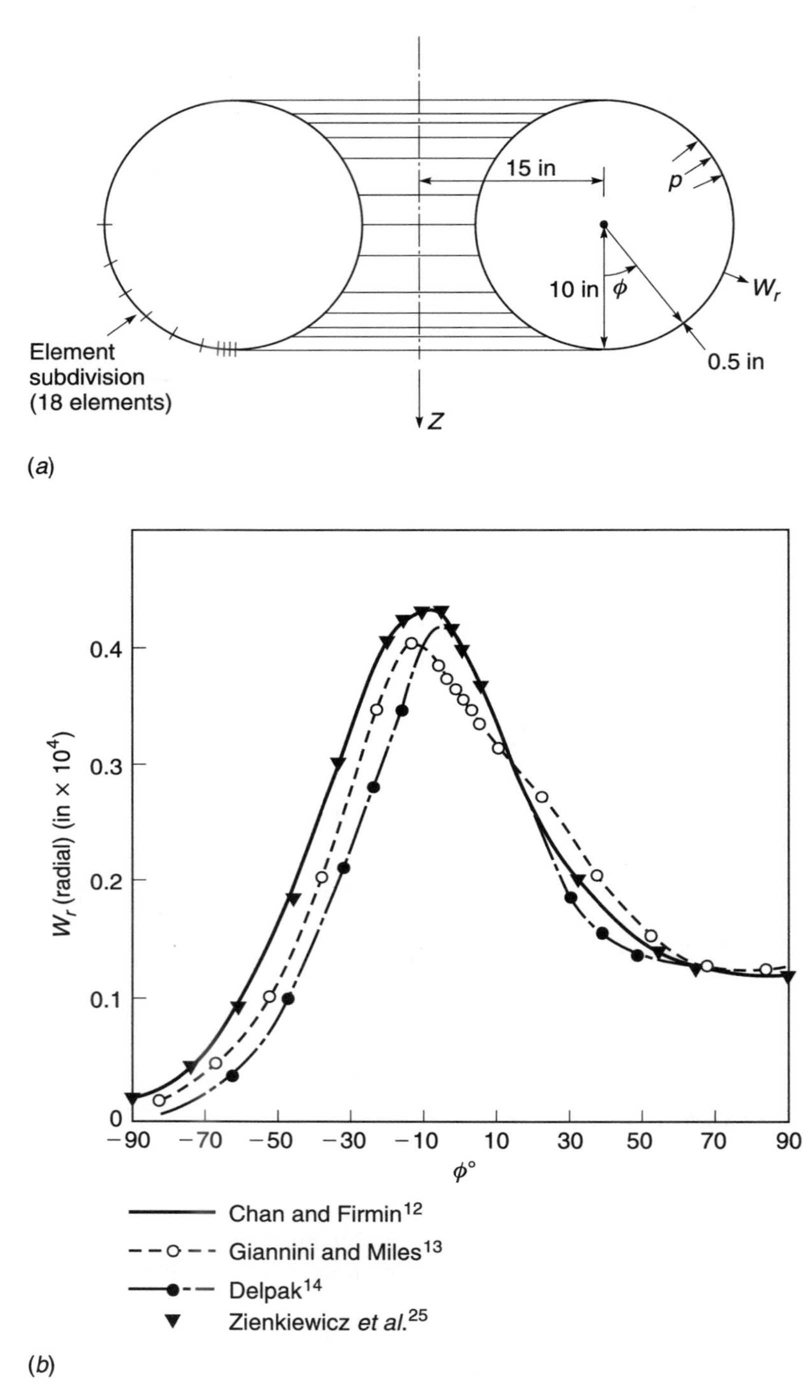

Fig. 7.8 Toroidal shell under internal pressure: (a) element subdivision; (b) radial displacements.

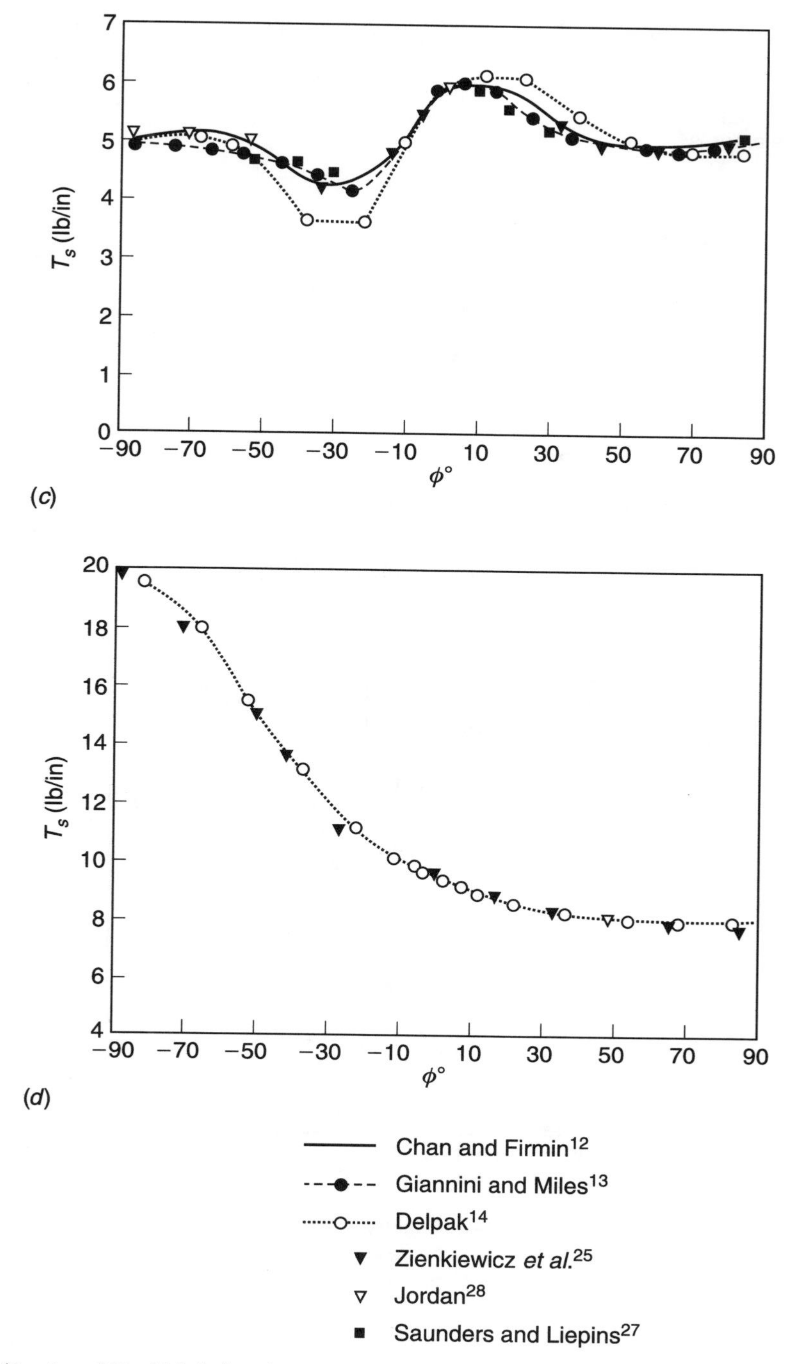

Fig. 7.8 (Continued) Toroidal shell under internal pressure: (c) in-plane stress resultants; (d) in-plane stress resultants.

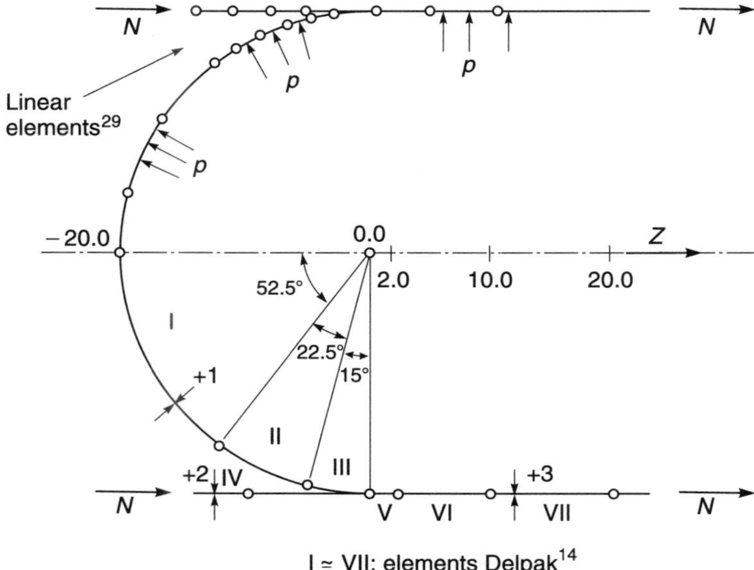

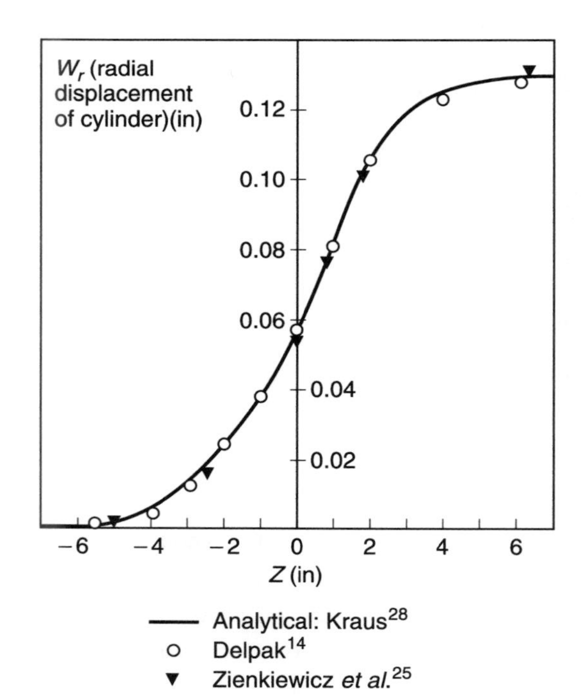

Fig. 7.9 Branching shell.

7.4 Independent slope–displacement interpolation with penalty functions (thick or thin shell formulations)

In Chapter 5 we discussed the use of independent slope and displacement interpolation in the context of beams and plates. Continuity was assured by the introduction of the shear force as an independent *mixed* variable which was defined within each element. The elimination of the shear variable led to a penalty-type formulation in which the shear rigidity played the role of the penalty parameter. The equivalence of the number of parameters used in defining the shear variation and the number of integration points used in evaluating the penalty terms was demonstrated there (and also in Chapter 11 of Volume 1) in special cases, and this justified the success of *reduced* integration methods. This equivalence is not exact in the case of the axisymmetric problem in which the radius, r, enters the integrals, and hence slightly different results can be expected from the use of the mixed form and simple use of reduced integration. The differences become greatest near the axis of rotation and disappear completely when $r \rightarrow \infty$ where the axisymmetric form results in an equivalent beam (or cylindrical bending plate) element.

Although in general the use of the mixed form yields a superior result, for simplicity we shall here derive only the reduced integration form, leaving the former to the reader as an exercise accomplished following the rules of Chapter 5.

In what follows we shall develop in detail the simplest possible element of this class. This is a direct descendant of the linear beam and plate elements.[25,29] (We note, however, that the plate element formulated in this way has singular modes and can on occasion give completely erroneous results; no such deficiency is present in the beam or axisymmetric shell.)

Consider the strain expressions of Eq. (7.1) for a straight element. When using these the need for C_1 continuity was implied by the second derivative of w existing there. If now we use

$$\frac{\mathrm{d}\bar{w}}{\mathrm{d}s} = \beta \tag{7.38}$$

the strain expression becomes

$$\boldsymbol{\varepsilon} = \begin{Bmatrix} \varepsilon_s \\ \varepsilon_\theta \\ \chi_s \\ \chi_\theta \end{Bmatrix} = \begin{Bmatrix} \mathrm{d}\bar{u}/\mathrm{d}s \\ [\bar{u}\cos\phi - \bar{w}\sin\phi]/r \\ -\mathrm{d}\beta/\mathrm{d}s \\ -\beta\cos\phi/r \end{Bmatrix} \tag{7.39}$$

As β can vary independently, a constraint has to be imposed:

$$C(\bar{w}, \beta) \equiv \frac{\mathrm{d}\bar{w}}{\mathrm{d}s} - \beta = 0 \tag{7.40}$$

This can be done by using the energy functional with a penalty multiplier α. We can thus write

$$\Pi = \pi \int_L \boldsymbol{\varepsilon}^{\mathrm{T}} \mathbf{D} \boldsymbol{\varepsilon} r \,\mathrm{d}s + \pi \int_L \alpha \left(\frac{\mathrm{d}w}{\mathrm{d}s} - \beta \right)^2 r \,\mathrm{d}s + \Pi_{\mathrm{ext}} \tag{7.41}$$

where Π_{ext} is a potential for boundary and loading terms and $\boldsymbol{\varepsilon}$ and $\mathbf{D}$ are defined as in Eq. (7.3). Immediately, α can be identified as the shear rigidity:

$$\alpha = \kappa G t \qquad \text{where for a homogeneous shell} \qquad \kappa = 5/6 \tag{7.42}$$

The penalty functional (7.41) can be identified on purely physical grounds. Washizu[22] quotes this on pages 199–201, and the general theory indeed follows that earlier suggested by Naghdi[30] for shells with shear deformation.

With first derivatives occurring in the energy expression only C_0 continuity is now required for the interpolation of u, w, and β, and in place of Eqs (7.6)–(7.12) we can write directly

$$\bar{\mathbf{u}} = \begin{Bmatrix} \bar{u} \\ \bar{w} \\ \beta \end{Bmatrix} = \sum_{i=1}^{2} N_i(\xi)\mathbf{T}\mathbf{a}_i \qquad \mathbf{a}_i^{\mathrm{T}} = [u_i \quad w_i \quad \beta_i] \tag{7.43}$$

where for $N_i(\xi)$ we can use any of the one-dimensional C_0 interpolations in Chapter 8 of Volume 1. Once again, isoparametric transformation could be used for curvilinear elements with strains defined by Eq. (7.23), and a formulation that we shall discuss in Chapter 8 is but an alternative to this process. If linear elements are used, we can write the expression without consequent use of isoparametric transformation. Indeed, we can replace the interpolations in Eq. (7.8) and now simply use

$$\begin{aligned} u &= N_1(\xi)u_1 + N_2(\xi)u_2 \\ w &= N_1(\xi)w_1 + N_2(\xi)w_2 \\ \beta &= N_1(\xi)\beta_1 + N_2(\xi)\beta_2 \end{aligned} \tag{7.44}$$

and evaluate the integrals arising from expression (7.41) at one Gauss point, which is sufficient to maintain convergence and yet here does not give a singularity.

This extremely simple form will, of course, give very poor results with exact integration, even for thick shells, but now with reduced integration shows excellent performance. In Figs 7.7–7.9 we superpose results obtained with this simple, straight element, and the results speak for themselves.

For other examples the reader can consult reference 25, but in Fig. 7.10 we show a very simple example of a bending of a circular plate with use of different numbers of equal elements. This purely bending problem shows the type of results and convergence attainable.

Interpreting the single integrating point as a single shear variable and applying the patch test count of Chapter 5, the reader can verify that this simple formulation passes the test in assemblies of two or more elements. In a similar way it can be verified that a quadratic interpolation of displacements and the use of two quadrature points (or a linear shear force) also will result in a robust element of excellent performance.

One final word of caution when the element is used in transient analyses is in order. Here it is necessary to compute a mass matrix which can be deduced from

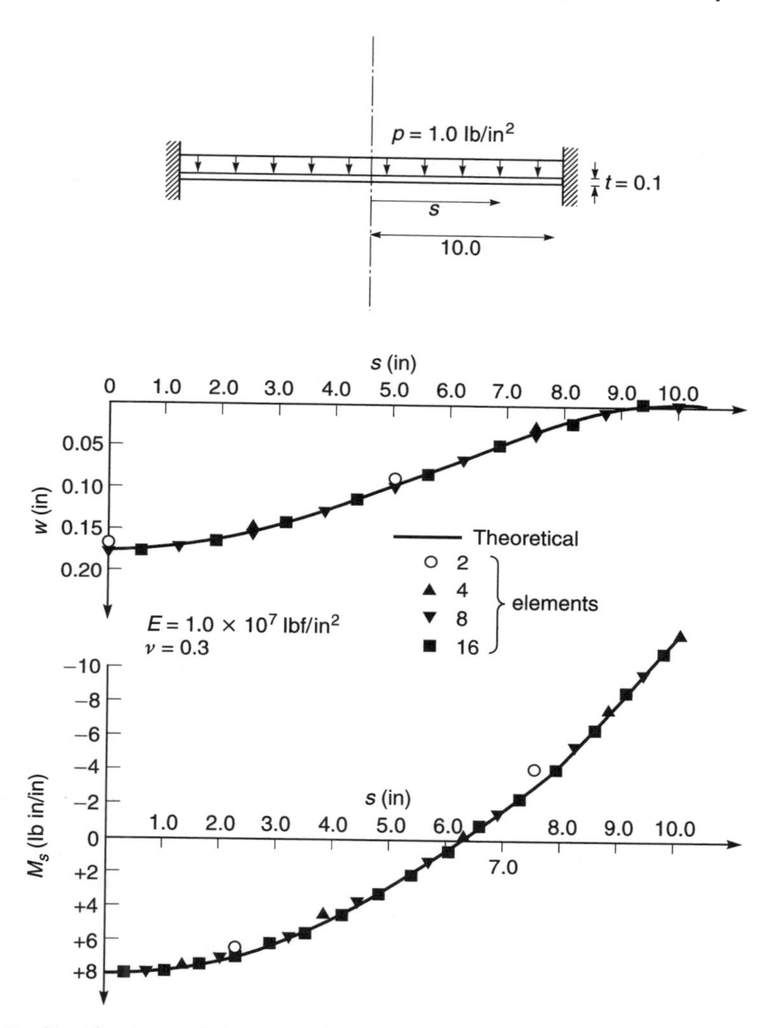

Fig. 7.10 Bending of a circular plate under uniform load; convergence study.

the term

$$\delta\Pi_{in} = 2\pi \int_L \left[\delta u \rho t \ddot{u} + \delta w \rho t \ddot{w} + \delta\beta \rho \frac{t^3}{12} \ddot{\beta} \right] r \, \mathrm{d}s \tag{7.45}$$

Evaluation of this integral with a single quadrature point will lead to a rank deficient mass matrix, which when used with any time stepping scheme can lead to large numerical errors (generally after many time steps have been computed). Accordingly, it is necessary to compute the mass matrix with at least two quadrature points (nodal quadrature giving immediately a diagonal 'lumped' mass).

References

1. P.E. Grafton and D.R. Strome. Analysis of axi-symmetric shells by the direct stiffness method. *Journal of AIAA*, **1**, 2342–7, 1963.
2. E.P. Popov, J. Penzien and Z.A. Liu. Finite element solution for axisymmetric shells. *Proc. Am. Soc. Civ. Eng.*, EM5, 119–45, 1964.
3. R.E. Jones and D.R. Strome. Direct stiffness method of analysis of shells of revolution utilizing curved elements. *Journal of AIAA*, **4**, 1519–25, 1966.
4. J.H. Percy, T.H.H. Pian, S. Klein and D.R. Navaratna. Application of matrix displacement method to linear elastic analysis of shells of revolution. *Journal of AIAA*, **3**, 2138–45, 1965.
5. S. Klein. A study of the matrix displacement method as applied to shells of revolution. In J.S. Przemienicki, R.M. Bader, W.F. Bozich, J.R. Johnson and W.J. Mykytow (eds), *Proc. 1st Conf. Matrix Methods in Structural Mechanics*, Volume AFFDL-TR-66-80, pp. 275–98, Air Force Flight Dynamics Laboratory, Wright Patterson Air Force Base, OH, October 1966.
6. R.E. Jones and D.R. Strome. A survey of analysis of shells by the displacement method. In J.S. Przemienicki, R.M. Bader, W.F. Bozich, J.R. Johnson and W.J. Mykytow (eds), *Proc. 1st Conf. Matrix Methods in Structural Mechanics*, Volume AFFDL-TR-66-80, pp. 205-29, Air Force Flight Dynamics Laboratory, Wright Patterson Air Force Base, OH, October 1966.
7. J.A. Stricklin, D.R. Navaratna and T.H.H. Pian. Improvements in the analysis of shells of revolution by matrix displacement method (curved elements). *Journal of AIAA*, **4**, 2069–72, 1966.
8. M. Khojasteh-Bakht. Analysis of elastic–plastic shells of revolution under axi-symmetric loading by the finite element method. Technical Report SESM 67-8, University of California, Berkeley, CA, 1967.
9. J.J. Webster. Free vibration of shells of revolution using ring elements. *J. Mech. Sci.*, **9**, 559–70, 1967.
10. S. Ahmad, B.M. Irons and O.C. Zienkiewicz. Curved thick shell and membrane elements with particular reference to axi-symmetric problems. In L. Berke, R.M. Bader, W.J. Mykytow, J.S. Przemienicki and M.H. Shirk (eds), *Proc. 2nd Conf. Matrix Methods in Structural Mechanics*, Volume AFFDL-TR-68-150, pp. 539–72, Air Force Flight Dynamics Laboratory, Wright Patterson Air Force Base, OH, October 1968.
11. E.A. Witmer and J.J. Kotanchik. Progress report on discrete element elastic and elastic–plastic analysis of shells of revolution subjected to axisymmetric and asymmetric loading. In L. Berke, R.M. Bader, W.J. Mykytow, J.S. Przemienicki and M.H. Shirk (eds), *Proc. 2nd Conf. Matrix Methods in Structural Mechanics*, Volume AFFDL-TR-68-150, pp. 1341–53, Air Force Flight Dynamics Laboratory, Wright Patterson Air Force Base, OH, October 1968.
12. A.S.L. Chan and A. Firmin. The analysis of cooling towers by the matrix finite element method. *Aeronaut. J.*, **74**, 826–35, 1970.
13. M. Giannini and G.A. Miles. A curved element approximation in the analysis of axi-symmetric thin shells. *Int. J. Num. Meth. Eng.*, **2**, 459–76, 1970.
14. R. Delpak. *Role of the Curved Parametric Element in Linear Analysis of Thin Rotational Shells*, PhD thesis, Department of Civil Engineering and Building, The Polytechnic of Wales, Pontypridd, 1975.
15. P.L. Gould and S.K. Sen. Refined mixed method finite elements for shells of revolution. In R.M. Bader, L. Berke, R.O. Meitz, W.J. Mykytow and J.S. Przemienicki (eds), *Proc. 3rd Conf. Matrix Methods in Structural Mechanics*, Volume AFFDL-TR-71-160, pp. 397–421, Air Force Flight Dynamics Laboratory, Wright Patterson Air Force Base, OH, 1972.

16. Z.M. Elias. Mixed finite element method for axisymmetric shells. *Int. J. Num. Meth. Eng.*, **4**, 261–72, 1972.
17. R.H. Gallagher. Analysis of plate and shell structures. In *Applications of Finite Element Method in Engineering*, pp. 155–205, ASCE, Vanderbilt University, Nashville, TN, 1969.
18. R.H. Gallagher. *Finite Element Analysis: Fundamentals*, Prentice-Hall, Englewood Cliffs, NJ, 1975.
19. J.A. Stricklin. Geometrically nonlinear static and dynamic analysis of shells of revolution. In *High Speed Computing of Elastic Structures*, pp. 383–411. University of Liège, Liège, 1976.
20. V.V. Novozhilov. *Theory of Thin Shells*, Noordhoff, Dordrecht, 1959 [English translation].
21. S.P. Timoshenko and S. Woinowski-Krieger. *Theory of Plates and Shells*, 2nd edition, McGraw-Hill, New York, 1959.
22. K. Washizu. *Variational Methods in Elasticity and Plasticity*, 3rd edition, Pergamon Press, New York, 1982.
23. E.L. Wilson. The static condensation algorithm. *Int. J. Num. Meth. Eng.*, **8**, 199–203, 1974.
24. W.E. Haisler and J.A. Stricklin. Rigid body displacements of curved elements in the analysis of shells by the matrix displacement method. *Journal of AIAA*, **5**, 1525–7, 1967.
25. O.C. Zienkiewicz, J. Bauer, K. Morgan and E. Oñate. A simple element for axi-symmetric shells with shear deformation. *Int. J. Num. Meth. Eng.*, **11**, 1545–58, 1977.
26. J.L. Sanders Jr. and A. Liepins. Toroidal membrane under internal pressure. *Journal of AIAA*, **1**, 2105–10, 1963.
27. F.F. Jordan. Stresses and deformations of the thin-walled pressurized torus. *J. Aero. Sci.*, **29**, 213–25, 1962.
28. H. Kraus. *Thin Elastic Shells*, John Wiley, New York, 1967.
29. T.J.R. Hughes, R.L. Taylor and W. Kanoknukulchai. A simple and efficient finite element for plate bending. *Int. J. Num. Meth. Eng.*, **11**, 1529–43, 1977.
30. P.M. Naghdi. Foundations of elastic shell theory. In I.N. Sneddon and R. Hill (eds), *Progress in Solid Mechanics*, Volume IV, Chapter 1, North-Holland, Amsterdam, 1963.

8

Shells as a special case of three-dimensional analysis – Reissner–Mindlin assumptions

8.1 Introduction

In the analysis of solids the use of isoparametric, curved two- and three-dimensional elements is particularly effective, as illustrated in Chapters 1 and 3 and presented in Chapters 9 and 10 of Volume 1. It seems obvious that use of such elements in the analysis of curved shells could be made directly simply by reducing their dimension in the thickness direction as shown in Fig. 8.1. Indeed, in an axisymmetric situation such an application is illustrated in the example of Fig. 9.25 of Volume 1. With a straightforward use of the three-dimensional concept, however, certain difficulties will be encountered.

In the first place the retention of 3 displacement degrees of freedom at each node leads to large stiffness coefficients from strains in the shell thickness direction. This presents numerical problems and may lead to ill-conditioned equations when the shell thickness becomes small compared with other dimensions of the element.

The second factor is that of economy. The use of several nodes across the shell thickness ignores the well-known fact that even for thick shells the 'normals' to the mid-surface remain practically straight after deformation. Thus an unnecessarily high number of degrees of freedom has to be carried, involving penalties of computer time.

In this chapter we present specialized formulations which overcome both of these difficulties. The constraint of straight 'normals' is introduced to improve economy and the strain energy corresponding to the stress perpendicular to the mid-surface is ignored to improve numerical conditioning.[1–3] With these modifications an efficient tool for analysing curved thick shells becomes available. The accuracy and wide range of applicability of the approach is demonstrated in several examples.

8.2 Shell element with displacement and rotation parameters

The reader will note that the two constraints introduced correspond precisely to the so-called Reissner–Mindlin assumptions already discussed in Chapter 5 to describe the

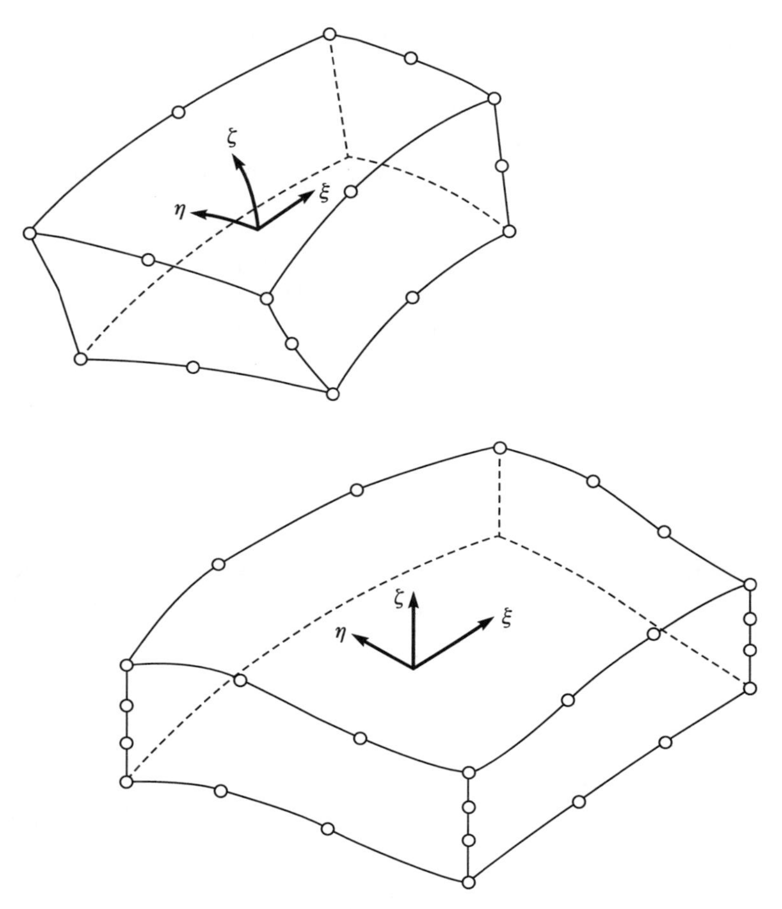

Fig. 8.1 Curved, isoparametric hexahedra in a direct approximation to a curved shell.

behaviour of thick plates. The omission of the third constraint associated with the thin plate theory (normals remaining normal to the mid-surface after deformation) permits the shell to experience transverse shear deformations – an important feature of thick shell situations.

The formulation presented here leads to additional complications compared with the straightforward use of a three-dimensional element. The elements developed here are in essence an alternative to the processes discussed in Chapter 5, for which an independent interpolation of slopes and displacement are used with a penalty function imposition of the continuity requirements. The use of reduced integration is useful if thin shells are to be dealt with – and, indeed, it was in this context that this procedure was first discovered.[4–7] Again the same restrictions for robust behaviour as those discussed in Chapter 5 become applicable and generally elements that perform well in plate situations will do well in shells.

8.2.1 Geometric definition of an element

Consider a typical shell element illustrated in Fig. 8.2. The external faces of the element are curved, while the sections across the thickness are generated by straight lines. Pairs of points, i_{top} and i_{bottom}, each with given cartesian coordinates, prescribe the shape of the element.

Let ξ, η be the two curvilinear coordinates in the mid-surface of the shell and let ζ be a linear coordinate in the thickness direction. If, further, we assume that ξ, η, ζ vary between -1 and 1 on the respective faces of the element we can write a relationship between the cartesian coordinates of any point of the shell and the curvilinear coordinates in the form

$$\begin{Bmatrix} x \\ y \\ z \end{Bmatrix} = \sum N_i(\xi, \eta) \left(\frac{1+\zeta}{2} \begin{Bmatrix} x_i \\ y_i \\ z_i \end{Bmatrix}_{\text{top}} + \frac{1-\zeta}{2} \begin{Bmatrix} x_i \\ y_i \\ z_i \end{Bmatrix}_{\text{bottom}} \right) \tag{8.1}$$

Here $N_i(\xi, \eta)$ is a standard two-dimensional shape function taking a value of unity at the top and bottom nodes i and zero at all other nodes (Chapter 9 of Volume 1). If the basic functions N_i are derived as 'shape functions' of a 'parent', two-dimensional

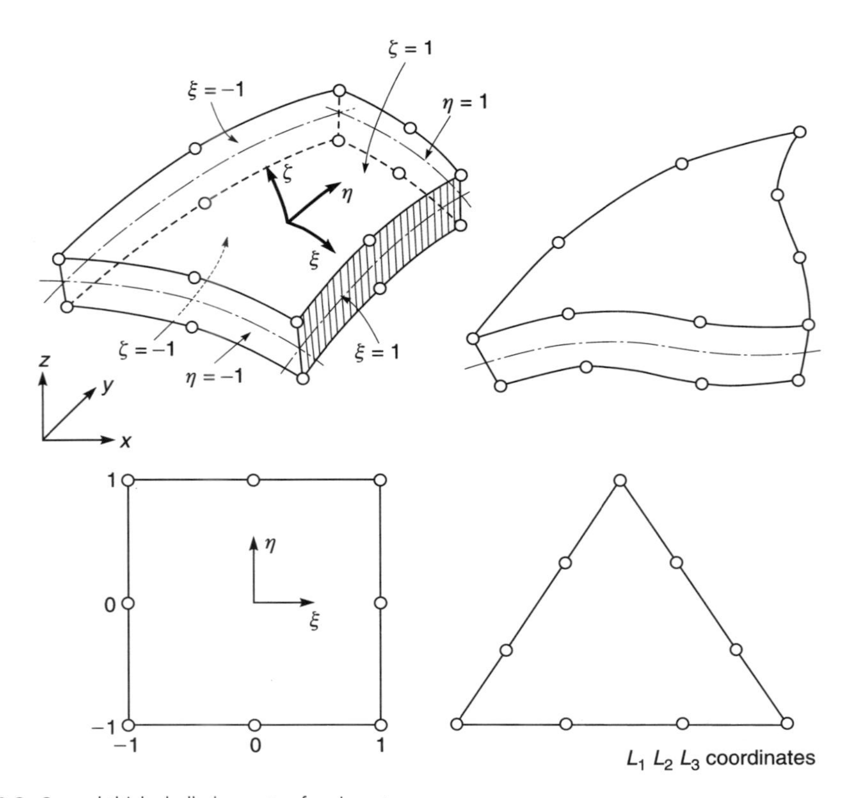

Fig. 8.2 Curved thick shell elements of various types.

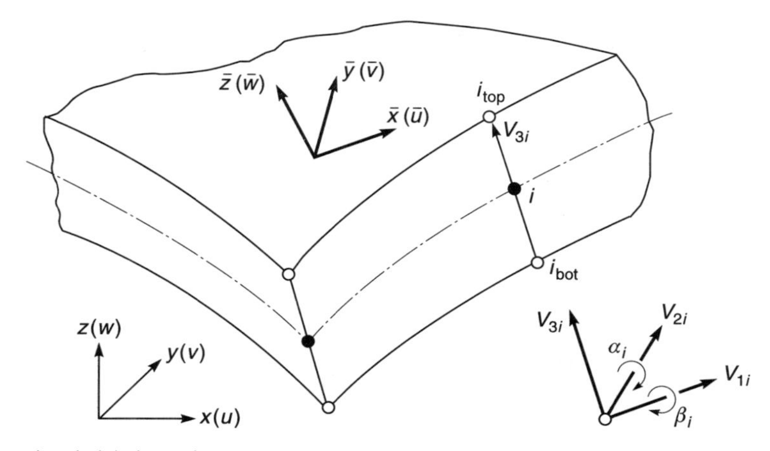

Fig. 8.3 Local and global coordinates.

element, square or triangular* in plan, and are so 'designed' that compatibility is achieved at interfaces, then the curved space elements will fit into each other. Arbitrary curved shapes of the element can be achieved by using shape functions of higher order than linear. Indeed, any of the two-dimensional shape functions of Chapter 8 of Volume 1 can be used here.

The relation between the cartesian and curvilinear coordinates is now established and it will be found desirable to operate with the curvilinear coordinates as the basis. It should be noted that often the coordinate direction ζ is *only approximately normal* to the mid-surface.

It is convenient to rewrite the relationship, Eq. (8.1), in a form specified by the 'vector' connecting the upper and lower points (i.e. a vector of length equal to the shell thickness t) and the mid-surface coordinates. Thus we can rewrite Eq. (8.1) as (Fig. 8.3)†

$$\begin{Bmatrix} x \\ y \\ z \end{Bmatrix} = \sum N_i(\xi,\eta)\left(\begin{Bmatrix} x_i \\ y_i \\ z_i \end{Bmatrix}_{\text{mid}} + \tfrac{1}{2}\zeta\mathbf{V}_{3i}\right) \tag{8.2}$$

where

$$\begin{Bmatrix} x_i \\ y_i \\ z_i \end{Bmatrix} = \tfrac{1}{2}\left(\begin{Bmatrix} x_i \\ y_i \\ z_i \end{Bmatrix}_{\text{top}} + \begin{Bmatrix} x_i \\ y_i \\ z_i \end{Bmatrix}_{\text{bottom}}\right) \quad \text{and} \quad \mathbf{V}_{3i} = \begin{Bmatrix} x_i \\ y_i \\ z_i \end{Bmatrix}_{\text{top}} - \begin{Bmatrix} x_i \\ y_i \\ z_i \end{Bmatrix}_{\text{bottom}} \tag{8.3}$$

with $\mathbf{V}_{3i}$ defining a vector whose length represents the shell thickness.

* Area coordinates L_j would be used in this case in place of ξ, η as in Chapter 8 of Volume 1.
† For details of vector algebra see Appendix F of Volume 1.

For relatively thin shells, it is convenient to replace the vector $\mathbf{V}_{3i}$ by a unit vector $\mathbf{v}_{3i}$ in the direction normal to the mid-surface. Now Eq. (8.2) is written simply as

$$\begin{Bmatrix} x \\ y \\ z \end{Bmatrix} = \sum N_i(\xi,\eta) \left(\begin{Bmatrix} x_i \\ y_i \\ z_i \end{Bmatrix}_{\text{mid}} + \tfrac{1}{2}\zeta t_i \mathbf{v}_{3i} \right)$$

where t_i is the shell thickness at the node i. Construction of a vector normal to the mid-surface is a simple process (see Sec. 6.4.2).

8.2.2 Displacement field

The displacement field is now specified for the element. As the strains in the direction normal to the mid-surface will be assumed to be negligible, the displacement throughout the element will be taken to be uniquely defined by the *three cartesian components* of the mid-surface node displacement and *two rotations* about two orthogonal directions normal to the nodal vector $\mathbf{V}_{3i}$. If these two orthogonal directions are denoted by unit vectors $\mathbf{v}_{1i}$ and $\mathbf{v}_{2i}$ with corresponding rotations α_i and β_i (see Fig. 8.3), we can write, similar to Eq. (8.2) but dropping the subscript 'mid' for simplicity,

$$\begin{Bmatrix} u \\ v \\ w \end{Bmatrix} = \sum N_i(\xi,\eta) \left(\begin{Bmatrix} u_i \\ v_i \\ w_i \end{Bmatrix} + \tfrac{1}{2}\zeta t_i [\mathbf{v}_{1i}, \quad -\mathbf{v}_{2i}] \begin{Bmatrix} \alpha_i \\ \beta_i \end{Bmatrix} \right) \tag{8.4}$$

from which the usual form is readily obtained as

$$\begin{Bmatrix} u \\ v \\ w \end{Bmatrix} = \mathbf{N}\mathbf{a}^e; \quad \mathbf{a}^e = \begin{Bmatrix} \mathbf{a}_i^e \\ \vdots \\ \mathbf{a}_j^e \end{Bmatrix} \qquad \text{with} \quad \mathbf{a}_i^e = \begin{Bmatrix} u_i \\ v_i \\ w_i \\ \alpha_i \\ \beta_i \end{Bmatrix} \tag{8.5}$$

where u, v and w are displacements in the directions of the global x, y and z axes.

As an infinity of vector directions normal to a given direction can be generated, a particular scheme has to be devised to ensure a *unique* definition. Some such schemes were discussed in Chapter 6. Here another unique alternative will be given,[2,4] but other possibilities are open.[7]

Here $\mathbf{V}_{3i}$ is the vector to which a normal direction is to be constructed. A coordinate vector in a Cartesian system may be defined by

$$\mathbf{x} = x\mathbf{i} + y\mathbf{j} + z\mathbf{k} \tag{8.6}$$

in which $\mathbf{i}$, $\mathbf{j}$ and $\mathbf{k}$ are three (orthogonal) base vectors. To find the first normal vector we find the minimum component of $\mathbf{V}_{3i}$ and construct a vector cross-product with the unit vector in this direction to define $\mathbf{V}_{1i}$. For example if the x component of $\mathbf{V}_{3i}$ is the smallest one we construct

$$\mathbf{V}_{1i} = \mathbf{i} \times \mathbf{V}_{3i} \tag{8.7}$$

where

$$\mathbf{i} = [1 \quad 0 \quad 0]^{\mathrm{T}}$$

is the form of the unit vector in the x direction. Now

$$\mathbf{v}_{1i} = \frac{\mathbf{V}_{1i}}{|\mathbf{V}_{1i}|} \qquad \text{where} \qquad |\mathbf{V}_{1i}| = \sqrt{\mathbf{V}_{1i}^{\mathrm{T}}\mathbf{V}_{1i}} \tag{8.8}$$

defines the first unit vector.

The second normal vector may now be computed from

$$\mathbf{V}_{2i} = \mathbf{V}_{3i} \times \mathbf{V}_{1i} \tag{8.9}$$

and normalized using the form in Eq. (8.8). We have thus three local, orthogonal axes defined by unit vectors

$$\mathbf{v}_{1i}, \mathbf{v}_{2i} \quad \text{and} \quad \mathbf{v}_{3i} \tag{8.10}$$

Once again if N_i are C_0 functions then displacement compatibility is maintained between adjacent elements.

The element coordinate definition is now given by the relation Eq. (8.2) and has more degrees of freedom than the definition of the displacements. The element is therefore of the 'superparametric' kind (see Chapter 9 of Volume 1) and the constant strain criteria are not automatically satisfied. Nevertheless, it will be seen from the definition of strain components involved that both rigid body motions and constant strain conditions are available.

Physically it has been assumed in the definition of Eq. (8.4) that no strains occur in the 'thickness' direction ζ. While this direction is not always exactly normal to the mid-surface it still represents a good approximation of one of the usual shell assumptions.

At each mid-surface node i of Fig 8.3 we now have the 5 basic degrees-of-freedom, and the connection of elements will follow precisely the patterns described in Chapter 6 (Secs 6.3 and 6.4).

8.2.3 Definition of strains and stresses

To derive the properties of a finite element the essential strains and stresses need first to be defined. The components in directions of *orthogonal axes* related to the surface ζ (constant) are essential if account is to be taken of the basic shell assumptions. Thus, if at any point in this surface we erect a normal $\bar{z}$ with two other orthogonal axes $\bar{x}$ and $\bar{y}$ tangential to it (Fig. 8.3), the strain components of interest are given simply by the three-dimensional relationships in Chapter 6 of Volume 1:

$$\bar{\boldsymbol{\varepsilon}} = \begin{Bmatrix} \varepsilon_{\bar{x}} \\ \varepsilon_{\bar{y}} \\ \gamma_{\bar{x}\bar{y}} \\ \gamma_{\bar{y}\bar{z}} \\ \gamma_{\bar{z}\bar{x}} \end{Bmatrix} = \begin{Bmatrix} \bar{u}_{,\bar{x}} \\ \bar{v}_{,\bar{y}} \\ \bar{u}_{,\bar{y}} + \bar{v}_{,\bar{x}} \\ \bar{v}_{,\bar{z}} + \bar{w}_{,\bar{y}} \\ \bar{w}_{,\bar{x}} + \bar{u}_{,\bar{z}} \end{Bmatrix} \tag{8.11}$$

with the strain in direction $\bar{z}$ neglected so as to be consistent with the usual shell assumptions. It must be noted that in general none of these directions coincide with those of the curvilinear coordinates ξ, η, ζ, although $\bar{x}$, $\bar{y}$ are in the $\xi\eta$ plane ($\zeta = \text{constant}$).*

The stresses corresponding to these strains are defined by a matrix $\bar{\boldsymbol{\sigma}}$ and for elastic behaviour are related to the usual elasticity matrix $\bar{\mathbf{D}}$. Thus

$$\bar{\boldsymbol{\sigma}} = \begin{Bmatrix} \sigma_{\bar{x}} \\ \sigma_{\bar{y}} \\ \tau_{\bar{x}\bar{y}} \\ \tau_{\bar{y}\bar{z}} \\ \tau_{\bar{z}\bar{x}} \end{Bmatrix} = \bar{\mathbf{D}}(\bar{\boldsymbol{\varepsilon}} - \bar{\boldsymbol{\varepsilon}}_0) + \bar{\boldsymbol{\sigma}}_0 \tag{8.12}$$

where $\bar{\boldsymbol{\varepsilon}}_0$ and $\bar{\boldsymbol{\sigma}}_0$ represent any 'initial' strains and stresses, respectively.

The 5×5 matrix $\bar{\mathbf{D}}$ can now include any anisotropic properties and indeed may be prescribed as a function of ζ if sandwich or laminated construction is used. For the present we shall define it only for an isotropic material. Here

$$\bar{\mathbf{D}} = \frac{E}{1-\nu^2} \begin{bmatrix} 1 & \nu & 0 & 0 & 0 \\ \nu & 1 & 0 & 0 & 0 \\ 0 & 0 & (1-\nu)/2 & 0 & 0 \\ 0 & 0 & 0 & \kappa(1-\nu)/2 & 0 \\ 0 & 0 & 0 & 0 & \kappa(1-\nu)/2 \end{bmatrix} \tag{8.13}$$

in which E and ν are Young's modulus and Poisson's ratio, respectively. The factor κ included in the last two shear terms is taken as 5/6 and its purpose is to improve the shear displacement approximations (see Chapter 4). From the displacement definition it will be seen that the shear distribution is approximately constant through the thickness, whereas in reality the shear distribution for elastic behaviour is approximately parabolic. The value $\kappa = 5/6$ is the ratio of relevant strain energies.

It is important to note that this matrix is *not* defined by deleting appropriate terms from the equivalent three-dimensional stress matrix. It must be derived by substituting $\sigma_{\bar{z}} = 0$ into Eqs (6.13) and (6.14) in Volume 1 and a suitable elimination so that this important shell assumption is satisfied. This is similar to the procedure for deriving plane stress behaviour in two-dimensional analyses.

8.2.4 Element properties and necessary transformations

The stiffness matrix – and indeed all other 'element' property matrices – involve integrals over the volume of the element, which are quite generally of the form

$$\int_{V^e} \mathbf{H}\, \mathrm{d}x\, \mathrm{d}y\, \mathrm{d}z \tag{8.14}$$

* Indeed, these directions will only approximately agree with the nodal directions $\mathbf{v}_{1i}$, $\mathbf{v}_{2i}$ previously derived, as in general the vector $\mathbf{v}_{3i}$ is only approximately normal to the mid-surface.

where the matrix **H** is a function of the coordinates. For instance, in the stiffness matrix

$$\mathbf{H} = \bar{\mathbf{B}}^{\mathrm{T}} \bar{\mathbf{D}} \bar{\mathbf{B}} \tag{8.15}$$

and with the usual definition of Chapter 2 of Volume 1,

$$\bar{\boldsymbol{\varepsilon}} = \bar{\mathbf{B}} \mathbf{a}^e \tag{8.16}$$

we have $\bar{\mathbf{B}}$ defined in terms of the displacement derivatives with respect to the local Cartesian coordinates $\bar{x}, \bar{y}, \bar{z}$ by Eq. (8.11). Now, therefore, *two sets of transformations* are necessary before the element can be integrated with respect to the curvilinear coordinates ξ, η, ζ.

First, by identically the same process as we used in Chapter 9 of Volume 1, the derivatives with respect to the x, y, z directions are obtained. As Eq. (8.4) relates the global displacements u, v, w to the curvilinear coordinates, the derivatives of these displacements with respect to the global x, y, z coordinates are given by a matrix relation:

$$\begin{bmatrix} u_{,x} & v_{,x} & w_{,x} \\ u_{,y} & v_{,y} & w_{,y} \\ u_{,z} & v_{,z} & w_{,z} \end{bmatrix} = \mathbf{J}^{-1} \begin{bmatrix} u_{,\xi} & v_{,\xi} & w_{,\xi} \\ u_{,\eta} & v_{,\eta} & w_{,\eta} \\ u_{,\zeta} & v_{,\zeta} & w_{,\zeta} \end{bmatrix} \tag{8.17}$$

In this, the Jacobian matrix is defined as

$$\mathbf{J} = \begin{bmatrix} x_{,\xi} & y_{,\xi} & z_{,\xi} \\ x_{,\eta} & y_{,\eta} & z_{,\eta} \\ x_{,\zeta} & y_{,\zeta} & z_{,\zeta} \end{bmatrix} \tag{8.18}$$

and calculated from the coordinate definitions of Eq. (8.2). Now, for every set of curvilinear coordinates the global displacement derivatives can be obtained numerically.

A second transformation to the local displacements $\bar{x}$, $\bar{y}$, $\bar{z}$ will allow the strains, and hence the $\bar{\mathbf{B}}$ matrix, to be evaluated. The directions of the local axes can be established from a vector normal to the $\xi\eta$ mid-surface ($\zeta = 0$). This vector can be found from two vectors $\mathbf{x}_{,\xi}$ and $\mathbf{x}_{,\eta}$ that are tangential to the mid-surface. Thus

$$\mathbf{V}_3 = \begin{bmatrix} x_{,\xi} \\ y_{,\xi} \\ z_{,\xi} \end{bmatrix} \times \begin{bmatrix} x_{,\eta} \\ y_{,\eta} \\ z_{,\eta} \end{bmatrix} = \begin{bmatrix} y_{,\xi} z_{,\eta} - y_{,\eta} z_{,\xi} \\ z_{,\xi} x_{,\eta} - z_{,\eta} x_{,\xi} \\ x_{,\xi} y_{,\eta} - x_{,\eta} y_{,\xi} \end{bmatrix} \tag{8.19}$$

We can now construct two perpendicular vectors $\mathbf{V}_1$ and $\mathbf{V}_2$ following the process given previously to describe the $\bar{x}$ and $\bar{y}$ directions, respectively. The three orthogonal vectors can be reduced to unit magnitudes to obtain a matrix of vectors in the $\bar{x}$, $\bar{y}$, $\bar{z}$ directions (which is in fact the direction cosine matrix) given as

$$\boldsymbol{\theta} = [\mathbf{v}_1, \quad \mathbf{v}_2, \quad \mathbf{v}_3] \tag{8.20}$$

The global derivatives of displacement u, v and w are now transformed to the local derivatives of the local orthogonal displacements by a standard operation

$$\begin{bmatrix} \bar{u}_{,\bar{x}} & \bar{v}_{,\bar{x}} & \bar{w}_{,\bar{x}} \\ \bar{u}_{,\bar{y}} & \bar{v}_{,\bar{y}} & \bar{w}_{,\bar{y}} \\ \bar{u}_{,\bar{z}} & \bar{v}_{,\bar{z}} & \bar{w}_{,\bar{z}} \end{bmatrix} = \boldsymbol{\theta}^{\mathrm{T}} \begin{bmatrix} u_{,x} & v_{,x} & w_{,x} \\ u_{,y} & v_{,y} & w_{,y} \\ u_{,z} & v_{,z} & w_{,z} \end{bmatrix} \boldsymbol{\theta} \tag{8.21}$$

From this the components of the $\bar{\mathbf{B}}$ matrix can now be found explicitly, noting that 5 degrees of freedom exist at each node:

$$\bar{\boldsymbol{\varepsilon}} = \bar{\mathbf{B}}\mathbf{a}^e \tag{8.22}$$

where the form of $\mathbf{a}^e$ is given in Eq. (8.5).

The infinitesimal volume is given in terms of the curvilinear coordinates as

$$\mathrm{d}x\,\mathrm{d}y\,\mathrm{d}z = \det|\mathbf{J}|\,\mathrm{d}\xi\,\mathrm{d}\eta\,\mathrm{d}\zeta = j\,\mathrm{d}\xi\,\mathrm{d}\eta\,\mathrm{d}\zeta \tag{8.23}$$

where $j = \det|\mathbf{J}|$. This standard expression completes the basic formulation.

Numerical integration within the appropriate limits is carried out in exactly the same way as for three-dimensional elements using the Gaussian quadrature formulae discussed in Chapter 9 of Volume 1. An identical process serves to define all the other relevant element matrices arising from body and surface loading, inertia matrices, etc.

As the variation of the strain quantities in the thickness, or ζ direction, is linear, two Gauss points in that direction are sufficient for homogeneous elastic sections, while three or four in the ξ, η directions are needed for parabolic and cubic shape functions N_i, respectively.

It should be remarked here that, in fact, the integration with respect to ζ can be performed explicitly if desired, thus saving computation time.[1,4]

8.2.5 Some remarks on stress representation

The element properties are now defined, and the assembly and solution are in standard form. It remains to discuss the presentation of the stresses, and this problem is of some consequence. The strains being defined in local direction, $\bar{\boldsymbol{\sigma}}$, are readily available. Such components are indeed directly of interest but as the directions of local axes are not easily visualized (and indeed may not be continuously defined between adjacent elements) it is sometimes convenient to transfer the components to the global system using the standard transformation

$$\begin{bmatrix} \sigma_x & \tau_{xy} & \tau_{xz} \\ \tau_{yx} & \sigma_y & \tau_{yz} \\ \tau_{zx} & \tau_{zy} & \sigma_z \end{bmatrix} = \boldsymbol{\theta} \begin{bmatrix} \sigma_{\bar{x}} & \tau_{\bar{x}\bar{y}} & \tau_{\bar{x}\bar{z}} \\ \tau_{\bar{y}\bar{x}} & \sigma_{\bar{y}} & \tau_{\bar{y}\bar{z}} \\ \tau_{\bar{z}\bar{x}} & \tau_{\bar{z}\bar{y}} & \sigma_{\bar{z}} \end{bmatrix} \boldsymbol{\theta}^{\mathrm{T}} \tag{8.24}$$

Such a transformation should be performed only for elements which belong to the approximation for the same smooth surface.

In a general shell structure, the stresses in a global system do not, however, give a clear picture of shell surface stresses. It is thus convenient always to compute the principal stresses (or invariants of stress) by a suitable transformation. Regarding the shell stresses more rationally, one may note that the shear components $\tau_{\bar{x}\bar{z}}$ and $\tau_{\bar{y}\bar{z}}$ are in fact zero on the top and bottom surfaces and this may be noted when making the transformation of Eq. (8.24) before converting to global components to ensure that the principal stresses lie on the surface of the shell. The values obtained directly for these shear components are the average values across the section. The maximum transverse shear on a solid cross-section occurs on the mid-surface and is equal to about 1.5 times the average value.

8.3 Special case of axisymmetric, curved, thick shells

For axisymmetric shells the formulation is simplified. Now the element mid-surface is defined by only two coordinates ξ, η and a considerable saving in computer effort is obtained.[1]

The element now is derived in a similar manner by starting from a two-dimensional definition of Fig. 8.4.

Equations (8.1) and (8.2) are now replaced by their two-dimensional equivalents defining the relation between coordinates as

$$\begin{Bmatrix} r \\ z \end{Bmatrix} = \sum N_i(\xi) \left(\frac{1+\eta}{2} \begin{Bmatrix} r_i \\ z_i \end{Bmatrix}_{\text{top}} + \frac{1-\eta}{2} \begin{Bmatrix} r_i \\ z_i \end{Bmatrix}_{\text{bottom}} \right)$$

$$= \sum N_i(\xi) \left(\begin{Bmatrix} r_i \\ z_i \end{Bmatrix}_{\text{mid}} + \tfrac{1}{2} \eta t_i \mathbf{v}_{3i} \right) \tag{8.25}$$

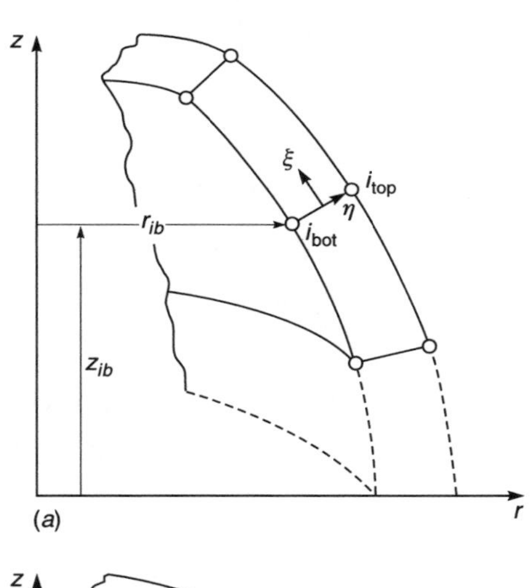

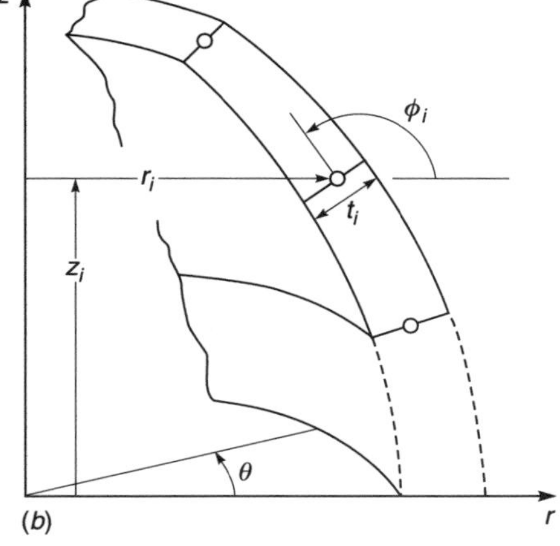

Fig. 8.4 Coordinates for an axisymmetric shell: (a) coordinate representation; (b) shell representation.

with

$$\mathbf{v}_{3i} = \begin{Bmatrix} \cos\phi_i \\ \sin\phi_i \end{Bmatrix}$$

in which ϕ_i is the angle defined in Fig. 8.4(b) and t_i is the shell thickness. Similarly, the displacement definition is specified by following the lines of Eq. (8.4).

Here we consider the case of axisymmetric loading only. Non-axisymmetric loading is addressed in Chapter 9 along with other schemes which permit treatment of problems in a reduced manner. Thus, we specify the two displacement components as

$$\begin{Bmatrix} u \\ w \end{Bmatrix} = \sum N_i \left(\begin{Bmatrix} u_i \\ w_i \end{Bmatrix} + \frac{\eta t_i}{2} \begin{Bmatrix} -\sin\phi_i \\ \cos\phi_i \end{Bmatrix} \beta_i \right) \tag{8.26}$$

In this β_i stands for the rotation illustrated in Fig. 8.5, and u_i, w_i stand for the displacement of the middle surface node.

Global strains are conveniently defined by the relationship[8]

$$\boldsymbol{\varepsilon} = \begin{Bmatrix} \varepsilon_r \\ \varepsilon_z \\ \varepsilon_\theta \\ \gamma_{rz} \end{Bmatrix} = \begin{Bmatrix} u_{,r} \\ w_{,z} \\ u/r \\ u_{,z} + w_{,r} \end{Bmatrix} \tag{8.27}$$

These strains are transformed to the local coordinates and the component normal to η (η = constant) is neglected.

All the transformations follow the pattern described in previous sections and need not be further commented on except perhaps to remark that they are now carried out only between sets of directions ξ,η, r,z, and $\bar{r},\bar{z}$, thus involving only two variables.

Similarly the integration of element properties is carried out numerically with respect to ξ and η only, noting, however, that the volume element is

$$\mathrm{d}x\,\mathrm{d}y\,\mathrm{d}z = \det|\mathbf{J}|\,\mathrm{d}\xi\,\mathrm{d}\eta\, r\,\mathrm{d}\theta = j\,r\,\mathrm{d}\xi\,\mathrm{d}\eta\,\mathrm{d}\theta \tag{8.28}$$

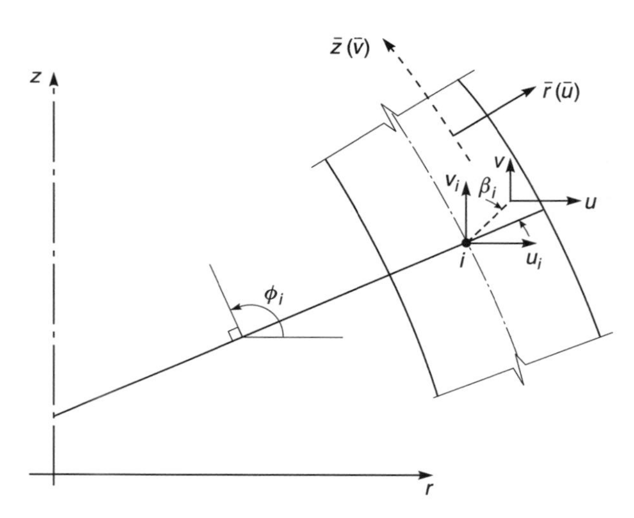

Fig. 8.5 Global displacements in an axisymmetric shell.

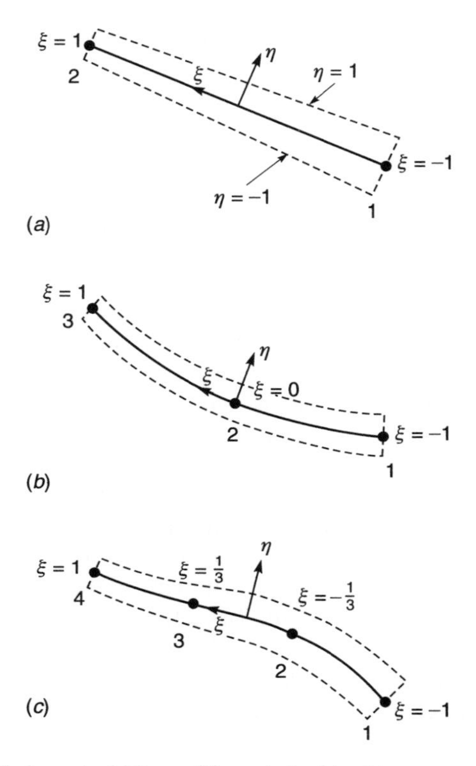

Fig. 8.6 Axisymmetric shell elements: (a) linear; (b) parabolic; (c) cubic.

By suitable choice of shape functions $N_i(\xi)$, straight, parabolic, or cubic shapes of variable thickness elements can be used as shown in Fig. 8.6.

8.4 Special case of thick plates

The transformations necessary in this chapter are somewhat involved and the programming steps are quite sophisticated. However, the application of the principle involved is available for thick plates and readers are advised to first test their comprehension on such a simple problem.

Here the following obvious simplifications arise.

1. $\zeta = 2z/t$ and unit vectors $\mathbf{v}_1$, $\mathbf{v}_2$ and $\mathbf{v}_3$ can be taken in the directions of the x, y, and z axes respectively.
2. α_i and β_i are simply the rotations θ_y and θ_x, respectively (see Chapter 5).
3. It is no longer necessary to transform stress and strain components to a local system of axes $\bar{x}$, $\bar{y}$, $\bar{z}$ and global definitions x, y, z can be used throughout. For elements of this type, numerical thickness integration can be avoided and, as an exercise, readers are encouraged to derive the stiffness matrices, etc., for, say, linear, rectangular elements. Forms will be found which are identical to those derived in Chapter 5 with an independent displacement and rotation interpolation

and using shear constraints. This demonstrates the essential identity of the alternative procedures.

8.5 Convergence

Whereas in three-dimensional analysis it is possible to talk about absolute convergence to the true exact solution of the elasticity problem, in equivalent plate and shell problems such a convergence cannot happen. As the element size decreases the so-called convergent solution of a plate bending problem approaches only to the exact solution of the approximate model implied in the formulation. Thus, here again convergence of the above formulation will only occur to the exact solution constrained by the requirement that straight 'normals' remain straight during deformation.

In elements of finite size it will be found that pure bending deformation modes are nearly always accompanied by some shear strains which in fact do not exist in the conventional thin plate or shell bending theory (although quite generally shear stresses may be deduced by equilibrium considerations on an element of the model, similar to the manner by which shear stresses in beams are deduced). Thus large elements deforming mainly under bending action (as would be the case of the shell element degenerated to a flat plate) tend to be appreciably too stiff. In such cases certain limits of the ratio of size of element to its thickness need to be imposed. However, it will be found that such restrictions often are relaxed by the simple expedient of *reducing the integration order*.[4]

Figure 8.7 shows, for instance, the application of the quadratic eight-node element to a square plate situation. Here results for integration with 3 × 3 and 2 × 2 Gauss points

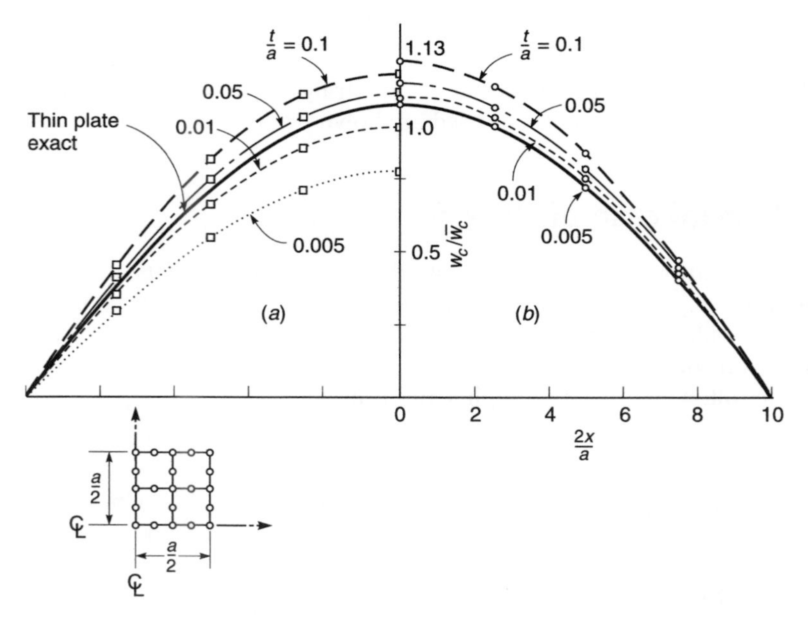

Fig. 8.7 A simply supported square plate under uniform load q_0: plot of central deflection w_c for eight-node elements with (a) 3 × 3 Gauss point integration and (b) with 2 × 2 (reduced) Gauss point integration. Central deflection is $\bar{w}_c$ for thin plate theory.

are given and results plotted for different thickness-to-span ratios. For reasonably thick situations, the results are similar and both give the additional shear deformation not available by thin plate theory. However, for thin plates the results with the more exact integration tend to diverge rapidly from the now correct thin plate results whereas the reduced integration still gives excellent results. The reasons for this improved performance are fully discussed in Chapter 2 and the reader is referred there for further plate examples using different types of shape functions.

8.6 Inelastic behaviour

All the formulations presented in this chapter can of course be used for all non-linear materials. The procedures are similar to those mentioned in Chapters 4 and 5 dealing with plates. Now it is only necessary to replace Eqs (8.12) and (8.13) by the appropriate constitutive equation and tangent operator, respectively. In this case it is necessary always to perform the through-thickness integration numerically since *a priori* knowledge of the behaviour will not be available. Any of the constitutive models described in Chapter 3 may be used for this purpose provided appropriate transformations are made to make $\sigma_{\bar{z}}$ zero.

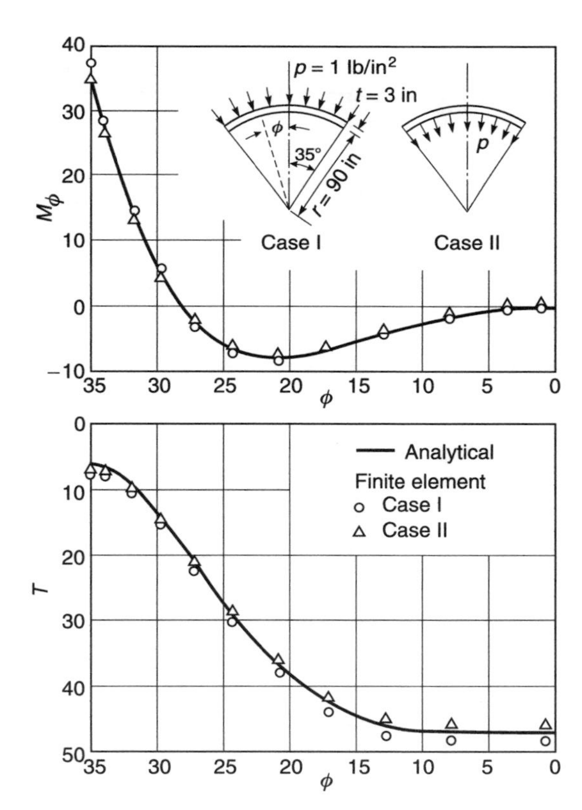

Fig. 8.8 Spherical dome under uniform pressure analysed with 24 cubic elements (first element subtends an angle of 0.1° from fixed end, others in arithmetic progression).

8.7 Some shell examples

A limited number of examples which show the accuracy and range of application of the axisymmetric shell formulation presented in this chapter will be given. For a fuller selection the reader is referred to references 1–7.

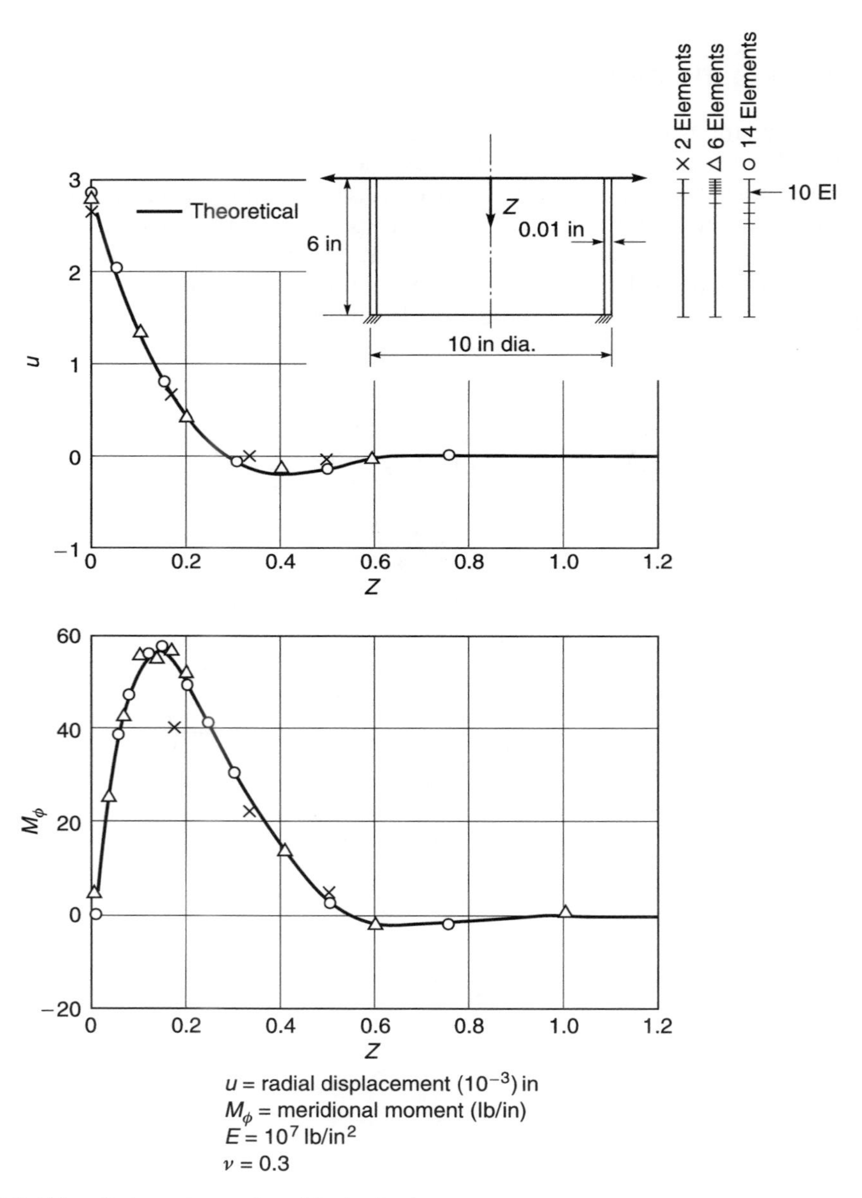

Fig. 8.9 Thin cylinder under a unit radial edge load.

8.7.1 Spherical dome under uniform pressure

The 'exact' solution of shell theory is known for this axisymmetric problem, illustrated in Fig. 8.8. Twenty-four cubic-type elements are used with graded size more closely spaced towards supports. Contrary to the 'exact' shell theory solution, the present formulation can distinguish between the application of pressure on the inner and outer surfaces as shown in the figure.

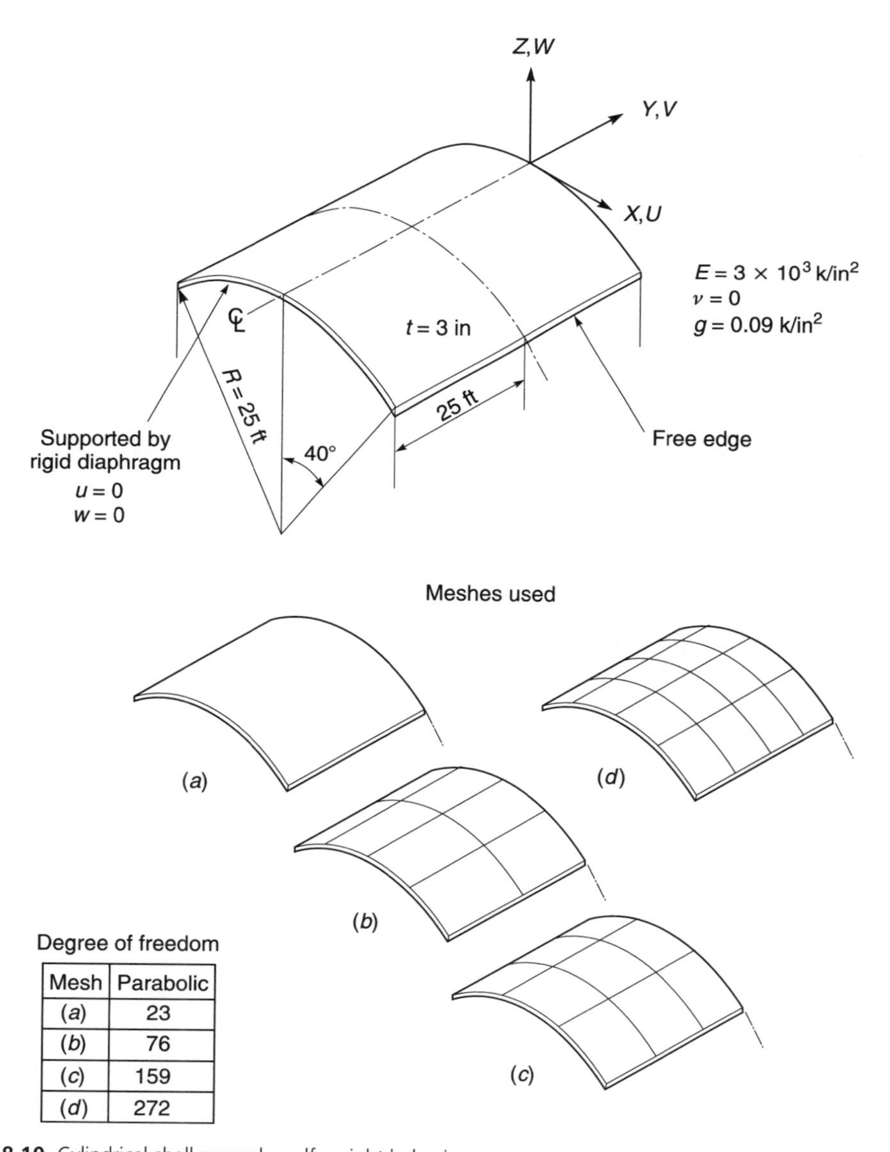

Mesh	Parabolic
(a)	23
(b)	76
(c)	159
(d)	272

Fig. 8.10 Cylindrical shell example: self-weight behaviour.

8.7.2 Edge loaded cylinder

A further axisymmetric example is shown in Fig. 8.9 to study the effect of subdivision. Two, six, or fourteen cubic elements of unequal length are used and the results for both of the finer subdivisions are almost coincident with the exact solution. Even the two-element solution gives reasonable results and departs only in the vicinity of the loaded edge.

Once again the solutions are basically identical to those derived with independent slope and displacement interpolation in the manner presented in Chapter 5.

8.7.3 Cylindrical vault

This is a test example of application of the full process to a shell in which bending action is dominant as a result of supports restraining deflection at the ends (see also Sec. 6.8.2).

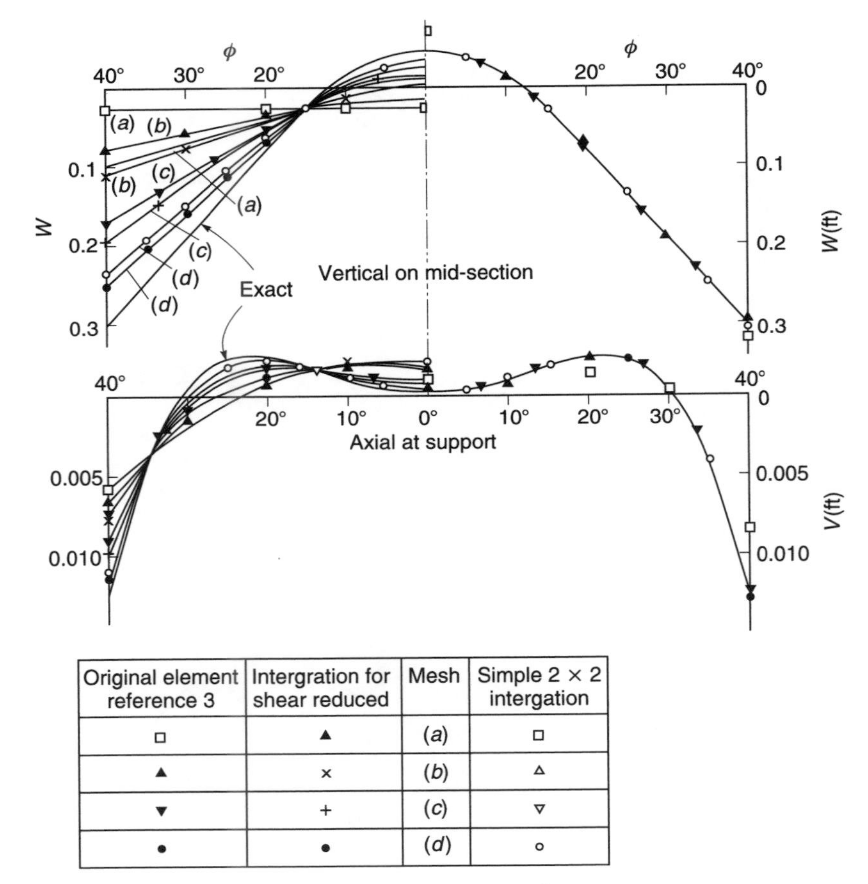

Original element reference 3	Intergration for shear reduced	Mesh	Simple 2 × 2 intergation
□	▲	(a)	□
▲	×	(b)	△
▼	+	(c)	▽
●	●	(d)	○

Fig. 8.11 Displacement (parabolic element), cylindrical shell roof.

In Fig. 8.10 the geometry, physical details of the problem, and subdivision are given, and in Fig. 8.11 the comparison of the effects of 3×3 and 2×2 integration using eight-node quadratic elements is shown on the displacements calculated. Both integrations result, as expected, in convergence. For the more exact integration, this is rather slow, but, with reduced integration order, very accurate results are obtained, even with one element. The improved convergence of displacements is matched by rapid convergence of stress components.

This example illustrates most dramatically the advantages of this simple expedient and is described more fully in references 4 and 6. The comparison solution for this problem is one derived along more conventional lines by Scordelis and Lo.[9]

8.7.4 Curved dams

All the previous examples were rather thin shells and indeed demonstrated the applicability of the process to these situations. At the other end of the scale, this formulation has been applied to the doubly curved dams illustrated in Chapter 9 of Volume 1 (Fig. 9.28). Indeed, exactly the same subdivision is again used and *results reproduce*

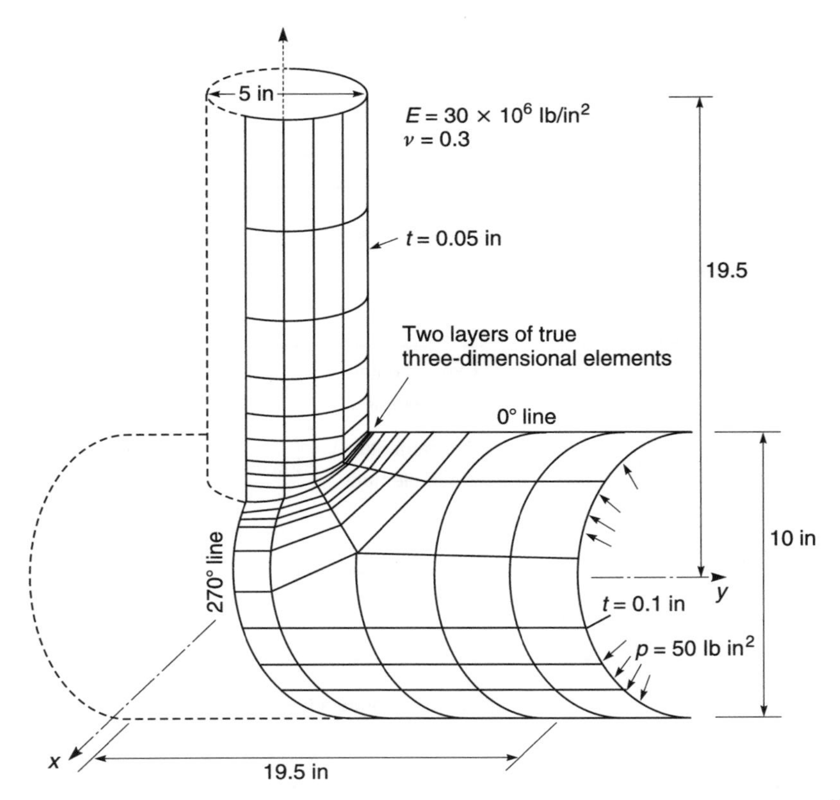

Fig. 8.12 An analysis of cylinder intersection by means of reduced integration shell-type elements.[10]

almost exactly those of the three-dimensional solution.[3] This remarkable result is achieved at a very considerable saving in both degrees of freedom and computer solution time.

Clearly, the range of application of this type of element is very wide.

8.7.5 Pipe penetration[10] and spherical cap[7]

The last two examples, shown in Figs 8.12–8.14, illustrate applications in which the irregular shape of elements is used. Both illustrate practical problems of some interest and show that with reduced integration a useful and very general shell element is available, even when the elements are quite distorted.

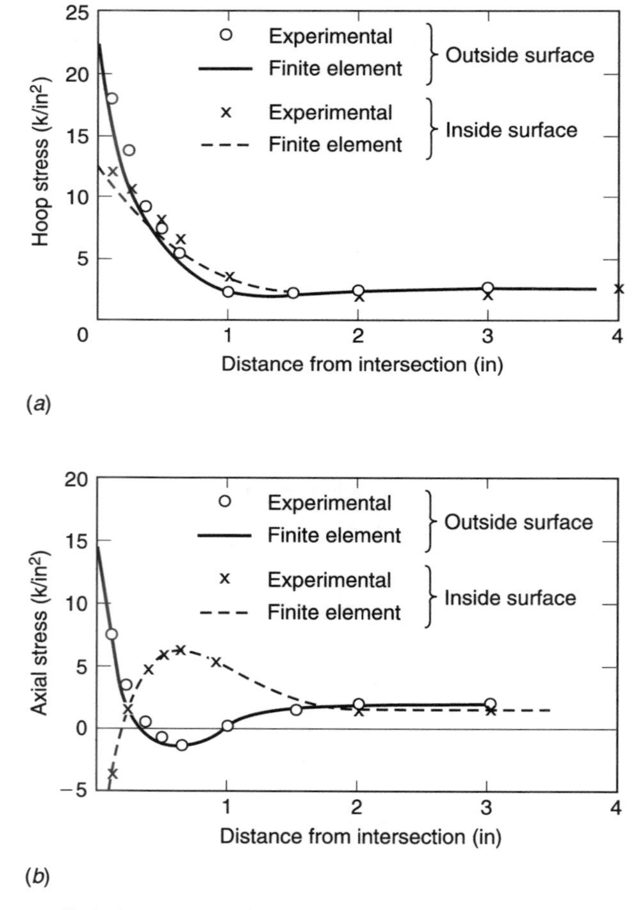

Fig. 8.13 Cylinder-to-cylinder intersections of Fig. 8.12: (a) hoop stresses near 0° line; (b) axial stresses near 0° line.

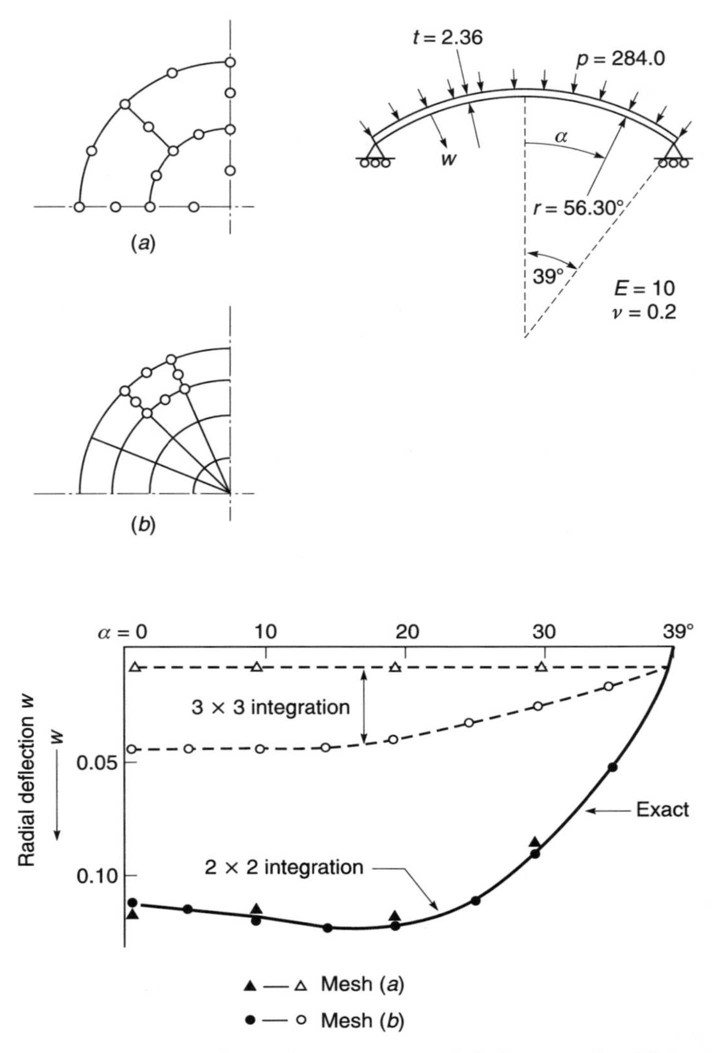

Fig. 8.14 A spherical cap analysis with irregular isoparametric shell elements using full 3×3 and reduced 2×2 integration.

8.8 Concluding remarks

The elements described in this chapter using degeneration of solid elements are shown in plate and axisymmetric problems to be nearly identical to those described in Chapters 5 and 7 where an independent slope and displacement interpolation is directly used in the middle plane. For the general curved shell the analogy is less obvious but clearly still exists. We should therefore expect that the conditions established in Chapter 5 for robustness of plate elements to be still valid. Further, it appears possible that other additional conditions on the various interpolations may have to be imposed in curved element forms. Both statements are true. The eight- and nine-node elements which we have shown in the previous section to perform well will fail under certain circumstances and for this reason many of the more successful plate elements also have been adapted to the shell problem.

The introduction of additional degrees of freedom in the interior of the eight-node serendipity element was first suggested by Cook[11,12] and later by Hughes[13–15] without, however, achieving complete robustness. The full lagrangian cubic interpolation as shown in Chapter 5 is quite effective and has been shown to perform well. However, the best results achieved to date appear to be those in which 'local constraints' are applied (see Sec. 5.5) and such elements as those due to Dvorkin and Bathe,[16] Huang and Hinton,[17] and Simo *et al.*[18,19] fall into this category.

While the importance of transverse shear strain constraints is now fully understood, the constraints introduced by the 'in-plane' (membrane) stress resultants are less amenable to analysis (although the elastic parameters Et associated with these are of the same order as those of shear Gt). It is well known that *membrane locking* can occur in situations that do not permit inextensional bending. Such locking has been thoroughly discussed[20–22] but to date the problem has not been rigorously solved and further developments are required.

Much effort is continuing to improve the formulation of the processes described in this chapter as they offer an excellent solution to the curved shell problem.[22–34]

References

1. S. Ahmad, B.M. Irons and O.C. Zienkiewicz. Curved thick shell and membrane elements with particular reference to axi-symmetric problems. In L. Berke, R.M. Bader, W.J. Mykytow, J.S. Przemienicki and M.H. Shirk (eds), *Proc. 2nd Conf. Matrix Methods in Structural Mechanics*, Volume AFFDL-TR-68-150, pp. 539–72, Air Force Flight Dynamics Laboratory, Wright Patterson Air Force Base, OH, October 1968.
2. S. Ahmad. *Curved Finite Elements in the Analysis of Solids, Shells and Plate Structures*, PhD thesis, Department of Civil Engineering, University of Wales, Swansea, 1969.
3. S. Ahmad, B.M. Irons and O.C. Zienkiewicz. Analysis of thick and thin shell structures by curved finite elements. *Int. J. Num. Meth. Eng.*, **2**, 419–51, 1970.
4. O.C. Zienkiewicz, J. Too and R.L. Taylor. Reduced integration technique in general analysis of plates and shells. *Int. J. Num. Meth. Eng.*, **3**, 275–90, 1971.
5. S.F. Pawsey and R.W. Clough. Improved numerical integration of thick slab finite elements. *Int. J. Num. Meth. Eng.*, **3**, 575–86, 1971.
6. S.F. Pawsey. *The Analysis of Moderately Thick to Thin Shells by the Finite Element Method*, PhD dissertation, Department of Civil Engineering, University of California, Berkeley, CA, 1970; also SESM Report 70–12.
7. J.J.M. Too. *Two-Dimensional, Plate, Shell and Finite Prism Isoparametric Elements and their Application*, PhD thesis, Department of Civil Engineering, University of Wales, Swansea, 1970.
8. I.S. Sokolnikoff. *The Mathematical Theory of Elasticity*, 2nd edition, McGraw-Hill, New York, 1956.
9. A.C. Scordelis and K.S. Lo. Computer analysis of cylindrical shells. *J. Am. Concr. Inst.*, **61**, 539–61, 1964.

10. S.A. Bakhrebah and W.C. Schnobrich. Finite element analysis of intersecting cylinders. Technical Report UILU-ENG-73-2018, Civil Engineering Studies, University of Illinois, Urbana, IL, 1973.
11. R.D. Cook. More on reduced integration and isoparametric elements. *Int. J. Num. Meth. Eng.*, **5**, 141–2, 1972.
12. R.D. Cook. *Concepts and Applications of Finite Element Analysis*, John Wiley, Chichester, Sussex, 1982.
13. T.J.R. Hughes and M. Cohen. The 'heterosis' finite element for plate bending. *Computers and Structures*, **9**, 445–50, 1978.
14. T.J.R. Hughes and W.K. Liu. Non linear finite element analysis of shells: Part I. *Comp. Meth. Appl. Mech. Eng.*, **26**, 331–62, 1981.
15. T.J.R. Hughes and W.K. Liu. Non linear finite element analysis of shells: Part II. *Comp. Meth. Appl. Mech. Eng.*, **27**, 331–62, 1981.
16. E.N. Dvorkin and K.-J. Bathe. A continuum mechanics based four node shell element for general non-linear analysis. *Engineering Computations*, **1**, 77–88, 1984.
17. E.C. Huang and E. Hinton. Elastic, plastic and geometrically non-linear analysis of plates and shells using a new, nine-noded element. In P. Bergan *et al.* (eds), *Finite elements for Non Linear Problems*, pp. 283–97, Springer-Verlag, Berlin, 1986.
18. J.C. Simo and D.D. Fox. On a stress resultant geometrically exact shell model. Part I, formulation and optimal parametrization. *Comp. Meth. Appl. Mech. Eng.*, **72**, 267–304, 1989.
19. J.C. Simo, D.D. Fox and M.S. Rifai. On a stress resultant geometrically exact shell model. Part II, the linear theory; computational aspects. *Comp. Meth. Appl. Mech. Eng.*, **73**, 53–92, 1989.
20. H. Stolarski and T. Belytschko. Membrane locking and reduced integration for curved elements. *J. Appl. Mech.*, **49**, 172–6, 1982.
21. H. Stolarski and T. Belytschko. Shear and membrane locking in curved C^0 elements. *Comp. Meth. Appl. Mech. Eng.*, **41**, 279–96, 1983.
22. R.V. Milford and W.C. Schnobrich. Degenerated isoparametric finite elements using explicit integration. *Int. J. Num. Meth. Eng.*, **23**, 133–54, 1986.
23. D. Bushnell. Computerized analysis of shells – governing equations. *Computers and Structures*, **18**, 471–536, 1984.
24. M.A. Cilia and N.G. Gray. Improved coordinate transformations for finite elements: the Lagrange cubic case. *Int. J. Num. Meth. Eng.*, **23**, 1529–45, 1986.
25. S. Vlachoutsis. Explicit integration for three dimensional degenerated shell finite elements. *Int. J. Num. Meth. Eng.*, **29**, 861–80, 1990.
26. H. Parisch. A continuum-based shell theory for nonlinear applications. *Int. J. Num. Meth. Eng.*, **38**, 1855–83, 1993.
27. U. Andelfinger and E. Ramm. EAS-elements for two-dimensional, three-dimensional, plate and shell structures and their equivalence to HR-elements. *Int. J. Num. Meth. Eng.*, **36**, 1311–37, 1993.
28. M. Braun, M. Bischoff and E. Ramm. Nonlinear shell formulations for complete three-dimensional constitutive laws include composites and laminates. *Comp. Mech.*, **15**, 1–18, 1994.
29. N. Büchter, E. Ramm and D. Roehl. Three-dimensional extension of nonlinear shell formulations based on the enhanced assumed strain concept. *Int. J. Num. Meth. Eng.*, **37**, 2551–68, 1994.
30. P. Betsch, F. Gruttmann and E. Stein. A 4-node finite shell element for the implementation of general hyperelastic 3d-elasticity at finite strains. *Comp. Meth. Appl. Mech. Eng.*, **130**, 57–79, 1996.
31. M. Bischoff and E. Ramm. Shear deformable shell elements for large strains and rotations. *Int. J. Num. Meth. Eng.*, **40**, 4427–49, 1997.

32. M. Bischoff and E. Ramm. Solid-like shell or shell-like solid formulation? A personal view. In W. Wunderlich, (ed.), *Proc. Eur. Conf. on Comp. Mech. (ECCM'99 on CD-ROM)*, Munich, September 1999.
33. E. Ramm. From Reissner plate theory to three dimensions in large deformations shell analysis. *Z. Angew. Math. Mech.*, **79**, 1–8, 1999.
34. H.T.Y. Yang, S. Saigal, A. Masud and R.K. Kapania. A survey of recent shell finite elements. *Int. J. Num. Meth. Eng.*, **47**, 101–27, 2000.

9

Semi-analytical finite element processes – use of orthogonal functions and 'finite strip' methods

9.1 Introduction

Standard finite element methods have been shown to be capable, in principle, of dealing with any two- or three- (or even four-)* dimensional situations. Nevertheless, the cost of solutions increases greatly with each dimension added and indeed, on occasion, overtaxes the available computer capability. It is therefore always desirable to search for alternatives that may reduce computational effort. One such class of processes of quite wide applicability will be illustrated in this chapter.

In many physical problems the situation is such that the *geometry* and *material properties* do not vary along one coordinate direction. However, the 'load' terms may still exhibit a variation in that direction, preventing the use of such simplifying assumptions as those that, for instance, permitted a two-dimensional plane strain or axisymmetric analysis to be substituted for a full three-dimensional treatment. In such cases it is possible still to consider a 'substitute' problem, not involving the particular coordinate (along which the geometry and properties do not vary), and to synthesize the true answer from a series of such simplified solutions.

The method to be described is of quite general use and, obviously, is not limited to structural situations. It will be convenient, however, to use the nomenclature of structural mechanics and to use potential energy minimization as an example.

We shall confine our attention to problems of minimizing a quadratic functional such as described in Chapters 2–6 of Volume 1. The interpretation of the process involved as the application of partial discretization in Chapter 3 of Volume 1 followed (or preceded) by the use of a Fourier series expansion should be noted.

Let (x, y, z) be the coordinates describing the domain (in this context these do not necessarily have to be the Cartesian coordinates). The last one of these, z, is the coordinate along which the geometry and material properties do not change and which is limited to lie between two values

$$0 \leqslant z \leqslant a$$

The boundary values are thus specified at $z = 0$ and $z = a$.

* See finite elements in the time domain in Chapter 18 of Volume 1.

We shall assume that the shape functions defining the variation of displacement $\mathbf{u}$ can be written in a product form as

$$\mathbf{u} = \mathbf{N}(x, y, z)\,\mathbf{a}^e = \sum_{l=1}^{L}\left(\bar{\mathbf{N}}(x, y)\cos\frac{l\pi z}{a} + \bar{\bar{\mathbf{N}}}(x, y)\sin\frac{l\pi z}{a}\right)(\mathbf{a}^l)^e$$
$$= \sum_{l=1}^{L}\mathbf{N}^l(x, y, z)\,(\mathbf{a}^l)^e \tag{9.1}$$

In this type of representation completeness is preserved in view of the capability of Fourier series to represent any continuous function within a given region (naturally assuming that the shape functions $\bar{\mathbf{N}}$ and $\bar{\bar{\mathbf{N}}}$ in the domain x, y satisfy the same requirements).

The loading terms will similarly be given a form

$$\mathbf{b} = \sum_{l=1}^{L}\left(\cos\frac{l\pi z}{a}\bar{\mathbf{b}}^l + \sin\frac{l\pi z}{a}\bar{\bar{\mathbf{b}}}^l\right) = \sum_{l=1}^{L}\mathbf{b}^l(x, y, z) \tag{9.2}$$

for body force, with similar form for concentrated loads and boundary tractions (see Chapter 2 of Volume 1). Indeed, initial strains and stresses, if present, would be expanded again in the above form.

Applying the standard processes of Chapter 2 of Volume 1 to the determination of the element contribution to the equation minimizing the potential energy, and limiting our attention to the contribution of body forces $\mathbf{b}$ only, we can write

$$\frac{\partial\Pi}{\partial\mathbf{a}^e} = \mathbf{K}^e\begin{Bmatrix}\mathbf{a}^{1e}\\ \vdots\\ \mathbf{a}^{Le}\end{Bmatrix} + \begin{Bmatrix}\mathbf{f}^{1e}\\ \vdots\\ \mathbf{f}^{Le}\end{Bmatrix} = \mathbf{0} \tag{9.3}$$

In the above, to avoid summation signs, the vectors $\mathbf{a}^e$, etc., are expanded, listing the contribution of each value of l separately.

Now a typical submatrix of $\mathbf{K}^e$ is

$$(\mathbf{K}^{lm})^e = \iiint_V (\mathbf{B}^l)^{\mathrm{T}}\mathbf{D}\mathbf{B}^m\,\mathrm{d}x\,\mathrm{d}y\,\mathrm{d}z \tag{9.4}$$

and a typical term of the 'force' vector becomes

$$(\mathbf{f}^l)^e = \iiint_V (\mathbf{N}^l)^{\mathrm{T}}\mathbf{b}^l\,\mathrm{d}x\,\mathrm{d}y\,\mathrm{d}z \tag{9.5}$$

Without going into details, it is obvious that the matrix given by Eq. (9.4) will contain the following integrals as products of various submatrices:

$$I_1 = \int_0^a \sin\frac{l\pi z}{a}\cos\frac{m\pi z}{a}\,\mathrm{d}z$$
$$I_2 = \int_0^a \sin\frac{l\pi z}{a}\sin\frac{m\pi z}{a}\,\mathrm{d}z \tag{9.6}$$
$$I_3 = \int_0^a \cos\frac{l\pi z}{a}\cos\frac{m\pi z}{a}\,\mathrm{d}z$$

These integrals arise from products of the derivatives contained in the definition of $\mathbf{B}^l$ and, owing to the well-known orthogonality property, give

$$I_2 = I_3 = 0 \qquad \text{for } l \neq m \tag{9.7}$$

when $l = 1, 2, \ldots$ and $m = 1, 2, \ldots$. The first integral I_1 is only zero when l and m are both even or odd numbers. The term involving I_1, however, vanishes in many applications because of the structure of $\mathbf{B}^l$. This means that the matrix $\mathbf{K}^e$ becomes a diagonal one and that the assembled final equations of the system have the form

$$\begin{bmatrix} \mathbf{K}^{11} & & & \\ & \mathbf{K}^{22} & & \\ & & \ddots & \\ & & & \mathbf{K}^{LL} \end{bmatrix} \begin{Bmatrix} \mathbf{a}^1 \\ \mathbf{a}^2 \\ \vdots \\ \mathbf{a}^L \end{Bmatrix} + \begin{Bmatrix} \mathbf{f}^1 \\ \mathbf{f}^2 \\ \vdots \\ \mathbf{f}^L \end{Bmatrix} \tag{9.8}$$

and *the large system of equations splits into L separate problems*:

$$\mathbf{K}^{ll}\mathbf{a}^l + \mathbf{f}^l = \mathbf{0} \tag{9.9}$$

in which

$$\mathbf{K}_{ij}^{ll} = \iiint_V (\mathbf{B}_i^l)^{\mathrm{T}} \mathbf{D} \mathbf{B}_j^l \, \mathrm{d}x \, \mathrm{d}y \, \mathrm{d}z \tag{9.10}$$

Further, from Eqs (9.5) and (9.2) we observe that owing to the orthogonality property of the integrals given by Eqs (9.6), the typical load term becomes simply

$$\mathbf{f}_i^l = \iiint_V (\mathbf{N}_i^l)^{\mathrm{T}} \mathbf{b}^l \, \mathrm{d}x \, \mathrm{d}y \, \mathrm{d}z \tag{9.11}$$

This means that the force term of the lth harmonic only affects the lth system of Eq. (9.9) and contributes nothing to the other equations. This extremely important property is of considerable practical significance for, *if the expansion of the loading factors involves only one term, only one set of equations need be solved.* The solution of this will tend to the exact one with increasing subdivision in the xy domain only. Thus, what was originally a three-dimensional problem now has been reduced to a two-dimensional one with consequent reduction of computational effort.

The preceding derivation was illustrated on a three-dimensional, elastic situation. Clearly, the arguments could equally well be applied for reduction of two-dimensional problems to one-dimensional ones, etc., and the arguments are not restricted to problems of elasticity. Any physical problem governed by a minimization of a quadratic functional (Chapter 3 of Volume 1) or by linear differential equations is amenable to the same treatment.

A word of warning should be added regarding the boundary conditions imposed on $\mathbf{u}$. For a complete decoupling to be possible these must be satisfied separately by each and every term of the expansion given by Eq. (9.1). Insertion of a zero displacement in the final reduced problem implies in fact a zero displacement fixed throughout all terms in the z direction by definition. Care must be taken not to treat the final matrix therefore as a simple reduced problem. Indeed, this is one of the limitations of the process described.

When the loading is complex and many Fourier components need to be considered the advantages of the approach outlined here reduce and the full solution sometimes becomes more efficient.

Other permutations of the basic definitions of the type given by Eq. (9.1) are obviously possible. For instance, two independent sets of parameters $\mathbf{a}^e$ may be specified with each of the trigonometric terms. Indeed, on occasion use of other orthogonal functions may be possible. The appropriate functions are often related to a reduction of the differential equation directly using separation of variables.[1]

As trigonometric functions will arise frequently it is convenient to remind the reader of the following integrals:

$$\begin{aligned} \int_0^a \sin\gamma_l z \cos\gamma_l z \,\mathrm{d}z = 0 \qquad & \text{when} \quad l = 0, 1, \ldots \\ \int_0^a \sin^2\gamma_l z \,\mathrm{d}z = \int_0^a \cos^2\gamma_l z \,\mathrm{d}z = \frac{a}{2} \qquad & \text{when} \quad l = 1, 2, \ldots \end{aligned} \tag{9.12}$$

where $\gamma_l = l\pi/a$.

9.2 Prismatic bar

Consider a prismatic bar, illustrated in Fig. 9.1, which is assumed to be held at $z = 0$ and $z = a$ in a manner preventing all displacements in the xy plane but permitting unrestricted motion in the z direction (traction $t_z = 0$). The problem is fully three-dimensional and three components of displacement u, v, and w have to be considered.

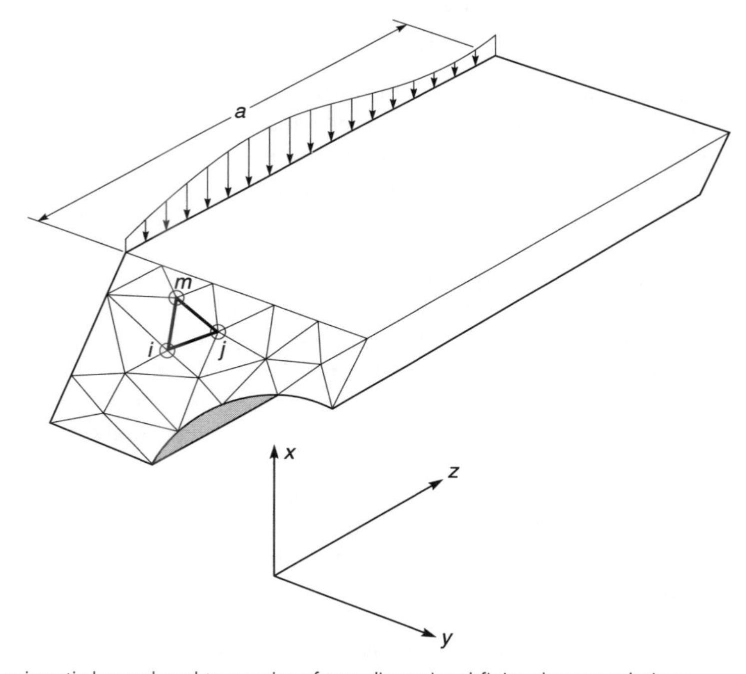

Fig. 9.1 A prismatic bar reduced to a series of two-dimensional finite element solutions.

Subdividing into finite elements in the xy plane we can prescribe the lth displacement components as

$$\mathbf{u}^l = \begin{Bmatrix} u^l \\ v^l \\ w^l \end{Bmatrix} = \sum_i N_i \begin{bmatrix} \sin\gamma_l z & 0 & 0 \\ 0 & \sin\gamma_l z & 0 \\ 0 & 0 & \cos\gamma_l z \end{bmatrix} \begin{Bmatrix} u_i^l \\ v_i^l \\ w_i^l \end{Bmatrix} \tag{9.13}$$

In this, N_i are simply the (scalar) shape functions appropriate to the elements used in the xy plane and again $\gamma_l = l\pi/a$. If, as shown in Fig. 9.1, simple triangles are used then the shape functions are given by Eqs (4.7) and (4.8) in Chapter 4 of Volume 1, but any of the more elaborate elements described in Chapter 8 of Volume 1 would be equally suitable (with or without the transformations given in Chapter 9 of Volume 1). The displacement expansion ensures zero u and v displacements at the ends and the zero t_z traction condition can be imposed in a standard manner.

As the problem is fully three-dimensional, the appropriate expression for strain involving all six components needs to be considered. This expression is given in Eq. (1.15) of Chapter 1. On substitution of the shape function given by Eq. (9.13) for a typical term of the $\mathbf{B}$ matrix we have

$$\mathbf{B}_I^l = \begin{bmatrix} N_{i,x}\sin\gamma_l z & 0 & 0 \\ 0 & N_{i,y}\sin\gamma_l z & 0 \\ 0 & 0 & -N_i\gamma_l\sin\gamma_l z \\ N_{i,y}\sin\gamma_l z & N_{i,x}\sin\gamma_l z & 0 \\ 0 & N_i\gamma_l\cos\gamma_l z & N_{i,y}\cos\gamma_l z \\ N_i\gamma_l\cos\gamma_l z & 0 & N_{i,x}\cos\gamma_l z \end{bmatrix} \tag{9.14}$$

It is convenient to separate the above as

$$\mathbf{B}_i^l = \bar{\mathbf{B}}_i^l \sin\gamma_l z + \bar{\bar{\mathbf{B}}}_i^l \cos\gamma_l z \tag{9.15}$$

In all of the above it is assumed that the parameters are listed in the usual order:

$$\mathbf{a}_i^l = \begin{bmatrix} u_i^l & v_i^l & w_i^l \end{bmatrix}^{\mathrm{T}} \tag{9.16}$$

and that the axes are as shown in Fig. 9.1.

The stiffness matrix can be computed in the usual manner, noting that

$$(\mathbf{K}_{ij}^{ll})^e = \iiint_{V^e} \mathbf{B}_i^{l\mathrm{T}} \mathbf{D} \mathbf{B}_j^l \, \mathrm{d}x\, \mathrm{d}y\, \mathrm{d}z \tag{9.17}$$

On substitution of Eq. (9.15), multiplying out, and noting the value of the integrals from Eq. (9.12), this reduces to

$$(\mathbf{K}_{ij}^{ll})^e = \frac{a}{2} \iint_{A^e} \left(\bar{\mathbf{B}}_i^{l\mathrm{T}} \mathbf{D} \bar{\mathbf{B}}_j^l + \bar{\bar{\mathbf{B}}}_i^{l\mathrm{T}} \mathbf{D} \bar{\bar{\mathbf{B}}}_j^l \right) \mathrm{d}x\, \mathrm{d}y \tag{9.18}$$

when $l = 1, 2, \ldots$. The integration is now simply carried out over the element *area*.*

* It should be noted that now, even for a single triangle, the integration is not trivial as some linear terms from N_i will remain in $\bar{\bar{\mathbf{B}}}$.

The contributions from distributed loads, initial stresses, etc., are found as the loading terms. To match the displacement expansions distributed body forces may be expanded in the Fourier series

$$\mathbf{b}_i^l = \int_0^a \begin{bmatrix} \sin\gamma_l z & 0 & 0 \\ 0 & \sin\gamma_l z & 0 \\ 0 & 0 & \cos\gamma_l z \end{bmatrix} \begin{Bmatrix} b_x(x,y,z) \\ b_y(x,y,z) \\ b_z(x,y,z) \end{Bmatrix} \mathrm{d}z \tag{9.19}$$

Similarly, concentrated line loads can be expressed directly as nodal forces

$$\mathbf{f}_i^l = \int_0^a \begin{bmatrix} \sin\gamma_l z & 0 & 0 \\ 0 & \sin\gamma_l z & 0 \\ 0 & 0 & \cos\gamma_l z \end{bmatrix} \begin{Bmatrix} f_x(x,y,z) \\ f_y(x,y,z) \\ f_z(x,y,z) \end{Bmatrix} \mathrm{d}z \tag{9.20}$$

in which $\mathbf{f}_i^l$ are intensities per unit length.

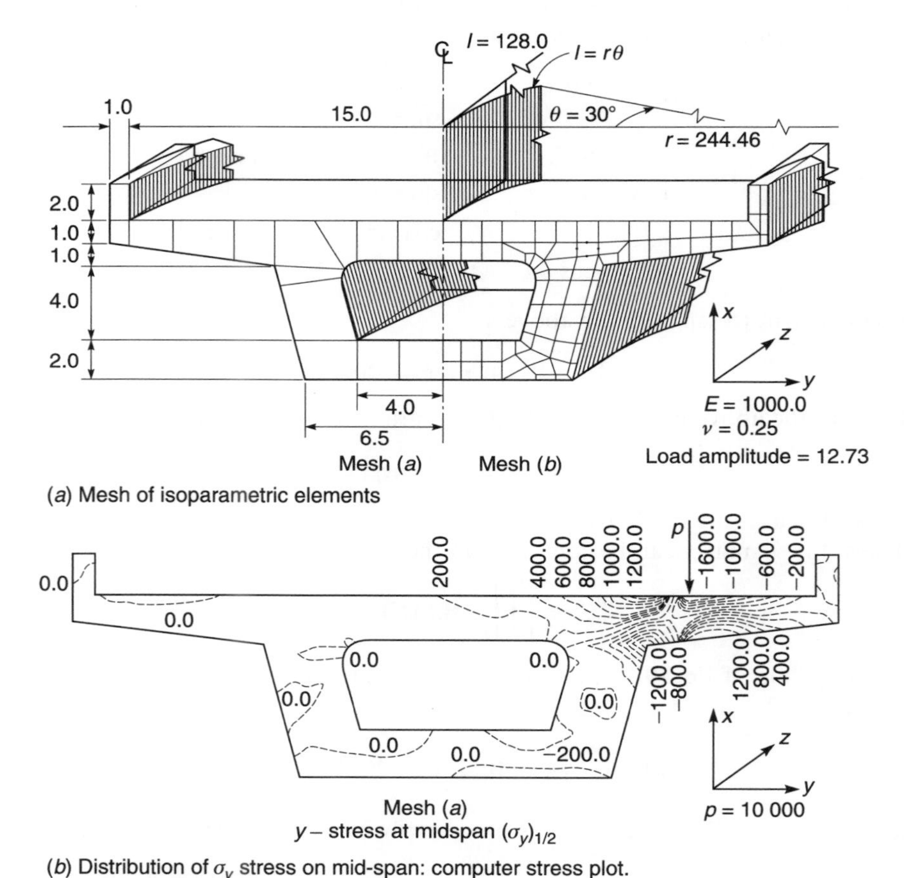

Fig. 9.2 A thick box bridge prism of straight or curved platform.

The boundary conditions used here have been of a type ensuring *simply supported* conditions for the prism. Other conditions can be inserted by suitable expansions.

The method of analysis outlined here can be applied to a range of practical problems – one of these being a popular type of box girder, concrete bridge, illustrated in Fig. 9.2. Here a particularly convenient type of element is the distorted, serendipity or lagrangian quadratic or cubic of Chapters 8 and 9 of Volume 1.[2] Finally, it should be mentioned that some restrictions placed on the general shapes defined by Eqs (9.1) or (9.13) can be removed by doubling the number of parameters and writing expansions in the form of two sums:

$$\mathbf{u} = \sum_{l=1}^{L} \bar{\mathbf{N}}(x, y) \cos \gamma_l z \, \mathbf{a}^{Al} + \sum_{l=1}^{L} \bar{\bar{\mathbf{N}}}(x, y) \sin \gamma_l z \, \mathbf{a}^{Bl} \tag{9.21}$$

Parameters $\mathbf{a}^{Al}$ and $\mathbf{a}^{Bl}$ are independent and for every component of displacement two values have to be found and two equations formed.

An alternative to the above process is to write the expansion as

$$\mathbf{u} = \sum [\mathbf{N}(x, y) \exp(i\gamma_l z)] \, \mathbf{a}^e$$

and to observe that both $\mathbf{N}$ and $\mathbf{a}$ are then complex quantities.

Complex algebra is available in standard programming languages and the identity of the above expression with Eq. (9.21) will be observed, noting that

$$\exp i\theta = \cos\theta + i \sin\theta$$

9.3 Thin membrane box structures

In the previous section a three-dimensional problem was reduced to that of two dimensions. Here we shall see how a somewhat similar problem can be reduced to one-dimensional elements (Fig. 9.3).

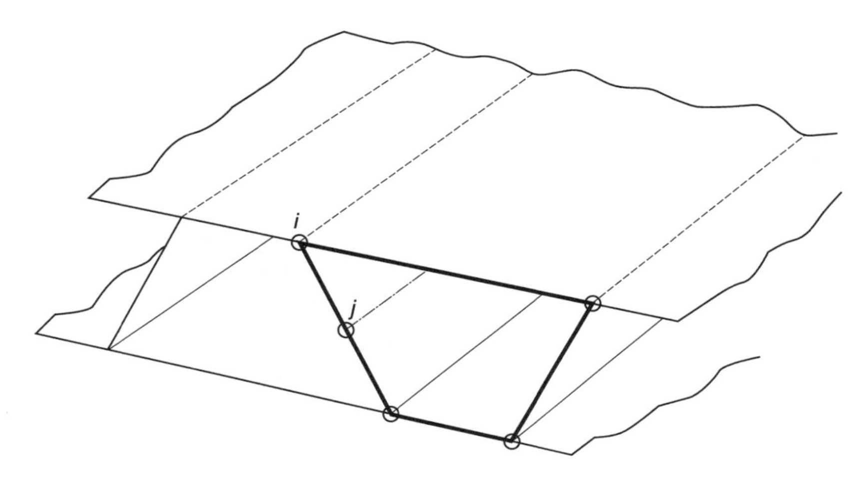

Fig. 9.3 A 'membrane' box with one-dimensional elements.

A box-type structure is made up of thin shell components capable of sustaining stresses only in its own plane. Now, just as in the previous case, three displacements have to be considered at every point and indeed similar variation can be prescribed for these. However, a typical element ij is 'one-dimensional' in the sense that integrations have to be carried out only along the line ij and only stresses in that direction be considered. Indeed, it will be found that the situation and solution are similar to that of a pin-jointed framework.

9.4 Plates and boxes with flexure

Consider now a rectangular plate simply supported at the ends and in which all strain energy is contained in flexure. Only one displacement, w, is needed to specify fully the state of strain (see Chapter 4).

For consistency of notation with Chapter 4, the direction in which geometry and material properties do not change is now taken as y (see Fig. 9.4). To preserve slope continuity the functions need to include now a 'rotation' parameter θ_i.

Use of simple beam functions (cubic Hermitian interpolations) is easy and for a typical element ij we can write (with $\gamma_l = l\pi/a$)

$$w^l = \bar{\mathbf{N}}(x) \sin \gamma_l y (\mathbf{a}^l)^e \tag{9.22}$$

ensuring *simply supported* end conditions. In this, the typical nodal parameters are

$$\mathbf{a}_i^l = \begin{Bmatrix} w_i^l \\ \theta_i^l \end{Bmatrix} \tag{9.23}$$

The shape functions of the cubic type are easy to write and are in fact identical to the Hermitian polynomials given in Sect. 4.14 and also those used for the asymmetric thin shell problem [Chapter 7, Eq. (7.9)].

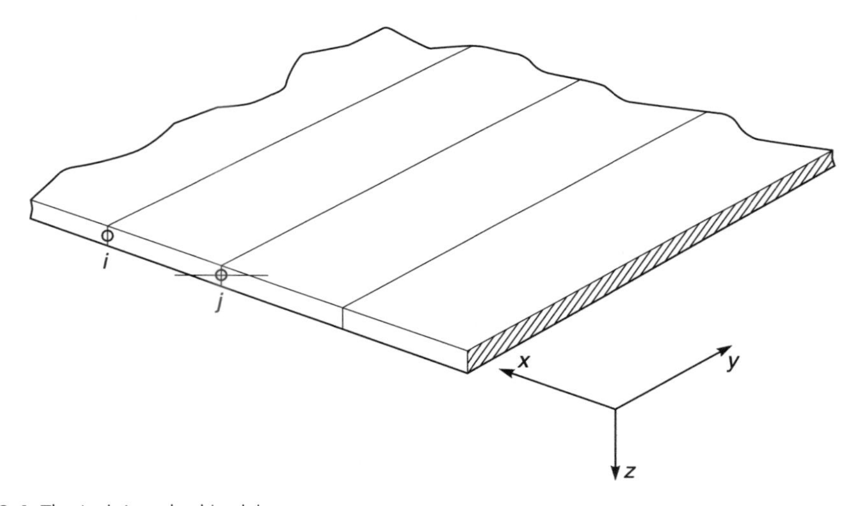

Fig. 9.4 The 'strip' method in slabs.

Table 9.1 Square plate, uniform load q; three sides simply supported one clamped (Poisson ratio = 0.3)

Term l	Central deflection	Central M_x	Maximum negative M_x
1	0.002832	0.0409	−0.0858
2	−0.000050	−0.0016	0.0041
3	0.002786	0.0396	−0.0007
Σ	0.002786	0.0396	−0.0824
Series	0.0028	0.039	−0.084
Multiplier	qa^4/D	qa^2	qa^2

Using all definitions of Chapter 4 the strains (curvatures) are found and the **B** matrices determined; now with C_1 continuity satisfied in a trivial manner, the problem of a two-dimensional kind has here been reduced to that of one dimension.

This application has been developed by Cheung and others,[3–17] named by him the 'finite strip' method, and used to solve many rectangular plate problems, box girders, shells, and various folded plates.

It is illuminating to quote an example from the above papers here. This refers to a square, uniformly loaded plate with three sides simply supported and one clamped. Ten strips or elements in the x direction were used in the solution, and Table 9.1 gives the results corresponding to the first three harmonics.

Not only is an accurate solution of each l term a simple one involving only some nineteen unknowns but the importance of higher terms in the series is seen to decrease rapidly.

Extension of the process to box structures in which both *membrane* and *bending effects* are present is almost obvious when this example is considered together with the one of the previous section.

In another paper Cheung[6] shows how functions other than trigonometric ones can be used to advantage, although only partial decoupling then occurs (see Sec. 9.7 below).

In the examples just quoted a thin plate theory using the single displacement variable w and enforcing C_1 compatibility in the x direction was employed. Obviously, any of the independently interpolated slope and displacement elements of Chapter 5 could be used here, again employing either reduced integration or mixed methods. Parabolic-type elements with reduced integration are employed in references 13 and 14, and linear interpolation with a single integration point is shown to be effective in reference 15.

Other applications for plate and box type structures abound and additional information is given in the text of reference 17.

9.5 Axisymmetric solids with non-symmetrical load

One of the most natural and indeed earliest applications of Fourier expansion occurs in axisymmetric bodies subject to non-axisymmetric loads. Now, not only the radial

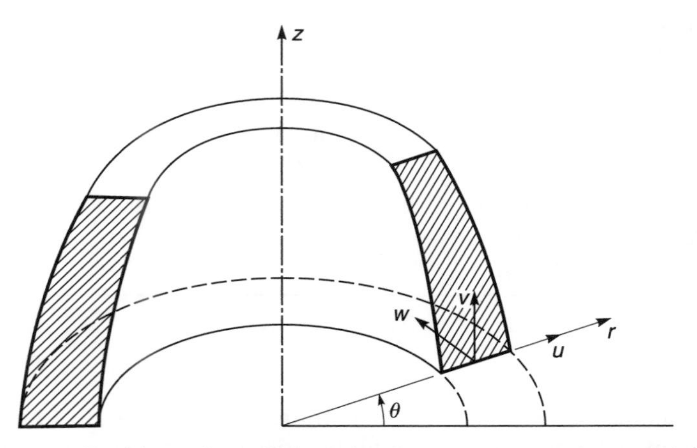

Fig. 9.5 An axisymmetric solid; coordinate displacement components in an axisymmetric body.

(u) and axial (w) displacement (as in Chapter 5 of Volume 1) will have to be considered but also a tangential component (v) associated with the tangential angular direction θ (Fig. 9.5). It is in this direction that the geometric and material properties do not vary and hence here that the elimination will be applied.

To simplify matters we shall consider first components of load which are symmetric about the $\theta = 0$ axis and later include those which are antisymmetric. Describing now only the nodal loads (with similar expansion holding for body forces, boundary conditions, initial strains, etc.) we specify forces per unit of circumference as

$$
\begin{aligned}
R &= \sum_{l=1}^{L} \bar{R}^l \cos l\theta \\
T &= \sum_{l=1}^{L} \bar{T}^l \sin l\theta \\
Z &= \sum_{l=1}^{L} \bar{Z}^l \cos l\theta
\end{aligned}
\tag{9.24}
$$

in the direction of the various coordinates for symmetric loads [Fig. 9.6(a)]. The apparently non-symmetric sine expansion is used for T, since to achieve symmetry the direction of T has to change for $\theta > \pi$.

The displacement components are described again in terms of the two-dimensional (r, z) shape functions appropriate to the element subdivision, and, observing symmetry, we write, as in Eq. (9.13),

$$
\mathbf{u}^l = \begin{Bmatrix} u^l \\ v^l \\ w^l \end{Bmatrix} = \sum_i N_i \begin{bmatrix} \cos\gamma_l\theta & 0 & 0 \\ 0 & \sin\gamma_l\theta & 0 \\ 0 & 0 & \cos\gamma_l\theta \end{bmatrix} \begin{Bmatrix} u_i^l \\ v_i^l \\ w_i^l \end{Bmatrix}
\tag{9.25}
$$

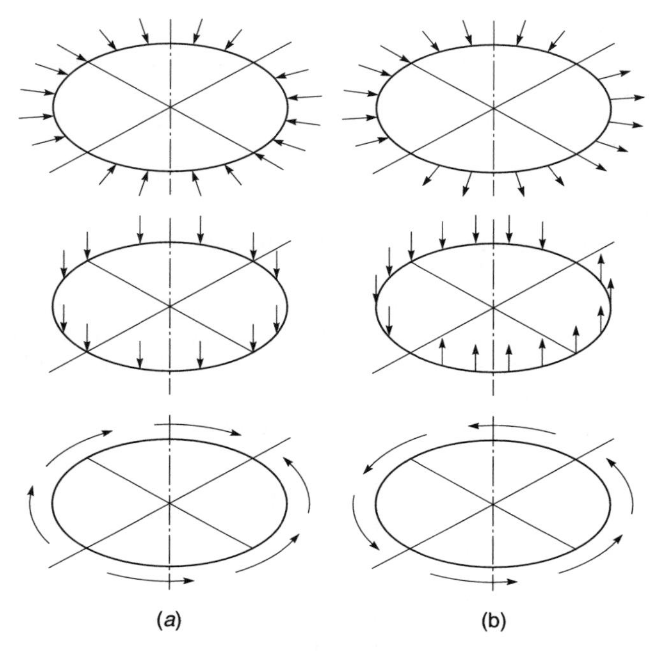

Fig. 9.6 Load and displacement components in an axisymmetric body: (a) symmetric; (b) antisymmetric.

To proceed further it is necessary to specify the general, three-dimensional expression for strains in cylindrical coordinates. These are given by[18]

$$\boldsymbol{\varepsilon} = \begin{Bmatrix} \varepsilon_r \\ \varepsilon_z \\ \varepsilon_\theta \\ \gamma_{rz} \\ \gamma_{z\theta} \\ \gamma_{\theta r} \end{Bmatrix} = \begin{Bmatrix} u_{,r} \\ w_{,z} \\ \left[u + v_{,\theta}\right]/r \\ u_{,z} + w_{,r} \\ v_{,z} + w_{,\theta}/r \\ u_{,\theta}/r + v_{,r} - v/r \end{Bmatrix} \tag{9.26}$$

We have on substitution of Eq. (9.25) into Eq. (9.26), and grouping the variables as in Eq. (9.16):

$$\mathbf{B}_i^l = \begin{bmatrix} N_{i,r}\cos l\theta & 0 & 0 \\ 0 & 0 & N_{i,z}\cos l\theta \\ N_i/r\cos l\theta & lN_i/r\cos l\theta & 0 \\ N_{i,z}\cos l\theta & 0 & N_{i,r}\cos l\theta \\ 0 & N_{i,z}\sin l\theta & -lN_i/r\sin l\theta \\ -lN_i/r\sin l\theta & \left(N_{i,r} - N_i/r\right)\sin l\theta & 0 \end{bmatrix} \tag{9.27}$$

A purely axisymmetric problem may be described for the complete zero harmonic ($l = 0$) and a further simplification arises in that the strains split into two problems: the first involves the displacement components u and w which appear only in the first four components of strain; and the second involves only the v displacement component and appears only in the last two shearing strains. This second problem is associated with a *torsion* problem on the axisymmetric body – with the first problem sometimes referred to as a *torsionless* problem. For an isotropic elastic material the stiffness matrix for these two problems completely decouples as a result of the structure of the **D** matrix, and they can be treated separately. However, for inelastic problems a coupling occurs whenever both torsionless and torsional loading are both applied as loading conditions on the same problem. Thus, it is often expedient to form the axisymmetric case including all three displacement components (as is necessary also for the other harmonics).

For the elastic case the remaining steps of the formulation follow precisely the previous derivations and can be performed by the reader as an exercise.

For the antisymmetric loading, of Fig. 9.6(b), we shall simply replace the sine by cosine and vice versa in Eqs (9.24) and (9.25).

The load terms in each harmonic are obtained by virtual work as

$$\mathbf{f}_i^l = \int_0^{2\pi} \begin{Bmatrix} \bar{R}^l \cos^2 l\theta \\ \bar{T}^l \sin^2 l\theta \\ \bar{Z}^l \cos^2 l\theta \end{Bmatrix} \mathrm{d}\theta = \begin{cases} \pi \begin{Bmatrix} \bar{R}^l \\ \bar{T}^l \\ \bar{Z}^l \end{Bmatrix} & \text{when } l = 1, 2, \ldots \\ 2\pi \begin{Bmatrix} \bar{R}^l \\ 0 \\ \bar{Z}^l \end{Bmatrix} & \text{when } l = 0 \end{cases} \tag{9.28}$$

for the symmetric case. Similarly, for the antisymmetric case

$$\mathbf{f}_i^l = \int_0^{2\pi} \begin{Bmatrix} \bar{R}^l \sin^2 l\theta \\ \bar{T}^l \cos^2 l\theta \\ \bar{Z}^l \sin^2 l\theta \end{Bmatrix} \mathrm{d}\theta = \begin{cases} \pi \begin{Bmatrix} \bar{R}^l \\ \bar{T}^l \\ \bar{Z}^l \end{Bmatrix} & \text{when } l = 1, 2, \ldots \\ 2\pi \begin{Bmatrix} 0 \\ \bar{T}^l \\ 0 \end{Bmatrix} & \text{when } l = 0 \end{cases} \tag{9.29}$$

We see from this and from an expansion of $\mathbf{K}^e$ that, as expected, for $l = 0$ the problem reduces to only two variables and the axisymmetric case is retrieved when symmetric terms only are involved. Similarly, when $l = 0$ only one set of equations

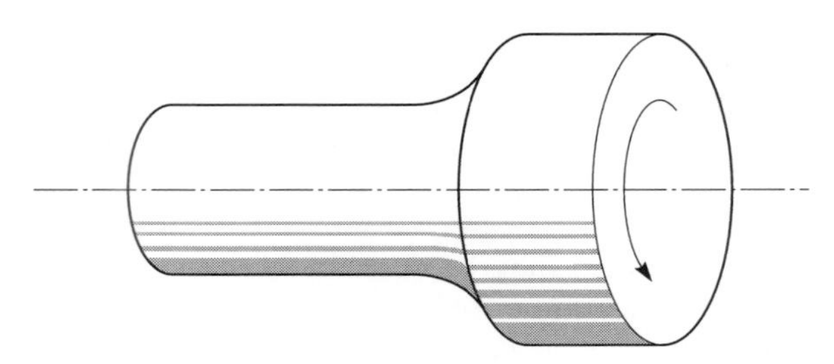

Fig. 9.7 Torsion of a variable section circular bar.

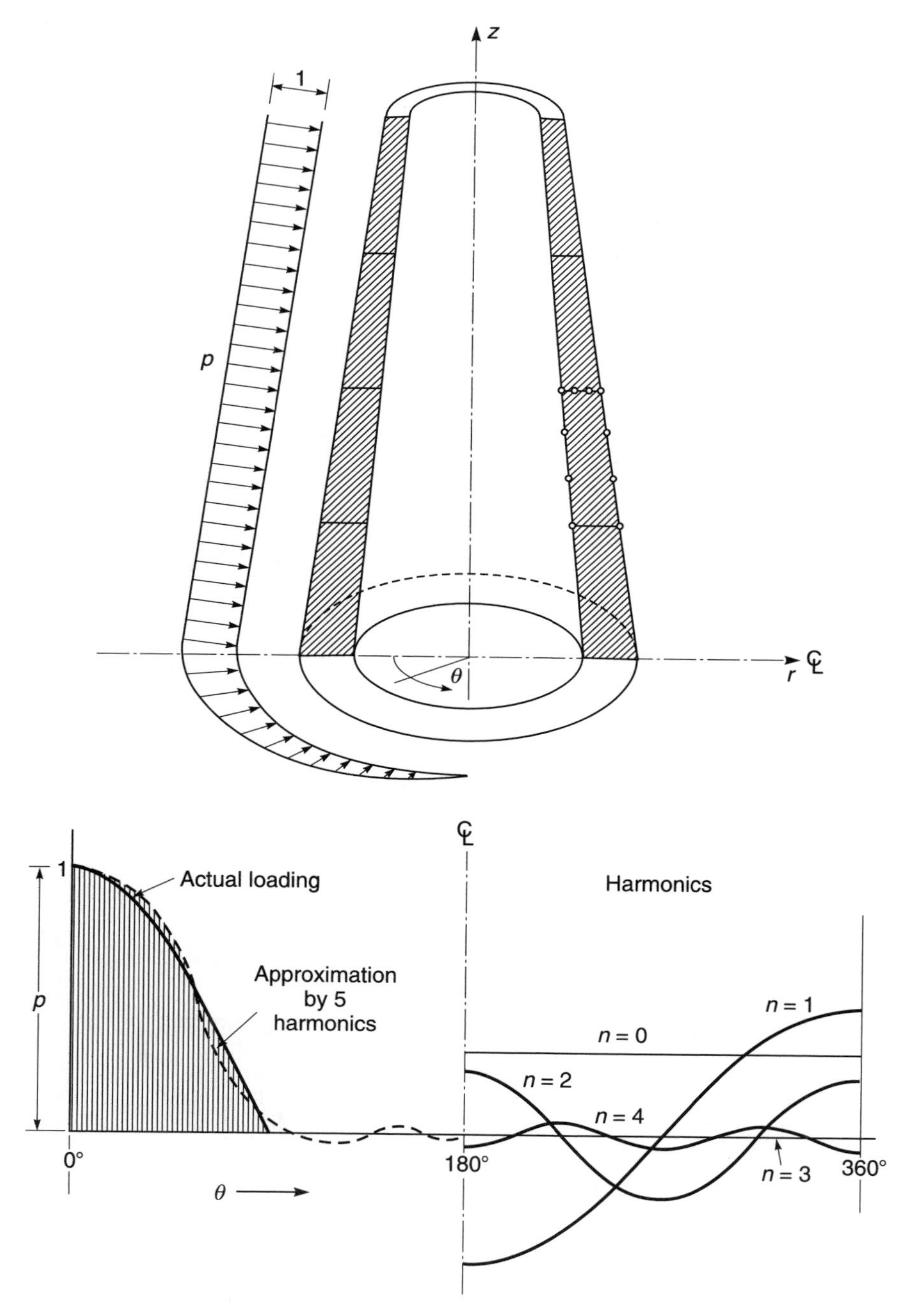

Fig. 9.8 (a) An axisymmetric tower under non-symmetric load; four cubic elements are used in the solution; the harmonics of load expansion used in the analysis are shown.

remains in the variable for v for the antisymmetric case. This corresponds to constant tangential traction and solves simply the torsion problem of shafts subject to known torques (Fig. 9.7). This problem is classically treated by the use of a stress function[19] and indeed in this way has been solved by using a finite element formulation.[20] Here, an alternative, more physical, approach is available.

The first application of the above concepts to the analysis of axisymmetric solids was made by Wilson.[21] A simple example illustrating the effects of various harmonics is shown in Figs 9.8(a) and 9.8(b).

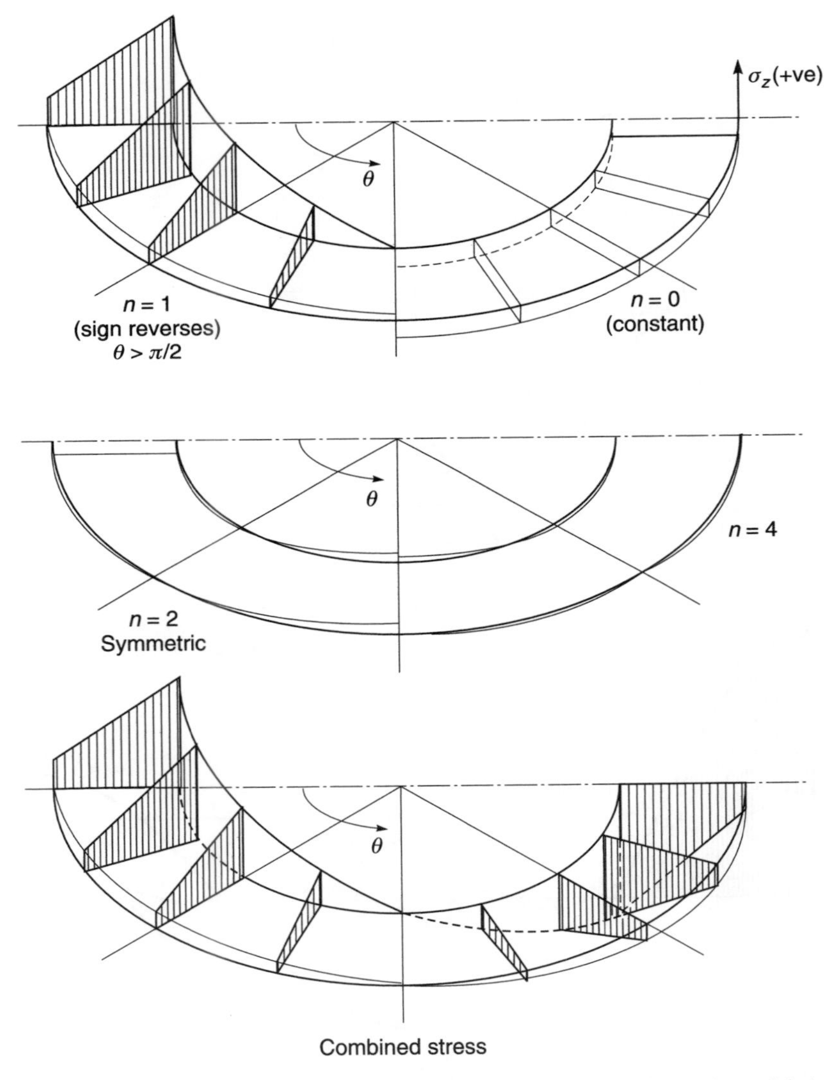

Fig. 9.8 (b) Distribution of σ_z, the vertical stress on base arising from various harmonics and their combination (third harmonic identically zero), the first two harmonics give practically the complete answer.

9.6 Axisymmetric shells with non-symmetrical load

9.6.1 Thin case – no shear deformation

The extension of analysis of axisymmetric thin shells as described in Chapter 7 to the case of non-axisymmetric loads is simple and will again follow the standard pattern.

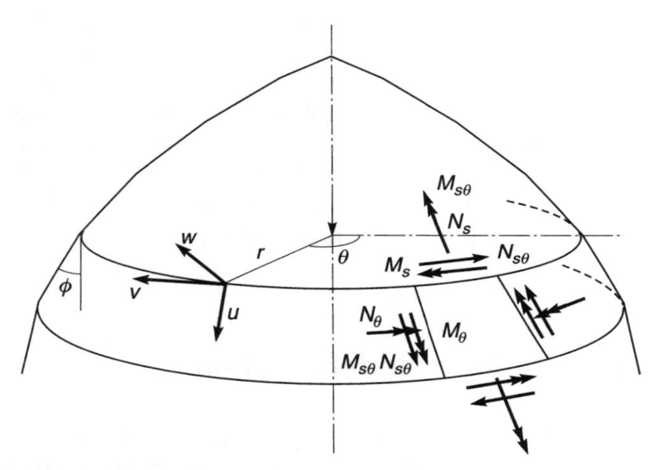

Fig. 9.9 Axisymmetric shell with non-symmetric load; displacements and stress resultants.

It is, however, necessary to extend the definition of strains to include now all three displacements and force components (Fig. 9.9). Three membrane and three bending effects are now present and, extending Eq. (7.1) involving straight generators, we now define strains as[22,23] *

$$\bar{\boldsymbol{\varepsilon}} = \begin{Bmatrix} \varepsilon_s \\ \varepsilon_\theta \\ \gamma_{s\theta} \\ \chi_s \\ \chi_\theta \\ \chi_{s\theta} \end{Bmatrix} = \begin{Bmatrix} \bar{u}_{,s} \\ \bar{v}_{,\theta}/r + (\bar{u}\cos\phi - \bar{w}\sin\phi)/r \\ \bar{u}_{,\theta}/r + \bar{v}_{,s} - \bar{v}\cos\phi/r \\ -\bar{w}_{,ss} \\ -\bar{w}_{,\theta\theta}/r^2 - \bar{w}_{,s}\cos\phi/r + \bar{v}_{,\theta}\sin\phi/r \\ 2\left(-\bar{w}_{,s\theta}/r + \bar{w}_{,\theta}\cos\phi/r^2 + \bar{v}_{,s}\sin\phi/r - \bar{v}\sin\phi\cos\phi/r^2\right) \end{Bmatrix} \tag{9.30}$$

* Various alternatives are available as a result of the multiplicity of shell theories. The one presented is quite commonly accepted.

The corresponding 'stress' matrix is

$$\boldsymbol{\sigma} = \begin{Bmatrix} N_s \\ N_\theta \\ N_{s\theta} \\ M_s \\ M_\theta \\ M_{s\theta} \end{Bmatrix} \tag{9.31}$$

with the three membrane and bending stresses defined as in Fig. 9.9.

Once again, symmetric and antisymmetric variation of loads and displacements can be assumed, as in the previous section. As the processes involved in executing this extension of the application are now obvious, no further description is needed here, but note again should be made of the more elaborate form of equations necessary when curved elements are involved [see Chapter 7, Eq. (7.23)].

The reader is referred to the original paper by Grafton and Strome[23] in which this problem is first treated and to the many later papers on the subject listed in Chapter 7.

9.6.2 Thick case – with shear deformation

The displacement definition for a shell which includes the effects of transverse shearing deformation is specified using the forms given in Eqs (8.4) and (8.26). For a case of loading which is symmetric about $\theta = 0$, the decomposition into global trigonometric components involves the three displacement components of the nth harmonic as

$$\begin{Bmatrix} u^n \\ v^n \\ w^n \end{Bmatrix} = \sum N_i \begin{bmatrix} \cos n\theta & 0 & 0 \\ 0 & \sin n\theta & 0 \\ 0 & 0 & \cos n\theta \end{bmatrix} \left(\begin{Bmatrix} u_i^n \\ v_i^n \\ w_i^n \end{Bmatrix} + \frac{\eta t_i}{2} \begin{bmatrix} -\sin\phi_i & 0 \\ 0 & 1 \\ \cos\phi_i & 0 \end{bmatrix} \begin{Bmatrix} \alpha_i^n \\ \beta_i^n \end{Bmatrix} \right) \tag{9.32}$$

In this u_i, w_i, and α_i stand for the displacements and rotation illustrated in Fig. 8.5, v_i is a displacement of the middle surface node in the tangential (θ) direction, and β_i is a rotation about the vector tangential to the mid-surface.

Global strains are conveniently defined by the relationship[18]

$$\boldsymbol{\varepsilon} = \begin{Bmatrix} \varepsilon_r \\ \varepsilon_z \\ \varepsilon_\theta \\ \gamma_{rz} \\ \gamma_{z\theta} \\ \gamma_{\theta r} \end{Bmatrix} = \begin{Bmatrix} u_{,r} \\ w_{,z} \\ [u + v_{,\theta}]/r \\ u_{,z} + w_{,r} \\ v_{,z} + w_{,\theta}/r \\ v_{,r} - v/r + u_{,\theta}/r \end{Bmatrix} \tag{9.33}$$

These strains are transformed to the local coordinates, and the component normal to η (η = constant) is neglected. As in the axisymmetric case described in Chapter 8, the $\bar{\mathbf{D}}$ matrix relating local stresses and strains takes a form identical to that defined by Eq. (8.13).

A purely axisymmetric problem may again be described for the complete zero harmonic problem and again, as in the non-symmetric loading of solids, the strains split into two problems defining a torsionless and a torsional state. However, for inelastic problems a coupling again occurs whenever both torsionless and torsional loading are both applied as loading conditions on the same problem. Thus, it is often expedient to form the axisymmetric case including all three displacement components.

9.7 Finite strip method – incomplete decoupling

In the previous discussion, orthogonal harmonic functions were used exclusively in the longitudinal/circumferential direction. However, the finite strip method developed by Cheung[17] can in fact be used to solve various structural problems involving different boundary conditions and arbitrary geometrical shapes at the expense of introducing a limited amount of coupling.

As already stated, the finite strip method calls for the use of displacement functions of the multiplicative type (similar to the use of separation of variables in solution of differential equations), in which simple, finite element polynomials are used in one direction, and continuously differentiable smooth series or spline functions in the other. The first type, similar to that previously discussed, is called the semi-analytical finite strip, and the series must be chosen in such a way that they satisfy *a priori* the boundary conditions at the ends of the strip. The second type is called the spline finite strip method, where usually cubic (B_3) spline functions are used and the boundary conditions are incorporated *a posteriori*. Here, for a strip, in which a two-dimensional problem is to be reduced to a one-dimensional one, the displacement previously defined by Eq. (9.22) now is assumed to be of the form

$$w^e = \sum_{n=1}^{r} \bar{\mathbf{N}}(x)\, Y_n(y)\, \mathbf{a}^e \qquad (9.34)$$

where $Y_n(y)$ are suitable continuous functions which are necessary to satisfy the boundary conditions.

Semi-analytical finite strips with orthogonal series Y_n have been developed for plates and shells with regular shapes. The method is a very good technique for solving single-span plates and prismatic thin-walled structures under arbitrary loading because of the uncoupling of the terms in the series. The method is also highly efficient for dynamic and stability analysis and for static analysis of multispan structures under uniformly distributed loads, because only a few coupled terms are required to yield a fairly accurate solution. Spline finite strips are better suited to plates with arbitrary shapes (parallelogram quadrilateral, S-shaped, etc.), for plates and shells with multispans, and for concentrated loading and point support conditions.

Displacement functions are of two types, the polynomial part made up of the shape functions $\bar{N}(x)$ of standard type and the $Y_n(y)$ series or spline function part.

The most commonly used series are the basic functions[24] (or eigenfunctions) which are derived from the solution of the beam vibration differential equation for a single span

$$\frac{d^4 Y}{dy} = \frac{\mu^4 Y}{a^4} \tag{9.35}$$

where a is length of the beam (strip) and μ is a parameter.

The general form of such basic functions is

$$Y_n(y) = C_1 \sin \eta_n + C_2 \cos \eta_n + C_3 \sinh \eta_n + C_4 \cosh \eta_n \tag{9.36}$$

where $\eta_n = \mu_n y/a$.

To a much more limited extent the buckling modes of a beam may be used for stability analysis,[24] and the series takes up the form

$$Y_n(y) = C_1 \sin \eta_n + C_2 \cos \eta_n + C_3 y + C_4 \tag{9.37}$$

The constants C_i are determined by the end conditions.

Another form of series solution that has been used for shear walls[25] is of the form

$$\begin{aligned} Y_1(y) &= \frac{y}{a} \\ Y_n(y) &= \sin \eta_n - \sinh \eta_n - [\cos \eta_n - \cosh \eta_n] \left[\frac{\sin \mu_n + \sinh \mu_n}{\cos \mu_n + \cosh \mu_n}\right] \\ &\quad \text{for} \quad n = 2, 3, \ldots, r \end{aligned} \tag{9.38}$$

where $\mu_n = 1.875, 4.694, \ldots, (2n-1)\pi/2$.

For multiple spans such as illustrated in Fig. 9.10 similar series can be used in each span with the constant appropriately adjusted to ensure continuity. However, spline functions are useful here.

Spline, which is originally the name of a small flexible wooden strip employed by a draftsman as a tool for drawing a continuous smooth curve segment by segment, became a mathematical tool after the seminal work of Schoenberg.[26] A variety of

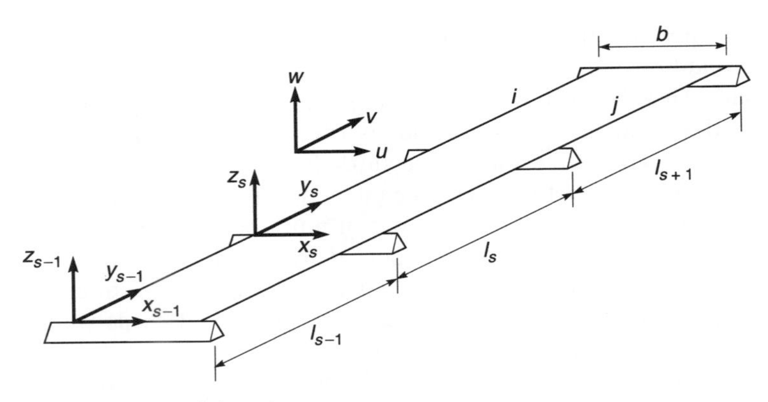

Fig. 9.10 A typical continuous finite strip.

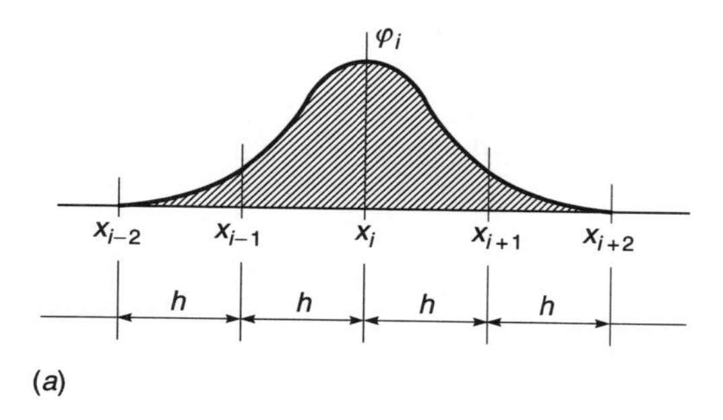

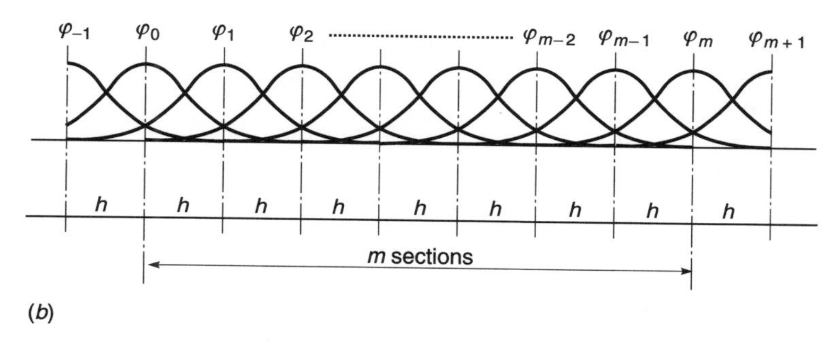

Fig. 9.11 Typical spline approximations: (a) typical B_3 spline; (b) basis functions for B_3 spline expression.

spline functions are available. The spline function chosen here (Fig. 9.11) to represent the displacement is the B_3 cubic spline of equal section length (B_3 splines of unequal section length have been discussed in the paper by Li *et al.*[27]) and is given as

$$Y(x) = \sum_{i=-1}^{m+1} \alpha_i \Psi_i(x) \tag{9.39}$$

in which each local B_3 spline Ψ_i has non-zero values over four consecutive sections with the section node $x = x_i$ as the centre and is defined by

$$\Psi_i = \frac{1}{6h^3}\begin{cases} 0, & x < x_{i-2} \\ (x - x_{i-2})^3, & x_{i-2} \leqslant x \leqslant x_{i-1} \\ h^3 + 3h^2(x - x_{i-1}) + 3h(x - x_{i-1})^2 - 3(x - x_{i-1})^3, & x_{i-1} \leqslant x \leqslant x_i \\ h^3 - 3h^2(x - x_{i+1}) + 3h(x - x_{i+1})^2 + 3(x - x_{i+1})^3, & x_i \leqslant x \leqslant x_{i+1} \\ (x_{i+2} - x)^3, & x_{i+1} \leqslant x \leqslant x_{i+2} \\ 0, & x > x_{i+2} \end{cases} \tag{9.40}$$

The use of B_3 splines offers certain distinct advantages when compared with the conventional finite element method and the semi-analytical finite strip method.

1. It is computationally efficient. When using B_3 splines as displacement functions, continuity is ensured up to the second order (C_2 continuity). However, to achieve

the same continuity conditions for conventional finite elements, it is necessary to have three times as many unknowns at the nodes (e.g. quintic Hermitian functions).
2. It is more flexible than the semi-analytical finite strip method in the boundary condition treatment. Only the local splines around the boundary point need to be amended to fit any specified boundary condition.
3. It has wider applications than the semi-analytical finite strip method. The spline finite strip method can be used to analyse plates with arbitrary shapes.[28] In this case any domain bounded by four curved (or straight) sides can be mapped into a rectangular one (see Chapter 9 of Volume 1) and all operations for one system (x, y) can be transformed to corresponding ones for the other system (ξ, η).

The finite strip methods have proved effective in a large number of engineering applications, many listed in the text by Cheung.[17] References 29–39 list some of the typical linear problems solved in statics, vibrations, and buckling analysis of structures. Indeed, non-linear problems of the type described in Sec. 4.20 have also been successfully tackled.[40,41]

Of considerable interest also is the extension of the procedures to the analysis of stratified (layered) media such as may be encountered in laminar structures or foundations.[42–44]

9.8 Concluding remarks

A fairly general process combining some of the advantages of finite element analysis with the economy of expansion in terms of generally orthogonal functions has been illustrated in several applications. Certainly, these only touch on the possibilities offered, but it should be borne in mind that the economy is achieved only in certain geometrically constrained situations and those to which the number of terms requiring solution is limited.

Similarly, other 'prismatic' situations can be dealt with in which only a segment of a body of revolution is developed (Fig. 9.12). Clearly, the expansion must now be taken in terms of the angle $l\pi\theta/\alpha$, but otherwise the approach is identical to that described previously.[2]

In the methods of this chapter it was assumed that material properties remain constant with one coordinate direction. This restriction can on occasion be lifted with the same general process maintained. An early example of this type was presented by Stricklin and DeAndrade.[45] Inclusion of inelastic behaviour has also been successfully treated.[46–49]

In Chapter 17 of Volume 1 dealing with semi-discretization we considered general classes of problems for time. All the problems we have described in this chapter could be derived in terms of similar semi-discretization. We would thus *first* semi-discretize, describing the problem in terms of an ordinary differential equation in z of the form

$$\mathbf{K}_1 \frac{\mathrm{d}^2\mathbf{a}}{\mathrm{d}z^2} + \mathbf{K}_2 \frac{\mathrm{d}\mathbf{a}}{\mathrm{d}z} + \mathbf{K}_3 \mathbf{a} + \mathbf{f} = \mathbf{0}$$

Second, the above equation system would be solved in the domain $0 < z < a$ by means of orthogonal functions that *naturally* enter the problem as solutions of ordinary

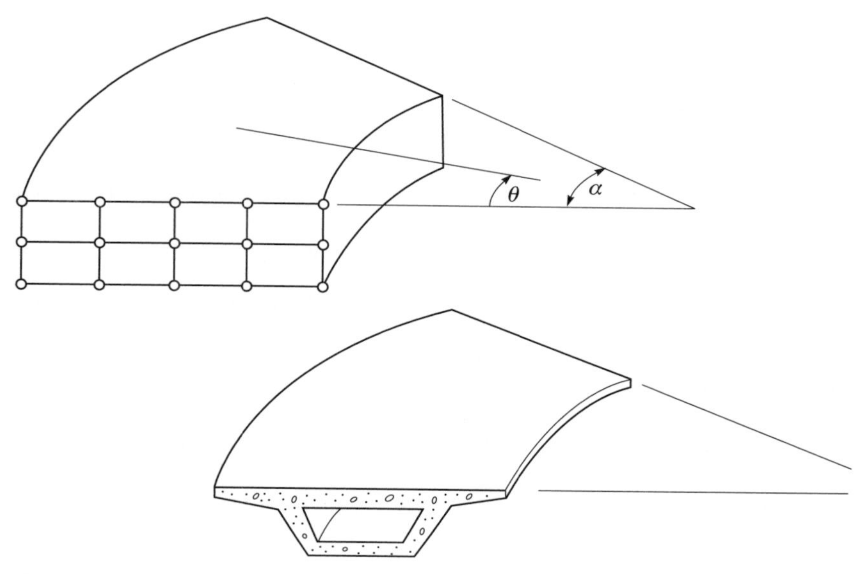

Fig. 9.12 Other segmental, prismatic situations.

differential equations *with constant coefficients*. This second solution step is most easily found by using a diagonalization process described in dynamic applications (see Chapter 17, Volume 1). Clearly, the final result of such computations would turn out to be identical with the procedures here described, but on occasion the above formulation is more self-evident.

References

1. P.M. Morse and H. Feshbach. *Methods of Theoretical Physics*, McGraw-Hill, New York, 1953.
2. O.C. Zienkiewicz and J.J.M. Too. The finite prism in analysis of thick simply supported bridge boxes. *Proc. Inst. Civ. Eng.*, **53**, 147–72, 1972.
3. Y.K. Cheung. The finite strip method in the analysis of elastic plates with two opposite simply supported ends. *Proc. Inst. Civ. Eng.*, **40**, 1–7, 1968.
4. Y.K. Cheung. Finite strip method of analysis of elastic slabs. *Proc. Am. Soc. Civ. Eng.*, **94**(EM6), 1365–78, 1968.
5. Y.K. Cheung. Folded plates by the finite strip method. *Proc. Am. Soc. Civ. Eng.*, **95**(ST2), 963–79, 1969.
6. Y.K. Cheung. The analysis of cylindrical orthotropic curved bridge decks. *Publ. Int. Ass. Struct. Eng.*, **29**-II, 41–52, 1969.
7. Y.K. Cheung, M.S. Cheung and A. Ghali. Analysis of slab and girder bridges by the finite strip method. *Building Sci.*, **5**, 95–104, 1970.
8. Y.C. Loo and A.R. Cusens. Development of the finite strip method in the analysis of cellular bridge decks. In K. Rockey *et al.* (eds), *Conf. on Developments in Bridge Design and Construction*, Crosby Lockwood, London, 1971.
9. Y.K. Cheung and M.S. Cheung. Static and dynamic behaviour of rectangular plates using higher order finite strips. *Building Sci.*, **7**, 151–8, 1972.

10. G.S. Tadros and A. Ghali. Convergence of semi-analytical solution of plates. *Proc. Am. Soc. Civ. Eng.*, **99**(EM5), 1023–35, 1973.
11. A.R. Cusens and Y.C. Loo. Application of the finite strip method in the analysis of concrete box bridges. *Proc. Inst. Civ. Eng.*, **57**-II, 251–73, 1974.
12. T.G. Brown and A. Ghali. Semi-analytic solution of skew plates in bending. *Proc. Inst. Civ. Eng.*, **57**-II, 165–75, 1974.
13. A.S. Mawenya and J.D. Davies. Finite strip analysis of plate bending including transverse shear effects. *Building Sci.*, **9**, 175–80, 1974.
14. P.R. Benson and E. Hinton. A thick finite strip solution for static, free vibration and stability problems. *Int. J. Num. Meth. Eng.*, **10**, 665–78, 1976.
15. E. Hinton and O.C. Zienkiewicz. A note on a simple thick finite strip. *Int. J. Num. Meth. Eng.*, **11**, 905–9, 1977.
16. H.C. Chan and O. Foo. Buckling of multilayer plates by the finite strip method. *Int. J. Mech. Sci.*, **19**, 447–56, 1977.
17. Y.K. Cheung. *Finite Strip Method in Structural Analysis*, Pergamon Press, Oxford, 1976.
18. I.S. Sokolnikoff. *The Mathematical Theory of Elasticity*, 2nd edition, McGraw-Hill, New York, 1956.
19. S.P. Timoshenko and J.N. Goodier. *Theory of Elasticity*, 3rd edition, McGraw-Hill, New York, 1969.
20. O.C. Zienkiewicz, P.L. Arlett and A.K. Bahrani. Solution of three-dimensional field problems by the finite element method. *The Engineer*, October 1967.
21. E.L. Wilson. Structural analysis of axi-symmetric solids. *Journal of AIAA*, **3**, 2269–74, 1965.
22. V.V. Novozhilov. *Theory of Thin Shells*, Noordhoff, Dordrecht, 1959 [English translation].
23. P.E. Grafton and D.R. Strome. Analysis of axi-symmetric shells by the direct stiffness method. *Journal of AIAA*, **1**, 2342–7, 1963.
24. O. Foo. *Application of Finite Strip Method in Structural Analysis with Particular Reference to Sandwich Plate Structure*, PhD thesis, The Queen's University of Belfast, Belfast, 1977.
25. Y.K. Cheung. Computer analysis of tall buildings. In *Proc. 3rd. Int. Conf. on Tall Buildings*, pp. 8–15, Hong Kong and Guangzhou, 1984.
26. I.J. Schoenberg. Contributions to the problem of approximation of equidistant data by analytic functions. *Q. Appl. Math.*, **4**, 45–99 and 112–114, 1946.
27. W.Y. Li, Y.K. Cheung and L.G. Tham. Spline finite strip analysis of general plates. *J. Eng. Mech., ASCE*, **112**(EM1), 43–54, 1986.
28. Y.K. Cheung, L.G. Tham and W.Y. Li. Free vibration and static analysis of general plates by spline finite strip. *Comp. Mech.*, **3**, 187–97, 1988.
29. Y.K. Cheung. Orthotropic right bridges by the finite strip method. In *Concrete Bridge Design*, pp. 812–905, Report SP-26, American Concrete Institute, Farmington, MI, 1971.
30. H.C. Chan and Y.K. Cheung. Static and dynamic analysis of multilayered sandwich plates. *Int. J. Mech. Sci.*, **14**, 399–406, 1972.
31. D.J. Dawe. Finite strip buckling of curved plate assemblies under biaxial loading. *Int. J. Solids Struct.*, **13**, 1141–55, 1977.
32. D.J. Dawe. Finite strip models for vibration of Mindlin plates. *J. Sound Vib.*, **59**, 441–52, 1978.
33. Y.K. Cheung and C. Delcourt. Buckling and vibration of thin, flat-walled structures continuous over several spans. *Proc. Inst. Civ. Eng.*, **64**-II, 93–103, 1977.
34. Y.K. Cheung and S. Swaddiwudhipong. Analysis of frame shear wall structures using finite strip elements. *Proc. Inst. Civ. Eng.*, **65**-II, 517–35, 1978.
35. D. Bucco, J. Mazumdax and G. Sved. Application of the finite strip method combined with the deflection contour method to plate bending problems. *J. Comp. Struct.*, **10**, 827–30, 1979.

36. C. Meyer and A.C. Scordelis. Analysis of curved folded plate structures. *Proc. Am. Soc. Civ. Eng.*, **97**(ST10), 2459–80, 1979.
37. Y.K. Cheung, L.G. Tham and W.Y. Li. Application of spline–finite strip method in the analysis of curved slab bridge. *Proc. Inst. Civ. Eng.*, **81**-II, 111–24, 1986.
38. W.Y. Li, L.G. Tham and Y.K. Cheung. Curved box-girder bridges. *Proc. Am. Soc. Civ. Eng.*, **114**(ST6), 1324–38, 1988.
39. Y.K. Cheung, W.Y. Li and L.G. Tham. Free vibration of singly curved shell by spline finite strip method. *J. Sound Vibr.*, **128**, 411–22, 1989.
40. Y.K. Cheung and D.S. Zhu. Large deflection analysis of arbitrary shaped thin plates. *J. Comp. Struct.*, **26**, 811–14, 1987.
41. D.S. Zhu and Y.K. Cheung. Postbuckling analysis of shells by spline finite strip method. *J. Comp. Struct.*, **31**, 357–64, 1989.
42. S.B. Dong and R.B. Nelson. On natural vibrations and waves in laminated orthotropic plates. *J. Appl. Mech.*, **30**, 739, 1972.
43. D.J. Guo, L.G. Tham and Y.K. Cheung. Infinite layer for the analysis of a single pile. *J. Comp. Geotechnics*, **3**, 229–49, 1987.
44. Y.K. Cheung, L.G. Tham and D.J. Guo. Analysis of pile group by infinite layer method. *Geotehnique*, **38**, 415–31, 1988.
45. J.A. Stricklin, J.C. DeAndrade, F.J. Stebbins and A.J. Cwertny Jr. Linear and non-linear analysis of shells of revolution with asymmetrical stiffness properties. In L. Berke, R.M. Bader, W.J. Mykytow, J.S. Przemienicki and M.H. Shirk (eds), *Proc. 2nd Conf. Matrix Methods in Structural Mechanics*, Volume AFFDL-TR-68-150, pp. 1231–51, Air Force Flight Dynamics Laboratory, Wright Patterson Air Force Base, OH, October 1968.
46. L.A. Winnicki and O.C. Zienkiewicz. Plastic or visco-plastic behaviour of axisymmetric bodies subject to non-symmetric loading; semi-analytical finite element solution. *Int. J. Num. Meth. Eng.*, **14**, 1399–412, 1979.
47. W. Wunderlich, H. Cramer and H. Obrecht. Application of ring elements in the nonlinear analysis of shells of revolution under nonaxisymmetric loading. *Comp. Meth. Appl. Mech. Eng.*, **51**, 259–75, 1985.
48. H. Obrecht, F. Schnabel and W. Wunderlich. Elastic–plastic creep buckling of circular cylindrical shells under axial compression. *Z. Angew. Math. Mech.*, **67**, T118–T120, 1987.
49. W. Wunderlich and C. Seiler. Nonlinear treatment of liquid-filled storage tanks under earthquake excitation by a quasistatic approach. In B.H.V. Topping (ed.), *Advances in Computational Structural Mechanics, Proceedings 4th International Conference on Computational Structures*, pp. 283–91, August 1998.

10

Geometrically non-linear problems – finite deformation

10.1 Introduction

In all our previous discussion we have assumed that deformations remained small so that linear relations could be used to represent the strain in a body. We now admit the possibility that deformations can become large during a loading process. In such cases it is necessary to distinguish between the *reference* configuration where initial shape of the body or bodies to be analysed is known and the *current* or *deformed* configuration after loading is applied. Figure 10.1 shows the two configurations and the coordinate frames which will be used to describe each one. We note that the deformed configuration of the body is unknown at the start of an analysis and, therefore, must be determined as part of the solution process – a process that is inherently non-linear. The relationships describing the finite deformation behaviour of solids involve equations related to both the reference and the deformed configurations. We shall generally find that such relations are most easily expressed using the indicial notation introduced in Volume 1 (see Appendix B, Volume 1); however, after these indicial forms are developed we shall again return to a matrix form to construct the finite element approximations.

The chapter starts by describing the basic kinematic relations used in finite deformation solid mechanics. This is followed by a summary of different stress and traction measures related to the reference and deformed configurations, a statement of boundary and initial conditions, and an overview of material constitution for finite elastic solids. A variational Galerkin statement for the finite elastic material is then given in the reference configuration. Using the variational form the problem is then cast into a matrix form and a standard finite element solution process is indicated. The procedure up to this point is based on equations related to the reference configuration. A transformation to a form related to the current configuration is performed and it is shown that a much simpler statement of the finite element formulation process results – one which again permits separation into a form for treating nearly incompressible situations.

A mixed variational form is introduced and the solution process for problems which can have nearly incompressible behaviour is presented. This follows closely the developments for the small strain form given in Chapter 1. An alternative to the mixed form is also given in the form of an enhanced strain model (see

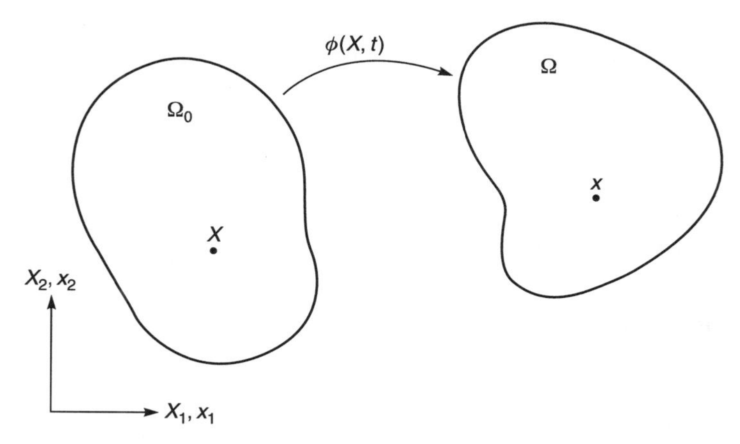

Fig. 10.1 Reference and deformed (current) configuration for finite deformation problems.

Chapter 11 of Volume 1). Here a fully mixed construction is shown and leads to a form which performs well in two- and three-dimensional problems.

In finite deformation problems, loads can be given relative to the deformed configuration. An example is a pressure loading which always remains normal to a deformed surface. Here we discuss this case and show that by using finite element type constructions a very simple result follows. Since the loading is no longer derivable from a potential function (i.e. conservative) the tangent matrix for the formulation is unsymmetric, leading in general to a requirement of an unsymmetric solver in a Newton–Raphson solution scheme.

We next consider the form of material constitutive models for finite deformation. This is a very complex subject and we present a discussion for only hyperelastic and isotropic elasto-plastic material forms. Thus, the reader undoubtedly will need to consult literature on the subject for additional types of models. We do give a rate model which can be used on some occasions to develop heuristic forms from small deformation concepts; however, such an approach should be used with caution and only when experimental data are available to verify the behaviour obtained.

In the last section of this chapter we consider the modelling of interaction between one or more bodies which come into contact with each other. Such *contact problems* are among the most difficult to model by finite elements and we summarize here only some of the approaches which have proved successful in practice. In general, the finite element discretization process itself leads to surfaces which are not smooth and, thus, when large sliding occurs the transition from one element to the next leads to discontinuities in the response – and in transient applications can induce non-physical inertial discontinuities also. For quasi-static response such discontinuity leads to difficulties in defining a unique solution and here methods of multisurface plasticity prove useful.

We include in the chapter some illustrations of performance for many of the formulations and problem classes discussed; however, the range is so broad that it is not possible to cover a comprehensive set. Here again the reader is referred to literature cited for additional insight and results.

The present chapter concentrates on continuum problems where finite elements are used to discretize the problem in all directions modelled. In the next chapter we consider forms for problems which have one (or more) small dimension(s) and thus can benefit from use of plate and shell formulations of the type discussed earlier in this volume for small deformation situations.

10.2 Governing equations

10.2.1 Kinematics and deformation

The basic equations for finite deformation solid mechanics may be found in standard references on the subject.[1–4] Here a summary of the basic equations in three dimensions is presented – two dimensional plane problems being a special case of these. A body B has material points whose positions are given by the vector $\mathbf{X}$ in a fixed reference configuration,* Ω, in a three-dimensional space. In Cartesian coordinates the position vector is described in terms of its components as:

$$\mathbf{X} = X_I \mathbf{E}_I; \qquad I = 1, 2, 3 \tag{10.1}$$

where $\mathbf{E}_I$ are unit orthogonal base vectors and summation convention is used for repeated indices of like kind (e.g. I). After the body is loaded each material point is described by its position vector, $\mathbf{x}$, in the *current* deformed configuration, ω. The position vector in the current configuration is given in terms of its Cartesian components as

$$\mathbf{x} = x_i \mathbf{e}_i; \qquad i = 1, 2, 3 \tag{10.2}$$

where $\mathbf{e}_i$ are unit base vectors for the current time, t, and again summation convention is used. In our discussion, common origins and directions of the reference and current coordinates are used for simplicity. Furthermore, in a Cartesian system base vectors do not change with position and all derivations may be made using components of tensors written in indicial form. Final equations are written in matrix form using standard transformations described in Chapter 1 and in Appendix B of Volume 1.

The position vector at the current time is related to the reference configuration position vector through the mapping

$$x_i = \phi_i(X_I, t) \tag{10.3}$$

Determination of ϕ_i is required as part of any solution and is analogous to the displacement vector, which we introduce next. When common origins and directions for the coordinate frames are used, a displacement vector may be introduced as the change between the two frames. Accordingly,

$$x_i = \delta_{iI}(X_I + U_I) \tag{10.4}$$

* As much as possible we adopt the notation that upper-case letters refer to quantities defined in the *reference* configuration and lower-case letters to quantities defined in the *current* deformed configuration. Exceptions occur when quantities are related to both the reference and the current configurations.

where summation convention is implied over indices of the same kind and δ_{iI} is a rank-two shifter tensor between the two coordinate frames, and is defined by a Kronnecker delta quantity such that

$$\delta_{iI} = \begin{cases} 1 & \text{if } i = I \\ 0 & \text{if } i \neq I \end{cases} \tag{10.5}$$

The shifter satisfies the relations

$$\delta_{iI}\,\delta_{iJ} = \delta_{IJ} \qquad \text{and} \qquad \delta_{iI}\,\delta_{jI} = \delta_{ij} \tag{10.6}$$

where δ_{IJ} and δ_{ij} are Kronnecker delta quantities in the reference and current configuration, respectively. Using the shifter, a displacement component may be written with respect to either the reference configuration or the current configuration and related through

$$u_i = \delta_{iI}\,U_I \qquad \text{and} \qquad U_I = \delta_{iI}\,u_i \tag{10.7}$$

and we observe that $u_1 = U_1$, etc. Thus, either may be used equally to develop finite element parameters.

A fundamental measure of deformation is described by the deformation gradient relative to X_I given by

$$F_{iI} = \frac{\partial \phi_i}{\partial X_I} \tag{10.8}$$

subject to the constraint

$$J = \det F_{iI} > 0 \tag{10.9}$$

to ensure that material volume elements remain positive. The deformation gradient is a direct measure which maps a differential line element in the reference configuration into one in the current configuration as (Fig. 10.1)

$$\mathrm{d}x_i = \frac{\partial \phi_i}{\partial X_I}\,\mathrm{d}X_I = F_{iI}\,\mathrm{d}X_I \tag{10.10}$$

Thus, it may be used to determine the change in length and direction of a differential line element. The determinant of the deformation gradient also maps a volume element in the reference configuration into one in the reference configuration, that is

$$\mathrm{d}v = J\,\mathrm{d}V \tag{10.11}$$

where $\mathrm{d}V$ is a volume element in the reference configuration and $\mathrm{d}v$ its corresponding form in the current configuration.

The deformation gradient may be expressed in terms of the displacement as

$$F_{iI} = \delta_{iI} + \frac{\partial u_i}{\partial X_I} = \delta_{iI} + u_{i,I} \tag{10.12}$$

and is a *two-point tensor* since it is referred to both the reference and the current configurations. Using F_{iI} directly complicates the development of constitutive equations and it is common to introduce deformation measures which are completely related to either the reference or the current configurations. For the reference configuration, the right Cauchy–Green deformation tensor, C_{IJ}, is introduced as

$$C_{IJ} = F_{iI} F_{iJ} \tag{10.13}$$

Alternatively the Green strain tensor, E_{IJ}, given as

$$E_{IJ} = \tfrac{1}{2}\left(C_{IJ} - \delta_{IJ}\right) \tag{10.14}$$

may be used. The Green strain may be expressed in terms of the reference displacements as

$$E_{IJ} = \frac{1}{2}\left[\frac{\partial U_I}{\partial X_J} + \frac{\partial U_J}{\partial X_I} + \frac{\partial U_K}{\partial X_I}\frac{\partial U_K}{\partial X_J}\right] = \frac{1}{2}\left[U_{I,J} + U_{J,I} + U_{K,I}U_{K,J}\right] \tag{10.15}$$

In the current configuration a common deformation measure is the left Cauchy–Green deformation tensor, b_{ij}, expressed as

$$b_{ij} = F_{iI} F_{jI} \tag{10.16}$$

The Almansi strain tensor, e_{ij}, is related to the inverse of b_{ij} as

$$e_{ij} = \tfrac{1}{2}\left(\delta_{ij} - b_{ij}^{-1}\right) \tag{10.17}$$

or inverting by

$$b_{ij} = \left(\delta_{ij} - 2e_{ij}\right)^{-1} \tag{10.18}$$

Generally, the Almansi strain tensor will not appear naturally in our constitutive equations and we often will use b_{ij} forms directly.

10.2.2 Stress and traction for reference and deformed states

Stress measures

Stress measures the amount of force per unit of area. In finite deformation problems care must be taken to describe the configuration to which a stress is measured. The Cauchy (true) stress, σ_{ij}, and the Kirchhoff stress, τ_{ij}, are symmetric measures of stress defined with respect to the current configuration. They are related through the determinant of the deformation gradient as

$$\tau_{ij} = J\sigma_{ij} \tag{10.19}$$

and usually are the stresses used to define general constitutive equations for materials. The second Piola–Kirchhoff stress, S_{IJ}, is a symmetric stress measure with respect to the reference configuration and is related to the Kirchhoff stress through the deformation gradient as

$$\tau_{ij} = F_{iI} S_{IJ} F_{jJ} \tag{10.20}$$

Finally, one can introduce the (unsymmetric) first Piola–Kirchhoff stress, P_{iI}, which is related to S_{IJ} through

$$P_{iI} = F_{iJ} S_{JI} \tag{10.21}$$

and to the Kirchhoff stress by

$$\tau_{ij} = P_{iI} F_{jI} \tag{10.22}$$

Traction measures

For the current configuration *traction* is given by

$$t_i = \sigma_{ij} n_j \tag{10.23}$$

where n_j are direction cosines of a unit outward pointing normal to a deformed surface. This form of the traction may be related to a reference surface quantity through *force* relations defined as

$$t_i \, \mathrm{d}s = \delta_{iI} T_I \, \mathrm{d}S \tag{10.24}$$

where $\mathrm{d}s$ and $\mathrm{d}S$ are surface area elements in the current and reference configurations, respectively, and T_I is traction on the reference configuration. Note that the direction of the traction component is preserved during the transformation and, thus, remains directly related to current configuration forces.

10.2.3 Equilibrium equations

Using quantities related to the current (deformed) configuration, the equilibrium equations for a solid subjected to finite deformation are nearly identical to those for small deformation. The local equilibrium equation (balance of linear momentum) is obtained as a force balance on a small differential volume of deformed solid and is given by[2–4]

$$\frac{\partial \sigma_{ij}}{\partial x_i} + \rho b_j^{(m)} = \rho \dot{v}_j \tag{10.25}$$

where ρ is mass density in the current configuration, $b_j^{(m)}$ is body force *per unit mass*, and v_j is the material velocity

$$v_j = \frac{\partial \phi_j}{\partial t} = \dot{x}_j = \dot{u}_j \tag{10.26}$$

The mass density in the current configuration may be related to the reference configuration (initial) mass density, ρ_0, using the balance-of-mass principle[2–4] and yields

$$\rho_0 = J\rho \tag{10.27}$$

Thus differences in the equilibrium equation from those of the small deformation case appear only in the body force and inertial force definitions.

Similarly, the moment equilibrium on a small differential volume element of the deformed solid gives the balance of angular momentum requirement for the

Cauchy stress as

$$\sigma_{ij} = \sigma_{ji} \tag{10.28}$$

which is identical to the result from the small deformation problem.

The equilibrium requirements may also be written for the reference configuration using relations between stress measures and the chain rule of differentiation.[2] We will show the form for the balance of linear momentum when discussing the variational form for the problem. Here, however, we comment on the symmetry requirements for stress resulting from angular momentum balance. Using symmetry of the Cauchy stress tensor and Eqs (10.19) and (10.22) leads to the requirement on the first Piola–Kirchhoff stress

$$F_{iI} P_{jI} = P_{iI} F_{jI} \tag{10.29}$$

and subsequently, using Eq. (10.21), to the symmetry of the second Piola–Kirchhoff stress tensor

$$S_{IJ} = S_{JI} \tag{10.30}$$

10.2.4 Boundary conditions

As described in Chapter 1 the basic boundary conditions for a continuum body consist of two types: displacement boundary conditions and traction boundary conditions. Boundary conditions generally are defined on each part of the boundary by specifying components with respect to a local coordinate system defined by the orthogonal basis, $\mathbf{e}'_i$, $i = 1, 2, 3$. Often one of the directions, say $\mathbf{e}_3$, coincides with the normal to the surface and the other two are in tangential directions along the surface. At each point on the boundary one (and only one) boundary condition must be specified for all three directions of the basis. These conditions can be all for displacements (fixed surface), all for tractions (stress or free surface), or a combination of displacements and tractions (mixed surface).

Displacement boundary conditions may be expressed for a component by requiring

$$x'_i = \bar{x}'_i \tag{10.31}$$

at each point on the displacement boundary, γ_u. A quantity with a superposed bar, such as $\bar{x}'_i$ again denotes a specified quantity. The boundary condition may also be expressed in terms of components of the displacement vector, u_i. Accordingly, on γ_u

$$u'_i = \bar{u}'_i \tag{10.32}$$

The second type of boundary condition is a traction boundary condition. Using the orthogonal basis described above, the traction boundary conditions may be given for each component by requiring

$$t'_i = \bar{t}'_i \tag{10.33}$$

at each point on the boundary, γ_t. The boundary condition may be non-linear for loadings such as pressure loads, as described later in Sec. 10.6.

10.2.5 Initial conditions

Initial conditions describe the state of a body at the start of an analysis. The conditions describe the initial kinematic and stress or strain states with respect to the reference configuration used to define the body. In addition, for constitutive equations with internal variables the initial values of terms which evolve in time must be given (e.g. initial plastic strain).

The initial conditions for the kinematic state consist of specifying the position and velocity at some initial time, commonly taken as zero. Accordingly,

$$x_i(X_I, 0) = \bar{\phi}_i(X_I, 0) \quad \text{or} \quad u_i(X_I, 0) = \bar{d}_i^0(X_I) \tag{10.34}$$

and

$$v_i(X_I, 0) = \dot{\phi}_i(X_I, 0) = \bar{v}_i^0(X_I) \tag{10.35}$$

are specified at each point in the body.

The initial conditions for stresses are specified as

$$\sigma_{ij}(X_I, 0) = \bar{\sigma}_{ij}^0(X_I) \tag{10.36}$$

at each point in the body. Finally, as noted above the internal variables in the stress–strain relations that evolve in time must have their initial conditions set. For a finite elastic model, generally there are no internal variables to be set unless initial stress effects are included.

10.3 Variational description for finite deformation

In order to construct finite element approximations for the solution of finite deformation problems it is necessary to write the formulation in a Galerkin (weak) or variational form as illustrated many times previously. Here again we can write these integral forms in either the reference configuration or in the current configuration. The simplest approach is to start from a reference configuration since here integrals are all expressed over *domains which do not change during the deformation process and thus are not affected by variation or linearization steps*. Later the results can be transformed and written in terms of the deformed configuration. Using the reference configuration form variations and linearizations can be carried out in an identical manner as was done in the small deformation case. Thus, all the steps outlined in Chapter 1 immediately can be extended to the finite deformation problem. We shall discover that the final equations obtained by this approach are very different from those of the small deformation problem. However, after all derivation steps are completed a transformation to expressions integrated over the current configuration will yield a form which is nearly identical to the small deformation problem and thus greatly simplifies the development of the final force and stiffness terms as well as programming steps.

To develop a finite element solution to the finite deformation problem we consider first the case of elasticity as a variational problem. Other material behaviour may be considered later by substitution of appropriate constitutive expressions for stress and tangent moduli – identical to the process used in Chapter 3 for the small deformation problem.

10.3.1 Reference configuration formulation

A variational theorem for finite elasticity may be written in the reference configuration as[4,5]

$$\Pi = \int_\Omega W(C_{IJ})\,\mathrm{d}V - \Pi_{\text{ext}} \tag{10.37}$$

in which $W(C_{IJ})$ is a stored energy function for a *hyperelastic* material from which the second Piola–Kirchhoff stress is computed using[4]

$$S_{IJ} = 2\frac{\partial W}{\partial C_{IJ}} = \frac{\partial W}{\partial E_{IJ}} \tag{10.38}$$

The simplest representation of the stored energy function is the Saint-Venant–Kirchhoff model given by

$$W(E_{IJ}) = \tfrac{1}{2} D_{IJKL} E_{IJ} E_{KL} \tag{10.39}$$

where D_{IJKL} are constant elastic moduli defined in a manner similar to the small deformation ones. Equation (10.38) then gives

$$S_{IJ} = D_{IJKL} E_{KL} \tag{10.40}$$

for the stress–strain relation. While this relation is simple it is not adequate to define the behaviour of elastic finite deformation states. It is useful, however, for the case where strains are small but displacements are large and we address this use further in the next chapter. Other models for representing elastic behaviour at large strain are considered in Sec. 10.7.

The potential for the external work is here assumed to be given by

$$\Pi_{\text{ext}} = \int_\Omega U_I \rho_0 b_I^{(m)}\,\mathrm{d}V + \int_{\Gamma_t} U_I \bar{T}_I\,\mathrm{d}S \tag{10.41}$$

where $\bar{T}_I$ denotes specified tractions in the reference configuration and Γ_t is the traction boundary surface in the reference configuration. Taking the variation of Eqs (10.37) and (10.41) we obtain

$$\delta\Pi = \int_\Omega \tfrac{1}{2}\,\delta C_{IJ} S_{IJ}\,\mathrm{d}V - \delta\Pi_{\text{ext}} = 0 \tag{10.42}$$

and

$$\delta\Pi_{\text{ext}} = \int_\Omega \delta U_I \rho_0 b_I^{(m)}\,\mathrm{d}V + \int_{\Gamma_t} \delta U_I \bar{T}_I\,\mathrm{d}S \tag{10.43}$$

where δU_I is a *variation* of the reference configuration displacement (i.e. a virtual displacement) which is arbitrary except at the kinematic boundary condition locations, Γ_u, where, for convenience, it vanishes. Since a virtual displacement is an arbitrary function, satisfaction of the variational equation implies satisfaction of the balance of linear momentum at each point in the body as well as the traction boundary conditions. We note that by using Eq. (10.38) and constructing the variation of C_{IJ}, the first term in the integrand of Eq. (10.42) can be expressed in alternate forms as

$$\tfrac{1}{2}\,\delta C_{IJ}\,S_{IJ} = \delta E_{IJ}\,S_{IJ} = \delta F_{iI}\,F_{iJ}\,S_{IJ} \tag{10.44}$$

where symmetry of S_{IJ} has been used. The variation of the deformation gradient may be expressed directly in terms of the current configuration displacement as

$$\delta F_{iI} = \frac{\partial \delta u_i}{\partial X_I} = \delta u_{i,I} \tag{10.45}$$

Using the above results, after integration by parts using Green's theorem (see Appendix G of Volume 1), the variational equation may be written as

$$\delta\Pi = -\int_\Omega \delta u_i \left[(F_{iJ}\,S_{IJ})_{,I} + \delta_{iI}\,\rho_0\,b_I^{(m)} \right] \mathrm{d}V + \int_{\Gamma_t} \delta u_i \left[F_{iJ}\,S_{IJ}\,N_I - \delta_{iI}\,\bar{T}_I \right] \mathrm{d}S = 0 \tag{10.46}$$

giving the Euler equations of (static) equilibrium in the reference configuration as

$$(F_{iJ}\,S_{IJ})_{,I} + \delta_{iI}\,\rho_0\,b_I^{(m)} = P_{iI,I} + \rho_0\,b_i^{(m)} = 0 \tag{10.47}$$

and the reference configuration traction boundary condition

$$S_{IJ}\,F_{iJ}\,N_I - \delta_{iI}\,\bar{T}_I = P_{iI}\,N_I - \delta_{iI}\,\bar{T}_I = 0 \tag{10.48}$$

The variational equation (10.42) is identical to a Galerkin method and, thus, can be used directly to formulate problems with constitutive models different from the hyperelastic behaviour above. In addition, direct use of the variational term (10.43) permits non-conservative loading forms, such as follower forces or pressures, to be introduced. We shall address such extensions in Section 10.6.

Matrix form

At this point we can again introduce matrix notation to represent the stress, strain, and variation of strain. For three-dimensional problems we define the matrix for the second Piola–Kirchhoff stress as

$$\mathbf{S} = [\, S_{11}, \quad S_{22}, \quad S_{33}, \quad S_{12}, \quad S_{23}, \quad S_{31} \,]^{\mathrm{T}} \tag{10.49}$$

and the Green strain as

$$\mathbf{E} = [\, E_{11}, \quad E_{22}, \quad E_{33}, \quad 2E_{12}, \quad 2E_{23}, \quad 2E_{31} \,]^{\mathrm{T}} \tag{10.50}$$

where, similar to the small strain problem, the shearing components are doubled to permit the reduction to six components. The variation of the Green strain is similarly given by

$$\delta \mathbf{E} = [\,\delta E_{11}, \quad \delta E_{22}, \quad \delta E_{33}, \quad 2\delta E_{12}, \quad 2\delta E_{23}, \quad 2\delta E_{31}\,]^{\mathrm{T}} \tag{10.51}$$

which permits Eq. (10.44) to be written as the matrix relation

$$\delta E_{IJ}\, S_{IJ} = \delta \mathbf{E}^{\mathrm{T}} \mathbf{S} \tag{10.52}$$

The variation of the Green strain is deduced from Eqs (10.13), (10.14) and (10.45) and written as

$$\delta E_{IJ} = \frac{1}{2}\left(\frac{\partial \delta u_i}{\partial X_I}\, F_{iJ} + \frac{\partial \delta u_i}{\partial X_J}\, F_{iI}\right) = \frac{1}{2}\left(\delta u_{i,I}\, F_{iJ} + \delta u_{i,J}\, F_{iI}\right) \tag{10.53}$$

Substitution of Eq. (10.53) into Eq. (10.51) we obtain

$$\delta \mathbf{E} = \begin{Bmatrix} F_{i1}\,\delta u_{i,1} \\ F_{i2}\,\delta u_{i,2} \\ F_{i3}\,\delta u_{i,3} \\ F_{i1}\,\delta u_{i,2} + F_{i2}\,\delta u_{i,1} \\ F_{i2}\,\delta u_{i,3} + F_{i3}\,\delta u_{i,2} \\ F_{i3}\,\delta u_{i,1} + F_{i1}\,\delta u_{i,3} \end{Bmatrix} \tag{10.54}$$

as the matrix form of the variation of the Green strain.

Finite element approximation

Using the isoparametric form developed in Chapters 8 and 9 of Volume 1 we represent the reference configuration coordinates as

$$X_I = \sum_{\alpha} N_{\alpha}(\boldsymbol{\xi})\, \tilde{X}_I^{\alpha} \tag{10.55}$$

where $\boldsymbol{\xi}$ are the natural coordinates ξ, η in two dimensions and ξ, η, ζ in three dimensions, N_{α} are shape standard functions (see Chapters 8 and 9 of Volume 1), and Greek symbols are introduced to identify uniquely the finite element nodal values from other indices. Similarly, we can approximate the displacement field in each element by

$$u_i = \sum_{\alpha} N_{\alpha}(\boldsymbol{\xi})\, \tilde{u}_i^{\alpha} \tag{10.56}$$

The reference system derivatives are constructed in an identical manner to that described in Chapter 9 of Volume 1. Thus,

$$u_{i,I} = N_{\alpha,I}\, \tilde{u}_i^{\alpha} \tag{10.57}$$

where explicit writing of the sum is omitted and summation convention for α is again invoked. The derivatives of the shape functions can be established by using standard routines to which the X_I^{α} coordinates of nodes attached to each element are supplied.

The deformation gradient and Green strain may now be computed with use of Eqs (10.12) and (10.15), respectively. Finally, the variation of the Green strain is given in matrix form as

$$
\delta \mathbf{E} = \begin{bmatrix}
F_{11}\, N_{\alpha,1} & F_{21}\, N_{\alpha,1} & F_{31}\, N_{\alpha,1} \\
F_{12}\, N_{\alpha,2} & F_{22}\, N_{\alpha,2} & F_{32}\, N_{\alpha,2} \\
F_{13}\, N_{\alpha,3} & F_{23}\, N_{\alpha,3} & F_{33}\, N_{\alpha,3} \\
F_{11}\, N_{\alpha,2} + F_{12}\, N_{\alpha,1} & F_{21}\, N_{\alpha,2} + F_{22}\, N_{\alpha,1} & F_{31}\, N_{\alpha,2} + F_{32}\, N_{\alpha,1} \\
F_{12}\, N_{\alpha,3} + F_{13}\, N_{\alpha,2} & F_{22}\, N_{\alpha,3} + F_{23}\, N_{\alpha,2} & F_{32}\, N_{\alpha,3} + F_{33}\, N_{\alpha,2} \\
F_{13}\, N_{\alpha,1} + F_{11}\, N_{\alpha,3} & F_{23}\, N_{\alpha,1} + F_{21}\, N_{\alpha,3} & F_{33}\, N_{\alpha,1} + F_{31}\, N_{\alpha,3}
\end{bmatrix}
\begin{Bmatrix} \delta \tilde{u}_1^\alpha \\ \delta \tilde{u}_2^\alpha \\ \delta \tilde{u}_3^\alpha \end{Bmatrix}
$$

$$
= \hat{\mathbf{B}}_\alpha\, \delta \tilde{\mathbf{u}}^\alpha \tag{10.58}
$$

where $\hat{\mathbf{B}}_\alpha$ replaces the form previously defined for the small deformation problem as $\mathbf{B}_\alpha$. Expressing the deformation gradient in terms of displacements it is also possible to split this matrix into two parts as

$$
\hat{\mathbf{B}}_\alpha = \mathbf{B}_\alpha + \mathbf{B}_\alpha^{\mathrm{NL}} \tag{10.59}
$$

in which $\mathbf{B}_\alpha$ is identical to the small deformation strain-displacement matrix and the remaining non-linear part is given by

$$
\mathbf{B}_\alpha^{\mathrm{NL}} = \begin{bmatrix}
u_{1,1}\, N_{\alpha,1} & u_{2,1}\, N_{\alpha,1} & u_{3,1}\, N_{\alpha,1} \\
u_{1,2}\, N_{\alpha,2} & u_{2,2}\, N_{\alpha,2} & u_{3,2}\, N_{\alpha,2} \\
u_{1,3}\, N_{\alpha,3} & u_{2,3}\, N_{\alpha,3} & u_{3,3}\, N_{\alpha,3} \\
u_{1,1}\, N_{\alpha,2} + u_{1,2}\, N_{\alpha,1} & u_{2,1}\, N_{\alpha,2} + u_{2,2}\, N_{\alpha,1} & u_{3,1}\, N_{\alpha,2} + u_{3,2}\, N_{\alpha,1} \\
u_{1,2}\, N_{\alpha,3} + u_{1,3}\, N_{\alpha,2} & u_{2,2}\, N_{\alpha,3} + u_{2,3}\, N_{\alpha,2} & u_{3,2}\, N_{\alpha,3} + u_{3,3}\, N_{\alpha,2} \\
u_{1,3}\, N_{\alpha,1} + u_{1,1}\, N_{\alpha,3} & u_{2,3}\, N_{\alpha,1} + u_{2,1}\, N_{\alpha,3} & u_{3,3}\, N_{\alpha,1} + u_{3,1}\, N_{\alpha,3}
\end{bmatrix} \tag{10.60}
$$

It is immediately evident that $\mathbf{B}_\alpha^{\mathrm{NL}}$ is zero in the reference configuration and therefore that $\hat{\mathbf{B}}_\alpha \equiv \mathbf{B}_\alpha$. We note, however, that in general no advantage results from this split over the single term expression given in Eq. (10.58).

The variational equation may now be written for the finite element problem by substituting Eqs (10.49) and (10.58) into Eq. (10.42) to obtain

$$
\delta \Pi = (\delta \tilde{\mathbf{u}}_\alpha)^{\mathrm{T}} \left(\int_\Omega \hat{\mathbf{B}}_\alpha^{\mathrm{T}} \mathbf{S}\, \mathrm{d}V - \mathbf{f}_\alpha \right) = 0 \tag{10.61}
$$

where the external forces are determined from $\delta \Pi_{\mathrm{ext}}$ as

$$
\mathbf{f}_\alpha = \int_\Omega N_\alpha \rho_0\, \mathbf{b}^{(m)}\, \mathrm{d}V + \int_{\Gamma_t} N_\alpha\, \bar{\mathbf{T}}\, \mathrm{d}S \tag{10.62}
$$

with $\mathbf{b}^{(m)}$ and $\bar{\mathbf{T}}$ the matrix form of the body and traction force vectors, respectively.

Using the d'Alembert principle we can introduce inertial forces through the body force as

$$
\mathbf{b}^{(m)} \rightarrow \mathbf{b}^{(m)} - \dot{\mathbf{v}} = \mathbf{b}^{(m)} - \ddot{\mathbf{x}} \tag{10.63}
$$

where $\mathbf{v}$ is the material velocity vector defined in Eq. (10.26). This adds an inertial term $\mathbf{M}_{\alpha\beta}\dot{\mathbf{v}}_\beta$ to the variational equation where the mass matrix is given in the reference configuration by

$$\mathbf{M}_{\alpha\beta} = \int_\Omega N_\alpha \rho_0 N_\beta \, \mathrm{d}V \mathbf{I} \tag{10.64}$$

For the transient problem we can introduce a Newton–Raphson type solution and replace Eq. (1.24) by

$$\boldsymbol{\Psi}_1 = \mathbf{f} - \int_\Omega \hat{\mathbf{B}}^\mathrm{T} \mathbf{S} \, \mathrm{d}V - \mathbf{M}\dot{\mathbf{v}} = \mathbf{0} \tag{10.65}$$

Here we consider further the Newton–Raphson solution process for a steady-state problem in which the inertial term $\mathbf{M}\dot{\mathbf{v}}$ is omitted. Extension to transient applications follows directly from the presentation given in Chapter 1. Applying the linearization process defined in Eq. (2.9) to Eq. (10.65) [without the inertia force] we obtain the tangent term

$$\mathbf{K}_\mathrm{T} = \int_\Omega \hat{\mathbf{B}}^\mathrm{T} \hat{\mathbf{D}}_\mathrm{T} \hat{\mathbf{B}} \, \mathrm{d}V + \int_\Omega \frac{\partial \hat{\mathbf{B}}^\mathrm{T}}{\partial \tilde{\mathbf{u}}} \mathbf{S} \, \mathrm{d}V - \frac{\partial \mathbf{f}}{\partial \tilde{\mathbf{u}}} = \mathbf{K}_\mathrm{M} + \mathbf{K}_\mathrm{G} + \mathbf{K}_\mathrm{L} \tag{10.66}$$

where the first term is the material tangent, $\mathbf{K}_\mathrm{M}$, in which $\hat{\mathbf{D}}_\mathrm{T}$ is the matrix form of the tangent moduli obtained from the derivative of constitution given in indicial form as

$$2\frac{\partial S_{IJ}}{\partial C_{KL}} = 4\frac{\partial^2 W}{\partial C_{IJ} \partial C_{KL}} = \frac{\partial^2 W}{\partial E_{IJ} \partial E_{KL}} = D_{IJKL} \tag{10.67}$$

and transformed to a matrix $\hat{\mathbf{D}}_\mathrm{T}$ (see Chapter 1 and Appendix B, Volume 1).

The second term, $\mathbf{K}_\mathrm{G}$, defines a tangent term arising from the non-linear form of the strain–displacement equations and is often called the *geometric stiffness*. The derivation of this term is most easily constructed from the indicial form written as

$$\int_\Omega \frac{\partial \delta E_{IJ}}{\partial \tilde{u}_j^\beta} S_{IJ} \, \mathrm{d}V \, d\tilde{u}_j^\beta = \delta \tilde{u}_i^\alpha \int_\Omega N_{\alpha,I} \delta_{ij} N_{\beta,J} S_{IJ} \, \mathrm{d}V \, d\tilde{u}_j^\beta = \delta \tilde{u}_i^\alpha (K_{ij}^{\alpha\beta})_\mathrm{G} \, d\tilde{u}_j^\beta \tag{10.68}$$

Thus, the geometric part of the tangent matrix is given by

$$\mathbf{K}_\mathrm{G}^{\alpha\beta} = G_{\alpha\beta} \mathbf{I} \tag{10.69}$$

where

$$G_{\alpha\beta} = \int_\Omega N_{\alpha,I} S_{IJ} N_{\beta,J} \, \mathrm{d}V \tag{10.70}$$

The last term in Eq. (10.66) is the tangent relating to loading which changes with deformation (e.g. follower forces, etc.). We assume for the present that the derivative of the force term $\mathbf{f}$ is zero so that $\mathbf{K}_\mathrm{L}$ vanishes.

10.3.2 Current configuration formulation

The form of the equations related to the reference configuration presented in the previous section follows from straightforward application of the variational procedures and finite element approximation methods introduced previously in this volume and throughout Volume 1. However, the form of the resulting equations leads to much more complicated strain–displacement matrices, $\hat{\mathbf{B}}$, than previously encountered. To implement such a form it is thus necessary to reprogram completely all the element routines. We will now show that if the equations given above are transformed to the current configuration a much simpler process results.

The transformations to the current configuration are made in two steps. In the first step we replace reference configuration terms by quantities related to the current configuration (e.g. we use Cauchy or Kirchhoff stress). In the second step we convert integrals over the undeformed body to ones in the current configuration.*

To transform from quantities in the reference configuration to ones in the current configuration we use the chain rule for differentiation to write

$$\frac{\partial(\cdot)}{\partial X_I} = \frac{\partial(\cdot)}{\partial x_i}\frac{\partial x_i}{\partial X_I} = \frac{\partial(\cdot)}{\partial x_i} F_{iI} \tag{10.71}$$

Using this relationship Eq. (10.53) may be transformed to

$$\delta E_{IJ} = \tfrac{1}{2}\left(\delta u_{i,j} + \delta u_{j,i}\right) F_{iI} F_{jJ} = \delta\varepsilon_{ij} F_{iI} F_{jJ} \tag{10.72}$$

where we have noted that the variation term is identical to the variation of the small deformation strain–displacement relations by again using the notation†

$$\delta\varepsilon_{ij} = \tfrac{1}{2}\left(\delta u_{i,j} + \delta u_{j,i}\right) \tag{10.73}$$

Equation (10.44) may now be written as

$$\delta E_{IJ} S_{IJ} = \delta\varepsilon_{ij} F_{iI} F_{iJ} S_{IJ} = \delta\varepsilon_{ij}\tau_{ji} = \delta\varepsilon_{ij}\sigma_{ij} J \tag{10.74}$$

and Eq. (10.42) as

$$\delta\Pi = \int_\Omega \delta\varepsilon_{ij}\sigma_{ij} J\,\mathrm{d}V - \delta\Pi_{\text{ext}} = 0 \tag{10.75}$$

The second step is now performed easily by noting the transformation of the volume element given in Eq. (10.11) to obtain finally

$$\delta\Pi = \int_\omega \delta\varepsilon_{ij}\sigma_{ij}\,\mathrm{d}v - \delta\Pi_{\text{ext}} = 0 \tag{10.76}$$

where ω is the domain in the current configuration.

The external potential Π_{ext} given in Eq. (10.43) may also be transformed to the current configuration using Eqs (10.24) and (10.27) to obtain

$$\delta\Pi_{\text{ext}} = \int_\omega \delta u_i \rho b_i^{(m)}\,\mathrm{d}v + \int_{\gamma_t} \delta u_i \bar{t}_I\,\mathrm{d}s \tag{10.77}$$

* This latter step need not be done to obtain advantage of the current configuration form of the integrand.
† We note that in finite deformation there is no meaning to ε_{ij} itself; only its variation, increment, or rate can appear in expressions.

The computation of the tangent matrix can similarly be transformed to the current configuration. The first term given in Eq. (10.66) is deduced from

$$\int_\Omega \delta E_{IJ}\, D_{IJKL}\, dE_{KL}\, \mathrm{d}V = \int_\Omega \delta\varepsilon_{ij}\, F_{iI}\, F_{jJ}\, D_{IJKL}\, F_{kK}\, F_{lL}\, d\varepsilon_{kl}\, \mathrm{d}V$$

$$= \int_\omega \delta\varepsilon_{ij}\, d_{ijkl}\, d\varepsilon_{kl}\, \mathrm{d}v \qquad (10.78)$$

where

$$J\, d_{ijkl} = F_{iI}\, F_{jJ}\, F_{kK}\, F_{lL}\, D_{IJKL} \qquad (10.79)$$

defines the moduli in the current configuration in terms of quantities in the reference state.

Finally, the geometric stiffness term in Eq. (10.66) may be written in the current configuration by transforming Eq. (10.70) to obtain

$$G_{\alpha\beta} = \int_\Omega N_{\alpha,I}\, S_{IJ}\, N_{\beta,J}\, \mathrm{d}V = \int_\omega N_{\alpha,i}\, \sigma_{ij}\, N_{\beta,j}\, \mathrm{d}v \qquad (10.80)$$

Thus, we obtain a form for the finite deformation problem which is identical to that of the small deformation problem except that a geometric stiffness term is added and integrals and derivatives are to be computed in the deformed configuration. Of course, another difference is the form of the constitutive equations which need to be given in an admissible finite deformation form.

Finite element formulation

The current configurational form of the variational problem is easily implemented in a finite element solution process. To obtain the shape functions and their derivatives it is necessary first to obtain the deformed Cartesian coordinates x_i by using Eq. (10.4). After this step standard shape function routines can be used to compute the derivatives of shape functions, $\partial N_\alpha / \partial x_i$. The terms in the variational equation can then be expressed in a form which is identical to that of the small deformation problem. Accordingly, the stress term is written as

$$\int_\omega \delta\varepsilon_{ij}\, \sigma_{ij}\, \mathrm{d}v = \delta\tilde{\mathbf{u}}^{\mathrm{T}} \int_\omega \mathbf{B}^{\mathrm{T}}\, \boldsymbol{\sigma}\, \mathrm{d}v \qquad (10.81)$$

where $\mathbf{B}$ is identical to the form of the small deformation strain–displacement matrix, and Cauchy stress is transformed to matrix form as

$$\boldsymbol{\sigma} = [\sigma_{11},\quad \sigma_{22},\quad \sigma_{33},\quad \sigma_{12},\quad \sigma_{23},\quad \sigma_{31}]^{\mathrm{T}} \qquad (10.82)$$

and involves only six independent components.

The residual for the static problem of a Newton–Raphson solution process is now given by

$$\boldsymbol{\Psi}_1 = \mathbf{f} - \int_\omega \mathbf{B}^{\mathrm{T}}\, \boldsymbol{\sigma}\, \mathrm{d}v = \mathbf{0} \qquad (10.83)$$

The linearization step of the Newton–Raphson solution process is performed by computing the tangent stiffness in matrix form. Transforming Eq. (10.78) to matrix form using the relations defined in Chapter 1, the material tangent is given by

$$\mathbf{K}_{\mathrm{M}}^{\alpha\beta} = \int_{\omega} \mathbf{B}_{\alpha}^{\mathrm{T}} \mathbf{D}_{\mathrm{T}} \mathbf{B}_{\beta} \, \mathrm{d}v \tag{10.84}$$

where now the material moduli $\mathbf{D}_{\mathrm{T}}$ are deduced by transforming the moduli in the current configuration, d_{ijkl}, to matrix form. The form for $G_{\alpha\beta}$ in Eq. (10.80) may be substituted into Eq. (10.69) to obtain the geometric tangent stiffness matrix. Thus, the total tangent matrix for the steady-state problem in the current configuration is given by

$$\mathbf{K}_{\mathrm{T}}^{\alpha\beta} = \int_{\omega} \mathbf{B}_{\alpha}^{\mathrm{T}} \mathbf{D}_{\mathrm{T}} \mathbf{B}_{\beta} \, \mathrm{d}v + G_{\alpha\beta} \mathbf{I} \tag{10.85}$$

and a Newton–Raphson iterate consists in solving

$$\mathbf{K}_{\mathrm{T}} \, \mathrm{d}\tilde{\mathbf{u}} = \mathbf{f} - \int_{\omega} \mathbf{B}^{\mathrm{T}} \boldsymbol{\sigma} \, \mathrm{d}v \tag{10.86}$$

where the external force is obtained from Eq. (10.77) as

$$\mathbf{f}_{\alpha} = \int_{\omega} N_{\alpha} \rho \mathbf{b}^{(m)} \, \mathrm{d}v + \int_{\gamma_t} N_{\alpha} \bar{\mathbf{t}} \, \mathrm{d}s \tag{10.87}$$

We can also transform the inertial force to a current configuration form by substituting Eqs (10.11) and (10.27) into Eq. (10.64) to obtain

$$\mathbf{M}_{\alpha\beta} = \int_{\Omega} N_{\alpha} \rho_0 N_{\beta} \, \mathrm{d}V \mathbf{I} = \int_{\omega} N_{\alpha} \rho N_{\beta} \, \mathrm{d}v \mathbf{I} \tag{10.88}$$

and thus, for the transient problem, the residual becomes

$$\boldsymbol{\Psi}_1 = \mathbf{f} - \int_{\omega} \mathbf{B}^{\mathrm{T}} \boldsymbol{\sigma} \, \mathrm{d}v - \mathbf{M}\dot{\mathbf{v}} = \mathbf{0} \tag{10.89}$$

Linearization of this term is identical to the small deformation problem and is not given here.

The development of *displacement-based finite element models* for two- and three-dimensional problems may be performed easily merely by adding a few modifications to a standard linear form. These modifications include the following steps.

1. Use current configuration coordinates x_i to compute shape functions and their derivatives. These are computed at nodes by adding current values of displacements $\tilde{u}_i^{\alpha}$ to reference configuration nodal coordinates $\tilde{X}_I^{\alpha}$.
2. Add a geometric stiffness matrix to the usual stiffness matrix as indicated in Eq. (10.85).
3. Use the appropriate material constitution for a finite deformation model.
4. Solve the problem by means of an appropriate strategy for non-linear problems.

It should be noted that the presence of the geometric stiffness and non-linear material behaviour may result in a tangent matrix which is no longer always positive definite (indeed, we shall discuss stability problems in the next chapter and this is a class of problems for which the tangent matrix can become singular as a result of the geometric stiffness term alone). Furthermore, use of displacement-based elements in finite deformation can lead to locking if the material has internal constraints, such as in nearly incompressible behaviour. It is then necessary again to resort to a mixed formulation to avoid such locking. The advantage of a properly constructed mixed form is that it may be used with equal accuracy for both the nearly incompressible problem as well as any compressible problem (see Chapter 12 of Volume 1).

10.4 A three-field mixed finite deformation formulation

A three-field, mixed variational form for the finite deformation hyperelastic problem is given by

$$\Pi = \int_{\Omega} [W(\bar{C}_{IJ}) + p(J - \theta)] \, \mathrm{d}V - \Pi_{\text{ext}} \tag{10.90}$$

where p is a mixed pressure in the current (deformed) configuration, J is the determinant of the deformation gradient F_{iI}, θ is the volume in the current configuration for a unit volume in the reference state, W is the stored energy function expressed in terms of a (mixed) right Cauchy–Green deformation tensor $\bar{C}_{IJ}$, and Π_{ext} is the functional for the body loading and boundary terms given in Eq. (10.41). This form of the variational problem has been used for problems formulated in principal stretches.[6] Here we use the form without referring to the specific structure of the stored energy function. In particular we wish to admit constitutive forms in which the volumetric and deviatoric parts are not split as in reference 6.

The (mixed) right Green deformation tensor is expressed as

$$\bar{C}_{IJ} = \bar{F}_{iI} \bar{F}_{iJ} \tag{10.91}$$

where

$$\bar{F}_{iI} = F^v_{ij} F^d_{jI} = (\theta^{1/3} \delta_{ij})(J^{-1/3} F_{jI}) = \left(\frac{\theta}{J}\right)^{1/3} F_{iI} \tag{10.92}$$

where F^v_{ij} is a volumetric and F^d_{jI} a deviatoric part. We also note that $\det F^d_{jI} = 1$ as required for a deviatoric (constant volume) state.

The variation of Eq. (10.90) is given by

$$\delta\Pi = \int_{\Omega} \left[\tfrac{1}{2} \, \delta\bar{C}_{IJ} \bar{S}_{IJ} + \delta p (J - \theta) + (\delta J - \delta\theta) p \right] \mathrm{d}V - \delta\Pi_{\text{ext}} \tag{10.93}$$

where a second Piola–Kirchhoff stress based on the mixed deformation tensor is defined as

$$\bar{S}_{IJ} = 2 \frac{\partial W}{\partial \bar{C}_{IJ}} = \frac{\partial W}{\partial \bar{E}_{IJ}} \quad \text{where} \quad \bar{E}_{IJ} = \tfrac{1}{2} (\bar{C}_{IJ} - \delta_{IJ}) \tag{10.94}$$

Using Eq. (10.91) the variation of the mixed deformation tensor is given by

$$\delta\bar{C}_{IJ} = \delta\bar{F}_{iI} \bar{F}_{iJ} + \delta\bar{F}_{iJ} \bar{F}_{iI} \tag{10.95}$$

and thus noting that[3,4]

$$\delta J = J F^{-1}_{jJ} \delta F_{jJ}$$

the first term of the integrand in Eq. (10.93) formally may be expanded as

$$\delta \bar{C}_{IJ} \bar{S}_{IJ} = \delta \bar{F}_{iI} \bar{F}_{iJ} \bar{S}_{IJ}$$

$$= \frac{1}{3} \frac{\delta\theta}{\theta} \bar{F}_{iI} \bar{F}_{iJ} \bar{S}_{IJ} + \left(\frac{\theta}{J}\right)^{1/3} \left[\delta F_{iI} - \frac{1}{3} \delta F_{jJ} F_{jJ}^{-1} F_{iI}\right] \bar{F}_{iJ} \bar{S}_{IJ} \tag{10.96}$$

This expression again may be greatly simplified by defining current configuration Kirchhoff and Cauchy stresses based on the mixed deformation gradient as

$$\bar{\tau}_{ij} = \bar{F}_{iI} \bar{S}_{IJ} \bar{F}_{jJ} = \theta \bar{\sigma}_{ij} \tag{10.97}$$

Also, we note from Eq. (10.71) that

$$\delta F_{jJ} F_{jJ}^{-1} = \delta u_{j,k} F_{kJ} F_{jJ}^{-1} = \delta u_{j,k} \delta_{kj} = \delta u_{j,j} \tag{10.98}$$

is the divergence of the variation in displacement. Thus, Eq. (10.96) simplifies to

$$\delta \bar{C}_{IJ} \bar{S}_{IJ} = \frac{1}{3} \left(\frac{\delta\theta}{\theta} - \delta u_{j,j}\right) \bar{\tau}_{ii} + \delta u_{i,j} \bar{\tau}_{ij} = \frac{1}{3} \frac{\delta\theta}{\theta} \bar{\tau}_{kk} + \frac{\partial \delta u_i}{\partial x_j} \left(\bar{\tau}_{ij} - \frac{1}{3} \delta_{ij} \bar{\tau}_{kk}\right) \tag{10.99}$$

Substituting relations deduced above into Eq. (10.93) and noting symmetry of the Kirchhoff stress, a formulation in terms of quantities related to the deformed position may be written as

$$\delta \Pi = \int_\Omega \delta\varepsilon_{ij} \left[\bar{\sigma}_{ij} + \delta_{ij} \left(\frac{J}{\theta} p - \bar{p}\right)\right] \theta \, \mathrm{d}V + \int_\Omega \delta\theta (\bar{p} - p) \, \mathrm{d}V$$

$$+ \int_\Omega \delta p (J - \theta) \, \mathrm{d}V - \delta \Pi_{\text{ext}} = 0 \tag{10.100}$$

where $\bar{p} = \bar{\sigma}_{ii}/3$ defines a mean stress based on the Cauchy stress deduced according to Eq. (10.97). This variational equation may be transformed to integrals over the current configuration by replacing $\mathrm{d}V$ by $\mathrm{d}v/J$; however, this step is not a necessary transformation to make the relations valid.

Finite elements: matrix notation

The finite element approximation of the mixed variational form is again given using deformation measures and stresses related to the current configuration. The development is very similar to that presented in Chapter 1 for the small deformation case.

The reference coordinate and displacement fields are approximated by isoparametric interpolations as indicated in Eqs (10.55) and (10.56), respectively. These are used to compute the deformation gradient by means of Eqs (10.12) and (10.57). The pressure and volume are interpolated in a manner which is identical to the small deformation case as

$$p = \mathbf{N}_p \tilde{\mathbf{p}} \qquad \text{and} \qquad \theta = \mathbf{N}_\theta \tilde{\boldsymbol{\theta}}$$

and for quadrilateral and brick elements are taken to be discontinuous between elements. A similar scheme with p being C_0 continuous can be used to develop triangular and tetrahedral elements.[7]

Using the above approximation, Eq. (10.100) may be expressed in matrix form as

$$\delta\Pi = \delta\tilde{\mathbf{u}}^{\mathrm{T}} \int_{\Omega} \mathbf{B}^{\mathrm{T}} \breve{\boldsymbol{\sigma}} \theta \, \mathrm{d}V + \delta\tilde{\mathbf{p}}^{\mathrm{T}} \int_{\Omega} \mathbf{N}_p^{\mathrm{T}} (J - \theta) \, \mathrm{d}V$$
$$+ \delta\tilde{\theta}^{\mathrm{T}} \int_{\Omega} \mathbf{N}_\theta^{\mathrm{T}} (\bar{p} - p) \, \mathrm{d}V + \delta\Pi_{\mathrm{ext}} \qquad (10.101)$$

In this form of the finite deformation problem $\mathbf{B}$ again is identical to the small deformation strain–displacement matrix and a modified mixed stress is defined as

$$\breve{\boldsymbol{\sigma}} = \bar{\boldsymbol{\sigma}} + (\breve{p} - \bar{p})\mathbf{m} \qquad \text{where} \qquad \breve{p} = \frac{J}{\theta} p \qquad (10.102)$$

Using these interpolations Eq. (10.101) gives the equations

$$\begin{aligned} \mathbf{P} + \mathbf{M}\dot{\tilde{\mathbf{v}}} &= \mathbf{f} \\ \mathbf{P}_p - \mathbf{K}_{\theta p}\tilde{\mathbf{p}} &= \mathbf{0} \\ -\mathbf{K}_{p\theta}\tilde{\boldsymbol{\theta}} + \mathbf{E}_J &= \mathbf{0} \end{aligned} \qquad (10.103)$$

where the arrays are given as

$$\begin{aligned} \mathbf{P} &= \int_{\Omega} \mathbf{B}^{\mathrm{T}} \breve{\boldsymbol{\sigma}} \theta \, \mathrm{d}V & \mathbf{P}_p &= \tfrac{1}{3} \int_{\Omega} \mathbf{N}_\theta^{\mathrm{T}} \breve{\boldsymbol{\sigma}} \theta \, \mathrm{d}V \\ \mathbf{K}_{\theta p} &= \int_{\Omega} \mathbf{N}_\theta^{\mathrm{T}} \mathbf{N}_p \, \mathrm{d}V = \mathbf{K}_{p\theta} & \mathbf{E}_J &= \int_{\Omega} \mathbf{N}_p^{\mathrm{T}} J \, \mathrm{d}V \end{aligned} \qquad (10.104)$$

and force **f** and mass **M** are identical to the terms appearing in the displacement model presented previously.

We can observe that the mixed model reduces to the displacement form if $\theta = J$ and $p = \bar{p}$ at every point in the element. This would occur if our approximations for θ and p contained all the terms appearing in results computed from deformations and, thus, again establishes the principle of limitation.[8] Moreover, if this occurred, any locking tendency in the displacement form would again occur in the mixed approach also.

To obtain a formulation free of locking it is again necessary to select approximations for pressure and volume which satisfy the mixed patch test count conditions as described in Chapters 11 and 12 of Volume 1. Here, to approximate p and θ in each element we assume that $\mathbf{N}_\theta = \mathbf{N}_p$ and for four-noded quadrilateral and eight-noded brick elements of linear order use constant (unit) interpolation. In nine-noded quadrilateral and 27-noded brick elements of quadratic order we assume linear interpolation. Linear interpolation in ξ, η, ζ or X_1, X_2, X_3 can be used; however, x_1, x_2, x_3 should not be used since then the interpolation becomes non-linear (x_i depend on u_i) and the solution complexity is greatly increased from that indicated above.

The second and third expressions in Eqs (10.103) are linear in $\tilde{\mathbf{p}}$ and $\tilde{\boldsymbol{\theta}}$, respectively, and also are completely formed in a single element. Moreover, the coefficient matrix $\mathbf{K}_{p\theta} = \mathbf{K}_{\theta p}$ is symmetric positive definite when $\mathbf{N}_\theta = \mathbf{N}_p$. Thus, a partial solution can be achieved in each element as

$$\begin{aligned} \tilde{\mathbf{p}} &= \mathbf{K}_{\theta p}^{-1} \mathbf{P}_p \\ \tilde{\boldsymbol{\theta}} &= \mathbf{K}_{p\theta}^{-1} \mathbf{E}_J \end{aligned} \qquad (10.105)$$

An explicit method in time (see Chapter 18, Volume 1) may be employed to solve the momentum equation: as was indeed used to solve examples shown at the end of Chapter 1. However, here we only consider further an implicit scheme which is applicable to either transient or static problems (see Chapters 1 and 2). A Newton–Raphson scheme may be employed to solve Eq. (10.101). To construct the tangent matrix $\mathbf{K}_T$ it is necessary to linearize Eq. (10.93). In indicial form, the Newton–Raphson linearization may be assembled as

$$\begin{aligned} d(\delta\Pi) = & \int_\Omega \left[\delta\bar{C}_{IJ}\,\bar{D}_{IJKL}\,d\bar{C}_{KL} + \tfrac{1}{2}\,d(\delta\bar{C}_{IJ})\,\bar{S}_{IJ}\right] \mathrm{d}V + \int_\Omega p\,d(\delta J)\,\mathrm{d}V \\ & + \int_\Omega \delta p\,(dJ - d\theta)\,\mathrm{d}V + \int_\Omega dp\,(\delta J - \delta\theta)\,\mathrm{d}V + d(\delta\Pi_{\text{ext}}) \end{aligned} \tag{10.106}$$

where $d\bar{C}_{KL}$, dp, etc., denote incremental quantities and material tangent moduli in the reference configuration are denoted by

$$2\frac{\partial \bar{S}_{IJ}}{\partial \bar{C}_{KL}} = \bar{D}_{IJKL} \tag{10.107}$$

The above integrals may also be expressed in quantities terms of current configuration terms in an identical manner as for the displacement model presented in Sec. 10.3.2. In this case the reference configuration moduli are transformed to the current configuration using

$$\bar{d}_{ijkl} = \frac{1}{\theta}\bar{F}_{iI}\,\bar{F}_{jJ}\,\bar{F}_{kK}\,\bar{F}_{lL}\,\bar{D}_{IJKL} \tag{10.108}$$

Using standard transformations from indicial to matrix form the moduli for the current configuration may be written in matrix form as $\bar{\mathbf{D}}_T$.

We can now write Eq. (10.106) in matrix form and obtain the set of equations which determine the parameters $d\tilde{\mathbf{u}}$, $d\tilde{\boldsymbol{\theta}}$ and $d\tilde{\mathbf{p}}$ as

$$\begin{bmatrix} \mathbf{K}_{uu} & \mathbf{K}_{u\theta} & \mathbf{K}_{up} \\ \mathbf{K}_{\theta u} & \mathbf{K}_{\theta\theta} & -\mathbf{K}_{\theta p} \\ \mathbf{K}_{pu} & -\mathbf{K}_{p\theta} & \mathbf{0} \end{bmatrix} \begin{Bmatrix} d\tilde{\mathbf{u}} \\ d\tilde{\boldsymbol{\theta}} \\ d\tilde{\mathbf{p}} \end{Bmatrix} = \begin{Bmatrix} \mathbf{f} - \mathbf{P} \\ \mathbf{0} \\ \mathbf{0} \end{Bmatrix} \tag{10.109}$$

where

$$\begin{aligned} \mathbf{K}_{uu} &= \int_\Omega \mathbf{B}^{\mathrm{T}}\bar{\mathbf{D}}_{11}\mathbf{B}\theta\,\mathrm{d}V + \mathbf{K}_G, \qquad & \mathbf{K}_{u\theta} &= \int_\Omega \mathbf{B}^{\mathrm{T}}\bar{\mathbf{D}}_{12}\mathbf{N}_\theta\,\mathrm{d}V = \mathbf{K}_{\theta u}^{\mathrm{T}} \\ \mathbf{K}_{up} &= \int_\Omega \mathbf{B}^{\mathrm{T}}\mathbf{m}\mathbf{N}_p J\,\mathrm{d}V = \mathbf{K}_{pu}^{\mathrm{T}}, \qquad & \mathbf{K}_{\theta\theta} &= \int_\Omega \mathbf{N}_\theta^{\mathrm{T}}\bar{\mathbf{D}}_{22}\mathbf{N}_\theta\,\frac{1}{\theta}\,\mathrm{d}V \end{aligned} \tag{10.110}$$

in which

$$\begin{aligned} \bar{\mathbf{D}}_{11} &= \mathbf{I}_d\bar{\mathbf{D}}_{\mathrm{T}}\mathbf{I}_d - \tfrac{2}{3}\left(\mathbf{m}\bar{\boldsymbol{\sigma}}_d^{\mathrm{T}} + \bar{\boldsymbol{\sigma}}_d\mathbf{m}^{\mathrm{T}}\right) + 2(\bar{p} - \tilde{p})\mathbf{I}_0 - \left(\tfrac{2}{3}\bar{p} - \tilde{p}\right)\mathbf{m}\mathbf{m}^{\mathrm{T}} \\ \bar{\mathbf{D}}_{12} &= \tfrac{1}{3}\,\mathbf{I}_d\bar{\mathbf{D}}_{\mathrm{T}}\,\mathbf{m} + \tfrac{2}{3}\,\bar{\boldsymbol{\sigma}}_d = \bar{\mathbf{D}}_{21}^{\mathrm{T}} \\ \bar{\mathbf{D}}_{22} &= \tfrac{1}{9}\,\mathbf{m}^{\mathrm{T}}\bar{\mathbf{D}}_{\mathrm{T}}\,\mathbf{m} - \tfrac{1}{3}\,\bar{p} \end{aligned} \tag{10.111}$$

where $\mathbf{I}_0$ is as defined in Eq. (1.37). Note also that the right-hand side is zero in the second and third rows of Eq. (10.109) since the solution for pressure and volume parameters was determined exactly using Eq. (10.105).

The geometric tangent term is given by

$$\mathbf{K}_G^{\alpha\beta} = \bar{G}_{\alpha\beta}\mathbf{I} \quad \text{where} \quad \bar{G}_{\alpha\beta} = \int_\Omega N_{\alpha,i}\,\bar{\sigma}_{ij}\,N_{\beta,j}\,\mathrm{d}V \tag{10.112}$$

A solution to Eq. (10.109) may be formed by solving the third and second rows as

$$\begin{aligned} d\tilde{\boldsymbol{\theta}} &= \mathbf{K}_{\theta p}^{-1}\,\mathbf{K}_{pu}\,d\tilde{\mathbf{u}} \\ d\tilde{\mathbf{p}} &= \mathbf{K}_{\theta p}^{-1}\,\mathbf{K}_{\theta u}\,d\tilde{\mathbf{u}} + \mathbf{K}_{\theta p}^{-1}\,\mathbf{K}_{\theta\theta}\,d\tilde{\boldsymbol{\theta}} \\ &= \left(\mathbf{K}_{\theta p}^{-1}\,\mathbf{K}_{\theta u} + \mathbf{K}_{\theta p}^{-1}\,\mathbf{K}_{\theta\theta}\,\mathbf{K}_{p\theta}^{-1}\,\mathbf{K}_{pu}\right) d\tilde{\mathbf{u}} \end{aligned} \tag{10.113}$$

and substituting the result into the first row to obtain

$$\mathbf{K}_{\mathrm{T}}\,d\tilde{\mathbf{u}} = [\mathbf{K}_{uu} + \mathbf{K}_{u\theta}\mathbf{K}_{p\theta}^{-1}\,\mathbf{K}_{pu} + \mathbf{K}_{up}\,\mathbf{K}_{\theta p}^{-1}\,\mathbf{K}_{\theta u} + \mathbf{K}_{up}\,\mathbf{K}_{\theta p}^{-1}\,\mathbf{K}_{\theta\theta}\,\mathbf{K}_{p\theta}^{-1}\,\mathbf{K}_{pu}]\,d\tilde{\mathbf{u}} = \mathbf{f} - \mathbf{P} \tag{10.114}$$

This result is obtained by inverting only the symmetric positive definite matrix $\mathbf{K}_{p\theta}$, which we also note is independent of any specific constitutive model.

10.5 A mixed–enhanced finite deformation formulation

An alternative method to that just discussed is the fully mixed method in which strain approximations are *enhanced*. The key idea of the mixed–enhanced formulation is the parameterization of the deformation gradient in terms of a mixed and an enhanced deformation gradient from which a consistent formulation is derived. This methodology allows for a formulation which has standard-order quadrature and variationally recoverable stresses, hence circumventing difficulties which arise in other *enhanced strain* methods.[9–14]

There is no need to separate any deformation gradient terms into deviatoric and mean parts as was necessary for the mixed approach discussed in the previous section. The mixed–enhanced formulation discussed here uses a three-field variational form for finite deformation hyperelasticity expressed as

$$\Pi = \int_\Omega [W(\hat{F}_{iI}) + \hat{P}_{iI}(F_{iI} - \hat{F}_{iI})]\,\mathrm{d}V - \Pi_{\mathrm{ext}} \tag{10.115}$$

where F_{iI} is the deformation gradient, $\tilde{F}_{iI}$ is the mixed deformation gradient, $\hat{P}_{iI}$ is the mixed first Piola–Kirchhoff stress, W is an objective stored energy function in terms of $\hat{F}_{iI}$, and Π_{ext} is the loading term given by Eq. (10.41).

The stationary point of Π is obtained by setting to zero the first variation of Eq. (10.115) with respect to the three independent fields. Accordingly,

$$\delta\Pi = \int_\Omega \left[\delta F_{iI}\hat{P}_{iI} + \delta\hat{F}_{iI}\left(\frac{\partial W}{\partial \hat{F}_{iI}} - \hat{P}_{iI}\right) + \delta\hat{P}_{iI}\left(F_{iI} - \hat{F}_{iI}\right)\right]\mathrm{d}V - \delta\Pi_{\mathrm{ext}} = 0 \tag{10.116}$$

where $\hat{F}_{iI}$ and $\hat{P}_{iI}$ are *mixed* variables to be approximated directly. The reader will note that we now use the deformation gradient directly instead of the usual C_{IJ},

E_{IJ}, or b_{ij} symmetric forms. We often will use constitutive models which are expressed in these symmetric quantities; however, we note that they are also implicitly functions of the deformation gradient through the definitions given in Sec. 10.2. Once again, at this point we may substitute a first Piola–Kirchhoff stress from any constitutive model in place of the derivative of the stored energy function $\partial W / \partial \hat{F}_{iI}$ in Eq. (10.116). Thus, the present form can be used in a general context.

Finite element approximations to the mixed deformation gradient and first Piola–Kirchhoff stress are constructed directly in terms of local coordinates of the parent element using standard tensor transformation concepts. Accordingly, we take

$$\hat{F}_{iI} = \bar{F}_{iA} \bar{J}_{\alpha A} \bar{J}_{\beta I} \mathcal{F}_{\alpha\beta}(\boldsymbol{\xi}) \tag{10.117}$$

and

$$\hat{P}_{iI} = \bar{F}_{iA}^{-1} \bar{J}_{\alpha A}^{-1} \bar{J}_{\beta I}^{-1} \mathcal{P}_{\alpha\beta}(\boldsymbol{\xi}) \tag{10.118}$$

where $\boldsymbol{\xi}$ denotes the natural coordinates ξ, η, ζ, the Greek subscripts are now associated with the natural coordinates (i.e. they are not here the finite element node numbers), and $\mathcal{P}_{\alpha\beta}$ and $\mathcal{F}_{\alpha\beta}$ are the first Piola–Kirchhoff stress and deformation gradient approximations in the isoparametric coordinate space, respectively.* The arrays $\bar{J}_{\alpha A}$ and $\bar{F}_{iI}$ used above are average quantities over the element volume, Ω_e. The average quantity $\bar{J}_{\alpha A}$ is defined as

$$\bar{J}_{\alpha A} = \frac{1}{\Omega_e} \int_{\Omega_e} J_{\alpha A} \, \mathrm{d}V \qquad \text{and} \qquad J_{\alpha A} = \frac{\partial X_A}{\partial \xi_a} \tag{10.119}$$

where $J_{\alpha A}$ is the standard Jacobian matrix as defined in Eq. (9.10) of Volume 1 (but now written for the reference coordinates), and $\bar{F}_{iI}$ is defined as

$$\bar{F}_{iI} = \frac{1}{\Omega_e} \int_{\Omega_e} F_{iI} \, \mathrm{d}V \tag{10.120}$$

The above form of approximation will ensure direct inclusion of constant states as well as minimize the order of quadrature needed to evaluate the finite element arrays and eliminate some sensitivity associated with initially distorted elements.

The form given in Eqs (10.117) and (10.118) are constructed so that the energy term of the physical and isoparametric pairs are equal. Accordingly, we observe that

$$\mathcal{P}_{\alpha\beta} \mathcal{F}_{\alpha\beta} = \hat{P}_{iI} \hat{F}_{iI} \tag{10.121}$$

This greatly simplifies the integrations needed to construct the terms in Eq. (10.116).

To construct the approximations we note that the tensor transformations for the mixed deformation gradient may be written in matrix form as

$$\hat{\mathbf{F}} = \mathbf{A} \boldsymbol{\mathcal{F}} \tag{10.122}$$

and

$$\hat{\mathbf{P}} = \mathbf{A}^{-1} \boldsymbol{\mathcal{P}} \tag{10.123}$$

* Note the resulting transformed arrays are objective under a superposed rigid body motion.[4]

Table 10.1 Matrix–tensor transformation for the nine-component form

Row or column	1	2	3	4	5	6	7	8	9
i or α	1	2	3	1	2	3	2	3	1
I or β	1	2	3	2	3	1	1	2	3

where **A** is a transformation to matrix form of the fourth-rank tensor given as

$$A_{iI\alpha\beta} = \bar{F}_{iA} \bar{J}_{\alpha A} \bar{J}_{\beta I} \tag{10.124}$$

The ordering for the matrix–tensor transformation for all the variables is described in Table 10.1.

The approximations for the mixed deformation gradient may now be written as

$$\hat{\mathbf{F}} = \tilde{\boldsymbol{\gamma}}^0 + \frac{1}{j} \mathbf{A} [\mathbf{E}_1(\boldsymbol{\xi}) \tilde{\boldsymbol{\gamma}} + \mathbf{E}_2(\boldsymbol{\xi}) \tilde{\boldsymbol{\alpha}}] \tag{10.125}$$

and

$$\hat{\mathbf{P}} = \tilde{\boldsymbol{\beta}}^0 + \mathbf{A}^{-1} \left[\mathbf{E}_1(\boldsymbol{\xi}) \tilde{\boldsymbol{\beta}}\right] \tag{10.126}$$

where $j = \det J_{\alpha A}$ and $\mathbf{E}_1$, $\mathbf{E}_2$ define the functions to be selected in terms of natural coordinates. The functions suggested in reference 15 are given in Table 10.2. The terms $\boldsymbol{\beta}^0$ and $\boldsymbol{\gamma}^0$ ensure that constant stress and strain are available in the element.

The above construction is similar to that used in Sec. 11.4.4 of Volume 1 to construct the Pian–Sumihara plane elastic element[16] and also in Sec. 5.6 to construct the shear and bubble interpolation for the thick-plate element Q4S2B2.

The enhanced parameters $\boldsymbol{\alpha}$ are added to the normal strains in Table 10.2 such that the resulting strain components are complete polynomials in natural coordinates. This is done to provide the necessary equations to enforce an incompressibility constraint without loss of rank in the resulting finite element arrays. In addition, the enhanced parameters improve coarse mesh accuracy in bending dominated regimes.

Finite elements: matrix notation

By isolating the equations associated with the first variation of the first Piola–Kirchhoff stress tensor $\delta\hat{\mathbf{P}}$ in Eq. (10.116) some of the element parameters of the

Table 10.2 Three-dimensional interpolations

α	β	$\mathbf{E}_1 \boldsymbol{\gamma}$	$\mathbf{E}_2 \boldsymbol{\alpha}$
1	1	$\xi_2\gamma_1 + \xi_3\gamma_2 + \xi_2\xi_3\gamma_3$	$\xi_1\alpha_1 + \xi_1\xi_2\alpha_2 + \xi_1\xi_3\alpha_3$
2	2	$\xi_1\gamma_4 + \xi_3\gamma_5 + \xi_1\xi_3\gamma_6$	$\xi_2\alpha_4 + \xi_2\xi_3\alpha_5 + \xi_1\xi_2\alpha_6$
3	3	$\xi_1\gamma_7 + \xi_2\gamma_8 + \xi_1\xi_2\gamma_9$	$\xi_3\alpha_7 + \xi_2\xi_3\alpha_8 + \xi_1\xi_3\alpha_9$
1	2	$\xi_3\gamma_{10}$	0
2	3	$\xi_1\gamma_{12}$	0
3	1	$\xi_2\gamma_{14}$	0
2	1	$\xi_3\gamma_{11}$	0
3	2	$\xi_1\gamma_{13}$	0
1	3	$\xi_2\gamma_{15}$	0

mixed–enhanced deformation gradient $\hat{\mathbf{F}}$ may be obtained as

$$\gamma^0 = \frac{1}{\Omega_e} \int_{\Omega_e} \mathbf{F} \, \mathrm{d}V = \bar{\mathbf{F}} \tag{10.127}$$

and

$$\gamma = \left(\int_{\square} \mathbf{E}_1^{\mathrm{T}} \mathbf{E}_1 \, \mathrm{d}\square \right)^{-1} \int_{\Omega_e} \mathbf{E}_1^{\mathrm{T}} \mathbf{A}^{-1} (\mathbf{F} - \bar{\mathbf{F}}) \, \mathrm{d}V \tag{10.128}$$

where the box denotes integration over the element region defined by the isoparametric coordinates ξ_I. We note that the construction for $\mathbf{E}_1$ and $\mathbf{E}_2$ are such that integrals have the property

$$\int_{\square} \mathbf{E}_1 \, \mathrm{d}\square = \int_{\square} \mathbf{E}_2 \, \mathrm{d}\square = \int_{\square} \mathbf{E}_1^{\mathrm{T}} \mathbf{E}_2 \, \mathrm{d}\square \equiv 0$$

This greatly simplifies the construction of the partial solution given above.

Use of the above definitions for $\hat{\mathbf{F}}$ and $\hat{\mathbf{P}}$ also makes the second term in the integrand of Eq. (10.115) zero, hence the modified functional $\hat{\Pi}$ is expressed as

$$\hat{\Pi} = \int_{\Omega} W(\tilde{F}_{iI}) \, \mathrm{d}V + \Pi_{\mathrm{ext}} \tag{10.129}$$

The stationary condition of $\hat{\Pi}$ yields a reduced set of nonlinear equations, in terms of the nodal displacements, $\tilde{\mathbf{u}}$, and the enhanced parameters, $\tilde{\boldsymbol{\alpha}}$, expressed as

$$\delta\hat{\Pi} = \int_{\Omega} \frac{\partial W}{\partial \tilde{F}_{iI}} \delta\tilde{F}_{iI} \, \mathrm{d}V - \delta\Pi_{\mathrm{ext}} = \begin{Bmatrix} \delta\tilde{\mathbf{u}}^{\mathrm{T}} & \delta\tilde{\boldsymbol{\alpha}}^{\mathrm{T}} \end{Bmatrix} \begin{Bmatrix} \mathbf{P}_{\mathrm{int}}(\tilde{\mathbf{u}}, \tilde{\boldsymbol{\alpha}}) - \mathbf{f} \\ \mathbf{P}_{\mathrm{enh}}(\tilde{\mathbf{u}}, \tilde{\boldsymbol{\alpha}}) \end{Bmatrix} = 0 \tag{10.130}$$

where $\mathbf{P}_{\mathrm{int}}$ is the internal force vector, $\mathbf{P}_{\mathrm{enh}}$ is the enhanced force vector, and $\mathbf{f}$ is the usual force vector computed from Π_{ext}. Noting that the variations $\delta\tilde{\mathbf{u}}$ and $\delta\tilde{\boldsymbol{\alpha}}$ in Eq. (10.130) are arbitrary the finite element residual vectors are given by

$$\boldsymbol{\Psi}_u = \mathbf{f} - \mathbf{P}_{\mathrm{int}}(\tilde{\mathbf{u}}, \tilde{\boldsymbol{\alpha}}) = \mathbf{0} \tag{10.131}$$

$$\boldsymbol{\Psi}_\alpha = -\mathbf{P}_{\mathrm{enh}}(\tilde{\mathbf{u}}, \tilde{\boldsymbol{\alpha}}) = \mathbf{0} \tag{10.132}$$

A solution to these equations may now be constructed in the standard manner discussed in Chapter 2. Using a Newton–Raphson scheme to linearize Eq. (10.130) we obtain

$$\begin{aligned} d(\delta\hat{\Pi}) &= \int_{\Omega} \delta\hat{F}_{iI} \frac{\partial^2 W}{\partial \hat{F}_{iI} \partial \hat{F}_{jJ}} \mathrm{d}\hat{F}_{jJ} + \frac{\partial W}{\partial \hat{F}_{iI}} d(\delta\hat{F}_{iI}) \, \mathrm{d}V \\ &\equiv \begin{Bmatrix} \delta\tilde{\mathbf{u}}^{\mathrm{T}} & \delta\tilde{\boldsymbol{\alpha}}^{\mathrm{T}} \end{Bmatrix} \begin{bmatrix} \hat{\mathbf{K}}_{uu} & \hat{\mathbf{K}}_{u\alpha} \\ \hat{\mathbf{K}}_{\alpha u} & \hat{\mathbf{K}}_{\alpha\alpha} \end{bmatrix} \begin{Bmatrix} d\tilde{\mathbf{u}} \\ d\tilde{\boldsymbol{\alpha}} \end{Bmatrix} \end{aligned} \tag{10.133}$$

where $\hat{\mathbf{K}}_{uu}$, etc., are obtained by evaluating all the terms in the integrals in a standard manner, and the process is by now so familiar to the reader we leave it as an exercise.

Using Eqs (10.131)–(10.133) we obtain the system of equations

$$\begin{bmatrix} \mathbf{K}_{uu} & \mathbf{K}_{u\alpha} \\ \mathbf{K}_{\alpha u} & \mathbf{K}_{\alpha\alpha} \end{bmatrix} \begin{Bmatrix} d\tilde{\mathbf{u}} \\ d\tilde{\boldsymbol{\alpha}} \end{Bmatrix} = \begin{Bmatrix} \boldsymbol{\Psi}_u \\ \boldsymbol{\Psi}_\alpha \end{Bmatrix} \tag{10.134}$$

where we note that the parameters $\boldsymbol{\alpha}$ are associated with individual elements. We have encountered such forms in many previous situations (e.g. bubble modes in plates) and used static condensation[17] to perform a partial solution at the element level. Here the situation is slightly different in that the equations are non-linear. Thus, it is necessary to use the static condensation process in an iterative manner. Accordingly, given a solution $\tilde{\mathbf{u}}$ for some iterate in a Newton–Raphson process we can isolate the part for each $\tilde{\boldsymbol{\alpha}}$ and consider a local solution for the equation set

$$d\tilde{\boldsymbol{\alpha}}^{(k)} = \hat{\mathbf{K}}_{\alpha\alpha}^{-1} \boldsymbol{\Psi}_{\alpha}^{(k)} \tag{10.135}$$

Iteration continues until $\boldsymbol{\Psi}_\alpha$ is zero with updates

$$\tilde{\boldsymbol{\alpha}}^{(k+1)} = \tilde{\boldsymbol{\alpha}}^{(k)} + d\tilde{\boldsymbol{\alpha}}^{(k)} \tag{10.136}$$

which is performed on each element separately.

Utilizing the final solution from Eq. (10.135) an equivalent displacement model involving only the nodal displacement parameters is obtained as

$$\hat{\mathbf{K}}_{\mathrm{T}}\, d\tilde{\mathbf{u}} = \boldsymbol{\Psi}_u \tag{10.137}$$

where

$$\mathbf{K}_{\mathrm{T}}^{(k)} = [\hat{\mathbf{K}}_{uu} - \hat{\mathbf{K}}_{u\alpha} \left(\hat{\mathbf{K}}_{\alpha\alpha}\right)^{-1} \hat{\mathbf{K}}_{\alpha u}]$$

The system of Eq. (10.137) is solved and the nodal displacements are updated in the usual manner for any displacement problem (see Chapter 1). Additional details and many example solutions using the above formulation, and its specialization to small deformations, may be found in references 15 and 18.

10.6 Forces dependent on deformation – pressure loads

In the derivations presented in the previous sections it was assumed that the forces $\mathbf{f}$ were not themselves dependent on the deformation. In some instances this is not true. For instance, pressure loads on a deforming structure are in this category. Aerodynamic forces are an example of such pressure loads and can induce flutter.

If forces vary with displacement then in relation (10.66) the variation of the forces with respect to the displacements has to be considered. This leads to the introduction of the *load correction matrix* $\mathbf{K}_{\mathrm{L}}$ as originally suggested by Oden[19] and Hibbitt *et al.*[20]

Here we consider the case where pressure acts on the current configuration and remains normal throughout the deformation history. If the pressure is given by $\bar{p}$ then the surface traction term in $\delta\Pi_{\mathrm{ext}}$ is given by

$$\int_{\gamma_t} \delta u_i \bar{t}_i \,\mathrm{d}s = \int_{\gamma_t} \delta u_i \bar{p} n_i \,\mathrm{d}s \tag{10.138}$$

where n_i are the direction cosines of an outward pointing normal to the deformed surface. The computation of the nodal forces and tangent matrix terms is most conveniently computed by transforming the above expression to the surface approximated by finite elements.[21–23] In this case we have the approximation to Eq. (10.138) for a three-dimensional problem given in matrix notation by (where once again we use Greek subscripts to denote the finite element node numbers)

$$\int_{\gamma_t} \delta u_i \bar{t}_i \, \mathrm{d}s = \delta \tilde{\mathbf{u}}_\alpha \int_{-1}^{1} \int_{-1}^{1} N_\alpha \bar{p}(\xi, \eta) \left[\left(N_{\gamma,\xi} \mathbf{x}_\gamma \right) \times \left(N_{\delta,\eta} \mathbf{x}_\delta \right) \right] \mathrm{d}\xi \, \mathrm{d}\eta \tag{10.139}$$

where ξ, η are natural coordinates of a two-dimensional finite element surface interpolation, $\bar{p}(\xi, \eta)$ is a specified nodal pressure at each point on the surface, $\mathbf{x}_\gamma$ are nodal coordinates of the deformed surface, and we have used the relation transforming surface area given in Sec. 7.5 of Volume 1. A cross-product may be written in the alternate forms

$$\mathbf{x}_\gamma \times \mathbf{x}_\delta = \hat{\mathbf{x}}_\gamma \mathbf{x}_\delta = -\hat{\mathbf{x}}_\delta \mathbf{x}_\gamma = \hat{\mathbf{x}}_\delta^{\mathrm{T}} \mathbf{x}_\gamma \tag{10.140}$$

where here $\hat{\mathbf{x}}$ denotes a *skew symmetric* matrix given as

$$\hat{\mathbf{x}} = \begin{bmatrix} 0 & -x_3 & x_2 \\ x_3 & 0 & -x_1 \\ -x_2 & x_1 & 0 \end{bmatrix} \tag{10.141}$$

Using the above relations the nodal forces for the 'follower' surface loading are given by

$$\mathbf{f}_\alpha = \int_{-1}^{1} \int_{-1}^{1} N_\alpha \bar{p}(\xi, \eta) N_{\gamma,\xi} N_{\delta,\eta} \hat{\mathbf{x}}_\gamma \mathbf{x}_\delta \, \mathrm{d}\xi \, \mathrm{d}\eta \tag{10.142}$$

Since the nodal forces involve the nodal coordinates in the current configuration explicitly, it is necessary to compute a tangent matrix $\mathbf{K}_\mathrm{L}$ for use in a Newton–Raphson solution scheme. Linearizing Eq. (10.142) we obtain the tangent as

$$\mathbf{K}_\mathrm{L}^{\alpha\beta} = -\frac{\partial \mathbf{f}_\alpha}{\partial \mathbf{u}_\beta} = \int_{-1}^{1} \int_{-1}^{1} N_\alpha \bar{p}(\xi, \eta) \left[N_{\beta,\xi} N_{\gamma,\eta} - N_{\gamma,\xi} N_{\beta,\eta} \right] \hat{\mathbf{x}}_\gamma \, \mathrm{d}\xi \, \mathrm{d}\eta \tag{10.143}$$

In general the tangent expression is unsymmetric; however, if the pressure loading is applied over a closed surface and is constant the final assembled terms are symmetric.

For cases where the pressure varies over the surface the pressure may be computed by using an interpolation

$$\bar{p}(\xi, \eta) = N_\alpha(\xi, \eta) \tilde{p}_\alpha \tag{10.144}$$

in which $\tilde{p}_\alpha$ are values of the known pressure at the nodes. Of course, these could also arise from solution of a problem which generates pressures on the contiguous surfaces and thus lead to the need to solve a coupled problem as discussed in Chapter 19 of Volume 1.

The form for two-dimensional plane problems simplifies considerably since in this case Eq. (10.139) becomes

$$\int_{\gamma_t} \delta u_i \bar{t}_i \, \mathrm{d}s = \delta \tilde{\mathbf{u}}_\alpha \int_{-1}^{1} N_\alpha \bar{p}(\xi) \left(N_{\gamma,\xi} \mathbf{x}_\gamma \right) \times \mathbf{e}_3 \, \mathrm{d}\xi \tag{10.145}$$

where ξ is a one-dimensional natural coordinate for the surface side, $\mathbf{e}_3$ is the unit vector normal to the plane of deformation (which is constant), and $\bar{p}(\xi)$ is now the force per unit length of surface side. For this case the nodal forces for the follower pressure load are given explicitly by

$$\mathbf{f}_\alpha = \int_{-1}^{1} N_\alpha \bar{p}(\xi) \begin{Bmatrix} -x_{2,\xi} \\ x_{1,\xi} \end{Bmatrix} \mathrm{d}\xi \tag{10.146}$$

where $x_{i,\xi}$ are derivatives computed from the one-dimensional finite element interpolation used to approximate the element side. The case for axisymmetry involves additional terms and the reader is referred to reference 21 for details.

10.7 Material constitution for finite deformation

In order to complete any finite element development it is necessary to describe how the material behaves when subjected to deformation or deformation histories. In the discussion above we considered elastic behaviour without introducing details on how to model specific material behaviour. Clearly, restriction to elastic behaviour is inadequate to model the behaviour of many engineering materials as we have already shown in many previous applications. The modelling of engineering materials at finite strain is a subject of much research and any complete summary on the state of the art is clearly outside the scope of what can be presented here. In this chapter we present only some classical methods which may be used to model elastic and elasto-plastic type behaviours. The reader is directed to literature for details on other constitutive models (e.g. see references 3 and 24).

We first consider some methods which may be used to describe the behaviour of isotropic elastic materials which undergo finite deformation. In this section we restrict attention to those materials in which a stored energy function is used. Later we will extend this to permit the use of plasticity models and show that much of the material presented in Chapter 3 is here again useful. Finally, to permit the modelling of materials which are not isotropic or cannot be expressed as an extension to elastic behaviour (e.g. generalized plasticity models of Chapter 3) we introduce a rate form – here again many options are possible.

10.7.1 Isotropic elasticity – formulation in invariants

We consider a finite deformation form for *hyperelasticity* in which a stored energy density function, W, is used to compute stresses. For a stored energy density expressed in terms of right Cauchy–Green deformation tensor, C_{IJ}, the second Piola–Kirchhoff stress is computed by using Eq. (10.38). Through standard transformation we can also obtain the Kirchhoff stress as[2–4]

$$\tau_{ij} = 2 b_{ik} \frac{\partial W}{\partial b_{kj}} \tag{10.147}$$

and thus, by using Eq. (10.19), also obtain directly the Cauchy stress.

For an isotropic material the stored energy density depends only on three invariants of the deformation. Here we consider the three invariants (noting they also are equal to those for b_{ij}) expressed as[3,4]

$$I = C_{KK} = b_{kk} \tag{10.148}$$

$$II = \tfrac{1}{2}\,(I^2 - C_{KL}C_{LK}) = \tfrac{1}{2}\,(I^2 - b_{kl}b_{lk}) \tag{10.149}$$

and

$$III = \det C_{KL} = \det b_{kl} = J^2 \quad \text{where} \quad J = \det F_{kL} \tag{10.150}$$

and write the strain energy density as

$$W(C_{KL}) = W(b_{kl}) \equiv W(I, II, J) \tag{10.151}$$

where we select J instead of III as the measure of the volume change. Thus, the second Piola–Kirchhoff stress is computed as

$$S_{IJ} = 2\left[\frac{\partial W}{\partial I}\frac{\partial I}{\partial C_{IJ}} + \frac{\partial W}{\partial II}\frac{\partial II}{\partial C_{IJ}} + \frac{\partial W}{\partial J}\frac{\partial J}{\partial C_{IJ}}\right] \tag{10.152}$$

The derivatives of the invariants may be evaluated as (see Appendix A)

$$\frac{\partial I}{\partial C_{IJ}} = \delta_{IJ}, \qquad \frac{\partial II}{\partial C_{IJ}} = I\,\delta_{IJ} - C_{IJ}, \qquad \frac{\partial J}{\partial C_{IJ}} = \tfrac{1}{2}\,JC_{IJ}^{-1} \tag{10.153}$$

Thus, the stress is given by

$$S_{IJ} = 2[\,\delta_{IJ} \quad (I\,\delta_{IJ} - C_{IJ}) \quad \tfrac{1}{2}\,J\,C_{IJ}^{-1}\,]\begin{Bmatrix} \dfrac{\partial W}{\partial I} \\ \dfrac{\partial W}{\partial II} \\ \dfrac{\partial W}{\partial J} \end{Bmatrix} \tag{10.154}$$

The second Piola–Kirchhoff stress may be transformed to the Cauchy stress by using Eq. (10.20), and gives

$$\sigma_{ij} = \frac{2}{J}\,[\,b_{ij} \quad (I\,b_{ij} - b_{im}b_{mj}) \quad \tfrac{1}{2}\,J\,\delta_{ij}\,]\begin{Bmatrix} \dfrac{\partial W}{\partial I} \\ \dfrac{\partial W}{\partial II} \\ \dfrac{\partial W}{\partial J} \end{Bmatrix} \tag{10.155}$$

Use of a Newton–Raphson type solution process requires computation of the elastic moduli for the finite elasticity model. The elastic moduli with respect to the reference configuration are deduced from[3,4]

$$D_{IJKL} = 4\,\frac{\partial^2 W}{\partial C_{IJ}\partial C_{KL}} = 2\,\frac{\partial S_{IJ}}{\partial C_{KL}} \tag{10.156}$$

Using Eq. (10.154) the general form for the elastic moduli of an isotropic material is obtained from

$$D_{IJKL} = 4[\,\delta_{IJ},\ \ (I\,\delta_{IJ} - C_{IJ}),\ \ \tfrac{1}{2}J\,C_{IJ}^{-1}\,]\begin{bmatrix} \dfrac{\partial^2 W}{\partial I^2} & \dfrac{\partial^2 W}{\partial I \partial II} & \dfrac{\partial^2 W}{\partial I \partial J} \\ \dfrac{\partial^2 W}{\partial II \partial I} & \dfrac{\partial^2 W}{\partial II^2} & \dfrac{\partial^2 W}{\partial II \partial J} \\ \dfrac{\partial^2 W}{\partial J \partial I} & \dfrac{\partial^2 W}{\partial J \partial II} & \dfrac{\partial^2 W}{\partial J^2} \end{bmatrix} \begin{Bmatrix} \delta_{KL} \\ (I\,\delta_{KL} - C_{KL}) \\ \tfrac{1}{2}J\,C_{KL}^{-1} \end{Bmatrix}$$

$$+\,[\,\delta_{IJ}\delta_{KL} - \tfrac{1}{2}(\delta_{IK}\delta_{JL} + \delta_{IL}\delta_{JK}),\ \ J[C_{IJ}^{-1}C_{KL}^{-1} - 2\mathcal{C}_{IJKL}^{-1}]]\begin{Bmatrix} 4\dfrac{\partial W}{\partial II} \\ \dfrac{\partial W}{\partial J} \end{Bmatrix} \qquad (10.157)$$

where

$$\mathcal{C}_{IJKL}^{-1} = \tfrac{1}{2}\left[C_{IK}^{-1}C_{JL}^{-1} + C_{IL}^{-1}C_{JK}^{-1}\right] \qquad (10.158)$$

The spatial elasticities related to the Cauchy stress are obtained by the *push forward* transformation

$$J\,d_{ijkl} = F_{iI}\,F_{jJ}\,F_{kK}\,F_{lL}\,D_{IJKL} \qquad (10.159)$$

which, applied to Eq. (10.157), gives

$$J\,d_{ijkl} = 4[\,b_{ij},\ \ (I\,b_{ij} - b_{im}\,b_{mj}),\ \ \tfrac{1}{2}J\,\delta_{ij}\,]\begin{bmatrix} \dfrac{\partial^2 W}{\partial I^2} & \dfrac{\partial^2 W}{\partial I \partial II} & \dfrac{\partial^2 W}{\partial I \partial J} \\ \dfrac{\partial^2 W}{\partial II \partial I} & \dfrac{\partial^2 W}{\partial II^2} & \dfrac{\partial^2 W}{\partial II \partial J} \\ \dfrac{\partial^2 W}{\partial J \partial I} & \dfrac{\partial^2 W}{\partial J \partial II} & \dfrac{\partial^2 W}{\partial J^2} \end{bmatrix} \begin{Bmatrix} b_{kl} \\ (I\,b_{kl} - b_{km}\,b_{ml}) \\ \tfrac{1}{2}J\,\delta_{kl} \end{Bmatrix}$$

$$+\,\left[\,b_{ij}\,b_{kl} - \tfrac{1}{2}\left(b_{ik}\,b_{jl} + b_{il}\,b_{jk}\right),\ \ J\left[\delta_{kl}\delta_{kl} - 2\mathcal{I}_{ijkl}\right]\right]\begin{Bmatrix} 4\dfrac{\partial W}{\partial II} \\ \dfrac{\partial W}{\partial J} \end{Bmatrix} \qquad (10.160)$$

where

$$\mathcal{I}_{ijkl} = \tfrac{1}{2}\left[\delta_{ik}\delta_{jl} + \delta_{il}\delta_{jk}\right] \qquad (10.161)$$

The above expressions describe completely the necessary equations to construct a finite element model for any isotropic hyperelastic material. All that remains is to select a specific form for the stored energy function W. Here, many options exist and we include below only a very simple model. For others the reader is referred to literature on the subject.

Example: compressible neo-Hookean material

As an example, we consider the case of a neo-Hookean material[25] that includes a compressibility effect. The stored energy density is expressed as

$$W(I,J) = \tfrac{1}{2}\,\mu(I - 3 - 2\ln J) + \tfrac{1}{2}\,\lambda(J-1)^2 \tag{10.162}$$

where the material constants λ and μ are selected to give the same response in small deformations as a linear elastic material using Lamé parameters.[3] Substitution into Eq. (10.154) gives

$$S_{IJ} = \mu(\delta_{IJ} - C_{IJ}^{-1}) + \lambda J\,(J-1)\,C_{IJ}^{-1} \tag{10.163}$$

which may be transformed to give the Cauchy stress

$$\sigma_{ij} = \frac{\mu}{J}\,(b_{ij} - \delta_{ij}) + \lambda(J-1)\,\delta_{ij} \tag{10.164}$$

For the neo-Hookean model the material moduli with respect to the reference configuration are given as

$$D_{IJKL} = \lambda J\,(2J-1)\,C_{IJ}^{-1}C_{KL}^{-1} + 2\,[\mu - \lambda J\,(J-1)]\,\mathcal{C}_{IJKL}^{-1} \tag{10.165}$$

Transformation to spatial configuration moduli gives

$$d_{ijkl} = \lambda(2J-1)\,\delta_{ij}\delta_{kl} + 2\left[\frac{\mu}{J} - \lambda(J-1)\right]\mathcal{I}_{ijkl} \tag{10.166}$$

We note that when $J \approx 1$ the small deformation result

$$d_{ijkl} = \lambda\,\delta_{ij}\delta_{kl} + 2\,\mu\,\mathcal{I}_{ijkl} \tag{10.167}$$

is obtained and thus matches the usual linear elastic relations. This permits the finite deformation formulation to be used directly for analyses in which the small strain assumptions hold as well as for situations in which deformations are large. The above model may also be used with the mixed forms described above for situations where the ratio λ/μ is large (i.e. nearly incompressible behaviour). Indeed this was an early use of the model.

10.7.2 Isotropic elasticity – formulation in principal stretches

Other forms of elastic constitutive equations may be introduced by using appropriate expansions of the stored energy density function. As an alternative, an elastic formulation expressed in terms of principal stretches (which are the square root of the eigenvalues of C_{IJ} or b_{ij}) may be introduced. This approach has been presented by Ogden[26] and by Simo and Taylor.[6]

We first consider a change of coordinates given by (see Appendix B, Volume 1)

$$x_i = \Lambda_{m'i}\,x_{m'} \tag{10.168}$$

where $\Lambda_{m'i}$ are direction cosines between the two Cartesian systems. The transformation equations for a second-rank tensor, say b_{ij}, may then be written in the form

$$b_{ij} = \Lambda_{m'i}\,b_{m'n'}\,\Lambda_{n'j} \tag{10.169}$$

To compute specific relations for the transformation array we consider the solution of the eigenproblem

$$b_{ij} q_j^{(n)} = q_j^{(n)} b_n; \quad n = 1, 2, 3 \qquad \text{with} \qquad q_k^{(m)} q_k^{(n)} = \delta_{mn} \tag{10.170}$$

where b_n are the *principal values* of b_{ij}, and $q_i^{(n)}$ are direction cosines for the principal directions. The principal values of b_{ij} are equal to the square of the *principal stretches*, λ_n, that is,

$$b_n = \lambda_n^2 \tag{10.171}$$

If we assign the direction cosines in the transformation equation (10.169) as

$$\Lambda_{n'j} \equiv q_j^{(n)} \tag{10.172}$$

the spectral representation of the deformation tensor results and may be expressed as

$$b_{ij} = \sum_m \lambda_m^2 q_i^{(m)} q_j^{(m)} \tag{10.173}$$

An advantage of a spectral form is that other forms of the tensor may easily be represented. For example,

$$b_{ik} b_{kj} = \sum_m \lambda_m^4 q_i^{(m)} q_j^{(m)} \qquad \text{and} \qquad b_{ik}^{-1} = \sum_m \lambda_m^{-2} q_i^{(m)} q_j^{(m)} \tag{10.174}$$

Also, we note that an identity tensor may be represented as

$$\delta_{ij} = \sum_m q_i^{(m)} q_j^{(m)} \tag{10.175}$$

From Eq. (10.155) we can immediately observe that Cauchy and Kirchhoff stresses have the same principal directions as the left Cauchy–Green tensor. Thus, for example, the Kirchhoff stress has the representation

$$\tau_{ij} = \sum_m \tau_m q_i^{(m)} q_j^{(m)} \tag{10.176}$$

where τ_m denote principal values.

If we now represent the stored energy function in terms of principal stretch values as $\hat{w}(\lambda_1, \lambda_2, \lambda_3)$ the principal values of the Kirchhoff stress may be deduced from[24,26]

$$\tau_m = \lambda_m \frac{\partial \hat{w}}{\partial \lambda_m} \tag{10.177}$$

The reader is referred to the literature for a more general discussion on formulations in principal stretches for use in general elasticity problems.[6,24,26] Here we wish to consider one form which is useful to develop solution algorithms for finite elasto-plastic behaviour of isotropic materials in which elastic strains are quite small. Such a form is useful, for example, in modelling metal plasticity.

Logarithmic principal stretch form

A particularly simple result is obtained by writing the stored energy function in terms of logarithmic principal stretches. Accordingly, we take

$$\hat{w}(\lambda_1, \lambda_2, \lambda_3) = w(\varepsilon_1, \varepsilon_2, \varepsilon_3) \quad \text{where} \quad \varepsilon_m = \log(\lambda_m) \tag{10.178}$$

From Eq. (10.177) it follows that

$$\tau_m = \frac{\partial w}{\partial \varepsilon_m} \tag{10.179}$$

which is now identical to the form from linear elasticity, but expressed in principal directions. It also follows that the elastic moduli may be written as[24,26] (summation convention is not used to write this expression)

$$\begin{aligned} J\, d_{ijkl} &= \sum_{m=1}^{3}\sum_{n=1}^{3} [c_{mn} - 2\tau_m \delta_{mn}]\, q_i^{(m)} q_j^{(m)} q_k^{(n)} q_l^{(n)} \\ &+ \frac{1}{2}\sum_{m=1}^{3}\sum_{\substack{n=1\\ n\neq m}}^{3} g_{mn}[q_i^{(m)} q_j^{(n)} q_k^{(m)} q_l^{(n)} + q_i^{(m)} q_j^{(n)} q_k^{(n)} q_l^{(m)}] \end{aligned} \tag{10.180}$$

where

$$c_{mn} = \frac{\partial^2 w}{\partial \varepsilon_m \partial \varepsilon_n} \quad \text{and} \quad g_{mn} = \begin{cases} \dfrac{\tau_m \lambda_n^2 - \tau_n \lambda_m^2}{\lambda_m^2 - \lambda_n^2}; & \lambda_m \neq \lambda_n \\ \dfrac{\partial(\tau_m - \tau_n)}{\partial \varepsilon_m}; & \lambda_m = \lambda_n \end{cases} \tag{10.181}$$

In practice the equal root form is used whenever differences are less than a small tolerance (say 10^{-8}).

Use of a quadratic form for w given by

$$w = \tfrac{1}{2}(K - \tfrac{2}{3}G)[\varepsilon_1 + \varepsilon_2 + \varepsilon_3]^2 + G\,[\varepsilon_1^2 + \varepsilon_2^2 + \varepsilon_3^2] \tag{10.182}$$

yields principal Kirchhoff stresses given by

$$\begin{Bmatrix} \tau_1 \\ \tau_2 \\ \tau_3 \end{Bmatrix} = \begin{bmatrix} K + \frac{4}{3}G & K - \frac{2}{3}G & K - \frac{2}{3}G \\ K - \frac{2}{3}G & K + \frac{4}{3}G & K - \frac{2}{3}G \\ K - \frac{2}{3}G & K - \frac{2}{3}G & K + \frac{4}{3}G \end{bmatrix} \begin{Bmatrix} \varepsilon_1 \\ \varepsilon_2 \\ \varepsilon_3 \end{Bmatrix} \tag{10.183}$$

in which the 3×3 elasticity matrix is given by a *constant* coefficient matrix which is identical to the usual linear elastic expression in terms of bulk and shear moduli. We also note that when roots are equal

$$\frac{\partial(\tau_m - \tau_n)}{\partial \varepsilon_m} (K + \tfrac{4}{3}G) - (K - \tfrac{2}{3}G) = 2G \tag{10.184}$$

which defines the usual shear modulus form in isotropic linear elasticity.

10.7.3 Plasticity models

For isotropic materials, the modelling of elasto-plastic behaviour in which the total deformations are large may be performed by an extension of a hyperelastic formulation. In this case the deformation gradient is decomposed in a *product form* (instead of the additive form assumed in Chapter 3) written as[27–29]

$$F_{iI} = F^{\mathrm{e}}_{i\hat{J}} F^{\mathrm{p}}_{\hat{J}I} \tag{10.185}$$

where $F^{\mathrm{e}}_{i\hat{J}}$ is the elastic part and $F^{\mathrm{p}}_{\hat{J}I}$ the plastic part. The deformation picture is often shown as three parts, a reference state, a deformed state, and an *intermediate* state. The intermediate state is assumed to be the state of a point in a stress-free condition.* From this decomposition deformation tensors may be defined as

$$b^{\mathrm{e}}_{ij} = F^{\mathrm{e}}_{i\hat{K}} F^{\mathrm{e}}_{j\hat{K}} \quad \text{and} \quad C^{\mathrm{p}}_{IJ} = F^{\mathrm{p}}_{\hat{K}I} F^{\mathrm{p}}_{\hat{K}J} \tag{10.186}$$

which when combined with Eq. (10.185) give the alternate representation

$$b^{\mathrm{e}}_{ij} = F_{iI} \left(C^{\mathrm{p}}_{IJ}\right)^{-1} F_{jJ} \tag{10.187}$$

An incremental setting may now be established that obtains a solution for a time t_{n+1} given the state at time t_n. The steps to establish the algorithm are too lengthy to include here and the interested reader is referred to literature for details.[5,24,30,31]

The components $(b^{\mathrm{e}}_{ij})_n$ denote values of the converged elastic deformation tensor at time t_n. We assume at the start of a new load step a *trial value* of the elastic tensor is determined from

$$(b^{\mathrm{e}}_{ij})^{\mathrm{tr}}_{n+1} = f_{ik} \, (b^{\mathrm{e}}_{kl})_n \, f_{jl} \tag{10.188}$$

where an incremental deformation gradient is computed as

$$f_{ij} = (F_{iK})_{n+1} (F^{-1}_{jK})_n \tag{10.189}$$

A spectral representation of the trial tensor is then determined by using Eq. (10.173) giving

$$(b^{\mathrm{e}}_{ij})_{n+1} = \sum_m (\lambda^{\mathrm{e}}_m)^2_{n+1} \, q_i^{(m),\mathrm{tr}} \, q_j^{(m),\mathrm{tr}} \tag{10.190}$$

Owing to isotropy $q_i^{(m),\mathrm{tr}}$ can be shown to equal the final directions $q_i^{(m)}$.[24]

Trial logarithmic strains are computed as

$$(\varepsilon^{\mathrm{tr}}_m)_{n+1} = \log(\lambda^{\mathrm{e}}_m)_{n+1} \tag{10.191}$$

and used with the stored energy function $W(b^{\mathrm{e}}_{ij})$ to compute trial values of the principal Kirchhoff stress $(\tau^{\mathrm{tr}}_m)_{n+1}$. This may be used in conjunction with the return map algorithm (see Section 3.4.2) and a yield function written in principal stresses τ_m to compute a final stress state and any internal hardening variables. This part of the algorithm is identical to the small strain form and needs no additional description except to emphasize that only the normal stress is included in the calculation of yield and flow directions. We note in particular that any of the yield functions for isotropic materials which we discussed in Chapter 3 may be used. The use of the return map algorithm also yields the consistent elasto-plastic tangent in principal space which can be transformed by means of Eq. (10.180) for subsequent use in the finite element matrix form.

* The intermediate state is not a configuration, as it is generally discontinuous across interfaces between elastic and inelastic response.

The last step in the algorithm is to compute the final elastic deformation tensor. This is accomplished from the spectral form and final elastic logarithmic strains resulting from the return map solution as

$$(b^{e}_{ij})_{n+1} = \sum_{m=1}^{3} \exp[2\,(\varepsilon^{e}_{m})_{n+1}]\, q_i^{(m)}\, q_j^{(m)} \tag{10.192}$$

The advantages of the above algorithm are numerous. The form again permits a consistent linearization of the algorithm resulting in optimal performance when used with the Newton–Raphson solution scheme. Most important, all the steps previously developed for the small deformation case are here used. For example, although not discussed here, extension to viscoplastic and generalized plastic forms for isotropic materials is again given by results contained in Secs 3.6.2 and 3.9. The primary difficulty is an inability to treat materials which are anisotropic. Here recourse to a rate form of the constitutive equation is possible, as discussed next.

10.7.4 Rate constitutive models

The construction of a rate form for elastic constitutive equations deduced from a stored energy function is easily performed in the reference configuration by taking a time derivative of Eq. (10.38), which gives

$$\dot{S}_{IJ} = D_{IJKL}\, \dot{E}_{KL} \tag{10.193}$$

where, as before, D_{IJKL} are moduli given by Eq. (10.156). The above result follows naturally from the notion of a derivative since

$$\dot{S}_{IJ} = \lim_{\eta \to 0} \frac{S_{IJ}(t+\eta) - S_{IJ}(t)}{\eta} \tag{10.194}$$

Such a definition is clearly not appropriate for the Cauchy or Kirchhoff stress since they are related to different configurations at time $t+\eta$ and t and thus would not satisfy the requirements of objectivity.[4,32] A definition of an *objective time derivative* may be computed for the Kirchhoff stress by using Eq. (10.20) and is sometimes referred to as the Truesdell rate[33] or equivalently a Lie derivative form.[34] Accordingly, we note that the objective time derivative is given by

$$\overset{\circ}{\tau}_{ij} = F_{iI}\, \dot{S}_{IJ}\, F_{jJ} + \dot{F}_{iI}\, S_{IJ}\, F_{jJ} + F_{iI}\, S_{IJ}\, \dot{F}_{jJ} \tag{10.195}$$

Introducing the rate of deformation tensor l_{ij} defined as

$$\dot{F}_{iI} = \dot{x}_{i,I} = \dot{x}_{i,j}\, x_{j,I} = l_{ij}\, F_{jI} \tag{10.196}$$

the stress rate may now be written as

$$\overset{\circ}{\tau}_{ij} = F_{iI}\, \dot{S}_{IJ}\, F_{jJ} + l_{ik}\, \tau_{kj} + \tau_{ik}\, l_{kj} \tag{10.197}$$

The rate of the second Piola–Kirchhoff stress may be transformed by noting

$$\dot{E}_{KL} = \tfrac{1}{2}\left(F_{kL}\, \dot{F}_{kK} + F_{kK}\, \dot{F}_{kL}\right) = \tfrac{1}{2}\left(F_{lK}\, F_{kL}\, l_{kl} + F_{kK}\, F_{lL}\, l_{kl}\right) = F_{kK}\, F_{lL}\, \dot{\varepsilon}_{kl} \tag{10.198}$$

where

$$\dot{\varepsilon}_{kl} = \frac{1}{2}(l_{kl} + l_{lk}) = \frac{1}{2}\left(\frac{\partial v_k}{\partial x_l} + \frac{\partial v_l}{\partial x_k}\right) \tag{10.199}$$

in which $v_k = \dot{x}_k = \dot{u}_k$ is the velocity vector. The form $\dot{\varepsilon}_{kl}$ is identical to the rate of small strain form. Furthermore we have upon grouping terms the rate of stress expression

$$\overset{\circ}{\tau}_{ij} = J\, d_{ijkl}\, \dot{\varepsilon}_{kl} + l_{ik}\, \tau_{kj} + \tau_{ik}\, l_{kj} \tag{10.200}$$

in which d_{ijkl} is computed now by means of Eq. (10.79). Incremental forms may be deduced for a rate equation which involve objective approximations for the Lie derivative.[5,24] For example an approximation to the 'strain rate' may be computed from[5]

$$(\dot{\varepsilon}_{ij})_{n+1/2} \approx \frac{1}{\Delta t}(f_{ik}^{-1})_{n+1/2}\, \Delta E_{kl}\, (f_{jl}^{-1})_{n+1/2} \tag{10.201}$$

$$\Delta E_{kl} = \tfrac{1}{2}\left[(f_{km})_{n+1} f_{lm})_{n+1} - \delta_{kl}\right] \tag{10.202}$$

where

$$(f_{ij})_{n+\alpha} = \delta_{ij} + \alpha\, \frac{\partial \Delta(u_i)_{n+1}}{\partial (x_j)_{\mathrm{n}}} \tag{10.203}$$

with $\Delta(u_i)_{n+1} = (u_i)_{n+1} - (u_i)_n$. Similarly, an approximation to the Lie derivative of Kirchhoff stress may be taken as

$$(\overset{\circ}{\tau}_{ij})_{n+1/2} \approx \frac{1}{\Delta t}(f_{ik})_{n+1/2}\left[(f_{km}^{-1})_{n+1}(\tau_{mn})_{n+1}(f_{ln}^{-1})_{n+1} - (\tau_{kl})_n\right](f_{jl})_{n+1/2} \tag{10.204}$$

Other approximations may be used; however, the above are quite convenient. In the approximation a modulus array d_{ijkl} must also be obtained. Here there is no simple form which is always consistent with the tangent needed for a full Newton–Raphson solution scheme and, often, a constant array is used based on results from linear elasticity.

Extension of the above to include general material constitution may be performed by replacing the strain rate by an additive form given as

$$\dot{\varepsilon} = \dot{\varepsilon}^{\mathrm{e}} + \dot{\varepsilon}^{\mathrm{p}} \tag{10.205}$$

Once again we can use all the constitutive equations discussed in Chapter 3 (including those which are not isotropic) to construct a finite element model for the large strain problem. Here, since approximations not consistent with a Newton–Raphson scheme are generally used for the moduli, convergence generally does not achieve an asymptotic quadratic rate. Use of quasi-Newton schemes and line search, as described in Chapter 2, can improve the convergence properties and leads to excellent performance in most situations.

Many other stress rates may be substituted for the Lie derivative. For example, the Jaumann–Zaremba stress rate form given as

$$\overset{\circ}{\tau}_{ij} = J\, d_{ijkl}\, \dot{\varepsilon}_{kl} + \dot{\omega}_{ik}\, \tau_{kj} - \dot{\omega}_{jk}\, \tau_{ik} \tag{10.206}$$

may be used. This form is deduced by noting that the rate of deformation tensor may be split into a symmetric and skew-symmetric form as

$$l_{ij} = \dot{\varepsilon}_{ij} + \dot{\omega}_{ij} \tag{10.207}$$

where $\dot{\omega}_{ij}$ is the rate of spin or vorticity. This form was often used in early developments of finite element solutions to large strain problems and enjoys considerable popularity even today.

10.8 Contact problems

In many problems situations are encountered where the points on a boundary of one body come into contact with points on the boundary of the same or another object. Such problems are commonly referred to as *contact problems*. Finite element methods have been used for many years to solve contact problems.[35–64] The patch test has also been extended to test consistency of contact developments.[65] Contact problems are inherently non-linear since, prior to contact, boundary conditions are given by traction conditions (often the traction being simply zero) whereas during 'contact' kinematic constraints must be imposed which prevent penetration of one boundary through the other, called the *impenetrability condition*.

The solution of a contact problem involves first identifying which points on a boundary interact and second the insertion of appropriate conditions to prevent the penetration. Figure 10.2 shows a typical situation in which one body is being pressed into a second body. In Fig. 10.2(a) the two objects are not in contact and the boundary conditions are specified by zero traction conditions for both bodies. In Fig. 10.2(b) the two objects are in contact along a part of the boundary segment

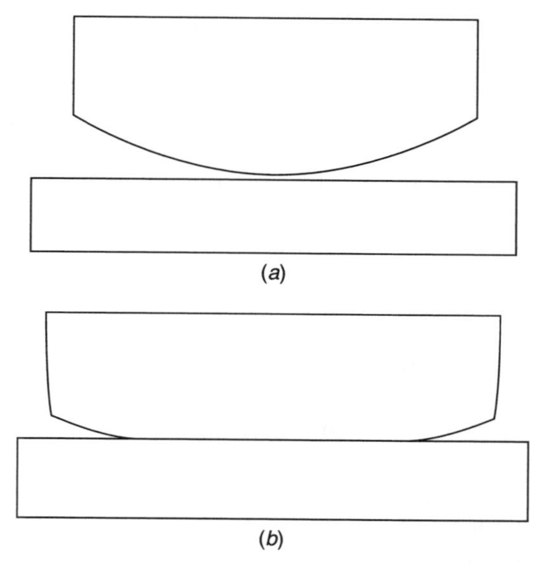

Fig. 10.2 Contact between two bodies: (a) no contact condition; (b) contact state.

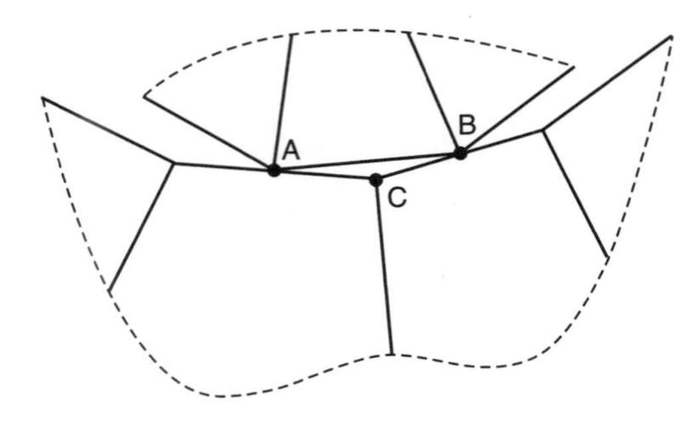

Fig. 10.3 Contact by finite elements.

and here conditions must be inserted to ensure that penetration does not occur. Along this boundary different types of contact can be modelled, the simplest being a *frictionless* condition in which the only non-zero contact traction is normal to the contact surface. A more complex condition occurs in which traction tangential to the surface can be generated by *frictional* conditions. The simplest model for a frictional condition is Coulomb friction where

$$|t_s| \leqslant \mu |t_n| \tag{10.208}$$

in which μ is a positive frictional parameter, t_n is the magnitude of the normal traction, and t_s is the tangential traction. If the magnitude of t_s is less than the limit condition the points on the surface are assumed to *stick*; whereas if the magnitude is at the limit condition *slip* occurs with an imposed tangential traction on each surface opposite to the direction of slip and equal to $\mu |t_n|$.

In modelling contact problems by finite element methods immediate difficulties arise. First, it is not possible to model contact at every point along a boundary. This is primarily because of the fact that the finite element representation of the boundary is not *smooth*. For example in the two-dimensional case in which boundaries of individual elements are straightline segments as shown in Fig. 10.3 nodes A and B are in contact with the lower body but the segment between the nodes is not in contact. Second, finite element modelling results in non-unique representation of a normal between the two bodies and, again because of finite element discretization, the normals are not continuous between elements. This is illustrated also in Fig. 10.3 where it is evident that the normal to the segment between nodes A and B is not the same as the negative normal of the facets around node C (which indeed are not unique at node C).

10.8.1 Geometric modelling

Node–node contact – Hertzian contact

For applications in which displacements on the contact boundary are *small* it is sometimes possible to model the contact by means of nodes. For this to be possible, the finite element mesh must be constructed such that boundary nodes on one body, here referred to as *slave* nodes, match the location of the boundary nodes for the other body, referred to as *master* nodes, to within conditions acceptable for

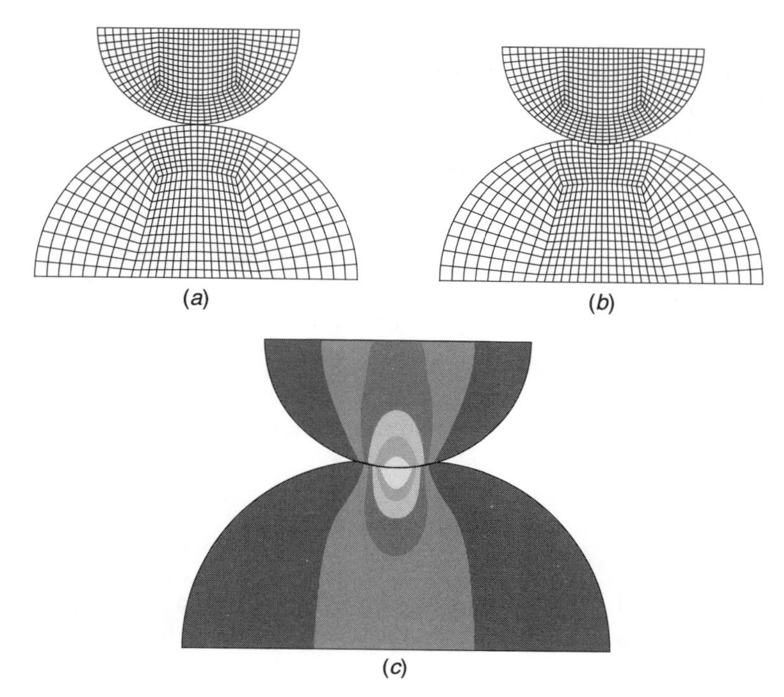

Fig. 10.4 Contact between semicircular discs: node–node solution. (a) Undeformed mesh; (b) deformed mesh; (c) vertical stress contours.

small deformation analysis. Such conditions may also be extended for cases where the boundary of one body is treated as flat and rigid (unilateral contact). A problem in which such conditions may be used is the interaction between two half discs (or hemispheres) which are pressed together along the line of action between their centres. A simple finite element model for such a problem is shown in Fig. 10.4(a) where it is observed that the horizontal alignment of potential contact nodes on the boundary of each disc are identical. The solution after pressing the bodies together is indicated in Fig. 10.4(b) and contours for the vertical normal stress are shown in Fig. 10.4(c). It is evident that the contours do not match perfectly along the vertical axis owing to lack of alignment of the nodes in the deformed position. However, the mismatch is not severe, and useful engineering results are possible. Later we will consider methods which give a more accurate representation; however, before doing so we consider the methods available to prevent penetration.

The determination of which nodes are in contact for such a problem can be monitored simply by comparing the vertical position of each node pair, which may be treated as a simple two-node element. Thus denoting the upper disc as slave body 's' and the lower one as master body 'm' we can monitor the vertical gap

$$g = x_2^{(s)} - x_2^{(m)} = [X_2^{(s)} + u_2^{(s)}] - [X_2^{(m)} + u_2^{(m)}] \tag{10.209}$$

If $g > 0$ no contact exists, whereas if $g \leqslant 0$ contact or penetration has occurred. (We note that penetration can exist for any iteration in which no modification of the formulation has been inserted.) Thus, the next step is to insert a constraint condition for any nodal pair (element) in which the gap g is negative or zero (here some tolerance usually is necessary to define 'zero'). There are many approaches which may be used to insert the constraint. Here we discuss use of a Lagrange multiplier form, a penalty approach, and an augmented lagrangian approach.[41]

Lagrange multiplier form

A Lagrange multiplier approach is given simply by multiplying the gap condition given in Eq. (10.209) by the multiplier. Accordingly, we can write for each nodal pair for which contact has been assigned a variational term

$$\Pi_c = \lambda g \tag{10.210}$$

and add its first variation to the variational equations being used to solve the problem. The first variation to Eq. (10.210) is given as

$$\delta\Pi_c = \delta\lambda g + [\delta u_2^{(s)} - \delta u_2^{(m)}]\lambda \tag{10.211}$$

and thus we identify λ as a 'force' applied to each node to prevent penetration. Linearization of Eq. (10.211) produces a tangent matrix term for use in a Newton–Raphson solution process. The final tangent and residual for the nodal contact element may be written as

$$\begin{bmatrix} 0 & 0 & 1 \\ 0 & 0 & -1 \\ 1 & -1 & 0 \end{bmatrix} \begin{Bmatrix} du_2^{(s)} \\ du_2^{(m)} \\ d\lambda \end{Bmatrix} = \begin{Bmatrix} -\lambda \\ \lambda \\ -g \end{Bmatrix} \tag{10.212}$$

and is added into the equations in a manner identical to any element assembly process. It is evident that the equations in this form introduce a new unknown for each contact pair. Also, as for any Lagrange multiplier approach, the equations are not positive definite and indeed have a zero diagonal for each multiplier term, thus, special care is needed in the solution process to avoid divisions by the zero diagonal.

Penalty function form

An approach which avoids equation solution difficulties of a Lagrange multiplier method is the penalty method, as described many times in Volume 1. In this the contact term is given by

$$\Pi = \kappa g^2 \tag{10.213}$$

where κ is a penalty parameter. The matrix equation for a nodal pair is now given by

$$\begin{bmatrix} \kappa & -\kappa \\ -\kappa & \kappa \end{bmatrix} \begin{Bmatrix} du_2^{(s)} \\ du_2^{(m)} \end{Bmatrix} = \begin{Bmatrix} -\kappa g \\ \kappa g \end{Bmatrix} \tag{10.214}$$

In a penalty approach the final gap will not be zero but becomes a small number depending on the value of the parameter κ selected. Thus, the advantage of the penalty method is somewhat offset by a need to identify the value of the parameter that gives an acceptable answer. Indeed, in a complex problem this is not a trivial task, especially for problems involving contact between beam, plate, or shell elements and solid elements.

Augmented lagrangian form

A compromise between the penalty method and the Lagrange multiplier method may be achieved by using an iterative update for the multiplier combined with a penalty-like form. We discussed this for incompressibility problems in Sec. 12.6 of Volume 1 and here indicate briefly how it applies equally to the contact problem. Based on results from Volume 1 we may write the augmented form as

$$\begin{bmatrix} \kappa & -\kappa \\ -\kappa & \kappa \end{bmatrix} \begin{Bmatrix} du_2^{(s)} \\ du_2^{(m)} \end{Bmatrix} = \begin{Bmatrix} -\lambda_k - \kappa g \\ \lambda_k + \kappa g \end{Bmatrix} \tag{10.215}$$

where an update to the Lagrange multiplier is computed by using[41]

$$\lambda_{k+1} = \lambda_k + \kappa g \tag{10.216}$$

Such an update may be computed after each Newton–Raphson iteration or in an added iteration loop after convergence of the Newton–Raphson iteration. In either case a loss of quadratic convergence results for the simple augmented strategy shown. Improvements to superlinear convergence are possible as shown by Zavarise and Wriggers,[66] and a more complex approach which restores the quadratic convergence rate may be introduced at the expense of retaining an added variable.[54] In general, however, use of a fairly large value of the penalty parameter in the simple scheme shown above is sufficient to achieve good solutions with few added iterations.

Node–surface contact

The simplest form for contact between bodies in which nodes on surfaces of one body do not interact directly with nodes on a second body is defined by a *node–surface* treatment. A two-dimensional treatment for this case is shown in Fig. 10.5 where a node, called the *slave node*, with deformed position $\mathbf{x}_s$ can contact a segment, called the *master surface*, defined for simplicity in two dimensions by an interpolation

$$\mathbf{x} = N_\alpha(\xi)\,\mathbf{x}_\alpha \tag{10.217}$$

This interpolation may be treated either as the usual interpolation along the edge facets of elements describing the target body as shown in Fig. 10.5(a) or by an interpolation which smooths the slope discontinuity between adjacent element surface facets as shown in Fig. 10.5(b).

A contact between the two bodies occurs when the *gap* g_n shown in Fig. 10.5 becomes zero. The determination of a contact requires a search to find which target facet is a potential contact surface and computation of the associated gap for each one. If the gap is positive no contact condition exists and, thus, no modification to the governing equations is required. If the gap is negative a 'penetration' of the two bodies has occurred and it is necessary to modify the equilibrium equations to reflect the contact forces which occur.

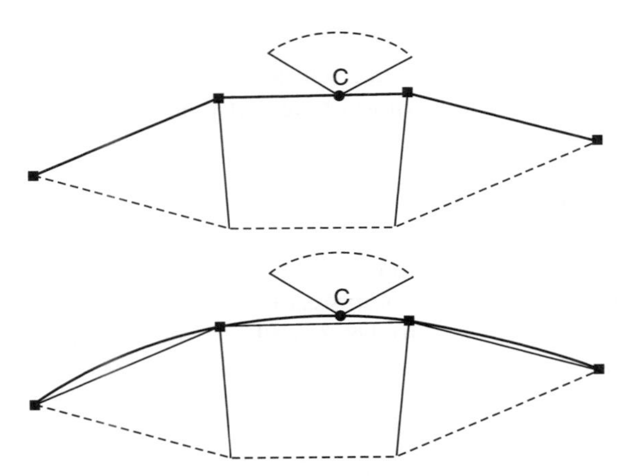

Fig. 10.5 Node-to-surface contact: (a) contact using element interpolations; (b) contact using 'smoothed' interpolations.

To determine the gap it is necessary to find the point on the target (master) facet which is closest to the slave node. This can be accomplished by expressing points on the facet by using Eq. (10.217) and finding the value of ξ which minimizes the function

$$f(\xi) = \tfrac{1}{2}\left(\mathbf{x}_s^T - \mathbf{x}^T\right)(\mathbf{x}_s - \mathbf{x}) = \text{minimum} \tag{10.218}$$

Here again a Newton–Raphson solution method may be used to find a solution. Linearizing, we solve for iterates from

$$[\mathbf{x}_{,\xi}^T \mathbf{x}_{,\xi} - (\mathbf{x}_s - \mathbf{x}(\xi_i))^T \mathbf{x}_{,\xi\xi}]\, d\xi_i = \mathbf{x}_{,\xi}^T(\mathbf{x}_s - \mathbf{x}(\xi_i)) = -\frac{df}{d\xi} = R \tag{10.219}$$

with updates

$$\xi_{i+1} = \xi_i + d\xi_i$$

until $R = 0$ is satisfied to within a specified tolerance. For the two-dimensional problem in which linear interpolation is used to define $N_\alpha(\xi)$, the expression for R is linear in ξ and, thus, convergence is achieved in one iteration. Denoting the solution as ξ_c, the location of the closest point on the target facet becomes $\mathbf{x}_c$ as shown in Fig. 10.5 and, using Eq. (10.217), is given by

$$\mathbf{x}_c = N_\alpha(\xi_c)\,\mathbf{x}_\alpha \tag{10.220}$$

For *frictionless* contact only normal tractions are involved on the surfaces between the two bodies; thus sliding can occur without generation of tangential forces and the traction is given by

$$\mathbf{t} = \lambda_n\,\mathbf{n}_c \tag{10.221}$$

where λ_n is the magnitude of a normal traction applied to the contact target and $\mathbf{n}_c$ is a unit normal to the master facet at the point ξ_c. This case can be included by appending the variation of a Lagrange multiplier term to the Galerkin (weak) form describing equilibrium of the problem for each contact slave node. This term may be expressed as

$$\Pi_c = \left(\lambda_n \mathbf{n}_c^T\right)\left(g_n \mathbf{n}_c\right) A_c \tag{10.222}$$

where at the solution point ξ_c

$$g_n \mathbf{n}_c = \mathbf{x}_s - \mathbf{x}(\xi_c) \tag{10.223}$$

and A_c is a surface area associated with the slave node.* In the development summarized here the surface area term is based on the reference configuration and kept constant during the analysis. Thus, the traction measure λ_n is a reference surface measure which must be scaled by the current surface area ratio to obtain the magnitude of the traction in the deformed state.

Use of the Lagrange multiplier form introduces an additional unknown λ_n for each master–slave contact pair. Since a contact traction interacts with both bodies it must be determined as part of the solution of the global equilibrium equations. Of course, we again can eliminate the contact tractions by using a *penalty* form for the constraint in a manner similar to that used for treating node–node contact [Eq. (10.213)]. However, even then the problem is more complex as we do not know *a priori* which master facet a contact node will interact with. This implies that the non-zero structure of the global tangent matrix will change during the solution of any contact problem and continual updates are required to describe the profile or sparse structure.

The variation of the potential given in Eq. (10.222) may be expressed as

$$\delta\Pi_c = \left[\delta\left(\lambda_n \mathbf{n}_c^T\right) g_n \mathbf{n}_c + \left(\delta\mathbf{x}_s^T - \delta\mathbf{x}_c^T\right)\mathbf{n}_c \lambda_n\right] A_c \tag{10.224}$$

where the variation of the coordinate $\mathbf{x}_c$ is given by

$$\delta\mathbf{x}_c = N_\alpha \delta\mathbf{x}_\alpha + \mathbf{x}_{,\xi}\,\delta\xi_c \tag{10.225}$$

with $\mathbf{x}_{,\xi}$ computed by differentiation of the interpolation functions in Eq. (10.217) and then evaluated at ξ_c. Noting now that

$$\mathbf{n}_c^T \mathbf{n}_c = 1 \qquad \text{and} \qquad \mathbf{n}_c^T \delta\mathbf{n}_c = 0 \tag{10.226}$$

the added contact term may be written in matrix form as

$$\delta\Pi_c = \left[\delta\lambda_n, \quad \delta\mathbf{x}_s^T, \quad \delta\mathbf{x}_\alpha^T, \quad \delta\xi_c\right] \begin{Bmatrix} g_n A_c \\ \lambda_n A_c \mathbf{n}_c \\ -\lambda_n A_c N_\alpha \mathbf{n}_c \\ -\lambda_n A_c \mathbf{x}_{,\xi}^T \mathbf{n}_c \end{Bmatrix} \tag{10.227}$$

where we note that the entry multiplying $\delta\xi_c$ vanishes at $\xi = \xi_c$. At a solution g_n should also be zero; however, in some iterations it may be non-zero as a result of penetrations occurring before the contact term is inserted (also for penalty type methods it will never be exactly zero).

* One is tempted to use $\mathbf{n}_c^T \mathbf{n}_c = 1$ to simplify Eq. (10.222); however, doing this the advantages of Eq. (10.223) are then lost and computation of the tangent becomes more complex.

Using a Newton–Raphson solution procedure generates a tangent matrix expressed formally by the term

$$d(\delta\Pi) = \{(\delta\mathbf{x}_s^T - \delta\mathbf{x}_c^T)\, d(\lambda_n \mathbf{n}_c) + \delta(\lambda_n \mathbf{n}_c^T)(d\mathbf{x}_s - d\mathbf{x}_c) - \lambda_n\, d(\delta\mathbf{x}_c^T)\mathbf{n}_c + g_n \mathbf{n}^T d[\delta(\lambda_n \mathbf{n}_c)]\} A_c \tag{10.228}$$

The linearization is straightforward except for the terms involving $\mathbf{n}_c$. It is necessary here to use a form which does not divide by g_n [which would result in using Eq. (10.223) directly]. This may be achieved by introducing a surface unit tangent vector

$$\mathbf{t}_c = \frac{\mathbf{x}_{,\xi}}{||\mathbf{x}_{,\xi}||} \tag{10.229}$$

and using

$$\mathbf{n}_c = \mathbf{t}_c \times \mathbf{e}_3 \qquad \text{and} \qquad \delta\mathbf{n}_c = \delta\mathbf{t}_c \times \mathbf{e}_3 \tag{10.230}$$

where $\mathbf{e}_3$ is a unit vector normal to the plane of deformation. Now

$$\delta\mathbf{t}_c = \frac{1}{||\mathbf{x}_{,\xi}||}\left[\mathbf{I} - \mathbf{t}_c\mathbf{t}_c^T\right]\delta\mathbf{x}_{,\xi} = \frac{1}{||\mathbf{x}_{,\xi}||}\mathbf{n}_c\mathbf{n}_c^T\delta\mathbf{x}_{,\xi} \tag{10.231}$$

which upon using Eq. (10.230) gives

$$\delta\mathbf{n}_c = -\frac{\mathbf{x}_{,\xi}\mathbf{n}_c^T\,\delta\mathbf{x}_{,\xi}}{||\mathbf{x}_{,\xi}||^2} \tag{10.232}$$

Performing all linearizations results in the tangent term

$$\left[\delta\lambda_n,\quad \delta\mathbf{x}_s^T,\quad \delta\mathbf{x}_\alpha^T,\quad \delta\xi_c\right]\begin{bmatrix} 0 & A_c\mathbf{n}_c^T & -N_\beta A_c\mathbf{n}_c^T & 0 \\ A_c\mathbf{n}_c & \mathbf{0} & \mathbf{G}_{s\beta} & \mathbf{G}_{s\xi} \\ -N_\alpha A_c\mathbf{n}_c & \mathbf{G}_{\alpha s} & \mathbf{G}_{\alpha\beta} & \mathbf{G}_{\alpha\xi} \\ 0 & \mathbf{G}_{\xi s} & \mathbf{G}_{\xi\beta} & G_{\xi\xi}\end{bmatrix}\begin{Bmatrix} d\lambda_n \\ d\mathbf{x}_s \\ d\mathbf{x}_\alpha \\ d\xi_c\end{Bmatrix} \tag{10.233}$$

where

$$\begin{aligned}
\mathbf{G}_{s\beta} &= -\frac{\lambda_n A_c N_{\beta,\xi}}{||\mathbf{x}||^2}\mathbf{x}_{,\xi}\mathbf{n}_c = \mathbf{G}_{\beta s}^T \\
\mathbf{G}_{s\xi} &= -\frac{\lambda_n A_c \kappa_c}{||\mathbf{x}||^2}\mathbf{x}_{,\xi} = \mathbf{G}_{\xi s}^T \\
\mathbf{G}_{\alpha\beta} &= \frac{\lambda_n A_c}{||\mathbf{x}||^2}\left[N_\alpha N_{\beta,\xi}\mathbf{x}_{,\xi}\mathbf{n}_c^T + N_{\alpha,\xi}N_\beta\mathbf{n}_c\mathbf{x}_{,\xi}^T - g_n N_{\alpha,\xi}N_{\beta,\xi}\mathbf{n}_c\mathbf{n}_c^T\right] \\
\mathbf{G}_{\alpha\xi} &= \frac{\lambda_n A_c \kappa_c}{||\mathbf{x}||^2}\left[N_\alpha\mathbf{x}_{,\xi} - g_n N_{\alpha,\xi}\mathbf{n}_c\right] \\
G_{\xi\xi} &= \lambda_n A_c\left[1 - \frac{g_n\kappa_c}{||\mathbf{x}_{,\xi}||^2}\right]
\end{aligned} \tag{10.234}$$

in which $\kappa_c = \mathbf{n}_c^T \mathbf{x}_{,\xi\xi}$ is related to the target facet curvature. Clearly, for a straight contact facet with linear interpolation defining $\mathbf{x}(\xi)$ many of the terms in Eq. (10.234) vanish and the tangent is greatly simplified. In such a case, however, special attention must be devoted to the case where a slave node is very near a master node and oscillations on which facet to use occur during subsequent Newton–Raphson iterations. Here an expedient solution is to use concepts from multi-surface plasticity to define a 'continuous' approximation for the normal. This leads to additional considerations which are not given here and are left for the reader to develop (see reference 55).

The tangent matrix may be reduced by eliminating the dependence on $d\xi_c$. The reduced tangent then depends only on the Lagrange multiplier λ_n and geometry terms computed from current values of nodal parameters.

Extension to three-dimensional problems is straightforward and involves addition of a second natural coordinate η and replacing $\mathbf{e}_3$ by a second surface tangent vector deduced from $\mathbf{x}_{,\eta}$. Extension to include frictional effects may also be performed and the reader is referred to the literature for additional details.[67–81]

10.9 Numerical examples

10.9.1 Node–surface contact between discs

As a first example here we consider the contact problem previously solved using a node–node approach. In that case we observed a small but significant discontinuity between the contours of vertical stress between the bodies, indicating that traction is not correctly transmitted across the section. Here we use the node–surface method given above in which the contact area of each body is taken as the boundary of elements. The solution is achieved by using a penalty method and a *two-pass* solution procedure where on the first pass one body is the slave and the other the master and on the second pass the designation is reversed. This approach has been shown to be necessary in order to satisfy the mixed patch test for contact.[65] The results using this approach are shown in Fig. 10.6. For the solution, the two-dimensional plane

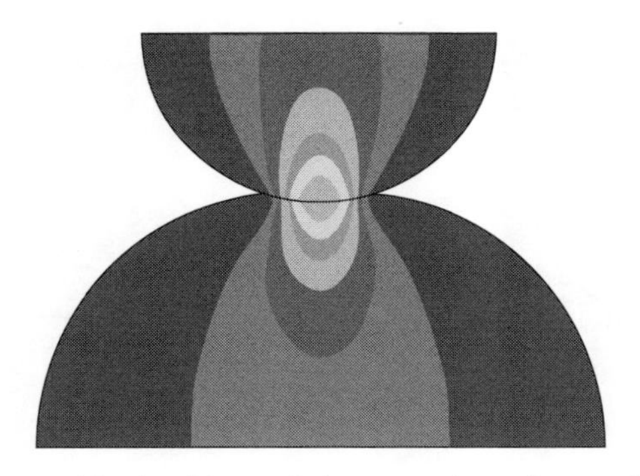

Fig. 10.6 Contact between semicircular disks: vertical contours for node-to-surface solution.

strain finite deformation displacement element described in Sec. 10.3.2 is used with material behaviour given by the neo-Hookean hyperelastic model described in Sec. 10.7.1. Properties are: $E = 100\,000$ for the upper body and $E = 1000$ for the lower body. A Poisson ratio of $\nu = 0.25$ is used to compute Lamé parameters λ and μ. As can readily be seen in the figure the results obtained are significantly better than those from the node-to-node analysis.

10.9.2 Upsetting of a cylindrical billet

To illustrate performance in highly strained regimes, we consider large compression of a three-dimensional cylindrical billet. The initial configuration is a cylinder

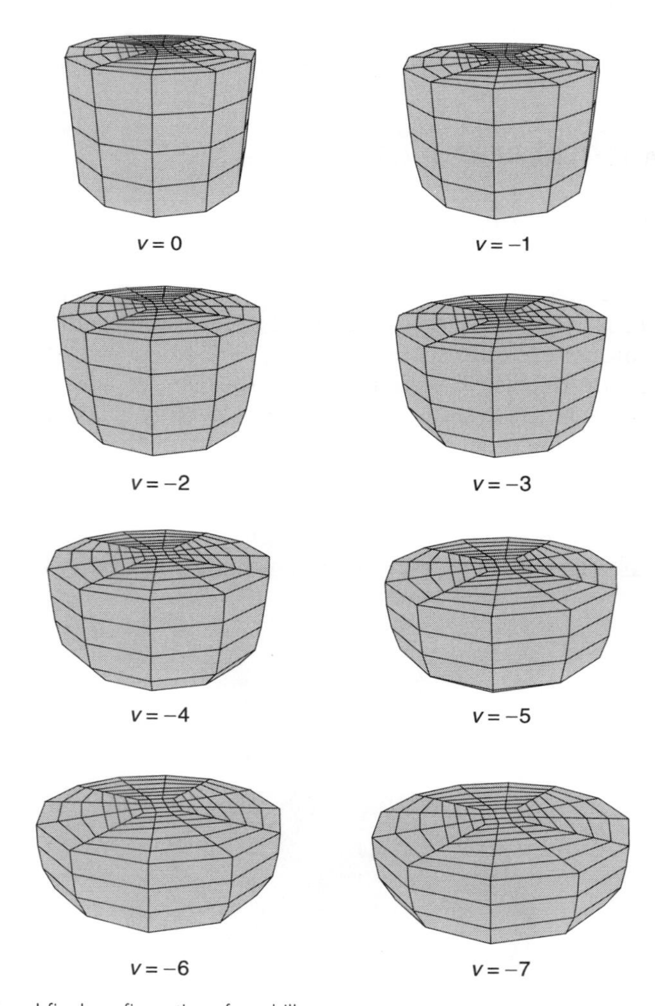

Fig. 10.7 Initial and final configurations for a billet.

with radius, $r = 10$, and height, $h = 15$. The mesh consists of 459 eight-noded hexahedral elements based on the mixed-enhanced formulation presented in Sec. 10.5. The billet is loaded via displacement control on the upper surface, while the lower edge is fully restrained. A full Newton–Raphson solution process is used in which the upper displacement is increased by increments of displacement equal to 0.25 units.

To prevent penetration with the rigid base during large deformations a simple node-on-node penalty formulation with a penalty parameter $k = 10^6$ is defined for nodes on the lower part of the cylindrical boundary. A neo-Hookean material model with $\lambda = 10^4$ and $\mu = 10$ is used for the simulation. Figure 10.7 depicts the initial mesh and progression of deformation.

10.9.3 Necking of circular bar

The last example we include is a large deformation plasticity problem. Here we consider the three-dimensional behaviour of a cylindrical bar subjected to tension. In the presence of plastic deformation an unstable plastic *necking* will occur at some location along a bar of mild steel, or similar elasto-plastic behaving material. This is easily observed from the tension test of a cylindrical specimen which tapers by a small amount to a central location to ensure that the location of necking will occur in a specified location. A finite element model is constructed having the same taper, and here only one-eighth of the bar need be modelled as shown in Fig. 10.8(a). In Fig. 10.8(b) we show the half-bar model which is projected by symmetry and reflection and on which the behaviour will be illustrated.

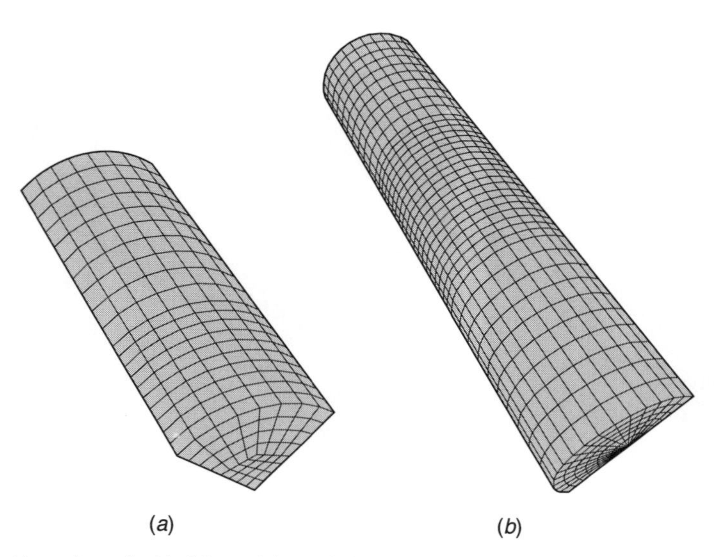

Fig. 10.8 Necking of a cylindrical bar: eight-noded elements. (a) Finite element model; (b) half-bar by symmetry.

This problem has been studied by several authors and here the properties are taken as described by Simo and co-workers.[5,9,24] The one-eighth quadrant model consists of 960 eight-noded hexahedra of the mixed type discussed in Sec. 10.5. The radius at the loading end is taken as $R = 6.413$ and a uniform taper to a central radius of $R_c = 0.982 \times R$ is used. The total length of the bar is $L = 53.334$ (giving a half length of 25.667). The mesh along the length is uniform between the centre (0) and a distance of 10, and again from 10 to the end. A blending function mesh generation is used (see Sec. 9.12, Volume 1) to ensure that exterior nodes lie exactly on the circular radius. This ensures that, as much as possible for the discretization employed, the response will be axisymmetric.

The finite deformation plasticity model based on the logarithmic stretch elastic behaviour from Sec. 10.7.2 and the finite plasticity as described in Sec. 10.7.3 is used for the analysis. The material properties used are as follows: elastic properties are $K = 164.21$ and $G = 80.1938$; a J_2 plasticity model in terms of principal Kirchhoff stresses τ_i with an initial yield in tension of $\tau_y = 0.45$ is used. Only isotropic hardening is included and a saturation type model defined by

$$\kappa = H_i\,\varepsilon^{\mathrm{p}} + \left[\tau_y^{\infty} - \tau_y\right]\left(1 - \exp\left[-\bar{\delta}\varepsilon^{\mathrm{p}}\right]\right)$$

with the parameters

$$H_i = 0.12924, \qquad \tau_y^{\infty} = 0.715, \quad \text{and} \quad \bar{\delta} = 16.93$$

is employed. An alternative to this is a piecewise linear behavior as suggested by some authors; however, the above model is very easy to implement and gives a smooth behaviour with increase in the accumulated plastic strain ε^{p} as the hardening parameter κ.

In Fig. 10.9 we show the deformed configuration of the bar at an elongation of 22.5 per cent (elongation = 6 units). Figure 10.9(a) has the contours of the first

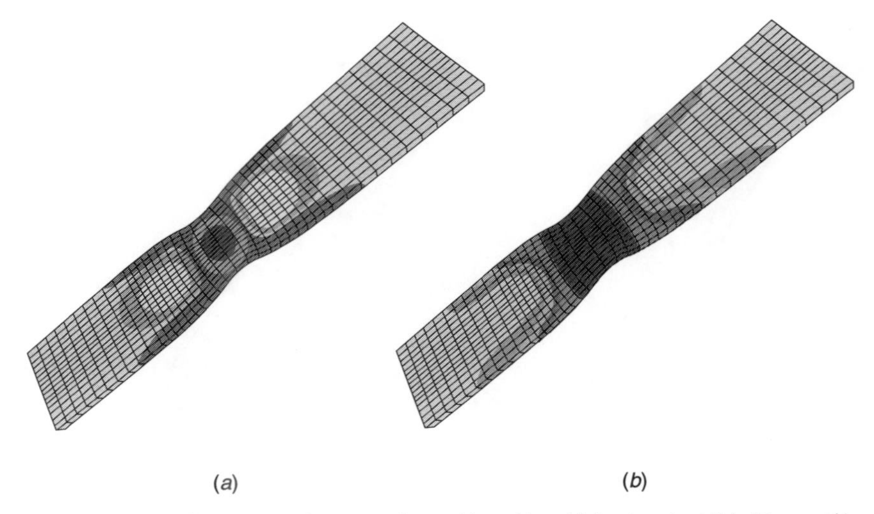

(*a*) (*b*)

Fig. 10.9 Deformed configuration and contours for necking of bar: (a) first invariant (I_1); (b) second invariant (J_2).

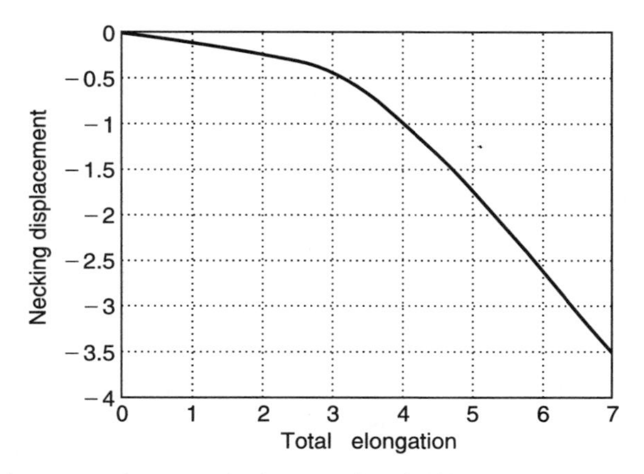

Fig. 10.10 Neck radius versus elongation displacement for a half-bar.

invariant of Cauchy stress superposed and Fig. 10.9(b) those for the second invariant of the deviator stresses. It is apparent that considerable variation in pressure (first invariant divided by 3) occurs in the necked region, whereas the values of the second deviator invariant vary more smoothly in this region. A plot of the radius of the bar at the centre is shown Fig. 10.10 for different elongation values.

This example is quite sensitive to solve as the response involves an unstable behaviour of the necking process. Use of a full Newton–Raphson scheme was generally ineffective in this regime and here a modified Newton–Raphson scheme together with a BFGS (Broyden–Fletcher–Goldfarb–Shanno) secant update was employed (see Sec. 2.2.4). When near convergence was achieved the algorithm was then switched to a full Newton–Raphson process and during the last iterations quadratic convergence was obtained when used with an algorithmic consistent tangent matrix as described in Sections 10.7.2 and 10.7.3.

10.10 Concluding remarks

This chapter presents a unified approach for all finite deformation problems. The various procedures for solution of the resulting non-linear algebraic system have followed those presented in Chapter 2. Although not discussed extensively in the chapter, the extension to consider transient (dynamic) situations is easily accomplished. The long-term integration of dynamic problems occasionally presents difficulties using the time integration procedures discussed in Volume 1. Here schemes which conserve momentum and energy for hyperelastic materials can be considered as alternatives, and the reader is referred to literature on the subject for additional details.[56,82–87]

We have also presented some mixed forms for developing elements which perform well at finite strains and with materials which can exhibit nearly incompressible behaviour. These elements are developed in a form which allow the introduction of finite elastic and inelastic material models without difficulty. Indeed, we have shown that there is no need to decouple the constitutive behaviour between volumetric and deviatoric response as often assumed in many presentations. We usually find that transformation to a current configuration form in which either the Kirchhoff stress or the Cauchy stress is used directly will lead to a form which admits a simple extension of existing small deformation finite element procedures for developing the necessary residual (force) and stiffness matrices. An exception here is the presentation of the mixed–enhanced form in which all basic development is shown using the deformation gradient and first Piola–Kirchhoff stress. Here we could express final answers in a current configuration form also, but we leave these steps for the reader to perform.

In this chapter we have concentrated on developments for continuum problems in which the full two- or three-dimensional equations are modelled by finite elements. In the next chapter we address problems which can be represented using beam (rod), plate, or shell models and thus permit a reduction of the discretization space to one or two dimensions.

References

1. P. Chadwick. *Continuum Mechanics*, John Wiley, New York, 1976.
2. M.E. Gurtin. *An Introduction to Continuum Mechanics*, Academic Press, New York, 1981.
3. I.H. Shames and F.A. Cozzarelli. *Elastic and Inelastic Stress Analysis*, revised edition, Taylor & Francis. Washington, DC, 1997.
4. J. Bonet and R.D. Wood. *Nonlinear Continuum Mechanics for Finite Element Analysis*, Cambridge University Press, Cambridge, 1997.
5. J.C. Simo and T.J.R. Hughes. *Interdisciplinary Applied Mathematics, Volume 7: Computational Inelasticity*, Springer-Verlag, Berlin, 1998.
6. J.C. Simo and R.L. Taylor. Quasi-incompressible finite elasticity in principal stretches: continuum basis and numerical algorithms. *Comp. Meth. Appl. Mech. Eng.*, **85**, 273–310, 1991.
7. R.L. Taylor. A mixed–enhanced formulation for tetrahedral finite elements. *Int. J. Num. Meth. Eng.*, **47**, 205–27, 2000.
8. B. Fraeijs de Veubeke. Displacement and equilibrium models in finite element method. In O.C. Zienkiewicz and G.S. Holister (eds), *Stress Analysis*, pp. 145–97. John Wiley, Chichester, Sussex, 1965.
9. J.C. Simo and F. Armero. Geometrically non-linear enhanced strain mixed methods and the method of incompatible modes. *Int. J. Num. Meth. Eng.*, **33**, 1413–49, 1992.
10. U. Andelfinger, E. Ramm and D. Roehl. 2d- and 3d-enhanced assumed strain elements and their application in plasticity. In D. Owen, E. Oñate and E. Hinton (eds), *Proceedings of the 4th International Conference on Computational Plasticity*, pp. 1997–2007, Pineridge Press, Swansea, 1992.
11. J.C. Simo, F. Armero and R.L. Taylor. Improved versions of assumed enhanced strain tri-linear elements for 3d finite deformation problems. *Comp. Meth. Appl. Mech. Eng.*, **110**, 359–86, 1993.
12. P. Wriggers and G. Zavarise. Application of augmented Lagrangian techniques for non-linear constitutive laws in contact interfaces. *Communications in Numerical Methods in Engineering*, **9**, 813–24, 1993.
13. S. Glaser and F. Armero. On the formulation of enhanced strain finite elements in finite deformation. *Engineering Computations*, **14**, 759–91, 1996.

14. M. Bischoff, E. Ramm and D. Braess. A class of equivalent enhanced assumed strain and hybrid stress finite elements. *Computational Mechanics*, **22**, 443–9, 1999.
15. E.P. Kasper and R.L. Taylor. A mixed–enhanced strain method: part 1 – geometrically linear problems. *Computers and Structures*, **75**(3), 237–50, 2000.
16. T.H.H. Pian and K. Sumihara. Rational approach for assumed stress finite elements. *Int. J. Num. Meth. Eng.*, **20**, 1685–95, 1985.
17. E.L. Wilson. The static condensation algorithm. *Int. J. Num. Meth. Eng.*, **8**, 199–203, 1974.
18. E.P. Kasper and R.L. Taylor. A mixed–enhanced strain method: part 2 – geometrically nonlinear problems. *Computers and Structures*, **75**(3), 215–60, 2000.
19. J.T. Oden. Discussion on 'Finite element analysis of non-linear structures' by R.H. Mallett and P.V. Marcal. *Proc. Am. Soc. Civ. Eng.*, **95**(ST6), 1376–81, 1969.
20. H.D. Hibbitt, P.V. Marcal and J.R. Rice. A finite element formulation for problems of large strain and large displacement. *International Journal of Solids and Structures*, **6**, 1069–86, 1970.
21. J.C. Simo, R.L. Taylor and P. Wriggers. A note on finite element implementation of pressure boundary loading. *Communications in Applied Numerical Methods*, **7**, 513–25, 1991.
22. K. Schweizerhof. *Nitchlineare Berechnung von Tragwerken unter verformungsabhanpiger belastung mit finiten Elementen*, PhD dissertation, University of Stuttgart, Stuttgart, 1982.
23. K. Schweizerhof and E. Ramm. Displacement dependent pressure loads in non-linear finite element analysis. *Computers and Structures*, **18**(6), 1099–114, 1984.
24. J.C. Simo. Topics on the numerical analysis and simulation of plasticity. In P.G. Ciarlet and J.L. Lions (eds), *Handbook of Numerical Analysis*, Volume III, pp. 183–499, Elsevier Science Publisher, Amsterdam, 1999.
25. M. Mooney. A theory of large elastic deformation. *J. Appl. Physics*, **1**, 582–92, 1940.
26. R.W. Ogden. *Non-linear Elastic Deformations*, Ellis Horwood (reprinted by Dover, 1997), Chichester, Sussex, 1984.
27. J. Mandel. Contribution theorique a l'etude de l'ecrouissage et des lois de l'ecoulement plastique. In *Proc. 11th Int. Cong. of Appl. Mech.*, pp. 502–9, 1964.
28. E.H. Lee and D.T. Liu. finite strain elastic–plastic theory particularly for plane wave analysis. *J. Appl. Phys.*, 38, 1967.
29. E.H. Lee. Elastic–plastic deformations at finite strains. *J. Appl. Mech.*, **36**, 1–6, 1969.
30. J.C. Simo. Algorithms for multiplicative plasticity that preserve the form of the return mappings of the infinitesimal theory. *Comp. Meth. Appl. Mech. Eng.*, **99**, 61–112, 1992.
31. F. Auricchio and R.L. Taylor. A return-map algorithm of general associative isotropic elastoplastic materials in large deformation regimes. *Int. J. Plasticity*, **15**, 1359–78, 1999.
32. L.E. Malvern. *Introduction to the Mechanics of a Continuous Medium*, Prentice-Hall, Englewood Cliffs, NJ, 1969.
33. C. Truesdell and W. Noll. The non-linear field theories of mechanics. In S. Flügge (ed.), *Handbuch der Physik III/3*, Springer-Verlag, Berlin, 1965.
34. J.E. Marsden and T.J.R. Hughes. *Mathematical Foundations of Elasticity*, Dover, New York, 1994.
35. S.K. Chan and I.S. Tuba. A finite element method for contact problems of solid bodies – part I. Theory and validation. *Int. J. Mech. Sci.*, **13**, 615–25, 1971.
36. S.K. Chan and I.S. Tuba. A finite element method for contact problems of solid bodies – part II. Application to turbine blade fastenings. *Int. J. Mech. Sci.*, **13**, 627–39, 1971.
37. J.J. Kalker and Y. van Randen. A minimum principle for frictionless elastic contact with application to non-Hertzian half-space contact problems. *J. Engr. Math.*, **6**, 193–206, 1972.
38. A. Francavilla and O.C. Zienkiewicz. A note on numerical computation of elastic contact problems. *Int. J. Num. Meth. Eng.*, **9**, 913–24, 1975.

39. T.J.R. Hughes, R.L. Taylor, J.L. Sackman, A. Curnier and W. Kanoknukulchai. A finite element method for a class of contact–impact problems. *Comp. Meth. Appl. Mech. Eng.*, **8**, 249–76, 1976.
40. J.O. Hallquist, G.L. Goudreau and D.J. Benson. Sliding interfaces with contact–impact in large scale Lagrangian computations. *Comp. Meth. Appl. Mech. Eng.*, **51**, 107–37, 1985.
41. J.A. Landers and R.L. Taylor. An augmented Lagrangian formulation for the finite element solution of contact problems. Technical Report SESM 85/09, University of California, Berkeley, CA, 1985.
42. P. Wriggers and J.C. Simo. A note on tangent stiffness for fully nonlinear contact problems. *Comm. Appl. Num. Meth.*, **1**, 199–203, 1985.
43. J.C. Simo, P. Wriggers and R.L. Taylor. A perturbed Lagrangian formulation for the finite element solution of contact problems. *Comp. Meth. Appl. Mech. Eng.*, **50**, 163–80, 1985.
44. J.C. Simo, P. Wriggers, K.H. Schweizerhof and R.L Taylor. Finite deformation post-buckling analysis involving inelasticity and contact constraints. *Int. J. Num. Meth. Eng.*, **23**, 779–800, 1986.
45. K.-J. Bathe and A.B. Chaudhary. A solution method for planar and axisymmetric contact problems. *Int. J. Num. Meth. Eng.*, **21**, 65–88, 1985.
46. K.-J. Bathe and P.A. Bouzinov. On the constraint function method for contact problems. *Computers and Structures*, **64**(5/6), 1069–85, 1997.
47. J.J. Kalker. Contact mechanical algorithms. *Comm. Appl. Num. Meth.*, **4**, 25–32, 1988.
48. N. Kikuchi and J.T. Oden. *Contact Problems in Elasticity: A Study of Variational Inequalities and Finite Element Methods*, Volume 8, SIAM, Philadelphia, PA, 1988.
49. H. Parisch. A consistent tangent stiffness matrix for three dimensional non-linear contact analysis. *Int. J. Num. Meth. Eng.*, **28**, 1803–12, 1989.
50. D.J. Benson and J.O. Hallquist. A single surface contact algorithm for the post-buckling analysis of shell structures. *Comp. Meth. Appl. Mech. Engr.*, **78**, 141–63, 1990.
51. T. Belytschko and M.O. Neal. Contact–impact by the pinball algorithm with penalty and Lagrangian methods. *Int. J. Num. Meth. Eng.*, **31**, 547–72, 1991.
52. R.L. Taylor and P. Papadopoulos. On a finite element method for dynamic contact–impact problems. *Int. J. Num. Meth. Eng.*, pp. 2123–39, 1992.
53. Z. Zhong and J. Mackerle. Static contact problems – a review. *Engineering Computations*, **9**, 3–37, 1992.
54. J.-H. Heegaard and A. Curnier. An augmented Lagrangian method for discrete large-slip contact problems. *Int. J. Num. Meth. Eng.*, **36**, 569–93, 1993.
55. T.A. Laursen and S. Govindjee. A note on the treatment of frictionless contact between non-smooth surfaces in fully non-linear problems. *Communications in Numerical Methods in Engineering*, **10**, 869–78, 1994.
56. T.A. Laursen and V. Chawla. Design of energy conserving algorithms for frictionless dynamic contact problems. *Int. J. Num. Meth. Eng.*, **40**, 863–86, 1997.
57. P. Papadopoulos and R.L. Taylor. A mixed formulation for the finite element solution of contact problems. *Comp. Meth. Appl. Mech. Eng.*, **94**, 373–89, 1994.
58. P. Papadopoulos, R.E. Jones and J.M. Solberg. A novel finite element formulation for frictionless contact problems. *Int. J. Num. Meth. Eng.*, **38**, 2603–17, 1995.
59. P. Papadopoulos and J.M. Solberg. A Lagrange multiplier method for the finite element solution of frictionless contact problems. *Math. Comput. Modelling*, **28**, 373–84, 1998.
60. J.M. Solberg and P. Papadopoulos. A finite element method for contact/impact. *Finite Elements in Analysis and Design*, **30**, 297–311, 1998.
61. E. Bittencourt and G.J. Creus. Finite element analysis of three-dimensional contact and impact in large deformation problems. *Computers and Structures*, **69**, 219–34, 1998.

62. M. Cuomo and G. Ventura. Complementary energy approach to contact problems based on consistent augmented Lagrangian formulation. *Mathematical and Computer Modelling*, **28**, 185–204, 1998.
63. C. Kane, E.A. Repetto, M. Ortiz and J.E. Marsden. Finite element analysis of nonsmooth contact. *Comp. Meth. Appl. Mech. Eng.*, **180**, 1–26, 1999.
64. I. Paczelt, B.A. Szabo and T. Szabo. Solution of contact problem using the *hp*-version of the finite element method. *Computers and Mathematics with Applications*, **38**, 49–69, 1999.
65. R.L. Taylor and P. Papadopoulos. A patch test for contact problems in two dimensions. In P. Wriggers and W. Wagner (eds), *Nonlinear Computational Mechanics*, pp. 690–702, Springer, Berlin, 1991.
66. G. Zavarise and P. Wriggers. A superlinear convergent augmented Lagrangian procedure for contact problems. *Engineering Computations*, **16**, 88–119, 1999.
67. A.B. Chaudhary and K.-J. Bathe. A solution method for static and dynamic analysis of three-dimensional contact problems with friction. *Computers and Structures*, **24**, 855–73, 1986.
68. J.-W. Ju and R.L. Taylor. A perturbed Lagrangian formulation for the finite element solution of nonlinear frictional contact problems. *Journal de Mecanique Theorique et Appliquée*, **7**(Supplement, 1), 1–14, 1988.
69. A. Curnier and P. Alart. A generalized Newton method for contact problems with friction. *Journal de Mecanique Theorique et Appliquée*, **7**, 67–82, 1988.
70. P. Wriggers, T. Vu Van and E. Stein. Finite element formulation of large deformation impact–contact problems with friction. *Computers and Structures*, **37**, 319–31, 1990.
71. P. Alart and A. Curnier. A mixed formulation for frictional contact problems prone to Newton like solution methods. *Comp. Meth. Appl. Mech. Eng.*, **92**, 353–75, 1991.
72. J.C. Simo and T.A. Laursen. An augmented Lagrangian treatment of contact problems involving friction. *Computers and Structures*, **42**, 97–116, 1992.
73. T.A. Laursen and J.C. Simo. A continuum-based finite element formulation for the implicit solution of multibody, large-deformation, frictional, contact problems. *Int. J. Num. Meth. Eng.*, **36**, 3451–86, 1993.
74. T.A. Laursen and J.C. Simo. Algorithmic symmetrization of Coulomb frictional problems using augmented Lagrangians. *Comp. Meth. Appl. Mech. Eng.*, **108**, 133–46, 1993.
75. T.A. Laursen and V.G. Oancea. Automation and assessment of augmented lagrangian algorithms for frictional contact problems. *J. Appl. Mech*, **61**, 956–63, 1994.
76. A. Heege and P. Alart. A frictional contact element for strongly curved contact problems. *Int. J. Num. Meth. Eng.*, **39**, 165–84, 1996.
77. C. Agelet de Saracibar. A new frictional time integration algorithm for large slip multibody frictional contact problems. *Comp. Meth. Appl. Mech. Eng.*, **142**, 303–34, 1997.
78. W. Ling and H.K. Stolarski. On elasto-plastic finite element analysis of some frictional contact problems with large sliding. *Engineering Computations*, **14**, 558–80, 1997.
79. F. Jourdan, P. Alart and M. Jean. A Gauss–Seidel like algorithm to solve frictional contact problems. *Comp. Meth. Appl. Mech. Eng.*, **155**, 31–47, 1998.
80. C. Agelet de Saracibar. Numerical analysis of coupled thermomechanical frictional contact: computational model and applications. *Archives of Computational Methods in Engineering*, **5**(3), 243–301, 1998.
81. G. Pietrzak and A. Curnier. Large deformation frictional contact mechanics: continuum formulation and augmented Lagrangian treatment. *Comp. Meth. Appl. Mech. Eng.*, **177**, 351–81, 1999.
82. J.C. Simo and N. Tarnow. The discrete energy–momentum method: conserving algorithm for nonlinear elastodynamics. *Zeitschrift für Angewandte Mathematik und Physik*, **43**, 757–93, 1992.

83. J.C. Simo and N. Tarnow. Exact energy–momentum conserving algorithms and symplectic schemes for nonlinear dynamics. *Comp. Meth. Appl. Mech. Eng.*, **100**, 63–116, 1992.
84. N. Tarnow. *Energy and Momentum Conserving Algorithms for Hamiltonian Systems in the Nonlinear Dynamics of Solids*, PhD thesis, Department of Civil Engineering, Stanford University, Stanford, CA, 1993.
85. O. Gonzalez. *Design and Analysis of Conserving Integrators for Nonlinear Hamiltonian Systems with Symmetry*, PhD thesis, Department of Civil Engineering, Stanford University, Stanford, CA, 1996.
86. M.A. Crisfield and J. Shi. An energy conserving co-rotational procedure for non-linear dynamics with finite elements. *Nonlinear Dynamics*, **9**, 37–52, 1996.
87. U. Galvanetto and M.A. Crisfield. An energy-conserving co-rotational procedure for the dynamics of planar beam structures. *Int. J. Num. Meth. Eng.*, **39**, 2265–82, 1996.

11

Non-linear structural problems – large displacement and instability

11.1 Introduction

In the previous chapter the question of finite deformations and non-linear material behaviour was discussed and methods were developed to allow the standard linear forms to be used in an iterative way to obtain solutions. In the present chapter we consider the more specialized problem of large displacements but with strains restricted to be small. Generally, we shall assume that 'small strain' stress–strain relations are adequate but for accurate determination of the displacements geometric non-linearity needs to be considered. Here, for instance, stresses arising from membrane action, usually neglected in plate flexure, may cause a considerable decrease of displacements as compared with the linear solution discussed in Chapters 4 and 5, even though displacements remain quite small. Conversely, it may be found that a load is reached where indeed a state may be attained where load-carrying capacity *decreases* with continuing deformation. This classic problem is that of structural stability and obviously has many practical implications. The applications of such an analysis are clearly of considerable importance in aerospace and automotive engineering applications, design of telescopes, wind loading on cooling towers, box girder bridges with thin diaphrams and other relatively 'slender' structures.

In this chapter we consider the above class of problems applied to beam, plate, and shell systems by examining the basic non-linear equilibrium equations. Such considerations lead also to the formulation of classical initial stability problems. These concepts are illustrated in detail by formulating the large deflection and initial stability problems for beams and flat plates. A lagrangian approach is adopted throughout in which displacements are referred to the original (reference) configuration.

11.2 Large displacement theory of beams

11.2.1 Geometrically exact formulation

In Sec. 2.10 of Volume 1 we briefly described the behaviour for the bending of a beam for the small strain theory. Here we present a form for cases in which large

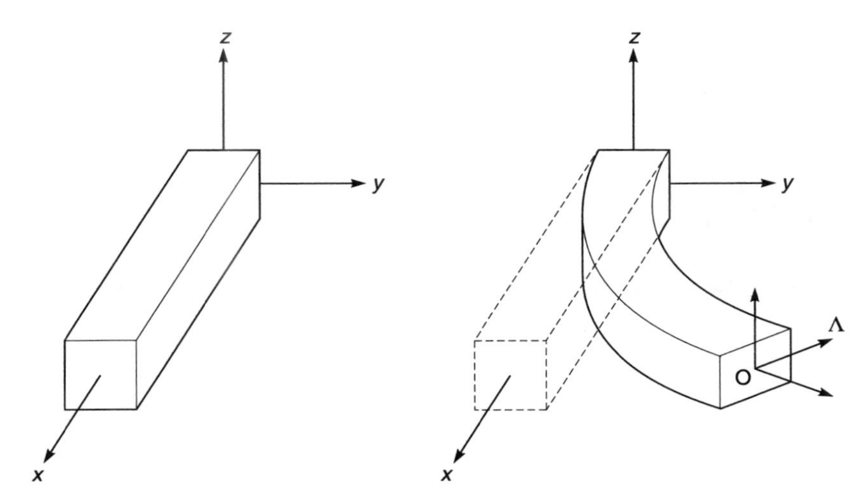

Fig. 11.1 Finite motion of three-dimensional beams.

displacements with finite rotations occur. We shall, however, assume that the strains which result are small. A two-dimensional theory of beams (rods) was developed by Reissner[1] and was extended to a three-dimensional dynamic form by Simo.[2] In these developments the normal to the cross-section is followed, as contrasted to following the tangent to the beam axis, by an orthogonal frame. Here we consider an initially straight beam for which the orthogonal triad of the beam cross-section is denoted by the vectors $\mathbf{a}_i$ (Fig. 11.1). The motion for the beam can then be written as

$$\phi_i \equiv x_i = x_i^0 + \Lambda_{iI} Z_I \tag{11.1}$$

where the orthogonal matrix is related to the $\mathbf{a}_i$ vectors as

$$\boldsymbol{\Lambda} = [\mathbf{a}_1 \quad \mathbf{a}_2 \quad \mathbf{a}_3] \tag{11.2}$$

If we assume that the reference coordinate X_1 (X) is the beam axis and X_2, X_3 (Y, Z) are the axes of the cross-section the above motion may be written in matrix form as

$$\begin{Bmatrix} x_1 \\ x_2 \\ x_3 \end{Bmatrix} = \begin{Bmatrix} x \\ y \\ z \end{Bmatrix} = \begin{Bmatrix} X \\ 0 \\ 0 \end{Bmatrix} + \begin{Bmatrix} u \\ v \\ w \end{Bmatrix} + \begin{bmatrix} \Lambda_{11} & \Lambda_{12} & \Lambda_{13} \\ \Lambda_{21} & \Lambda_{22} & \Lambda_{23} \\ \Lambda_{31} & \Lambda_{32} & \Lambda_{33} \end{bmatrix} \begin{Bmatrix} 0 \\ Y \\ Z \end{Bmatrix} \tag{11.3}$$

where $u(X)$, $v(X)$, and $w(X)$ are displacements of the beam reference axis and where $\Lambda(X)$ is the rotation of the beam cross-section which does not necessarily remain normal to the beam axis and thus admits the possibility of transverse shearing deformations.

The derivation of the deformation gradient for Eq. (11.3) requires computation of the derivatives of the displacements and the rotation matrix. The derivative of the rotation matrix is given by[2,3]

$$\mathbf{\Lambda}_{,X} = \hat{\boldsymbol{\theta}}_{,X}\mathbf{\Lambda} \tag{11.4}$$

where $\hat{\boldsymbol{\theta}}_{,X}$ denotes a skew symmetric matrix for the derivatives of a rotation vector $\boldsymbol{\theta}$ and is expressed by

$$\hat{\boldsymbol{\theta}} = \begin{bmatrix} 0 & -\theta_{Z,X} & \theta_{Y,X} \\ \theta_{Z,X} & 0 & -\theta_{X,X} \\ -\theta_{Y,X} & \theta_{X,X} & 0 \end{bmatrix} \tag{11.5}$$

Here we consider in detail the two-dimensional case where the motion is restricted to the X–Z plane. The orthogonal matrix may then be represented as ($\theta_Y = \beta$)

$$\mathbf{\Lambda} = \begin{bmatrix} \cos\beta & 0 & \sin\beta \\ 0 & 1 & 0 \\ -\sin\beta & 0 & \cos\beta \end{bmatrix} \tag{11.6}$$

Inserting this in Eq. (11.3) and expanding, the deformed position then is described compactly by

$$\begin{aligned} x &= X + u(X) + Z\sin\beta(X) \\ y &= Y \\ z &= w(X) + Z\cos\beta(X) \end{aligned} \tag{11.7}$$

This results in the deformed configuration for a beam shown in Fig. 11.2. It is a two-dimensional specialization of the theory presented by Simo and co-workers[2,4,5] and is called *geometrically exact* since no small-angle approximations are involved. The deformation gradient for this displacement is given by the relation

$$F_{iI} = \begin{bmatrix} [1 + u_{,X} + Z\beta_{,X}\cos\beta] & 0 & \sin\beta \\ 0 & 1 & 0 \\ [w_{,X} - Z\beta_{,X}\sin\beta] & 0 & \cos\beta \end{bmatrix} \tag{11.8}$$

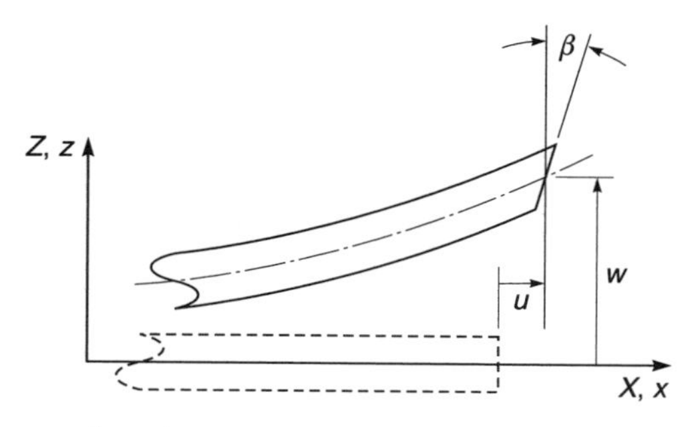

Fig. 11.2 Deformed beam configuration.

Using Eq (10.15) and computing the Green–Lagrange strain tensor, two non-zero components are obtained which, ignoring a quadratic term in Z, are expressed by

$$\begin{aligned} E_{XX} &= u_{,X} + \tfrac{1}{2}\,(u_{,X}^2 + w_{,X}^2) + Z\Lambda\beta_{,X} = E^0 + ZK^{\mathrm{b}} \\ 2E_{XZ} &= (1+u_{,X})\sin\beta + w_{,X}\cos\beta = \Gamma \end{aligned} \tag{11.9}$$

where E^0 and Γ are strains which are constant on the cross-section and K^{b} measures change in rotation (curvature) of the cross-sections and

$$\Lambda = (1+u_{,X})\cos\beta - w_{,X}\sin\beta \tag{11.10}$$

A variational equation for the beam can be written now by introducing second Piola–Kirchhoff stresses as described in Chapter 10 to obtain

$$\delta\Pi = \int_\Omega (\delta E_{XX}\,S_{XX} + 2\delta E_{XZ}\,S_{XZ})\,\mathrm{d}V - \delta\Pi_{\mathrm{ext}} \tag{11.11}$$

where $\delta\Pi_{\mathrm{ext}}$ denotes the terms from end forces and loading along the length. If we separate the volume integral into one along the length times an integral over the beam cross-sectional area A and define force resultants as

$$T^{\mathrm{p}} = \int_A S_{XX}\,\mathrm{d}A, \quad S^{\mathrm{p}} = \int_A S_{XZ}\,\mathrm{d}A \quad \text{and} \quad M^{\mathrm{b}} = \int_A S_{XX}\,Z\,\mathrm{d}A \tag{11.12}$$

the variational equation may be written compactly as

$$\delta\Pi = \int_{\mathrm{L}} (\delta E^0\,T^{\mathrm{p}} + \delta\Gamma\,S^{\mathrm{p}} + \delta K^{\mathrm{b}}\,M^{\mathrm{b}})\,\mathrm{d}X - \delta\Pi_{\mathrm{ext}} \tag{11.13}$$

where virtual strains for the beam are given by

$$\begin{aligned} \delta E^0 &= (1+u_{,X})\,\delta u_{,X} + w_{,X}\,\delta w_{,X} \\ \delta\Gamma &= \sin\beta\,\delta u_{,X} + \cos\beta\,\delta w_{,X} + \Lambda\,\delta\beta \\ \delta K^{\mathrm{b}} &= \Lambda\,\delta\beta_{,X} + \Gamma\,\delta\beta + \cos\beta\,\delta u_{,X} + \sin\beta\,\delta w_{,X} \end{aligned} \tag{11.14}$$

A finite element approximation for the displacements may be introduced in a manner identical to that used in Sec. 7.4 for axisymmetric shells. Accordingly, we can write

$$\begin{Bmatrix} u \\ w \\ \beta \end{Bmatrix} = N_\alpha(X) \begin{Bmatrix} u_\alpha \\ w_\alpha \\ \beta_\alpha \end{Bmatrix} \tag{11.15}$$

where the shape functions for each variable are the same. Using this approximation the virtual work is computed as

$$\delta\Pi = [\,\delta u_\alpha \quad \delta w_\alpha \quad \delta\beta_\alpha\,] \int_{\mathrm{L}} \mathbf{B}_\alpha^{\mathrm{T}} \begin{Bmatrix} T^{\mathrm{p}} \\ S^{\mathrm{p}} \\ M^{\mathrm{b}} \end{Bmatrix} \mathrm{d}X - \delta\Pi_{\mathrm{ext}} \tag{11.16}$$

where

$$\mathbf{B}_\alpha = \begin{bmatrix} (1+u_{,X})N_{\alpha,X} & w_{,X}N_{\alpha,X} & 0 \\ \sin\beta N_{\alpha,X} & \cos\beta N_{\alpha,X} & \Lambda N_\alpha \\ \beta_{,X}\cos\beta N_{\alpha,X} & -\beta_{,X}\sin\beta N_{\alpha,X} & (\Lambda N_{\alpha,X} - \Gamma \beta_{,X} N_\alpha) \end{bmatrix} \tag{11.17}$$

Just as for the axisymmetric shell described in Sec. 7.4 this interpolation will lead to 'shear locking' and it is necessary to compute the integrals for stresses by using a 'reduced quadrature'. For a two-noded beam element this implies use of one quadrature point for each element. Alternatively, a mixed formulation where Γ and S^{p} are assumed constant in each element can be introduced as was done in Sec. 5.6 for the bending analysis of plates using the T6S3B3 element.

The non-linear equilibrium equation for a quasi-static problem that is solved at each load level (or time) is given by

$$\boldsymbol{\Psi}_{n+1} = \mathbf{f}_{n+1} - \int_{\mathrm{L}} \mathbf{B}_\alpha^{\mathrm{T}} \begin{Bmatrix} N_{n+1}^{\mathrm{p}} \\ S_{n+1}^{\mathrm{p}} \\ M_{n+1}^{\mathrm{b}} \end{Bmatrix} \mathrm{d}X = \mathbf{0} \tag{11.18}$$

For a Newton–Raphson-type solution the tangent stiffness matrix is deduced by a linearization of Eq. (11.18). To give a specific relation for the derivation we assume, for simplicity, the strains are small and the constitution may be expressed by a linear elastic relation between the Green–Lagrange strains and the second Piola–Kirchhoff stresses. Accordingly, we take

$$S_{XX} = E E_{XX} \quad \text{and} \quad S_{XZ} = 2G E_{XZ} \tag{11.19}$$

where E is a Young's modulus and G a shear modulus. Integrating Eq. (11.12) the elastic behaviour of the beam resultants becomes

$$T^{\mathrm{p}} = EA\,E^0, \quad S^{\mathrm{p}} = \kappa GA\Gamma \quad \text{and} \quad M^{\mathrm{b}} = EI\,K^{\mathrm{b}}$$

in which A is the cross-sectional area, I is the moment of inertia about the centroid, and κ is a shear correction factor to account for the fact that S_{XZ} is not constant on the cross-section. Using these relations the linearization of Eq. (11.18) gives the tangent stiffness

$$(\mathbf{K}_{\mathrm{T}})_{\alpha\beta} = \int_L \mathbf{B}_\alpha^{\mathrm{T}} \mathbf{D}_{\mathrm{T}} \mathbf{B}_\beta \,\mathrm{d}X + (\mathbf{K}_G)_{\alpha\beta} \tag{11.20}$$

where for the simple elastic relation Eq. (11.20)

$$\mathbf{D}_{\mathrm{T}} = \begin{bmatrix} EA & & \\ & \kappa GA & \\ & & EI \end{bmatrix} \tag{11.21}$$

and $\mathbf{K}_{\mathrm{G}}$ is the geometric stiffness resulting from linearization of the non-linear expression for $\mathbf{B}$. After some algebra the reader can verify that the geometric stiffness is given by

$$(\mathbf{K}_G)_{\alpha\beta} = \int_L \left(N_{\alpha,X} \begin{bmatrix} T^p & 0 & M^b \cos\beta \\ 0 & T^p & -M^b \sin\beta \\ M^b \cos\beta & -M^b \sin\beta & 0 \end{bmatrix} N_{\beta,X} + N_\alpha \begin{bmatrix} 0 & 0 & 0 \\ 0 & 0 & 0 \\ 0 & 0 & G_3 \end{bmatrix} N_\beta \right.$$

$$\left. + N_{\alpha,X} \begin{bmatrix} 0 & 0 & G_1 \\ 0 & 0 & G_2 \\ 0 & 0 & -M^b\Gamma \end{bmatrix} N_\beta + N_\alpha \begin{bmatrix} 0 & 0 & 0 \\ 0 & 0 & 0 \\ G_1 & G_2 & -M^b\Gamma \end{bmatrix} N_{\beta,X} \right) \mathrm{d}X \quad (11.22)$$

where

$$G_1 = S^p \cos\beta - M^b \beta_{,X} \sin\beta, \quad G_2 = -S^p \sin\beta - M^b \beta_{,X} \cos\beta,$$

and

$$G_3 = -S^p \Gamma - M^b \beta_{,X} \Lambda$$

11.2.2 Large displacement formulation with small rotations

In many applications the full non-linear displacement field with finite rotations is not needed; however, the behaviour is such that limitations of the small displacement theory are not appropriate. In such cases we can assume that rotations are small so that the trigonometric functions may be approximated as

$$\sin\beta \approx \beta \qquad \text{and} \qquad \cos\beta \approx 1$$

In this case the displacement approximations become

$$\begin{aligned} x &= X + u(X) + Z\beta(X) \\ y &= Y \\ z &= w(X) + Z \end{aligned} \quad (11.23)$$

which yield now the non-zero Green–Lagrange strain expressions

$$\begin{aligned} E_{XX} &= u_{,X} + \tfrac{1}{2}(u_{,X}^2 + w_{,X}^2) + Z\beta_{,X} = E^0 + ZK^b \\ 2E_{XZ} &= w_{,X} + \beta = \Gamma \end{aligned} \quad (11.24)$$

where terms in Z^2 as well as products of β with derivatives of displacements are ignored. With this approximation and again using Eq. (11.15) for the finite element representation of the displacements in each element we obtain the set of non-linear equilibrium equations given by Eq. (11.18) in which now

$$\mathbf{B}_\alpha = \begin{bmatrix} (1 + u_{,X}) N_{\alpha,X} & w_{,X} N_{\alpha,X} & 0 \\ 0 & N_{\alpha,X} & N_\alpha \\ 0 & 0 & N_{\alpha,X} \end{bmatrix} \quad (11.25)$$

This expression results in a much simpler geometric stiffness term in the tangent matrix given by Eq. (11.20) and may be written simply as

$$(\mathbf{K}_G)_{\alpha\beta} = \int_L N_{\alpha,X} \begin{bmatrix} T^p & 0 & 0 \\ 0 & T^p & 0 \\ 0 & 0 & 0 \end{bmatrix} N_{\beta,X} \, dX \tag{11.26}$$

It is also possible to reduce the theory further by assuming shear deformations to be negligible so that from $\Gamma = 0$ we have

$$\beta = -w_{,X} \tag{11.27}$$

Taking the approximations now in the form

$$\begin{aligned} u &= N^u_\alpha \tilde{u}_\alpha \\ w &= N^w_\alpha \tilde{w}_\alpha + N^\beta_\alpha \tilde{\beta}_\alpha \end{aligned} \tag{11.28}$$

in which $\tilde{\beta}_\alpha \equiv \tilde{w}_{\alpha,X}$ at nodes.

The equilibrium equation is now given by

$$\boldsymbol{\Psi}_{n+1} = \mathbf{f}_{n+1} - \int_L \mathbf{B}^T_\alpha \begin{Bmatrix} N^p_{n+1} \\ M^b_{n+1} \end{Bmatrix} dX = \mathbf{0} \tag{11.29}$$

where the strain displacement matrix is expressed as

$$\mathbf{B}_\alpha = \begin{bmatrix} (1+u_{,X}) N^u_{\alpha,X} & w_{,X} N^w_{\alpha,X} & w_{,X} N^\beta_{\alpha,X} \\ 0 & -N^w_{\alpha,XX} & -N^\beta_{\alpha,XX} \end{bmatrix} \tag{11.30}$$

The tangent matrix is given by Eq. (11.20) where the elastic tangent moduli involve only the terms from T^p and M^b as

$$\mathbf{D}_T = \begin{bmatrix} EA & 0 \\ 0 & EI \end{bmatrix} \tag{11.31}$$

and the geometric tangent is given by

$$(\mathbf{K}_G)_{\alpha\beta} = \int_L N_{\alpha,X} \begin{bmatrix} N^u_{\alpha,X} T^p N^u_{\beta,X} & 0 & 0 \\ 0 & N^w_{\alpha,X} T^p N^w_{\beta,X} & N^w_{\alpha,X} T^p N^\beta_{\beta,X} \\ 0 & N^\beta_{\alpha,X} T^p N^w_{\beta,X} & N^\beta_{\alpha,X} T^p N^\beta_{\beta,X} \end{bmatrix} dX \tag{11.32}$$

Example: a clamped–hinged arch

To illustrate the performance and limitations of the above formulations we consider the behaviour of a circular arch with one boundary clamped, the other boundary hinged and loaded by a single point load, as shown in Fig. 11.3(a). Here it is necessary to introduce a transformation between the axes used to define each beam element and the global axes used to define the arch. This follows standard procedures as used many times previously. The cross-section of the beam is a unit square with other properties as shown in the figure. An analytical solution to this problem has been obtained by da Deppo and Schmidt[6] and an early finite element solution by Wood

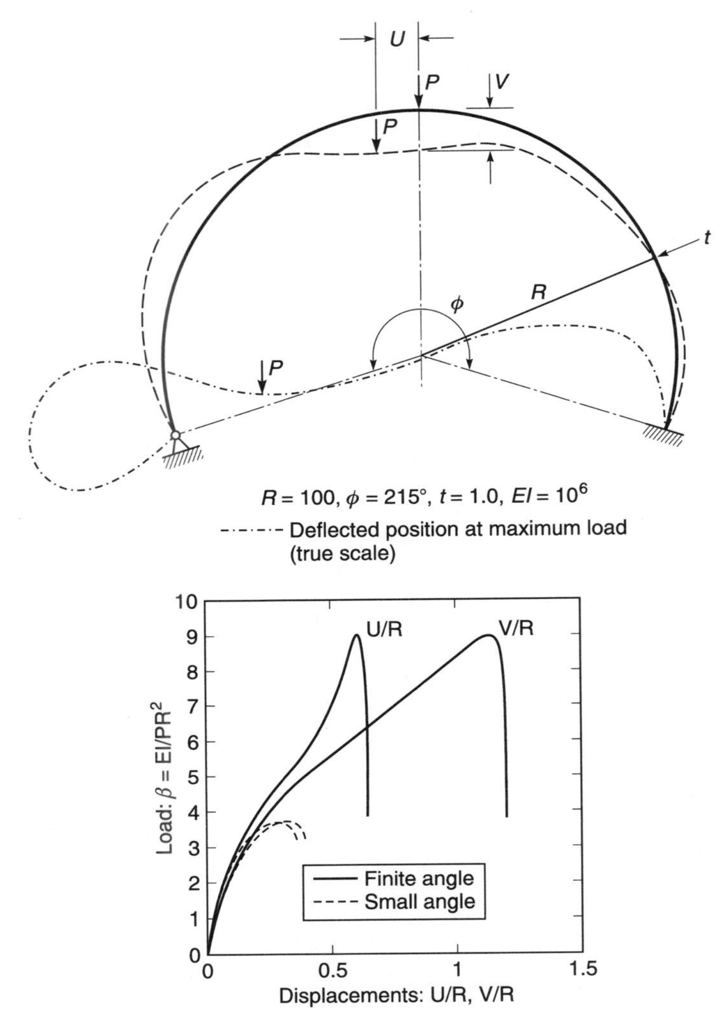

Fig. 11.3 Clamped–hinged arch: (a) problem definition; (b) load deflection.

and Zienkiewicz.[7] Here a solution is obtained using 40 two-noded elements of the types presented in this section. The problem produces a complex load displacement history with 'softening' behaviour that is traced using the arc-length method described in Sec. 2.2.6 [Fig. 11.3(b)]. It is observed from Fig. 11.3(b) that the assumption of small rotation produces an accurate trace of the behaviour only during the early parts of loading and also produces a limit state which is far from reality. This emphasizes clearly the type of discrepancies that can occur by misusing a formulation in which assumptions are involved.

Deformed configurations during the deformation history are shown for the load parameter $\beta = EI/PR^2$ in Fig. 11.4. In Fig. 11.4(a) we show the deformed configuration for five loading levels – three before the limit load is reached and two after

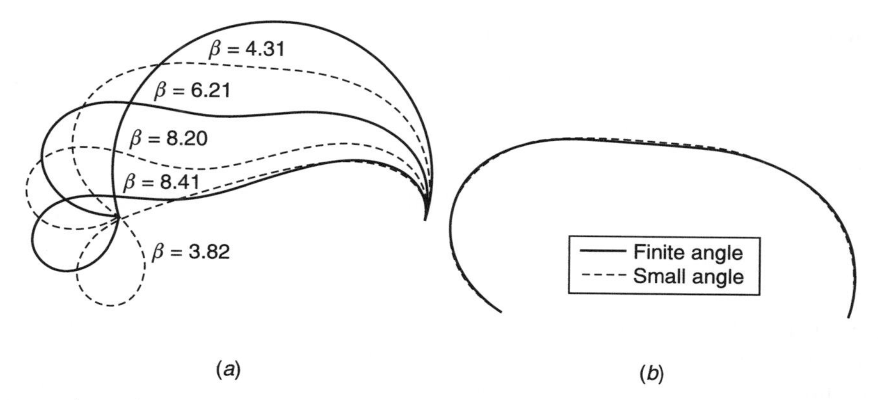

Fig. 11.4 Clamped–hinged arch: deformed shapes. (a) Finite-angle solution; (b) finite-angle form compared with small-angle form.

passing the limit load. It will be observed that continued loading would not lead to correct solutions unless a contact state is used between the support and the arch member. This aspect was considered by Simo *et al.*[8] and loading was applied much further into the deformation process. In Fig. 11.4(b) we show a comparison of the deformed shapes for $\beta = 3.0$ where the small-angle assumption is still valid.

11.3 Elastic stability – energy interpretation

The energy expression given in Eq. (10.37) and the equilibrium behaviour deduced from the first variation given by Eq. (10.42) may also be used to assess the *stability* of equilibrium.[9] For an equ'ilibrium state we always have

$$\delta\Pi = -\delta\tilde{\mathbf{u}}^{\mathrm{T}}\boldsymbol{\Psi} = 0 \tag{11.33}$$

that is, *the total potential energy is stationary* [which, ignoring inertia effects, is equivalent to Eq. (10.65)].

The second variation of Π is

$$\delta^2\Pi = \delta(\delta\Pi) = -\delta\tilde{\mathbf{u}}^{\mathrm{T}}\delta\boldsymbol{\Psi} = \delta\tilde{\mathbf{u}}^{\mathrm{T}}\mathbf{K}_{\mathrm{T}}\delta\tilde{\mathbf{u}} \tag{11.34}$$

The stability criterion is given by a positive value of this second variation and, conversely, instability by a negative value (as in the first case energy has to be added to the structure whereas in the second it contains surplus energy). In other words, *if* $\mathbf{K}_{\mathrm{T}}$ *is positive definite, stability exists*. This criterion is well known[9] and of considerable use when investigating stability during large deformation.[10,11] An alternative test is to investigate the sign of the determinant of $\mathbf{K}_{\mathrm{T}}$, a positive sign denoting stability.[12]

A limit on stability exists when the second variation is zero. We note from Eq. (10.66) that the stability test then can be written as (assuming $\mathbf{K}_{\mathrm{L}}$ is zero)

$$\delta\tilde{\mathbf{u}}^{\mathrm{T}}\mathbf{K}_{\mathrm{M}}\delta\tilde{\mathbf{u}} + \delta\tilde{\mathbf{u}}^{\mathrm{T}}\mathbf{K}_{\mathrm{G}}\delta\tilde{\mathbf{u}} = 0 \tag{11.35}$$

This may be written in the *Rayleigh quotient form*[13]

$$\frac{\delta\tilde{\mathbf{u}}^{\mathrm{T}}\mathbf{K}_{\mathrm{M}}\,\delta\tilde{\mathbf{u}}}{\delta\tilde{\mathbf{u}}^{\mathrm{T}}\mathbf{K}_{\mathrm{G}}\,\delta\tilde{\mathbf{u}}} = -\lambda \tag{11.36}$$

where we have

$$\lambda \begin{cases} < 1, & \text{stable} \\ = 1, & \text{stability limit} \\ > 1, & \text{unstable} \end{cases} \tag{11.37}$$

The limit of stability is sometimes called *neutral equilibrium* since the configuration may be changed by a small amount without affecting the value of the second variation (i.e. equilibrium balance). Several options exist for implementing the above test and the simplest is to let $\lambda = 1 + \Delta\lambda$ and write the problem in the form of a generalized linear eigenproblem given by

$$\mathbf{K}_{\mathrm{T}}\,\delta\mathbf{u} = \Delta\lambda\mathbf{K}_{\mathrm{G}}\,\delta\mathbf{u} \tag{11.38}$$

Here we seek the solution where $\Delta\lambda$ is zero to define a stability limit. This form uses the usual tangent matrix directly and requires only a separate implementation for the geometric term and availability of a general eigensolution routine. To maintain numerical conditioning in the eigenproblem near a buckling or limit state where $\mathbf{K}_{\mathrm{T}}$ is singular a *shift* may be used as described for the vibration problem in Chapter 17 of Volume 1.

Euler buckling – propped cantilever

As an example of the stability test we consider the buckling of a straight beam with one end fixed and the other on a roller support. We can also use this example to show the usefulness of the small angle beam theory.

An axial compressive load is applied to the roller end and the Euler buckling load computed. This is a problem in which the displacement prior to buckling is purely axial. The buckling load may be estimated relative to the small deformation theory by using the solution from the first tangent matrix computed. Alternatively, the buckling load can be computed by increasing the load until the tangent matrix becomes singular. In the case of a structure where the distribution of the internal forces does not change with load level and material is linear elastic there is no difference in the results obtained. Table 11.1 shows the results obtained for the propped cantilever using different numbers of elements. Here it is observed that accurate results for higher modes require use of more elements; however, both the finite rotation and small rotation formulations given above yield identical answers

Table 11.1 Linear buckling load estimates

Number of elements		
20	100	500
20.36	20.19	20.18
61.14	59.67	59.61
124.79	118.85	118.62

since no rotation is present prior to buckling. The properties used in the analysis are $E = 12 \times 10^6, A = 1, I = 1/12$, and length $L = 100$. The classical Euler buckling load is given by

$$P_{\text{cr}} = \alpha \frac{EI}{L^2} \tag{11.39}$$

with the lowest buckling load given as $\alpha = 20.18$.[14]

11.4 Large displacement theory of thick plates

11.4.1 Definitions

The small rotation form for beams described in Sec. 11.2.2 may be used to consider problems associated with deformation of plates subject to 'in-plane' and 'lateral' forces, when displacements are not infinitesimal but also not excessively large (Fig. 11.5). In this situation the 'change-in-geometry' effect is less important than the relative magnitudes of the linear and non-linear strain-displacement terms, and in fact for 'stiffening' problems the non-linear displacements are always less than the corresponding linear ones (see Fig. 11.6). It is well known that in such situations the lateral displacements will be responsible for development of 'membrane'-type strains and now the two problems of 'in-plane' and 'lateral' deformation can no longer be dealt with separately but are *coupled*.

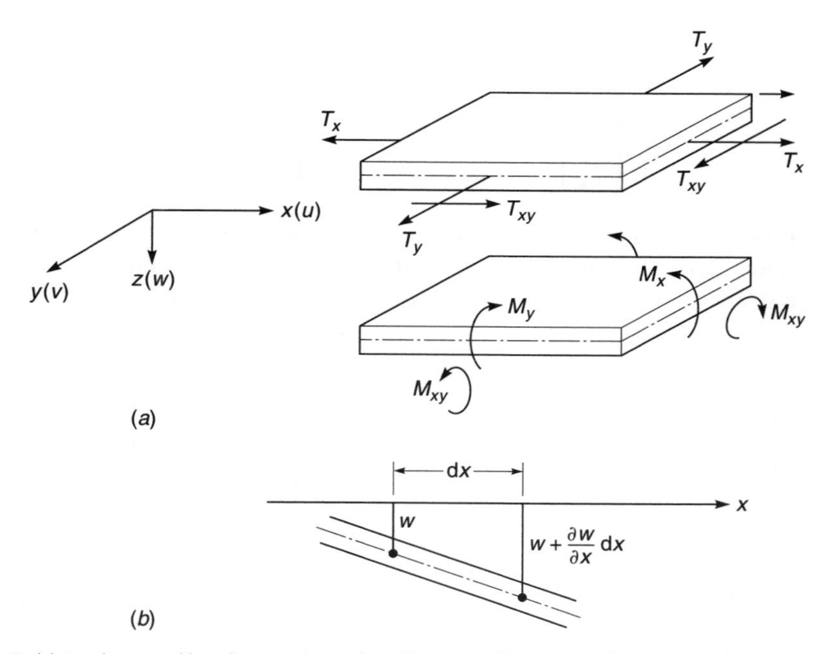

Fig. 11.5 (a) 'In-plane' and bending resultants for a flat plate; (b) increase of middle surface length owing to lateral displacement.

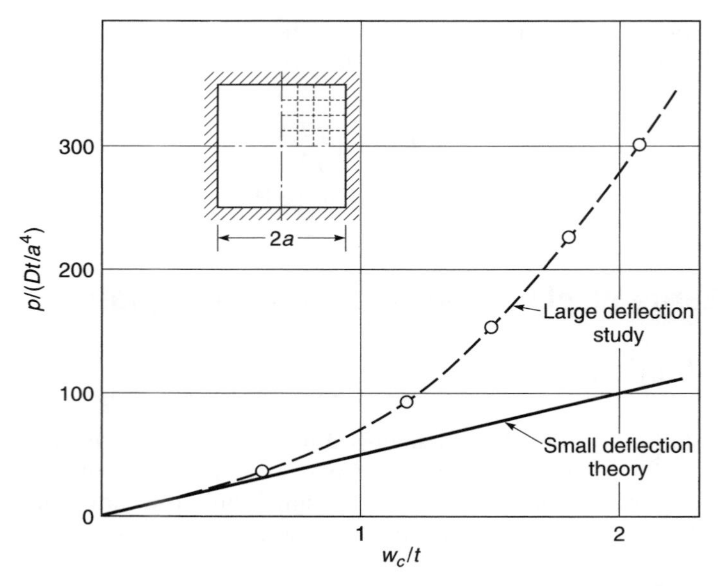

Fig. 11.6 Central deflection w_c of a clamped square plate under uniform load p;[12] $u = v = 0$ at edge.

Generally, for plates the rotation angles remain small unless in-plane strains also become large. To develop the equations for small rotations in which plate bending is modelled using the formulations discussed in Chapter 5 we generalize the displacement field given in Eq. (4.9) to include the effects of in-plane displacements. Accordingly, we write

$$\mathbf{u} = \begin{Bmatrix} u_1 \\ u_2 \\ u_3 \end{Bmatrix} = \begin{Bmatrix} u(X, Y) \\ v(X, Y) \\ w(X, Y) \end{Bmatrix} - Z \begin{Bmatrix} \theta_X(X, Y) \\ \theta_Y(X, Y) \\ 0 \end{Bmatrix} \tag{11.40}$$

where $\boldsymbol{\theta}$ are small rotations defined according to Fig. 4.3 and X, Y, Z denote positions in the reference configuration of the plate. Using these to compute the Green–Lagrange strains given by Eq. (10.15) we can write the non-zero terms as

$$\begin{Bmatrix} E_{XX} \\ E_{YY} \\ 2E_{XY} \\ 2E_{XZ} \\ 2E_{YZ} \end{Bmatrix} = \begin{Bmatrix} u_{,X} + \frac{1}{2}(w_{,X})^2 \\ v_{,Y} + \frac{1}{2}(w_{,Y})^2 \\ u_{,Y} + v_{,X} + w_{,X}w_{,Y} \\ -\theta_X + w_{,X} \\ -\theta_Y + w_{,Y} \end{Bmatrix} - Z \begin{Bmatrix} \theta_{X,X} \\ \theta_{Y,Y} \\ \theta_{X,Y} + \theta_{Y,X} \\ 0 \\ 0 \end{Bmatrix} \tag{11.41}$$

In these expressions we have used classical results[15] that ignore all square terms involving $\boldsymbol{\theta}$ and derivatives of u and v, as well as terms which contain quadratic powers of Z.

Generally, the position of the in-plane reference coordinates X and Y change very little during deformations and we can replace them with the current coordinates x and

y just as is done implicitly for the small strain case considered in Chapter 4. Thus, we can represent the Green–Lagrange strains in terms of the middle surface strains and changes in curvature as

$$\mathbf{E} = \begin{Bmatrix} u_{,x} + \frac{1}{2}(w_{,x})^2 \\ v_{,y} + \frac{1}{2}(w_{,y})^2 \\ u_{,y} + v_{,x} + w_{,x}w_{,y} \end{Bmatrix} - Z\begin{Bmatrix} \theta_{x,x} \\ \theta_{y,y} \\ \theta_{x,y} + \theta_{y,x} \end{Bmatrix} = \mathbf{E}^{\mathrm{p}} - Z\mathbf{K}^{\mathrm{b}} \tag{11.42}$$

where $\mathbf{E}^{\mathrm{p}}$ denotes the in-plane membrane strains and $\mathbf{K}^{\mathrm{b}}$ the change in curvatures owing to bending. In addition we have the transverse shearing strains given by

$$\boldsymbol{\Gamma}^{\mathrm{s}} = \begin{Bmatrix} -\theta_x + w_{,x} \\ -\theta_y + w_{,y} \end{Bmatrix} \tag{11.43}$$

The variations of the strains are given by

$$\delta\mathbf{E}^{\mathrm{p}} = \begin{Bmatrix} \delta u_{,x} \\ \delta v_{,y} \\ \delta u_{,y} + \delta v_{,x} \end{Bmatrix} + \begin{bmatrix} w_{,x} & 0 \\ 0 & w_{,y} \\ w_{,y} & w_{,x} \end{bmatrix} \begin{Bmatrix} \delta w_{,x} \\ \delta w_{,y} \end{Bmatrix} \tag{11.44}$$

$$\delta\mathbf{K}^{\mathrm{b}} = \begin{Bmatrix} \delta\theta_{x,x} \\ \delta\theta_{y,y} \\ \delta\theta_{x,y} + \delta\theta_{y,x} \end{Bmatrix} \quad \text{and} \quad \delta\boldsymbol{\Gamma}^{\mathrm{s}} = \begin{Bmatrix} -\delta\theta_x + \delta w_{,x} \\ -\delta\theta_y + \delta w_{,y} \end{Bmatrix} \tag{11.45}$$

Using these expressions the variation of the plate equations may be expressed as

$$\delta\Pi = \int_\Omega (\delta\mathbf{E}^{\mathrm{p}})^{\mathrm{T}}\mathbf{S}\,\mathrm{d}V + \int_\Omega (\delta\boldsymbol{\Gamma}^{\mathrm{s}})^{\mathrm{T}}\mathbf{S}^{\mathrm{s}}\,\mathrm{d}V + \int_\Omega (\delta\mathbf{K}^{\mathrm{b}})^{\mathrm{T}}\mathbf{S}Z\,\mathrm{d}V - \delta\Pi_{\mathrm{ext}} \tag{11.46}$$

Defining the integrals through the thickness in terms of the 'in-plane' membrane forces

$$\mathbf{T}^{\mathrm{p}} = \begin{Bmatrix} T_x \\ T_y \\ T_{xy} \end{Bmatrix} = \int_{-t/2}^{t/2} \mathbf{S}\,\mathrm{d}Z \equiv \int_{-t/2}^{t/2} \begin{Bmatrix} S_{XX} \\ S_{YY} \\ S_{XY} \end{Bmatrix} \mathrm{d}Z \tag{11.47}$$

transverse shears

$$\mathbf{T}^{\mathrm{s}} = \begin{Bmatrix} T_{xz} \\ T_{yz} \end{Bmatrix} = \int_{-t/2}^{t/2} \mathbf{S}^{\mathrm{s}}\,\mathrm{d}Z \equiv \int_{-t/2}^{t/2} \begin{Bmatrix} S_{XZ} \\ S_{YZ} \end{Bmatrix} \mathrm{d}Z \tag{11.48}$$

and bending forces

$$\mathbf{M}^{\mathrm{b}} = \begin{Bmatrix} M_{xx} \\ M_{yy} \\ M_{xy} \end{Bmatrix} = -\int_{-t/2}^{t/2} \mathbf{S}Z\,\mathrm{d}Z \equiv -\int_{-t/2}^{t/2} \begin{Bmatrix} S_{XX} \\ S_{YY} \\ S_{XY} \end{Bmatrix} Z\,\mathrm{d}Z \tag{11.49}$$

we obtain the virtual work expression for the plate, given as

$$\delta\Pi = \int_A [(\delta\mathbf{E}^{\mathrm{p}})^{\mathrm{T}}\mathbf{T}^{\mathrm{p}} + \delta(\boldsymbol{\Gamma}^{\mathrm{s}})^{\mathrm{T}}\mathbf{T}^{\mathrm{s}} + \delta(\mathbf{K}^{\mathrm{b}})^{\mathrm{T}}\mathbf{M}^{\mathrm{b}}]\,\mathrm{d}A - \delta\Pi_{\mathrm{ext}} \tag{11.50}$$

This may now be used to construct a finite element solution.

11.4.2 Finite element evaluation of strain–displacement matrices

For further evaluation it is necessary to establish expressions for the finite element $\mathbf{B}$ and $\mathbf{K}_{\mathrm{T}}$ matrices. Introducing the finite element approximations, we have

$$\begin{Bmatrix} u \\ v \\ w \end{Bmatrix} = \begin{bmatrix} N_\alpha & & \\ & N_\alpha & \\ & & N_\alpha^w \end{bmatrix} \begin{Bmatrix} \tilde{u}_\alpha \\ \tilde{v}_\alpha \\ \tilde{w}_\alpha \end{Bmatrix} \tag{11.51}$$

and

$$\begin{Bmatrix} \theta_x \\ \theta_y \end{Bmatrix} = N_\alpha^\theta \begin{Bmatrix} (\tilde{\theta}_x)_\alpha \\ (\tilde{\theta}_y)_\alpha \end{Bmatrix} \tag{11.52}$$

The expressions for the strain–displacement matrices are deduced from Eqs (11.44) and (11.45) as

$$\begin{aligned} \delta\mathbf{E}^{\mathrm{p}} = \mathbf{B}_\alpha^{\mathrm{p}}\,\delta\tilde{\mathbf{a}}_\alpha &= \begin{bmatrix} N_{\alpha,x} & 0 \\ 0 & N_{\alpha,y} \\ N_{\alpha,y} & N_{\alpha,x} \end{bmatrix} \begin{Bmatrix} \delta\tilde{u}_\alpha \\ \delta\tilde{v}_\alpha \end{Bmatrix} + \begin{bmatrix} w_{,x} & 0 \\ 0 & w_{,y} \\ w_{,y} & w_{,x} \end{bmatrix} \mathbf{G}_\alpha \begin{Bmatrix} \delta\tilde{w}_\alpha \\ \delta(\tilde{\theta}_x)_\alpha \\ \delta(\tilde{\theta}_y)_\alpha \end{Bmatrix} \\ &= \mathbf{B}_\alpha\,\delta\tilde{\mathbf{u}}_\alpha + \mathbf{B}_\alpha^{\mathrm{L}}\,\delta\tilde{\mathbf{w}}_\alpha \end{aligned} \tag{11.53}$$

$$\delta\boldsymbol{\Gamma}^{\mathrm{s}} = \mathbf{B}_\alpha^{\mathrm{s}}\,\delta\mathbf{w}_\alpha = \begin{bmatrix} N_{\alpha,x}^w & -N_\alpha^\theta & 0 \\ N_{\alpha,y}^w & 0 & -N_\alpha^\theta \end{bmatrix} \begin{Bmatrix} \delta\tilde{w}_\alpha \\ \delta(\tilde{\theta}_x)_\alpha \\ \delta(\tilde{\theta}_y)_\alpha \end{Bmatrix} \tag{11.54}$$

and

$$\delta\mathbf{K}^{\mathrm{b}} = \mathbf{B}_\alpha^{\mathrm{b}}\,\delta\tilde{\mathbf{w}}_\alpha = \begin{bmatrix} 0 & N_{\alpha,x}^\theta & 0 \\ 0 & 0 & N_{\alpha,y}^\theta \\ 0 & N_{\alpha,y}^\theta & N_{\alpha,x}^\theta \end{bmatrix} \begin{Bmatrix} \delta\tilde{w}_\alpha \\ \delta(\tilde{\theta}_x)_\alpha \\ \delta(\tilde{\theta}_y)_\alpha \end{Bmatrix} \tag{11.55}$$

where

$$\mathbf{G}_\alpha = \begin{bmatrix} N_{\alpha,x}^w & 0 & 0 \\ N_{\alpha,y}^w & 0 & 0 \end{bmatrix} \tag{11.56}$$

with nodal parameters defined by

$$\tilde{\mathbf{a}}_\alpha^{\mathrm{T}} = [\,\tilde{u}_\alpha \quad \tilde{v}_\alpha \quad \tilde{w}_\alpha \quad (\tilde{\theta}_x)_\alpha \quad (\tilde{\theta}_y)_\alpha\,] = [\,\mathbf{u}_\alpha^{\mathrm{T}} \quad \mathbf{w}_\alpha^{\mathrm{T}}\,]$$

$$\tilde{\mathbf{u}}_\alpha^{\mathrm{T}} = [\,\tilde{u}_\alpha \quad \tilde{v}_\alpha\,] \quad \text{and} \quad \tilde{\mathbf{w}}_\alpha^{\mathrm{T}} = [\,\tilde{w}_\alpha \quad (\tilde{\theta}_x)_\alpha \quad (\tilde{\theta}_y)_\alpha\,]$$

We here immediately recognize an in-plane term which is identical to the small strain (linear) plane stress (membrane) form and a term which is identical to the small strain bending and transverse shear form. The added nonlinear in-plane term results from the quadratic displacement terms in the membrane strains.

Using the above strain–displacement matrices we can now write Eq. (11.50) as

$$\delta\Pi = \delta\tilde{\mathbf{a}}_\alpha^{\mathrm{T}} \int_A (\mathbf{B}_\alpha^{\mathrm{p}})^{\mathrm{T}} \mathbf{T}^{\mathrm{p}} \, \mathrm{d}A + \delta\tilde{\mathbf{w}}_\alpha^{\mathrm{T}} \int_A (\mathbf{B}_\alpha^{\mathrm{s}})^{\mathrm{T}} \mathbf{T}^{\mathrm{s}} \, \mathrm{d}A + \delta\tilde{\mathbf{w}}_\alpha^{\mathrm{T}} \int_A (\mathbf{B}_\alpha^{\mathrm{b}})^{\mathrm{T}} \mathbf{M}^{\mathrm{b}} \, \mathrm{d}A - \delta\Pi_{\mathrm{ext}} = 0 \tag{11.57}$$

Grouping the force terms as

$$\bar{\boldsymbol{\sigma}} = \begin{Bmatrix} \mathbf{T}^{\mathrm{p}} \\ \mathbf{T}^{\mathrm{s}} \\ \mathbf{M}^{\mathrm{b}} \end{Bmatrix} \tag{11.58}$$

and the strain matrices as

$$\bar{\mathbf{B}}_\alpha = \begin{bmatrix} \mathbf{B}_\alpha & \mathbf{B}_\alpha^{\mathrm{L}} \\ \mathbf{0} & \mathbf{B}_\alpha^{\mathrm{s}} \\ \mathbf{0} & \mathbf{B}_\alpha^{\mathrm{b}} \end{bmatrix} \tag{11.59}$$

the virtual work expression may be written compactly as

$$\delta\Pi = \delta\tilde{\mathbf{a}}_\alpha^{\mathrm{T}} \int_A \bar{\mathbf{B}}_\alpha^{\mathrm{T}} \bar{\boldsymbol{\sigma}} \, \mathrm{d}A - \delta\Pi_{\mathrm{ext}} = 0 \tag{11.60}$$

The non-linear problem to be solved is thus expressed as

$$\boldsymbol{\Psi}_\alpha = \mathbf{f}_\alpha - \int_A \bar{\mathbf{B}}_\alpha^{\mathrm{T}} \bar{\boldsymbol{\sigma}} \, \mathrm{d}A = \mathbf{0} \tag{11.61}$$

This may be solved by using a Newton–Raphson process for which a tangent matrix is required.

11.4.3 Evaluation of tangent matrix

A tangent matrix for the non-linear plate formulation may be computed by a linearization of Eq. (11.60). Formally, this may be written as

$$d(\delta\Pi) = \delta\tilde{\mathbf{a}}_\alpha^{\mathrm{T}} \int_A \left[d(\bar{\mathbf{B}}_\alpha^{\mathrm{T}}) \bar{\boldsymbol{\sigma}} + \bar{\mathbf{B}}_\alpha^{\mathrm{T}} d(\bar{\boldsymbol{\sigma}}) \right] \mathrm{d}A - d(\delta\Pi_{\mathrm{ext}}) = 0 \tag{11.62}$$

We shall assume for simplicity that loading is conservative so that $d(\delta\Pi_{\mathrm{ext}}) = 0$ and hence the only terms to be linearized are the strain-displacement matrix and the stress–strain relation. If we assume linear elastic behaviour, the relation between the plate forces and strains may be written as

$$\begin{Bmatrix} \mathbf{T}^{\mathrm{p}} \\ \mathbf{T}^{\mathrm{s}} \\ \mathbf{M}^{\mathrm{b}} \end{Bmatrix} = \begin{bmatrix} \mathbf{D}^{\mathrm{p}} & \mathbf{0} & \mathbf{0} \\ \mathbf{0} & \mathbf{D}^{\mathrm{s}} & \mathbf{0} \\ \mathbf{0} & \mathbf{0} & \mathbf{D}^{\mathrm{b}} \end{bmatrix} \begin{Bmatrix} \mathbf{E}^{\mathrm{p}} \\ \boldsymbol{\Gamma}^{\mathrm{s}} \\ \mathbf{K}^{\mathrm{b}} \end{Bmatrix} \tag{11.63}$$

where for an isotropic homogeneous plate

$$\mathbf{D}^{\mathrm{p}} = \frac{Et}{1-\nu^2}\begin{bmatrix} 1 & \nu & 0 \\ \nu & 1 & 0 \\ 0 & 0 & (1-\nu)/2 \end{bmatrix}, \quad \mathbf{D}^{\mathrm{s}} = \frac{\kappa E t}{2(1+\nu)}\begin{bmatrix} 1 & 0 \\ 0 & 1 \end{bmatrix}, \quad \text{and} \quad \mathbf{D}^{\mathrm{b}} = \frac{t^2}{12}\mathbf{D}^{\mathrm{p}} \tag{11.64}$$

Again, κ is a shear correction factor which, for homogeneous plates, is usually taken as 5/6. Thus, the linearization of the constitution becomes

$$d(\bar{\boldsymbol{\sigma}}) = \begin{Bmatrix} d(\mathbf{T}^{\mathrm{p}}) \\ d(\mathbf{T}^{\mathrm{s}}) \\ d(\mathbf{M}^{\mathrm{b}}) \end{Bmatrix} = \begin{bmatrix} \mathbf{D}^{\mathrm{p}} & \mathbf{0} & \mathbf{0} \\ \mathbf{0} & \mathbf{D}^{\mathrm{s}} & \mathbf{0} \\ \mathbf{0} & \mathbf{0} & \mathbf{D}^{\mathrm{b}} \end{bmatrix} \begin{Bmatrix} d(\mathbf{E}^{\mathrm{p}}) \\ d(\boldsymbol{\Gamma}^{\mathrm{s}}) \\ d(\mathbf{K}^{\mathrm{b}}) \end{Bmatrix}$$

$$= \begin{bmatrix} \mathbf{D}^{\mathrm{p}} & \mathbf{0} & \mathbf{0} \\ \mathbf{0} & \mathbf{D}^{\mathrm{s}} & \mathbf{0} \\ \mathbf{0} & \mathbf{0} & \mathbf{D}^{\mathrm{b}} \end{bmatrix} \begin{bmatrix} \mathbf{B}_{\beta} & \mathbf{B}_{\beta}^{\mathrm{L}} \\ \mathbf{0} & \mathbf{B}_{\alpha}^{\mathrm{s}} \\ \mathbf{0} & \mathbf{B}_{\beta}^{\mathrm{b}} \end{bmatrix} \begin{Bmatrix} d(\tilde{\mathbf{u}}_{\beta}) \\ d(\tilde{\mathbf{w}}_{\beta}) \end{Bmatrix} \tag{11.65}$$

Using this result the material part of the tangent matrix is expressed as

$$(\mathbf{K}_{\mathrm{M}})_{\alpha\beta} = \int_A \begin{bmatrix} (\mathbf{B}_{\alpha}^{\mathrm{p}})^{\mathrm{T}} & \mathbf{0} & \mathbf{0} \\ (\mathbf{B}_{\alpha}^{\mathrm{L}})^{\mathrm{T}} & (\mathbf{B}_{\alpha}^{\mathrm{s}})^{\mathrm{T}} & (\mathbf{B}_{\alpha}^{\mathrm{b}})^{\mathrm{T}} \end{bmatrix} \begin{bmatrix} \mathbf{D}^{\mathrm{p}} & \mathbf{0} & \mathbf{0} \\ \mathbf{0} & \mathbf{D}^{\mathrm{s}} & \mathbf{0} \\ \mathbf{0} & \mathbf{0} & \mathbf{D}^{\mathrm{b}} \end{bmatrix} \begin{bmatrix} \mathbf{B}_{\beta}^{\mathrm{p}} & \mathbf{B}_{\beta}^{\mathrm{L}} \\ \mathbf{0} & \mathbf{B}_{\beta}^{\mathrm{s}} \\ \mathbf{0} & \mathbf{B}_{\beta}^{\mathrm{b}} \end{bmatrix} \mathrm{d}A$$

$$= \int_A \bar{\mathbf{B}}_{\alpha}^{\mathrm{T}} \mathbf{D}_{\mathrm{T}} \bar{\mathbf{B}}_{\beta} \,\mathrm{d}A = \begin{bmatrix} (\mathbf{K}_{\mathrm{M}}^{\mathrm{p}})_{\alpha\beta} & (\mathbf{K}_{\mathrm{M}}^{\mathrm{L}})_{\alpha\beta} \\ (\mathbf{K}_{\mathrm{M}}^{\mathrm{L}})_{\alpha\beta}^{\mathrm{T}} & (\mathbf{K}_{\mathrm{M}}^{\mathrm{b}})_{\alpha\beta} \end{bmatrix} \tag{11.66}$$

where $\mathbf{D}_{\mathrm{T}}$ is the coefficient matrix from Eq. (11.63), and the individual parts of the tangent matrix are

$$\begin{aligned} (\mathbf{K}_{\mathrm{M}}^{\mathrm{p}})_{\alpha\beta} &= \int_A (\mathbf{B}_{\alpha}^{\mathrm{p}})^{\mathrm{T}} \mathbf{D}^{\mathrm{p}} \mathbf{B}_{\beta} \,\mathrm{d}A \\ (\mathbf{K}_{\mathrm{M}}^{\mathrm{L}})_{\alpha\beta} &= \int_A (\mathbf{B}_{\alpha}^{\mathrm{p}})^{\mathrm{T}} \mathbf{D}^{\mathrm{p}} \mathbf{B}_{\beta}^{\mathrm{L}} \,\mathrm{d}A \\ (\mathbf{K}_{\mathrm{M}}^{\mathrm{b}})_{\alpha\beta} &= \int_A \left[(\mathbf{B}_{\alpha}^{\mathrm{s}})^{\mathrm{T}} \mathbf{D}^{\mathrm{s}} \mathbf{B}_{\beta}^{\mathrm{s}} + (\mathbf{B}_{\alpha}^{\mathrm{b}})^{\mathrm{T}} \mathbf{D}^{\mathrm{b}} \mathbf{B}_{\beta}^{\mathrm{b}} \right] \mathrm{d}A \end{aligned} \tag{11.67}$$

We immediately recognize that the material part of the tangent matrix consists of the same result as that of the small displacement analysis except for the added term $\mathbf{K}_{\mathrm{M}}^{\mathrm{L}}$ which establishes coupling between membrane and bending behaviour.

The remainder of the computation for the tangent involves the linearization of the non-linear part of the strain–displacement matrix, $\mathbf{B}_{\alpha}^{\mathrm{L}}$. As in the continuum problem discussed in Chapter 10 it is easiest to rewrite this term as

$$d(\mathbf{B}_\alpha^{\mathrm{L}})^{\mathrm{T}}\mathbf{T}^{\mathrm{b}} = \mathbf{G}_\alpha^{\mathrm{T}} \begin{bmatrix} d(w_{,x}) & 0 & d(w_{,y}) \\ 0 & d(w_{,y}) & d(w_{,x}) \end{bmatrix} \begin{Bmatrix} T_x^{\mathrm{p}} \\ T_y^{\mathrm{p}} \\ T_{xy}^{\mathrm{p}} \end{Bmatrix}$$

$$= \mathbf{G}_\alpha^{\mathrm{T}} \begin{bmatrix} T_x^{\mathrm{p}} & T_{xy}^{\mathrm{p}} \\ T_{xy}^{\mathrm{p}} & T_y^{\mathrm{p}} \end{bmatrix} \begin{Bmatrix} d(w_{,x}) \\ d(w_{,y}) \end{Bmatrix} \tag{11.68}$$

This may now be expressed in terms of finite element interpolations to obtain the geometric part of the tangent as

$$(\mathbf{K}_{\mathrm{G}}^{\mathrm{L}})_{\alpha\beta} = \int_A \mathbf{G}_\alpha^{\mathrm{T}} \begin{bmatrix} T_x^{\mathrm{p}} & T_{xy}^{\mathrm{p}} \\ T_{xy}^{\mathrm{p}} & T_y^{\mathrm{p}} \end{bmatrix} \mathbf{G}_\beta \, \mathrm{d}A \tag{11.69}$$

which is inserted into the total geometric tangent as

$$(\mathbf{K}_{\mathrm{G}})_{\alpha\beta} = \begin{bmatrix} \mathbf{0} & \mathbf{0} \\ \mathbf{0} & (\mathbf{K}_{\mathrm{G}}^{\mathrm{L}})_{\alpha\beta} \end{bmatrix} \tag{11.70}$$

This geometric matrix is also referred to in the literature as the *initial stress matrix* for plate bending.

11.5 Large displacement theory of thin plates

The above theory may be specialized to the thin plate formulation by neglecting the effects of transverse shearing strains as discussed in Chapter 4. Thus setting $E_{XZ} = E_{YZ} = 0$ in Eq. (11.41), this yields the result

$$\theta_X = w_{,X} \qquad \text{and} \qquad \theta_Y = w_{,Y} \tag{11.71}$$

The displacements of the plate middle surface may then be approximated as

$$\mathbf{u} = \begin{Bmatrix} u_1 \\ u_2 \\ u_3 \end{Bmatrix} = \begin{Bmatrix} u(X,Y) \\ v(X,Y) \\ w(X,Y) \end{Bmatrix} - Z \begin{Bmatrix} w_{,X}(X,Y) \\ w_{,Y}(X,Y) \\ 0 \end{Bmatrix} \tag{11.72}$$

Once again we can note that in-plane positions X and Y do not change significantly, thus permitting substitution of x and y in the strain expressions to obtain Green–Lagrange strains as

$$\mathbf{E} = \begin{Bmatrix} u_{,x} + \frac{1}{2}(w_{,x})^2 \\ v_{,y} + \frac{1}{2}(w_{,y})^2 \\ u_{,y} + v_{,x} + w_{,x}w_{,y} \end{Bmatrix} - Z \begin{Bmatrix} w_{,xx} \\ w_{,yy} \\ 2w_{,xy} \end{Bmatrix} = \mathbf{E}^{\mathrm{p}} - Z\mathbf{K}^{\mathrm{b}} \tag{11.73}$$

where we have once again neglected square terms involving derivatives of the in-plane displacements and terms in Z^2. We note now that introduction of Eq. (11.71) modifies the expression for change in curvature to the same form as that used for thin plates in Chapter 4.

11.5.1 Evaluation of strain–displacement matrices

For further formulation it is again necessary to establish expressions for the $\bar{\mathbf{B}}$ and $\mathbf{K}_\mathrm{T}$ matrices. The finite element approximations to the displacements now involve only u, v, and w. Here we assume these to be expressed in the form

$$\begin{Bmatrix} u \\ v \end{Bmatrix} = N_\alpha \begin{Bmatrix} \tilde{u}_\alpha \\ \tilde{v}_\alpha \end{Bmatrix} = N_\alpha \tilde{\mathbf{u}}_\alpha \tag{11.74}$$

and

$$w = N_\alpha^w \tilde{w}_\alpha + \mathbf{N}_\alpha^\theta \tilde{\boldsymbol{\theta}}_\alpha \tag{11.75}$$

where now the rotation parameters are defined as

$$\tilde{\boldsymbol{\theta}}_\alpha^\mathrm{T} = [\,(\tilde{\theta}_x)_\alpha \quad (\tilde{\theta}_y)_\alpha\,] = [\,(\tilde{w}_{,x})_\alpha \quad (\tilde{w}_{,y})_\alpha\,] \tag{11.76}$$

The expressions for $\mathbf{B}^\mathrm{p}$ and $\mathbf{B}^\mathrm{L}$ are identical to those given previously except for the definition of $\mathbf{G}$. Owing to the form of the interpolation for w, we now obtain

$$\mathbf{G}_\alpha = \begin{bmatrix} N_{\alpha,x}^w & N_{\alpha,x}^{\theta x} & N_{\alpha,x}^{\theta y} \\ N_{\alpha,y}^w & N_{\alpha,y}^{\theta x} & N_{\alpha,y}^{\theta y} \end{bmatrix} \tag{11.77}$$

The variation in curvature for the thin plate is given by

$$\begin{aligned} \delta\mathbf{K}^\mathrm{b} &= \begin{bmatrix} N_{\alpha,xx}^w & N_{\alpha,xx}^{\theta x} & N_{\alpha,xx}^{\theta y} \\ N_{\alpha,yy}^w & N_{\alpha,yy}^{\theta x} & N_{\alpha,yy}^{\theta y} \\ 2N_{\alpha,xy}^w & 2N_{\alpha,xy}^{\theta x} & 2N_{\alpha,xy}^{\theta y} \end{bmatrix} \begin{Bmatrix} \delta\tilde{w}_\alpha \\ (\delta\tilde{\theta}_x)_\alpha \\ (\delta\tilde{\theta}_x)_\alpha \end{Bmatrix} \\ &= \mathbf{B}_\alpha^\mathrm{b}\, \delta\tilde{\mathbf{w}}_\alpha \end{aligned} \tag{11.78}$$

Grouping the force terms, now without the shears $\mathbf{T}^\mathrm{s}$, as

$$\bar{\boldsymbol{\sigma}} = \begin{Bmatrix} \mathbf{T}^\mathrm{p} \\ \mathbf{M}^\mathrm{b} \end{Bmatrix} \tag{11.79}$$

and the strain matrices as

$$\bar{\mathbf{B}}_\alpha = \begin{bmatrix} \mathbf{B}_\alpha & \mathbf{B}_\alpha^\mathrm{L} \\ \mathbf{0} & \mathbf{B}_\alpha^\mathrm{b} \end{bmatrix} \tag{11.80}$$

the virtual work expression may be written in matrix form as

$$\delta\Pi = \delta\tilde{\mathbf{a}}_\alpha^\mathrm{T} \int_A \bar{\mathbf{B}}_\alpha^\mathrm{T} \bar{\boldsymbol{\sigma}}\, \mathrm{d}A - \delta\Pi_\mathrm{ext} = 0 \tag{11.81}$$

and once again a non-linear problem in the form of Eq. (11.61) is obtained.

11.5.2 Evaluation of tangent matrix

A tangent matrix for the non-linear plate formulation may be computed by a linearization of Eq. (11.60). If we again assume linear elastic behaviour, the relation between the plate forces and strains may be written as

$$\begin{Bmatrix} \mathbf{T}^{\mathrm{p}} \\ \mathbf{M}^{\mathrm{b}} \end{Bmatrix} = \begin{bmatrix} \mathbf{D}^{\mathrm{p}} & \mathbf{0} \\ \mathbf{0} & \mathbf{D}^{\mathrm{b}} \end{bmatrix} \begin{Bmatrix} \mathbf{E}^{\mathrm{p}} \\ \mathbf{K}^{\mathrm{b}} \end{Bmatrix} \tag{11.82}$$

where the elastic constants are given in Eq. (11.64). Thus, the linearization of the constitution becomes

$$d(\boldsymbol{\sigma}) = \begin{bmatrix} \mathbf{D}^{\mathrm{p}} & \mathbf{0} \\ \mathbf{0} & \mathbf{D}^{\mathrm{b}} \end{bmatrix} \begin{bmatrix} \mathbf{B}^{\mathrm{p}}_{\beta} & \mathbf{0} \\ \mathbf{B}^{\mathrm{L}}_{\beta} & \mathbf{B}^{\mathrm{b}}_{\beta} \end{bmatrix} \begin{Bmatrix} d(\tilde{\mathbf{u}}_{\beta}) \\ d(\tilde{\mathbf{w}}_{\beta}) \end{Bmatrix} \tag{11.83}$$

Using this result the material part of the tangent matrix is expressed as

$$\begin{aligned} (\mathbf{K}_{\mathrm{M}})_{\alpha\beta} &= \int_A \begin{bmatrix} (\mathbf{B}^{\mathrm{p}}_{\alpha})^{\mathrm{T}} & \mathbf{0} \\ (\mathbf{B}^{\mathrm{L}}_{\alpha})^{\mathrm{T}} & (\mathbf{B}^{\mathrm{b}}_{\alpha})^{\mathrm{T}} \end{bmatrix} \begin{bmatrix} \mathbf{D}^{\mathrm{p}} & \mathbf{0} \\ \mathbf{0} & \mathbf{D}^{\mathrm{b}} \end{bmatrix} \begin{bmatrix} \mathbf{B}^{\mathrm{p}}_{\beta} & \mathbf{B}^{\mathrm{L}}_{\beta} \\ \mathbf{0} & \mathbf{B}^{\mathrm{b}}_{\beta} \end{bmatrix} \mathrm{d}A \\ &= \begin{bmatrix} (\mathbf{K}^{\mathrm{p}}_{\mathrm{M}})_{\alpha\beta} & (\mathbf{K}^{\mathrm{L}}_{\mathrm{M}})_{\alpha\beta} \\ (\mathbf{K}^{\mathrm{L}}_{\mathrm{M}})^{\mathrm{T}}_{\alpha\beta} & (\mathbf{K}^{\mathrm{b}}_{\mathrm{M}})_{\alpha\beta} \end{bmatrix} \end{aligned} \tag{11.84}$$

where $\mathbf{K}^{\mathrm{p}}_{\mathrm{M}}$ and $\mathbf{K}^{\mathrm{L}}_{\mathrm{M}}$ are given as in Eq. (11.67), and $\mathbf{K}^{\mathrm{b}}_{\mathrm{M}}$ simplifies to

$$(\mathbf{K}^{\mathrm{b}}_{\mathrm{M}})_{\alpha\beta} = \int_A (\mathbf{B}^{\mathrm{b}}_{\alpha})^{\mathrm{T}} \mathbf{D}^{\mathrm{b}} \mathbf{B}^{\mathrm{b}}_{\beta} \, \mathrm{d}A \tag{11.85}$$

and now $\mathbf{B}^{\mathrm{b}}_{\alpha}$ is given by Eq. (11.78). Using Eq. (11.77) the geometric matrix has identical form to Eqs (11.69) and (11.70).

11.6 Solution of large deflection problems

All the ingredients necessary for computing the 'large deflection' plate problem are now available. Here we may use results from either the thick or thin plate formulations described above. Below we describe the process for the thin plate formulation.

As a first step displacements $\tilde{\mathbf{a}}^0$ are found according to the small displacement uncoupled solution. This is used to determine the actual strains by considering the non-linear relations for $\mathbf{E}^{\mathrm{p}}$ and the linear curvature relations for $\mathbf{K}^{\mathrm{b}}$ defined in Eq. (11.73). Corresponding stresses can be found by the elastic relations and a Newton–Raphson iteration process set up to solve Eq. (11.61) [which is obtained from Eq. (11.81)].

A typical solution which shows the stiffening of the plate with increasing deformation arising from the development of 'membrane' stresses was shown in Fig. 11.6.[12] The results show excellent agreement with an alternative analytical solution. The element properties were derived using for the in-plane deformation the simplest bilinear rectangle and for the bending deformation the non-conforming shape function for a rectangle (Sec. 4.3, Chapter 4).

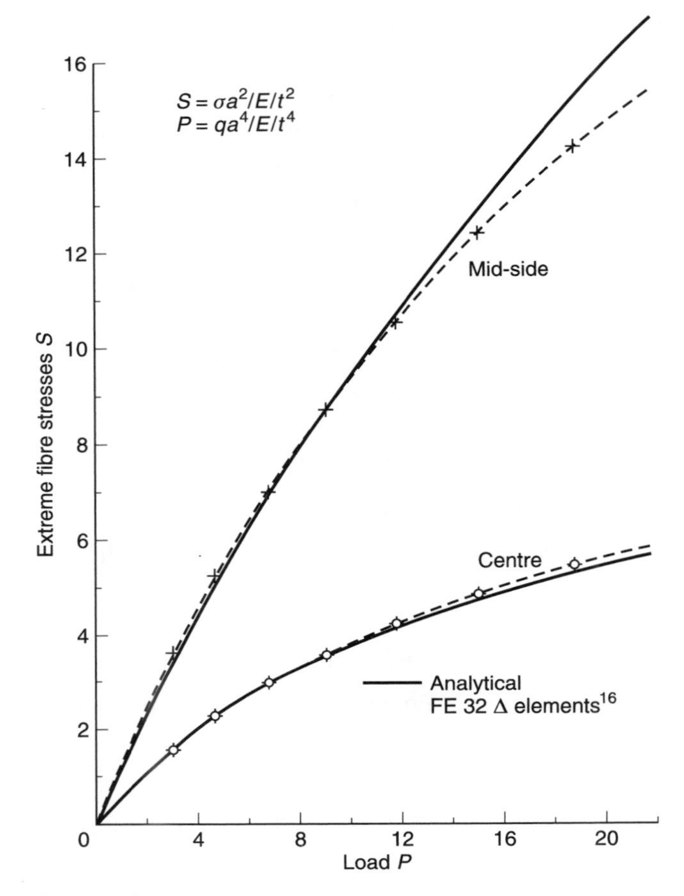

Fig. 11.7 Clamped square plate: stresses.

An example of the stress variation with loads for a clamped square plate under uniform dead load is shown in Fig. 11.7.[16] A quarter of the plate is analysed as above with 32 triangular elements, using the 'in-plane' triangular element given in Chapter 4 of Volume 1 together with a modified version of the non-conforming plate bending element of Chapter 4.[17] Many other examples of large plate deformation obtained by finite element methods are available in the literature.[18–23]

11.6.1 Bifurcation instability

In a few practical cases, as in the classical Euler problem, a bifurcation instability is possible similar to the case considered for straight beams in Sec. 11.3. Consider the situation of a plate loaded purely in its own plane. As lateral deflections, w, are not produced, the small deflection theory gives an exact solution. However, even with zero lateral displacements, the geometric stiffness (initial stress) matrix can be

Table 11.2 Values of C for a simply supported square plate and compressed axially by T_x

Elements in quarter plate	Non-compatible		Compatible	
	rectangle[26] 12 d.o..f.	triangle[27] 9 d.o.f.	rectangle[28] 116 d.o.f	quadrilateral[29] 16 d.o.f.
2×2		3.22		
4×4	3.77	3.72	4.015	4.029
8×8	3.93	3.90	4.001	4.002

Exact $C = 4.00$.[14]
d.o.f. = degrees-of-freedom.

found while $\mathbf{B}^{\rm L}$ remains zero. If the in-plane stresses are compressive this matrix will be such that real eigenvalues of the bending deformation can be found by solving the eigenproblem

$$\mathbf{K}_{\rm M}^{\rm b}\, d\tilde{\mathbf{w}} = -\lambda \mathbf{K}_{\rm G}^{\rm L}\, d\tilde{\mathbf{w}} \tag{11.86}$$

in which λ denotes a multiplying factor on the in-plane stresses necessary to achieve neutral equilibrium (limit stability), and $\delta\tilde{\mathbf{w}}$ is the eigenvector describing the shape that a 'buckling' mode may take.

At such an increased load incipient buckling occurs and lateral deflections can occur without any lateral load. The problem is simply formulated by writing only the bending equations with $\mathbf{K}_{\rm M}^{\rm b}$ determined as in Chapter 4 and with $\mathbf{K}_{\rm G}^{\rm L}$ found from Eq. (11.69).

Points of such incipient stability (buckling) for a variety of plate problems have been determined using various element formulations.[24–29] Some comparative results for a simple problem of a square, simply supported plate under a uniform compression T_x applied in one direction are given in Table 11.2. In this the buckling parameter is defined as

$$C = \frac{T_x a^2}{\pi^2 D}$$

where a is the side length of a square plate and D the bending rigidity.

The elements are all of the type described in Chapter 4 and it is of interest to note that all those that are slope compatible always overestimate the buckling factor. This result is obtained only for cases where the in-plane stresses $\mathbf{T}^{\rm p}$ are exact solutions to the differential equations; in cases where these are approximate solutions this bound property is not assured. The non-conforming elements in this case underestimate the load, although there is now no theoretical lower bound available.

Figure 11.8 shows a buckling mode for a geometrically more complex case.[27] Here again the non-conforming triangle was used.

Such incipient stability problems in plates are of limited practical importance. As soon as lateral deflection occurs a stiffening of the plate follows and additional loads can be carried. This stiffening was noted in the example of Fig. 11.6. Post-buckling behaviour thus should be studied by the large deformation process described in previous sections.[21,30]

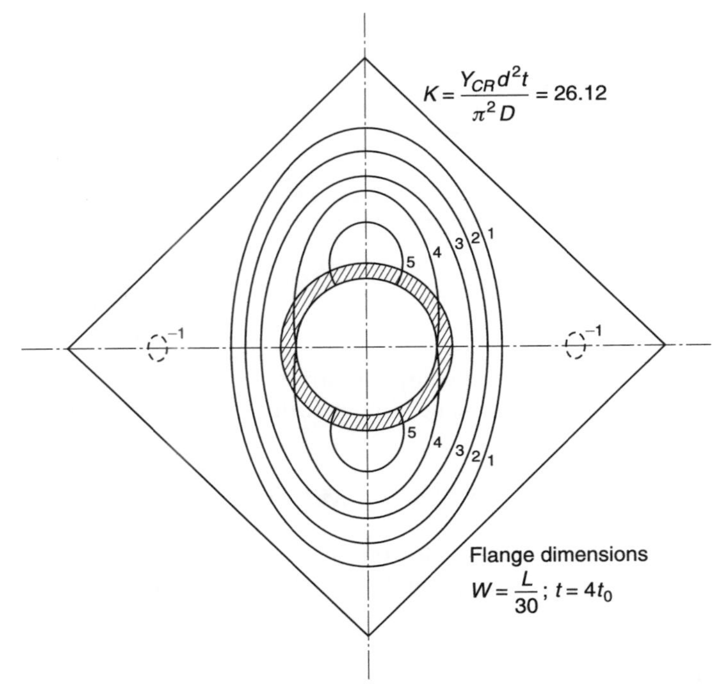

Fig. 11.8 Buckling mode of a square plate under shear: clamped edges, central hole stiffened by flange.[27]

11.7 Shells

In shells, non-linear response and stability problems are much more relevant than in plates. Here, in general, the problem is one in which the tangential stiffness matrix $\mathbf{K}_T$ should always be determined taking the actual displacements into account, as now the special case of uncoupled membrane and bending effects does not occur under load except in the most trivial cases. If the *initial stability* matrix $\mathbf{K}_G$ is determined for the elastic stresses it is, however, sometimes possible to obtain useful results concerning the stability factor λ, and indeed in the classical work on the subject of shell buckling this initial stability often has been considered. The true collapse load may, however, be well *below* the initial stability load and it is important to determine at least approximately the deformation effects.

If the shell is assumed to be built up of flat plate elements, the same transformations as given in Chapter 6 can be followed with the plate tangential stiffness matrix.[31] If curved shell elements are used it is important to revert to the equations of shell theory and to include in these the non-linear terms.[12,32–34] Alternatively, one may approach the problem from a degeneration of solids, as described in Chapter 7 for the small deformation case, suitably extended to the large deformation form. This approach was introduced by several authors and extensively developed in recent years.[35–46] A key to successful implementation of this approach is the treatment of finite rotations. For details on the complete formulation the reader is referred to the cited references.

11.7.1 Axisymmetric shells

Here we consider the extension for the beam presented above in Sec. 11.2 to treat axisymmetric shells. We limit our discussion to the extension of the small deformation case treated in Sec. 7.4 in which two-noded straight conical elements (see Fig. 7.2) and reduced quadrature are employed. Local axes on the shell segment may be defined by

$$\begin{aligned}\bar{R} &= \cos\phi\,(R - R_0) - \sin\phi\,(Z - Z_0)\\ \bar{z} &= \sin\phi\,(R - R_0) + \cos\phi\,(Z - Z_0)\end{aligned} \tag{11.87}$$

where R_0, Z_0 are centred on the element as

$$\begin{aligned}R_0 &= \tfrac{1}{2}(R_1 + R_2)\\ Z_0 &= \tfrac{1}{2}(Z_1 + Z_2)\end{aligned} \tag{11.88}$$

with R_I, Z_I nodal coordinates of the element. The deformed position with respect to the local axes may be written in a form identical to Eq. (11.7). Accordingly, we have

$$\begin{aligned}\bar{r} &= \bar{R} + \bar{u}(\bar{R}) + \bar{Z}\sin\beta(\bar{R})\\ \bar{z} &= \bar{w}(\bar{R}) + \bar{Z}\cos\beta(\bar{R})\end{aligned} \tag{11.89}$$

To consider the axisymmetric shell it is necessary to integrate over the volume of the shell and to include the axisymmetric hoop strain effects. Accordingly, we now consider a segment of shell in the R–Z plane (i.e. X is replaced by the radius R). The volume of the shell in the reference configuration is obtained by multiplying the beam volume element by the factor $2\pi R$. In axisymmetry the deformation gradient in the tangential (hoop) direction must be included. Accordingly, in the local coordinate frame the deformation gradient is given by

$$F_{iI} = \begin{bmatrix} [1 + \bar{u}_{,\bar{R}} + \bar{Z}(\cos\beta)\beta_{,\bar{R}}] & 0 & \sin\beta \\ 0 & r/R & 0 \\ [\bar{w}_{,\bar{R}} - \bar{Z}(\sin\beta)\beta_{,\bar{R}}] & 0 & \cos\beta \end{bmatrix} \tag{11.90}$$

Following the same procedures as indicated for the beam we obtain the expressions for Green–Lagrange strains as

$$\begin{aligned}E_{\bar{R}\bar{R}} &= \bar{u}_{,\bar{R}} + \tfrac{1}{2}(\bar{u}_{,\bar{R}}^2 + \bar{w}_{,\bar{R}}^2) + \bar{Z}\Lambda\beta_{,\bar{R}} = E^0_{\bar{R}\bar{R}} + \bar{Z}K^{\mathrm{b}}_{\bar{R}\bar{R}}\\ E_{TT} &= \frac{u}{R} + \frac{1}{2}\left(\frac{u}{R}\right)^2 + \bar{Z}\left(1 + \frac{u}{R}\right)\frac{\sin\beta}{R} = E^0_{TT} + ZK^{\mathrm{b}}_{TT}\\ 2E_{\bar{R}\bar{Z}} &= (1 + \bar{u}_{,\bar{R}})\sin\beta + \bar{w}_{,\bar{R}}\cos\beta = \Gamma\end{aligned} \tag{11.91}$$

where $\Lambda = (1 + \bar{u}_{,\bar{R}})\cos\beta - \bar{w}_{,\bar{R}}\sin\beta$, and the additional hoop strain results in two additional strain components, E^0_{TT} and K^{b}_{TT}.

With the above modifications, the virtual work expression for the shell now becomes

$$\delta\Pi = \int_\Omega (\delta E_{\bar{R}\bar{R}}S_{\bar{R}\bar{R}} + \delta E_{TT}S_{TT} + 2\delta E_{\bar{R}\bar{Z}}S_{\bar{R}\bar{Z}})\,\mathrm{d}V - \delta\Pi_{\mathrm{ext}} = 0 \tag{11.92}$$

in which S_{TT} is the hoop stress in the cylindrical direction. The remainder of the development follows the procedures presented in Sec. 11.2.1 and is left as an exercise for the reader. It is also possible to develop a small rotation theory following the methods described in Sec. 11.2.2.

Here we demonstrate the use of the axisymmetric shell theory by considering a shallow spherical cap subjected to an axisymmetric vertical ring load (Fig. 11.9). The case where the ring load is concentrated at the crown has been examined analytically by Biezeno[47] and Reissner.[48] Solutions using finite difference methods on the equations of Reissner are presented by Mescall.[49] Solutions by finite elements have been presented earlier by Zienkiewicz and co-workers.[7,50] Owing to the shallow nature of the shell, rotations remain small, and excellent agreement exists between the finite rotation and small rotation forms.

11.7.2 Shallow shells – co-rotational forms

In the case of shallow shells the transformations of Chapter 6 may conveniently be avoided by adopting a formulation based on Marguerre shallow shell theory.[23,51,52] A simple extension to a shallow shell theory for the formulation presented for thin plates may be obtained by replacing the displacements by

$$\begin{Bmatrix} u \\ v \\ w \end{Bmatrix} \rightarrow \begin{Bmatrix} u_0 + u \\ v_0 + v \\ w_0 + w \end{Bmatrix} \tag{11.93}$$

in which u_0, v_0, and w_0 describe the position of the shell reference configuration from the $X-Y$ plane. Now the current configuration of the shell (where, often, u_0 and v_0 are taken as zero) may be described by

$$\begin{aligned} x_1(t) &= X + u_0(X, Y) + u(X, Y, t) - Z[w_{0,X}(X, Y) + w_{,X}(X, Y, t)] \\ x_2(t) &= Y + v_0(X, Y) + v(X, Y, t) - Z[w_{0,Y}(X, Y) + w_{,Y}(X, Y, t)] \\ x_3(t) &= w_0(X, Y) + w(X, Y, t) \end{aligned} \tag{11.94}$$

where a time t is introduced to remind the reader that at time zero the reference configuration is described by

$$\begin{aligned} x_1(0) &= X + u_0(X, Y) - Z w_{0,X}(X, Y) \\ x_2(0) &= Y + v_0(X, Y) - Z w_{0,Y}(X, Y) \\ x_3(0) &= w_0(X, Y) \end{aligned} \tag{11.95}$$

where u, v, w vanish. Using these expressions we can compute the deformation gradient for the deformed configuration and for the reference configuration. Denoting these by F_{iI} and F_{iI}^0, respectively, we can deduce the Green–Lagrange strains from

$$E_{IJ} = \tfrac{1}{2}\left[F_{iI} F_{iJ} - F_{iI}^0 F_{iJ}^0\right] \tag{11.96}$$

The remainder of the derivations are straightforward and left as an exercise for the reader. This approach may be generalized and used also to deduce the equations

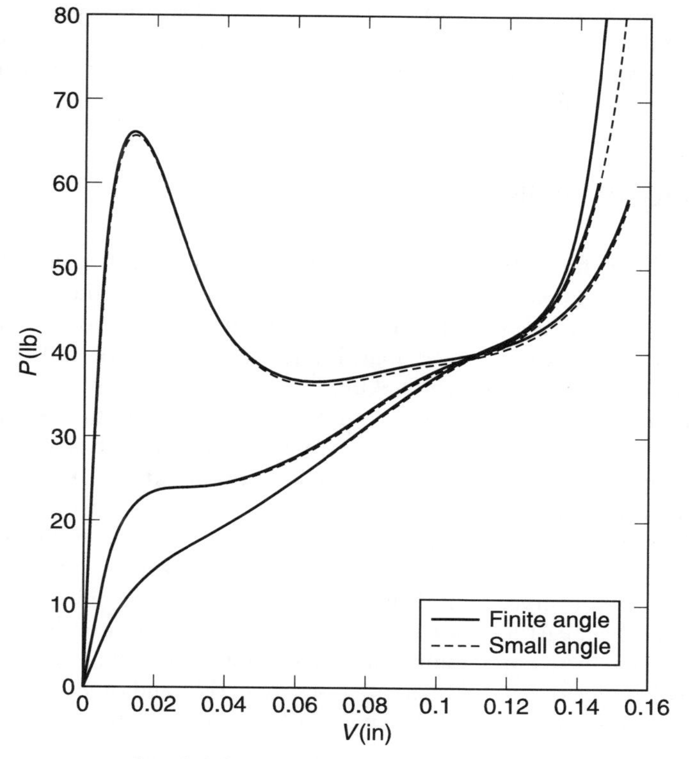

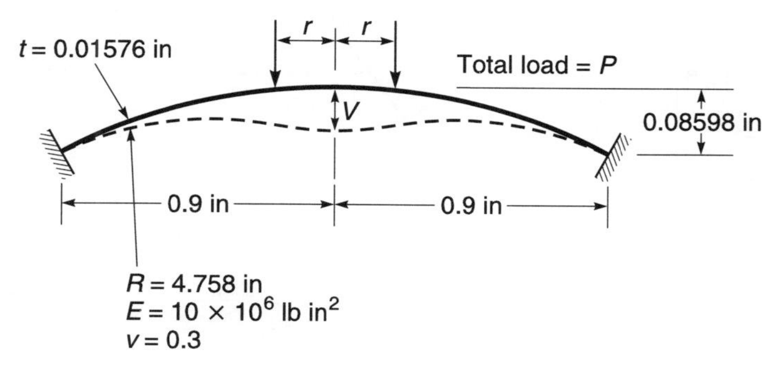

Fig. 11.9 Spherical cap under vertical ring load: (a) load–deflection curves for various ring loads. Spherical cap under vertical ring load: (b) geometry definition and deflected shape.

for deep shells.[42] Alternatively, we can note that as finite elements become small they are essentially shallow shells relative to a rotated plane. This observation led to the development of many general shells based on a concept named 'co-rotational'. Here the reader is referred to the literature for additional details.[53–67]

11.7.3 Stability of shells

It is extremely important to emphasize again that instability calculations are meaningful only in special cases and that they often overestimate the collapse loads considerably. For correct answers a full non-linear process has to be invoked. A progressive 'softening' of a shell under load is shown in Fig. 11.10 and the result is well below the one given by linearized buckling.[12] Figure 11.11 shows the progressive collapse of an arch at a load much below that given by the linear stability value. The solution from the finite rotation beam formulation is compared with an early solution obtained by Marcal[68] who employed small-angle approximations. Here again it is evident that use of finite angles is important.

The determination of the actual collapse load of a shell or other slender structure presents obvious difficulties (of a kind already discussed in Chapter 2 and encountered above for beams), as convergence of displacements cannot be obtained when load is 'increased' near the peak carrying capacity. In such cases one can proceed by prescribing displacement increments and computing the corresponding reactions if only one concentrated load is considered. By such processes, Argyris[69] and others[34,50] succeeded in following a complete snap-through behaviour of a shallow arch.

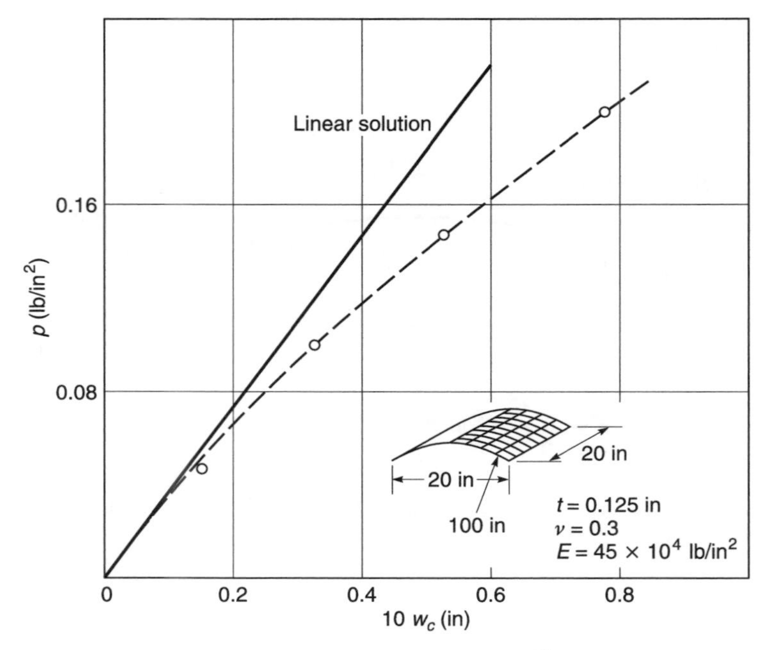

Fig. 11.10 Deflection of cylindrical shell at centre: all edges clamped.[12]

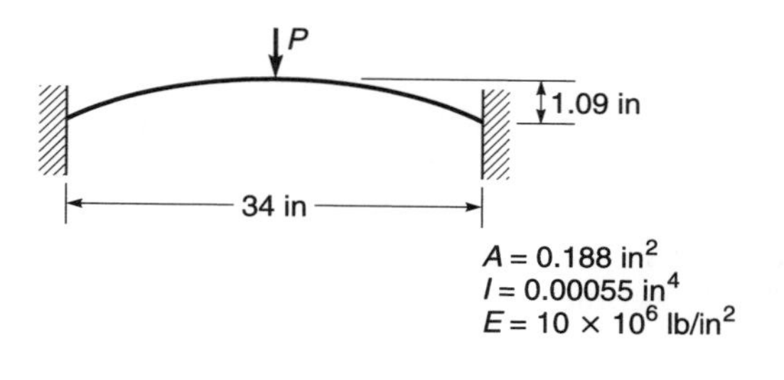

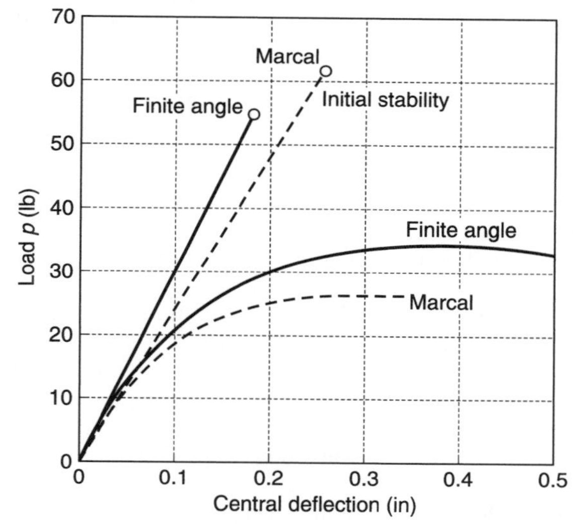

Fig. 11.11 'Initial stability' and incremental solution for large deformation of an arch under central load p.[68]

Pian and Tong[70] show how the process can be generalized simply when a system of proportional loads is considered. This and other 'arc-length' methods are considered in Sec. 2.2.6.

11.8 Concluding remarks

This chapter presents a summary of approaches that can be used to solve problems in structures composed of beams (rods), plates, and shells. The various procedures follow the general theory presented in Chapter 10 combined with solution methods for non-linear algebraic systems as presented in Chapter 2. Again we find that solution of a non-linear large displacement problem is efficiently approached by using a Newton–Raphson type approach in which a residual and a tangent matrix are used. We remind the reader, however, that use of modified approaches, such as use of a constant tangent matrix, is often as, or even more, economical than use of the full Newton–Raphson process.

If a full load deformation study is required it has been common practice to proceed with small load increments and treat, for each such increment, the problem by a form of the Newton–Raphson process. It is recommended that each solution step be accurately solved so as not to accumulate errors. We have observed that for problems which have a limit load, beyond which the system is stable, a full solution can be achieved only by use of an 'arc-length' method (except in the trivial case of one point load as noted above).

Extension of the problem to dynamic situations is readily accomplished by adding the inertial terms. In the geometrically exact approach in three dimensions one may encounter quite complex forms for these terms and here the reader should consult literature on the subject before proceeding with detailed developments.[2–5] For the small-angle assumptions the treatment of rotations is identical to the small deformation problem and no such difficulties arise.

References

1. E. Reissner. On one-dimensional finite strain beam theory: the plane problem. *J. Appl. Math. Phys.*, **23**, 795–804, 1972.
2. J.C. Simo. A finite strain beam formulation: the three-dimensional dynamic problem: part I. *Comp. Meth. Appl. Mech. Eng.*, **49**, 55–70, 1985.
3. A. Ibrahimbegovic and M. Al Mikdad. Finite rotations in dynamics of beams and implicit time-stepping schemes. *Int. J. Num. Meth. Eng.*, **41**, 781–814, 1998.
4. J.C. Simo and L. Vu-Quoc. A three-dimensional finite strain rod model. Part II: geometric and computational aspects. *Comp. Meth. Appl. Mech. Eng.*, **58**, 79–116, 1986.
5. J.C. Simo, N. Tarnow and M. Doblare. Non-linear dynamics of three-dimensional rods: exact energy and momentum conserving algorithms. *Int. J. Num. Meth. Eng.*, **38**, 1431–73, 1995.
6. D.A. da Deppo and R. Schmidt. Instability of clamped–hinged circular arches subjected to a point load. *Trans. Am. Soc. Mech. Eng.*, 894–6, 1975.
7. R.D. Wood and O.C. Zienkiewicz. Geometrically non-linear finite element analysis of beams–frames–circles and axisymmetric shells. *Computers and Structures*, **7**, 725–35, 1977.
8. J.C. Simo, P. Wriggers, K.H. Schweizerhof and R.L Taylor. Finite deformation post-buckling analysis involving inelasticity and contact constraints. *Int. J. Num. Meth. Eng.*, **23**, 779–800, 1986.
9. H.L. Langhaar. *Energy Methods in Applied Mechanics*, John Wiley, New York, 1962.
10. K. Marguerre. Über die Anwendung der energetishen Methode auf Stabilitätsprobleme. *Hohrb.*, **DVL**, 252–62, 1938.
11. B. Fraeijs de Veubeke. The second variation test with algebraic and differential constraints. In *Advanced Problems and Methods for Space Flight Optimization*, Pergamon Press, Oxford, 1969.
12. C.A. Brebbia and J. Connor. Geometrically non-linear finite element analysis. *Proc. Am. Soc. Civ. Eng.*, **95**(EM2), 463–83, 1969.
13. B.N. Parlett. *The Symmetric Eigenvalue Problem*, Prentice-Hall, Englewood Cliffs, NJ, 1980.
14. S.P. Timoshenko and J.M. Gere. *Theory of Elastic Stability*, McGraw-Hill, New York, 1961.
15. R. Szilard. *Theory and Analysis of Plates*, Prentice-Hall, Englewood Cliffs, NJ, 1974.
16. R.D. Wood. *The Application of Finite Element Methods to Geometrically Non-linear Analysis*, PhD thesis, Department of Civil Engineering, University of Wales, Swansea, 1973.

17. A. Razzaque. Program for triangular bending element with derivative smoothing. *Int. J. Num. Meth. Eng.*, **5**, 588–9, 1973.
18. M.J. Turner, E.H. Dill, H.C. Martin and R.J. Melosh. Large deflection of structures subjected to heating and external loads. *J. Aero. Sci.*, **27**, 97–106, 1960.
19. L.A. Schmit, F.K. Bogner and R.L. Fox. Finite deflection structural analysis using plate and cylindrical shell discrete elements. *Journal of AIAA*, **5**, 1525–7, 1968.
20. R.H. Mallett and P.V. Marcal. Finite element analysis of non-linear structures. *Proc. Am. Soc. Civ. Eng.*, **94**(ST9), 2081–105, 1968.
21. D.W. Murray and E.L. Wilson. Finite element post buckling analysis of thin plates. In L. Berke, R.M. Bader, W.J. Mykytow, J.S. Przemienicki and M.H. Shirk (eds), *Proc. 2nd Conf. Matrix Methods in Structural Mechanics*, Volume AFFDL-TR-68-150, Air Force Flight Dynamics Laboratory, Wright Patterson Air Force Base, OH, October 1968.
22. T. Kawai and N. Yoshimura. Analysis of large deflection of plates by finite element method. *Int. J. Num. Meth. Eng.*, **1**, 123–33, 1969.
23. P.G. Bergan and R.W. Clough. Large deflection analysis of plates and shallow shells using the finite element method. *Int. J. Num. Meth. Eng.*, **5**, 543–56, 1973.
24. R.H. Gallagher and J. Padlog. Discrete element approach to structural instability analysis. *Journal of AIAA*, **1**, 1537–9, 1963.
25. H.C. Martin. On the derivation of stiffness matrices for the analysis of large deflection and stability problems. In J.S. Przemienicki, R.M. Bader, W.F. Bozich, J.R. Johnson and W.J. Mykytow (eds), *Proc. 1st Conf. Matrix Methods in Structural Mechanics*, Volume AFFDI, TR-66-80, Air Force Flight Dynamics Laboratory, Wright Patterson Air Force Base, OH, October 1966.
26. K.K. Kapur and B.J. Hartz. Stability of thin plates using the finite element method. *Proc. Am. Soc. Civ. Eng.*, **92**(EM2), 177–95, 1966.
27. R.G. Anderson, B.M. Irons and O.C. Zienkiewicz. Vibration and stability of plates using finite elements. *International Journal of Solids and Structures*, **4**, 1033–55, 1968.
28. W.G. Carson and R.E. Newton. Plate buckling analysis using a fully compatible finite element. *Journal of AIAA*, **8**, 527–9, 1969.
29. Y.K. Chan and A.P. Kabaila. A conforming quadrilateral element for analysis of stiffened plates. Technical Report UNICIV, Report R-121, University of New South Wales, 1973.
30. K.C. Rockey and D.K. Bagchi. Buckling of plate girder webs under partial edge loadings. *Int. J. Mech. Sci.*, **12**, 61–76, 1970.
31. R.H. Gallagher, R.A. Gellately, R.H. Mallett and J. Padlog. A discrete element procedure for thin shell instability analysis. *Journal of AIAA*, **5**, 138–45, 1967.
32. R.H. Gallagher and H.T.Y. Yang. Elastic instability predictions for doubly curved shells. In L. Berke, R.M. Bader, W.J. Mykytow, J.S. Przemienicki and M.H. Shirk (eds), *Proc. 2nd Conf. Matrix Methods in Structural Mechanics*, Volume AFFDITR-68-150, Air Force Flight Dynamics Laboratory, Wright Patterson Air Force Base, OH, October 1968.
33. J.L. Batoz, A. Chattapadhyay and G. Dhatt. Finite element large deflection analysis of shallow shells. *Int. J. Num. Meth. Eng.*, **10**, 35–8, 1976.
34. T. Matsui and O. Matsuoka. A new finite element scheme for instability analysis of thin shells. *Int. J. Num. Meth. Eng.*, **10**, 145–70, 1976.
35. E. Ramm. Geometrishe nichtlineare Elastostatik und Finite elemente. Technical Report 76-2, Institut für Baustatik, Universität Stuttgart, Stuttgart, 1976.
36. H. Parisch. Efficient non-linear finite element shell formulation involving large strains. *Engineering Computations*, **3**, 121–8, 1986.
37. J.C. Simo and D.D. Fox. On a stress resultant geometrically exact shell model. Part I: formulation and optimal parametrization. *Comp. Meth. Appl. Mech. Eng.*, **72**, 267–304, 1989.

38. J.C. Simo, D.D. Fox and M.S. Rifai. On a stress resultant geometrically exact shell model. Part II: the linear theory; computational aspects. *Comp. Meth. Appl. Mech. Eng.*, **73**, 53–92, 1989.
39. J.C. Simo, S. Rifai and D.D. Fox. On a stress resultant geometrically exact shell model. Part IV: nonlinear plasticity: formulation and integration algorithms. *Comp. Meth. Appl. Mech. Eng.*, **81**, 91–126, 1990.
40. J.C. Simo and N. Tarnow. On a stress resultant geometrically exact shell model. Part VI: 5/6 DOF treatments. *Int. J. Num. Meth. Eng.*, **34**, 117–64, 1992.
41. H. Parisch. A continuum-based shell theory for nonlinear applications. *Int. J. Num. Meth. Eng.*, **38**, 1855–83, 1993.
42. K.-J. Bathe. *Finite Element Procedures*, Prentice-Hall, Englewood Cliffs, NJ, 1996.
43. P. Betsch, F. Gruttmann and E. Stein. A 4-node finite shell element for the implementation of general hyperelastic 3d-elasticity at finite strains. *Comp. Meth. Appl. Mech. Eng.*, **130**, 57–79, 1996.
44. M. Bischoff and E. Ramm. Shear deformable shell elements for large strains and rotations. *Int. J. Num. Meth. Eng.*, **40**, 4427–49, 1997.
45. E. Ramm. From Reissner plate theory to three dimensions in large deformation shell analysis. *Zeitschrift für Angewandte Mathematik und Mechanik*, **79**, 1–8, 1999.
46. M. Bischoff, E. Ramm and D. Braess. A class of equivalent enhanced assumed strain and hybrid stress finite elements. *Computational Mechanics*, **22**, 443–9, 1999.
47. C.B. Biezeno. Über die Bestimmung der Durchschlagkraft einer schwach-gekrummten kneinformigen Platte. *Zeitschrift für Angewandte Mathematik und Mechanik*, **15**, 10, 1935.
48. E. Reissner. On axisymmetric deformation of thin shells of revolution. In *Proc. Symp. in Appl. Math.*, p. 32, 1950.
49. J.F. Mescall. Large deflections of spherical caps under concentrated loads. *Trans. ASME, J. Appl. Mech*, **32**, 936–8, 1965.
50. O.C. Zienkiewicz and G.C. Nayak. A general approach to problems of plasticity and large deformation using isoparametric elements. In R.M. Bader, L. Berke, R.O. Meitz, W.J. Mykytow and J.S. Przemienicki (eds), *Proc. 3rd Conf. Matrix Methods in Structural Mechanics*, Volume AFFDITR-71-160, Air Force Flight Dynamics Laboratory, Wright Patterson Air Force Base, OH, 1972.
51. T.Y. Yang. A finite element procedure for the large deflection analysis of plates with initial imperfections. *Journal of AIAA*, **9**, 1468–73, 1971.
52. T.M. Roberts and D.G. Ashwell. The use of finite element mid-increment stiffness matrices in the post-buckling analysis of imperfect structures. *International Journal of Solids and Structures*, **7**, 805–23, 1971.
53. T. Belytschko and R. Mullen. Stability of explicit–implicit time domain solution. *Int. J. Num. Meth. Eng.*, **12**, 1575–86, 1978.
54. T. Belytschko, J.I. Lin and C.-S. Tsay. Explicit algorithms for the nonlinear dynamics of shells. *Comp. Meth. Appl. Mech. Eng.*, **42**, 225–51, 1984.
55. C.C. Rankin and F.A. Brogan. An element independent co-rotational procedure for the treatment of large rotations. *ASME, J. Press. Vessel Tech.*, **108**, 165–74, 1986.
56. K.-M. Hsiao and H.-C. Hung. Large-deflection analysis of shell structure by using co-rotational total Lagrangian formulation. *Comp. Meth. Appl. Mech. Eng.*, **73**, 209–25, 1989.
57. H. Stolarski, T. Belytschko and S.-H. Lee. A review of shell finite elements and co-rotational theories. *Computational Mechanics Advances*, **2**, 125–212, 1995.
58. E. Madenci and A. Barut. A free-formulation-based flat shell element for non-linear analysis of thin composite structures. *Int. J. Num. Meth. Eng.*, **30**, 3825–42, 1994.
59. E. Madenci and A. Barut. Dynamic response of thin composite shells experiencing non-linear elastic deformations coupled with large and rapid overall motions. *Int. J. Num. Meth. Eng.*, **39**, 2695–723, 1996.

60. T.M. Wasfy and A.K. Noor. Modeling and sensitivity analysis of multibody systems using new solid, shell and beam elements. *Comp. Meth. Appl. Mech. Eng.*, **138**, 187–211, 1996.
61. A. Barut, E. Madenci and A. Tessler. Nonlinear elastic deformations of moderately thick laminated shells subjected to large and rapid rigid-body motion. *Finite Elements in Analysis and Design*, **22**, 41–57, 1996.
62. M.A. Crisfield and G.F. Moita. A unified co-rotational framework for solids, shells and beams. *International Journal of Solids and Structures*, **33**, 2969–92, 1996.
63. M.A. Crisfield and J. Shi. An energy conserving co-rotational procedure for non-linear dynamics with finite elements. *Nonlinear Dynamics*, **9**, 37–52, 1996.
64. A.A. Barut, E. Madenci and A. Tessler. Nonlinear analysis of laminates through a Mindlin-type shear deformable shallow shell element. *Comp. Meth. Appl. Mech. Eng.*, **143**, 155–73, 1997.
65. J.L. Meek and S. Ristic. Large displacement analysis of thin plates and shells using a flat facet finite element formulation. *Comp. Meth. Appl. Mech. Eng.*, **145**(3–4), 285–99, 1997.
66. H.G. Zhong and M.A. Crisfield. An energy-conserving co-rotational procedure for the dynamics of shell structures. *Engineering Computations*, **15**, 552–76, 1998.
67. C. Pacoste. Co-rotational flat facet triangular elements for shell instability analyses. *Comp. Meth. Appl. Mech. Eng.*, **156**, 75–110, 1998.
68. P.V. Marcal. Effect of initial displacement on problem of large deflection and stability. Technical Report ARPA E54, Brown University, 1967.
69. J.H. Argyris. Continua and discontinue. In J.S. Przemienicki, R.M. Bader, W.F. Bozich, J.R. Johnson and W.J. Mykytow (eds), *Proc. 1st Conf. Matrix Methods in Structural Mechanics*, Volume AFFDL-TR-66-80, pp. 11–189, Air Force Flight Dynamics Laboratory, Wright Patterson Air Force Base, OH, October 1966.
70. T.H.H. Pian and P. Tong. Variational formulation of finite displacement analysis. In *Symp. on High Speed Electronic Computation of Structures*, Liège, 1970.

12

Pseudo-rigid and rigid–flexible bodies

12.1 Introduction

Many situations are encountered where treatment of the entire system as deformable bodies is neither necessary nor practical. For example, the frontal impact of a vehicle against a barrier requires a detailed modelling of the front part of the vehicle but the primary function of the engine and the rear part is to provide inertia, deformation being negligible for purposes of modelling the frontal impact. A second example, from geotechnical engineering, is the modelling of rock mass landslides or interaction between rocks on a conveyor belt where deformation of individual blocks is secondary. In this chapter we consider briefly the study of such systems.

The above problem classes divide themselves into two further sub-classes: one where it is necessary to include some simple mechanisms of deformation in each body (e.g. an individual rock piece) and the second in which the individual bodies have no deformation at all. The first class is called *pseudo-rigid* body deformation[1] and the second *rigid-body* behaviour.[2] Here we wish to illustrate how such behaviour can be described and combined in a finite element system. For the modelling of pseudo-rigid body analyses we follow closely the work of Cohen and Muncaster[1] and the numerical implementation proposed by Solberg and Papadopoulos.[3] The literature on rigid body analysis is extensive, and here we refer the reader to papers for additional details on methods and formulations beyond those covered here.[4–21]

12.2 Pseudo-rigid motions

In this section we consider the analysis of systems which are composed of many small bodies, each of which is assumed to undergo large displacements and a uniform deformation.* The individual bodies which we consider are of the types shown in Fig. 12.1. In particular, a faceted shape can be constructed directly from a finite element discretization in which the elements are designated as all belonging to a

* Higher-order approximations can be included using polynomial approximation for the deformation of each body.

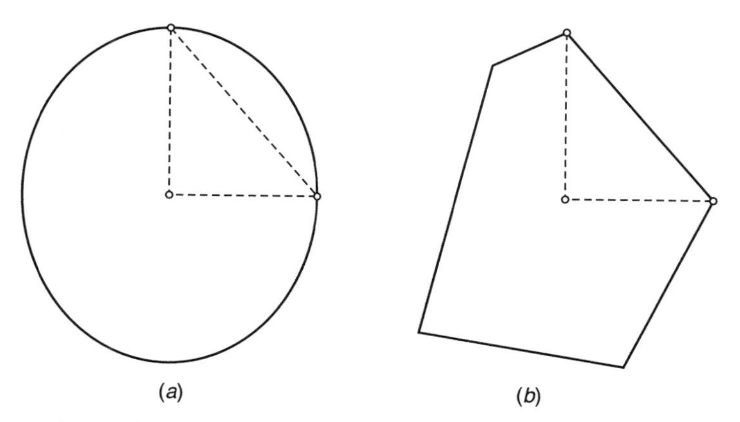

Fig. 12.1 Shapes for pseudo-rigid and rigid body analysis: (a) ellipsoid; (b) faceted body.

single solid object or the individual bodies can be described by simple geometric forms such as discs or ellipsoids.

A *homogeneous* motion of a body may be written as

$$\phi_i(X_I, t) = r_i(t) + F_{iI}(t)\,[X_I - R_I] \tag{12.1}$$

in which X_I is position, t is time, R_I is some reference point in the undeformed body, r_i is the position of the same point in the deformed body, and F_{iI} is a constant deformation gradient. We note immediately that at time zero the deformation gradient is the identity tensor (matrix) and Eq. (12.1) becomes

$$\phi_i(X_I, 0) = r_i(0) + \delta_{iI}\,[X_I - R_0] = r_0(0) + \delta_{iI}\,X_I - \delta_{iI}\,R_I \equiv \delta_{iI}\,X_I \tag{12.2}$$

where $r_i(0) = \delta_{iI} R_I$ by definition. The behaviour of solids which obey the above description is sometimes referred to as analysis of *pseudo-rigid bodies.*[1] A treatment by finite elements has been considered by Solberg and Papadopoulos,[3] and an alternative expression for motions restricted to incrementally linear behaviour has been developed by Shi, and the method is commonly called *discontinuous deformation analysis* (DDA).[22] The DDA form, while widely used in the geotechnical community, is usually combined with a simple linear elastic constitutive model and linear strain–displacement forms which can lead to large errors when finite rotations are encountered.

Once the deformation gradient is computed, the procedures for analysis follow the methods described in Chapter 10. It is, of course, necessary to include the inertial term for each body in the analysis. No difficulties are encountered once a shape of each body is described and a constitutive model is introduced. For elastic behaviour it is not necessary to use a complicated model, and here use of the Saint-Venant–Kirchhoff relation is adequate – indeed, if large deformations occur within an individual body the approximation of homogeneous deformation generally is not adequate to describe the solution. The primary difficulty for this class of problems is modelling the large number of interactions between bodies by contact phenomena and here the reader is referred to Chapter 10 and references on the subject for additional information on contact and other details.[22,23]

12.3 Rigid motions

The pseudo-rigid body form can be directly extended to rigid bodies by using the polar decomposition on the deformation tensor. The polar decomposition of the deformation gradient may be given as[24–26]

$$F_{iI} = \Lambda_{iJ} U_{JI} \qquad \text{where} \quad \Lambda_{iI}\Lambda_{iJ} = \delta_{IJ} \qquad \text{and} \qquad \Lambda_{iI}\Lambda_{jI} = \delta_{ij} \tag{12.3}$$

Here Λ_{iI} is a rigid rotation* and U_{IJ} is a stretch tensor (that has eigenvalues λ_m as defined in Chapter 10). In the case of rigid motions the stretches are all unity and U_{IJ} simply becomes an identity. Thus, a rigid body motion may be specified as

$$\phi_i(X_I, t) = r_i(t) + \Lambda_{iI}(t)\,[X_I - R_I] \tag{12.4}$$

or, in matrix form, as

$$\boldsymbol{\phi}(\mathbf{X}, t) = \mathbf{r}(t) + \boldsymbol{\Lambda}(t)\,[\mathbf{X} - \mathbf{R}] \tag{12.5}$$

Alternatively, we can express the rigid motion using Eq. (12.1) and impose constraints to make the stretches unity. For example, in two dimensions we can represent the motion in terms of the displacements of the vertices of a triangle and apply constraints that the lengths of the triangle sides are unchanged during deformation. The constraints may be added as Lagrange multipliers or other constraint methods and the analysis may proceed directly from a standard finite element representation of the triangle. Such an approach has been used in reference 27 with a penalty method used to impose the constraints. Here we do not pursue this approach further and instead consider direct use of rigid body motions to construct the formulation.

For subsequent use we note the form of the variation of a rigid motion and its incremental part. These may be expressed as

$$\delta\boldsymbol{\phi} = \delta\mathbf{r} + \widehat{\delta\boldsymbol{\theta}}\,\boldsymbol{\Lambda}\,[\mathbf{X} - \mathbf{R}]$$

$$d\boldsymbol{\phi} = d\mathbf{r} + \widehat{d\boldsymbol{\theta}}\,\boldsymbol{\Lambda}\,[\mathbf{X} - \mathbf{R}]$$

Using Eq. (12.5) these may be simplified to

$$\begin{aligned} \delta\boldsymbol{\phi} &= \delta\mathbf{r} - \hat{\mathbf{y}}\,\delta\boldsymbol{\theta} \quad \text{where} \quad \mathbf{y} = \mathbf{x} - \mathbf{r} \\ d\boldsymbol{\phi} &= d\mathbf{r} - \hat{\mathbf{y}}\,d\boldsymbol{\theta} \end{aligned} \tag{12.6}$$

where $d\boldsymbol{\phi}$ and $\delta\boldsymbol{\theta}$ are incremental and variational rotation vectors, respectively.

In a similar manner we obtain the velocity for the rigid motion as

$$\dot{\boldsymbol{\phi}} = \dot{\mathbf{r}} - \hat{\mathbf{y}}\,\boldsymbol{\omega} \tag{12.7}$$

in which $\dot{\mathbf{r}}$ is translational velocity and $\boldsymbol{\omega}$ angular velocity, both at the centre of mass. The angular velocity is obtained by solving

$$\dot{\boldsymbol{\Lambda}} = \hat{\boldsymbol{\omega}}\boldsymbol{\Lambda} \tag{12.8}$$

or

$$\dot{\boldsymbol{\Lambda}} = \boldsymbol{\Lambda}\hat{\boldsymbol{\Omega}} \tag{12.9}$$

* Often literature denotes this rotation as R_{iI}; however, here we use R_I as a position of a point in the body and to avoid confusion use Λ_{iI} to denote rotation.

where $\boldsymbol{\Omega}$ is the reference configuration angular velocity.[8] This is clearer by writing the equations in indicial form given by

$$\dot{\Lambda}_{iI} = \omega_{ij}\Lambda_{jI} = \Lambda_{iJ}\Omega_{JI} \tag{12.10}$$

where the velocity matrices are defined in terms of vector components and give the skew symmetric form

$$\omega_{ij} = \begin{bmatrix} 0 & -\omega_3 & \omega_2 \\ \omega_3 & 0 & -\omega_1 \\ -\omega_2 & \omega_1 & 0 \end{bmatrix} \tag{12.11}$$

and similarly for Ω_{IJ}. The above form allows for the use of either the material angular velocity or the spatial one. Transformation between the two is easily performed since the rigid rotation must satisfy the orthogonality conditions

$$\boldsymbol{\Lambda}^{\mathrm{T}}\boldsymbol{\Lambda} = \boldsymbol{\Lambda}\boldsymbol{\Lambda}^{\mathrm{T}} = \mathbf{I} \tag{12.12}$$

at all times. Using Eqs (12.8) and (12.9) we obtain

$$\hat{\boldsymbol{\omega}} = \boldsymbol{\Lambda}\hat{\boldsymbol{\Omega}}\boldsymbol{\Lambda}^{\mathrm{T}} \tag{12.13}$$

or by transforming in the opposite way

$$\hat{\boldsymbol{\Omega}} = \boldsymbol{\Lambda}^{\mathrm{T}}\hat{\boldsymbol{\omega}}\boldsymbol{\Lambda} \tag{12.14}$$

12.3.1 Equations of motion for a rigid body

If we consider a single rigid body subjected to concentrated loads $\mathbf{f}_a$ applied at points whose current position is $\mathbf{x}_a$ and locate the reference position for $\mathbf{R}$ at the centre of mass, the equations of equilibrium are given by conservation of linear momentum

$$\dot{\mathbf{p}} = \sum_a \mathbf{f}_a = \mathbf{f}; \qquad \mathbf{p} = m\dot{\mathbf{r}} \tag{12.15}$$

where $\mathbf{p}$ defines a *linear momentum*, $\mathbf{f}$ is a resultant force and total mass of the body is computed from

$$m = \int_\Omega \rho_0 \, \mathrm{d}V \tag{12.16}$$

and conservation of angular momentum

$$\dot{\boldsymbol{\pi}} = \sum_a (\mathbf{x}_a - \mathbf{r}) \times \mathbf{f}_a = \mathbf{m}; \qquad \boldsymbol{\pi} = \mathbb{I}\boldsymbol{\omega} \tag{12.17}$$

where $\boldsymbol{\pi}$ is the *angular momentum* of the rigid body, $\mathbf{m}$ is a resultant couple and $\mathbb{I}$ is the spatial inertia tensor.

The spatial inertia tensor (matrix) $\mathbb{I}$ is computed from

$$\mathbb{I} = \boldsymbol{\Lambda}\mathbb{J}\boldsymbol{\Lambda}^{\mathrm{T}} \tag{12.18}$$

where $\mathbb{J}$ is the inertia tensor (matrix) computed from an integral on the reference configuration and is given by

$$\mathbb{J} = \int_{\Omega} \rho_0 \left[(\mathbf{Y}^{\mathrm{T}}\mathbf{Y})\mathbf{I} - \mathbf{Y}\mathbf{Y}^{\mathrm{T}} \right] \mathrm{d}V \quad \text{where} \quad \mathbf{Y} = \mathbf{X} - \mathbf{R} \tag{12.19}$$

Thus, description of an individual rigid body requires locating the centre of mass $\mathbf{R}$ and computing the total mass m and inertia matrix $\mathbb{J}$. It is then necessary to integrate the equilibrium equations to define the position $\mathbf{r}$ and the orientation of the body $\boldsymbol{\Lambda}$.

12.3.2 Construction from a finite element model

If we model a body by finite elements, as described throughout the volumes of this book, we can define individual bodies or parts of bodies as being rigid. For each such body (or part of a body) it is then necessary to define the total mass, inertia matrix, and location of the centre of mass.

This may be accomplished by computing the integrals given by Eqs (12.16) and (12.19) together with the relation to determine the centre of mass given by

$$m\mathbf{R} = \int_{\Omega} \rho_0 \mathbf{X} \, \mathrm{d}V \tag{12.20}$$

In these expressions it is necessary only to define each point in the volume of an element by its reference position interpolation $\mathbf{X}$. For solid (e.g., brick or tetrahedral) elements such interpolation is given by Eq. (10.55) which in matrix form becomes (omitting the summation symbol)

$$\mathbf{X} = N_{\alpha} \tilde{\mathbf{X}}_{\alpha} \tag{12.21}$$

This interpolation may be used to determine the volume element necessary to carry out all the integrals numerically (see Chapter 9 of Volume 1).

The total mass may now be computed as

$$m = \sum_{e} \left(\int_{\Omega_e} \rho_0 \, \mathrm{d}V \right) \tag{12.22}$$

where Ω_e is the reference volume of each element e. Use of Eq. (12.21) in Eq. (12.20) to determine the centre of mass now gives

$$\mathbf{R} = \frac{1}{m} \sum_{e} \left(\int_{\Omega_e} \rho_0 N_{\alpha} \, \mathrm{d}V \right) \mathbf{X}_{\alpha} \tag{12.23}$$

and finally the reference inertia tensor (matrix) as

$$\mathbb{J} = \sum_{e} M_{\alpha\beta} \left[(\mathbf{Y}_{\alpha}^{\mathrm{T}} \mathbf{Y}_{\beta}) \mathbf{I} - \mathbf{Y}_{\alpha} \mathbf{Y}_{\beta}^{\mathrm{T}} \right]; \qquad \mathbf{Y}_{\alpha} = \mathbf{X}_{\alpha} - \mathbf{R} \tag{12.24}$$

where

$$M_{\alpha\beta}^{e} = \int_{\Omega_e} \rho_0 N_{\alpha} N_{\beta} \, \mathrm{d}V \tag{12.25}$$

The above definition of $\mathbf{Y}_\alpha$ tacitly assumes that $\sum_\alpha N_\alpha = 1$. If other interpolations are used to define the shape functions (e.g. hierarchical shape functions) it is necessary to modify the above procedure to determine the mass and inertia matrix.

12.3.3 Transient solutions

The integration of the translational rigid term $\mathbf{r}$ may be performed using any of the methods described in Chapter 18 of Volume 1 or indeed by other methods described in the literature. The integration of the rotational part can also be performed by many schemes, however, it is important that updates of the rotation produce discrete time values for rigid rotations which retain an orthonormal character, that is, the $\mathbf{\Lambda}_n$ must satisfy the orthogonality condition given by Eq. (12.12). One procedure to obtain this is to assume that the angular velocity within a time increment is constant, being measured as

$$\boldsymbol{\omega}(t) \approx \boldsymbol{\omega}_{n+\alpha} = \frac{1}{\Delta t}\boldsymbol{\theta} \qquad (12.26)$$

in which Δt is the time increment between t_n and t_{n+1}, $\boldsymbol{\theta}$ is the increment of rotation during the time step, and $0 \leqslant \alpha \leqslant 1$. The approximation

$$\boldsymbol{\omega}_{n+\alpha} = (1-\alpha)\,\boldsymbol{\omega}_n + \alpha\,\boldsymbol{\omega}_{n+1} \qquad (12.27)$$

is used to define intermediate values in terms of those at t_n and t_{n+1}. Equation (12.8) now becomes a constant coefficient ordinary differential equation which may be integrated exactly, yielding the solution

$$\mathbf{\Lambda}(t) = \exp[\hat{\boldsymbol{\theta}}(t-t_n)/\Delta t]\,\mathbf{\Lambda}_n \qquad t_n \leqslant t \leqslant t_{n+1} \qquad (12.28)$$

In particular at $t_{n+\alpha}$ we obtain

$$\mathbf{\Lambda}_{n+\alpha} = \exp[\alpha\hat{\boldsymbol{\theta}}]\,\mathbf{\Lambda}_n$$

This may also be performed using the material angular velocity $\mathbf{\Omega}$.[8] Many algorithms exist to construct the exponential of a matrix, and the *closed-form expression* given by the classical formula of Euler and Rodrigues (e.g. see Wittaker[28]) is quite popular. This is given by

$$\exp[\hat{\boldsymbol{\theta}}] = \mathbf{I} + \frac{\sin|\boldsymbol{\theta}|}{|\boldsymbol{\theta}|}\hat{\boldsymbol{\theta}} + \frac{1}{2}\frac{\sin^2|\boldsymbol{\theta}|/2}{[|\boldsymbol{\theta}|^2/2]}\hat{\boldsymbol{\theta}}^2 \quad \text{where} \quad |\boldsymbol{\theta}| = \left[\boldsymbol{\theta}^{\mathrm{T}}\boldsymbol{\theta}\right]^{1/2} \qquad (12.29)$$

This update may also be given in terms of quaternions and has been used for integration of both rigid body motions as well as for the integration of the rotations appearing in three-dimensional beam formulations (see Chapter 11).[8,29,30] Another alternative to the direct use of the exponential update is to use the Cayley transform to perform updates for $\mathbf{\Lambda}$ which remain orthonormal.

Once the form for the update of the rigid rotation is defined any of the integration procedures defined in Chapter 18 of Volume 1 may be used to advance the incremental rotation by noting that $\boldsymbol{\theta}$ or $\mathbf{\Theta}$ (the material counterpart) are in fact the change from time t_n to t_{n+1}. The reader also is referred to reference 8 for additional algorithms directly based on the GN11 and GN22 methods presented in Chapter 18 of Volume 1. Here forms for conservation of linear and angular momentum are of particular importance.

12.4 Connecting a rigid body to a flexible body

In some analyses the rigid body is directly attached to flexible body parts of the problem [Fig. 12.2(a)]. Consider a rigid body that occupies the part of the domain denoted as Ω_r and is 'bonded' to a flexible body with domain Ω_f. In such a case the formulation to 'bond' the surface may be performed in a concise manner using Lagrange multiplier constraints. We shall find that these multiplier constraints can be easily eliminated from the analysis by a local solution process, as opposed to the need to carry them to the global solution arrays as was the case in their use in contact problems (see Sec. 10.8).

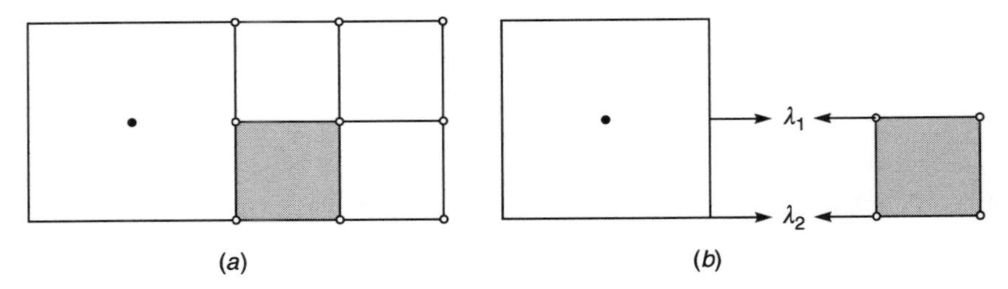

Fig. 12.2 Lagrange multiplier constraint between flexible and rigid bodies: (a) rigid–flexible body; (b) Lagrange multipliers.

12.4.1 Lagrange multiplier constraints

A simple two-dimensional rigid–flexible body problem is shown in Fig. 12.2(a) in which the interface will involve only three-nodal points. In Fig. 12.2(b) we show an exploded view between the rigid body and one of the elements which lies along the rigid–flexible interface. Here we need to enforce that the position of the two interface nodes for the element will have the same deformed position as the corresponding point on the rigid body. Such a constraint can easily be written using Eq. (12.4) as

$$\mathbf{C}_\alpha = \mathbf{r}(t) + \boldsymbol{\Lambda}(t)[\mathbf{X}_\alpha - \mathbf{R}] - \mathbf{x}_\alpha(t) = \mathbf{0} \tag{12.30}$$

in which the subscript α denotes a node number. We can now modify a functional to include the constraint using a classical Lagrange multiplier approach in which we add the term

$$\Pi_{rf} = \boldsymbol{\lambda}_\alpha \mathbf{C}_\alpha = \boldsymbol{\lambda}_\alpha [\mathbf{x}_\alpha(t) - \mathbf{r}(t) - \boldsymbol{\Lambda}(t)[\mathbf{X}_\alpha - \mathbf{R}]] \tag{12.31}$$

Taking the variation we obtain

$$\delta\Pi_{rf} = \delta\lambda_\alpha [\mathbf{x}_\alpha - \mathbf{r} - \boldsymbol{\Lambda}[\mathbf{X}_\alpha - \mathbf{R}]] + \lambda_\alpha [\delta\mathbf{x}_\alpha - \delta\mathbf{r} - \delta\boldsymbol{\theta}\,\boldsymbol{\Lambda}[\mathbf{X}_\alpha - \mathbf{R}]] \tag{12.32}$$

From this we immediately obtain the constraint equation and a modification to the equilibrium equations for each flexible node and the rigid body. Accordingly, the modified variational principle may now be written for a typical node α on the interface of the rigid body as

$$\delta\Pi + \delta\Pi_{rf} = \begin{bmatrix} \delta\mathbf{x}_\mu & \delta\mathbf{x}_\alpha & \delta\mathbf{r} & \delta\boldsymbol{\theta} & \delta\boldsymbol{\lambda}_\mu \end{bmatrix} \begin{Bmatrix} \mathbf{M}_{\mu\nu}\dot{\mathbf{v}}_\nu + \mathbf{M}_{\mu\beta}\dot{\mathbf{v}}_\beta + \mathbf{P}_\mu - \mathbf{f}_\mu \\ \mathbf{M}_{\alpha\nu}\dot{\mathbf{v}}_\nu + \mathbf{M}_{\alpha\beta}\dot{\mathbf{v}}_\beta + \mathbf{P}_\alpha - \mathbf{f}_\alpha + \boldsymbol{\lambda}_\alpha \\ \dot{\mathbf{p}} - \mathbf{f} - \boldsymbol{\lambda}_\alpha \\ \dot{\boldsymbol{\pi}} - \mathbf{m} - \hat{\mathbf{y}}_\alpha^{\mathrm{T}}\boldsymbol{\lambda}_\alpha \\ \mathbf{x}_\alpha - \mathbf{r} - \boldsymbol{\Lambda}[\mathbf{X}_\alpha - \mathbf{R}] \end{Bmatrix} = 0 \tag{12.33}$$

where $\mathbf{y}_\alpha = \mathbf{x}_\alpha - \mathbf{r}$ are the nodal values of $\mathbf{y}$, β are any other rigid body nodes connected to node α and μ, ν are flexible nodes connected to node α.

Since the parameters $\mathbf{x}_\alpha$ enter the equations in a linear manner we can use the constraint equation to eliminate their appearance in the equations. Accordingly, from the variation of the constraint equation we may write

$$\delta\mathbf{x}_\alpha = [\mathbf{I} - \hat{y}_\alpha] \tag{12.34}$$

which permits the remaining equations in Eq. (12.33) to be rewritten as

$$\delta\Pi + \delta\Pi_{rf} = \begin{bmatrix} \delta\mathbf{x}_\mu & \delta\mathbf{r} & \delta\boldsymbol{\theta} \end{bmatrix} \begin{Bmatrix} \mathbf{M}_{\mu\nu}\dot{\mathbf{v}}_\nu + \mathbf{M}_{\mu\beta}\dot{\mathbf{v}}_\beta + \mathbf{P}_\mu - \mathbf{f}_\mu \\ \dot{\mathbf{p}} - \mathbf{f}\mathbf{M}_{\alpha\nu}\dot{\mathbf{v}}_\nu + \mathbf{M}_{\alpha\beta}\dot{\mathbf{v}}_\beta + \mathbf{P}_\alpha - \mathbf{f}_\alpha \\ \dot{\boldsymbol{\pi}} - \mathbf{m} - \hat{\mathbf{y}}_\alpha^{\mathrm{T}}(\mathbf{M}_{\alpha\nu}\dot{\mathbf{v}}_\nu + \mathbf{M}_{\alpha\beta}\dot{\mathbf{v}}_\beta + \mathbf{P}_\alpha - \mathbf{f}_\alpha) \end{Bmatrix} = 0 \tag{12.35}$$

For use in a Newton–Raphson solution scheme it is necessary to linearize Eq. (12.35). This is easily achieved

$$d(\delta\Pi) + d(\delta\Pi_{rf}) = [\delta\mathbf{x}_\mu \quad \delta\mathbf{r} \quad \delta\boldsymbol{\theta}] \begin{bmatrix} (\mathbf{K}_{\mu\nu})_{\mathrm{T}} & (\mathbf{K}_{\mu\beta})_{\mathrm{T}} & \mathbf{0} & \mathbf{0} \\ (\mathbf{K}_{\alpha\nu})_{\mathrm{T}} & (\mathbf{K}_{\alpha\beta})_{\mathrm{T}} & \mathbf{K}_{\mathrm{T}}^{p} & \mathbf{0} \\ -\hat{\mathbf{y}}_\alpha^{\mathrm{T}}(\mathbf{K}_{\alpha\nu})_{\mathrm{T}} & -\hat{\mathbf{y}}_\alpha^{\mathrm{T}}(\mathbf{K}_{\alpha\beta})_{\mathrm{T}} & \mathbf{0} & \mathbf{K}_{\mathrm{T}}^{\theta} \end{bmatrix} \begin{Bmatrix} d\mathbf{x}_\nu \\ d\mathbf{x}_\beta \\ d\mathbf{r} \\ d\boldsymbol{\theta} \end{Bmatrix} \tag{12.36}$$

Once again this form may be reduced using the equivalent of Eq. (12.34) for an incremental $d\mathbf{x}_\beta$ to obtain

$$d(\delta\Pi) + d(\delta\Pi_{rf}) = [\delta\mathbf{x}_\mu \quad \delta\mathbf{r} \quad \delta\boldsymbol{\theta}] \begin{bmatrix} (\mathbf{K}_{\mu\nu})_{\mathrm{T}} & (\mathbf{K}_{\mu\beta})_{\mathrm{T}} & -(\mathbf{K}_{\mu\beta})_{\mathrm{T}}\hat{\mathbf{y}}_\beta \\ (\mathbf{K}_{\alpha\nu})_{\mathrm{T}} & \mathbf{K}_{\mathrm{T}}^{p} + (\mathbf{K}_{\alpha\beta})_{\mathrm{T}} & -(\mathbf{K}_{\alpha\beta})_{\mathrm{T}}\hat{\mathbf{y}}_\beta \\ -\hat{\mathbf{y}}_\alpha^{\mathrm{T}}(\mathbf{K}_{\alpha\nu})_{\mathrm{T}} & -\hat{\mathbf{y}}_\alpha^{\mathrm{T}}(\mathbf{K}_{\alpha\beta})_{\mathrm{T}} & [\mathbf{K}_{\mathrm{T}}^{\theta} + \hat{\mathbf{y}}_\alpha^{\mathrm{T}}(\mathbf{K}_{\alpha\beta})_{\mathrm{T}}\hat{\mathbf{y}}_\beta] \end{bmatrix} \times \begin{Bmatrix} d\mathbf{x}_\nu \\ d\mathbf{r} \\ d\boldsymbol{\theta} \end{Bmatrix} \tag{12.37}$$

Combining all the steps we obtain the set of equations for each rigid body as

$$\begin{bmatrix} (\mathbf{K}_{\mu\nu})_{\mathrm{T}} & (\mathbf{K}_{\mu\beta})_{\mathrm{T}} & -(\mathbf{K}_{\mu\beta})_{\mathrm{T}}\hat{\mathbf{y}}_\beta \\ (\mathbf{K}_{\alpha\nu})_{\mathrm{T}} & [\mathbf{K}^p_{\mathrm{T}} + (\mathbf{K}_{\alpha\beta})_{\mathrm{T}}] & -(\mathbf{K}_{\alpha\beta})_{\mathrm{T}}\hat{\mathbf{y}}_\beta \\ -\hat{\mathbf{y}}^{\mathrm{T}}_\alpha(\mathbf{K}_{\alpha\nu})_{\mathrm{T}} & -\hat{\mathbf{y}}^{\mathrm{T}}_\alpha(\mathbf{K}_{\alpha\beta})_{\mathrm{T}} & [\mathbf{K}^\theta_{\mathrm{T}} + \hat{\mathbf{y}}^{\mathrm{T}}_\alpha(\mathbf{K}_{\alpha\beta})_{\mathrm{T}}\hat{\mathbf{y}}_\beta] \end{bmatrix} \begin{Bmatrix} d\mathbf{x}_\nu \\ d\mathbf{r} \\ d\boldsymbol{\theta} \end{Bmatrix} = \begin{Bmatrix} \boldsymbol{\Psi}_\mu \\ \mathbf{f} - \dot{\mathbf{p}} + \boldsymbol{\Psi}_\alpha \\ \mathbf{m} - \dot{\boldsymbol{\pi}} + \hat{\mathbf{y}}^{\mathrm{T}}_\alpha\boldsymbol{\Psi}_\alpha \end{Bmatrix} \tag{12.38}$$

in which $\boldsymbol{\Psi}_\alpha$ and $\boldsymbol{\Psi}_\mu$ are the residuals from the finite element calculation at node α and μ, respectively. We recall from Chapter 10 that each is given by a form

$$\boldsymbol{\psi}_\alpha = \mathbf{f}_\alpha - \mathbf{P}_\alpha(\boldsymbol{\sigma}) - \mathbf{M}_{\alpha\nu}\dot{\mathbf{v}}_\nu - \mathbf{M}_{\alpha\beta}\dot{\mathbf{v}}_\beta \tag{12.39}$$

which is now not zero since total balance of momentum includes the addition of the $\boldsymbol{\lambda}_\alpha$.

The above steps to compute the residual and the tangent can be performed in each element separately by noting that

$$\boldsymbol{\lambda}_\alpha = \sum_e \boldsymbol{\lambda}^e_\alpha \tag{12.40}$$

where $\boldsymbol{\lambda}^e_\alpha$ denotes the contribution from element e. Thus, the steps to constrain a flexible body to a rigid body are once again a standard finite element assembly process and may easily be incorporated into a solution system.

The above discussion has considered the connection between a rigid body and a body which is modelled using solid finite elements (e.g. quadrilateral and hexahedral elements in two and three dimensions, respectively). It is also possible directly to connect beam elements which have nodal parameters of translation and rotation. This is easily performed if the rotation parameters of the beam are also defined in terms of the rigid rotation $\boldsymbol{\Lambda}$. In this case one merely transforms the rotation to be defined relative to the reference description of the rigid body rotation and assembles the result directly into the rotation terms of the rigid body. If one uses a rotation for both the beam and the rigid body which is defined in terms of the global Cartesian reference configuration no transformation is required. Shells can be similarly treated; however, it is best then to define the shell directly in terms of three rotation parameters instead of only two at points where connection is to be performed.[31,32]

12.5 Multibody coupling by joints

Often it is desirable to have two (or more) rigid bodies connected in some specified manner. For example, in Fig. 12.3 we show a disc connected to an arm. Both are treated as rigid bodies but it is desired to have the disc connected to the arm in such a way that it can rotate freely about the axis normal to the page. This type of motion is characteristic of many rotating machine connections and it as well as many other types of connections are encountered in the study of rigid body motions.[4,33] This type of interconnection is commonly referred to as a *joint*. In quite general terms joints may be constructed by a combination of two types of simple constraints: *translational constraints* and *rotational constraints*.

12.5.1 Translation constraints

The simplest type of joint is a spherical connection in which one body may freely rotate around the other but relative translation is prevented. Such a situation is shown in Fig. 12.3 where it is evident the spinning disc must stay attached to the rigid arm at its axle. Thus it may not translate relative to the arm in any direction (additional constraints are necessary to ensure it rotates only about the one axis – these are discussed in Sec. 12.5.2). If a full translation constraint is imposed a simple relation may be introduced as

$$\mathbf{C}_j = \mathbf{x}^{(a)} - \mathbf{x}^{(b)} = \mathbf{0} \tag{12.41}$$

where a and b denote two rigid bodies. Thus, addition of the Lagrange multiplier constraint

$$\Pi_j = \boldsymbol{\lambda}_j^{\mathrm{T}} [\mathbf{x}^{(a)} - \mathbf{x}^{(b)}] \tag{12.42}$$

imposes the spherical joint condition. It is necessary only to define the location for the spherical joint in the reference configuration. Denoting this as $\mathbf{X}_j$ (which is common to the two bodies) and introducing the rigid motion yields a constraint in terms of the rigid body positions as

$$\Pi_j = \boldsymbol{\lambda}_j^{\mathrm{T}} [\mathbf{r}^{(a)} + \boldsymbol{\Lambda}^{(a)}(\mathbf{X}_j - \mathbf{R}^{(a)}) - \mathbf{r}^{(b)} - \boldsymbol{\Lambda}^{(b)}(\mathbf{X}_j - \mathbf{R}^{(b)})] \tag{12.43}$$

The variation and subsequent linearization of this relation yields the contribution to the residual and tangent matrix for each body, respectively. This is easily performed using relations given above and is left as an exercise for the reader.

If the translation constraint is restricted to be in one direction with respect to, say, body a it is necessary to track this direction and write the constraint accordingly. To accomplish this the specific direction of the body a in the reference configuration is required. This may be computed by defining two points in space $\mathbf{X}_1$ and $\mathbf{X}_2$ from

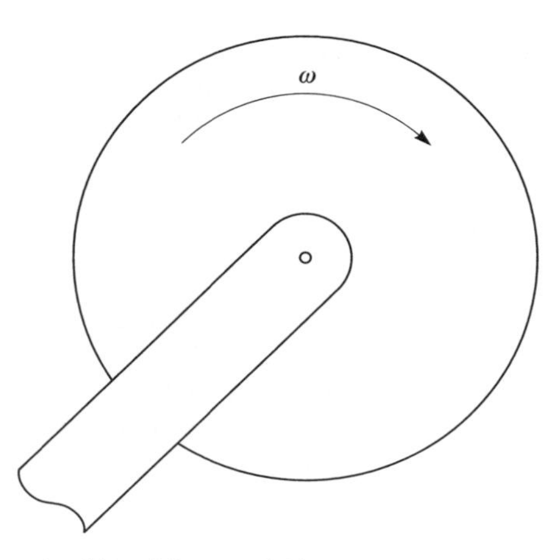

Fig. 12.3 Spinning disc constrained by a joint to a rigid arm.

which a unit vector $\mathbf{V}$ is defined by

$$\mathbf{V} = \frac{\mathbf{X}_2 - \mathbf{X}_1}{|\mathbf{X}_2 - \mathbf{X}_1|} \tag{12.44}$$

The direction of this vector in the current configuration, $\mathbf{v}$, may be obtained using the rigid rotation for body a

$$\mathbf{v} = \mathbf{\Lambda}^{(a)} \mathbf{V} \tag{12.45}$$

A constraint can now be introduced into the variational problem as

$$\Pi_j = \lambda_j \{\mathbf{V}^{\mathrm{T}} (\mathbf{\Lambda}^{(a)})^{\mathrm{T}} [\mathbf{r}^{(a)} + \mathbf{\Lambda}^{(a)}(\mathbf{X}_j - \mathbf{R}^{(a)}) - \mathbf{r}^{(b)} - \mathbf{\Lambda}^{(b)}(\mathbf{X}_j - \mathbf{R}^{(b)})]\} \tag{12.46}$$

where, owing to the fact there is only a single constraint direction, the Lagrange multiplier is a scalar λ_j and, again, $\mathbf{X}_j$ denotes the reference position where the constraint is imposed.

The above constraints may also be imposed by using a penalty function. The most direct form is to *perturb* each Lagrange multiplier form by a penalty term. Accordingly, for each constraint we write the variational problem as

$$\Pi_j = \lambda_j C_j - \frac{1}{2k_j} \lambda_j^2 \tag{12.47}$$

where it is immediately obvious that the limit $k_j \to \infty$ yields exact satisfaction of the constraint. Use of a large k_j and variation with respect to λ_j gives

$$\delta\lambda_j \left[C_j - \frac{1}{k_j} \lambda_j \right] = 0 \tag{12.48}$$

and may easily be solved for the Lagrange multiplier as

$$\lambda_j = k_j C_j \tag{12.49}$$

which when substituted back into Eq. (12.47) gives the classical form

$$\Pi_j = \frac{k_j}{2} [C_j]^2 \tag{12.50}$$

The reader will recognize that Eq. (12.47) is a mixed problem, whereas, Eq. (12.50) is irreducible. An augmented lagrangian form is also possible following the procedures introduced in Volume 1 and used in Chapter 10 for contact problems.

12.5.2 Rotation constraints

A second kind of constraint that needs to be considered relates to rotations. We have already observed in Fig. 12.3 that the disc is free to rotate around only one axis. Accordingly, constraints must be imposed which limit this type of motion. This may be accomplished by constructing an orthogonal set of unit vectors $\mathbf{V}_I$ in the reference configuration and tracking the orientation of the deformed set of axes for each body as

$$\mathbf{v}_i^{(c)} = \delta_{iI} \mathbf{\Lambda}^{(c)} \mathbf{V}_I \quad \text{for} \quad c = a, b \quad \mathbf{v}_I^{\mathrm{T}} \mathbf{V}_J = \delta_{IJ} \tag{12.51}$$

A rotational constraint which imposes that axis i of body a remain perpendicular to axis j of body b may then be written as

$$(\mathbf{v}_i^{(a)})^{\mathrm{T}}\mathbf{v}_j^{(b)} = \mathbf{V}_I^{\mathrm{T}}(\boldsymbol{\Lambda}^{(a)})^{\mathrm{T}}\boldsymbol{\Lambda}^{(b)}\mathbf{V}_J = 0 \qquad (12.52)$$

Example: revolute joint

As an example, consider the situation shown for the disc in Fig. 12.3 and define the axis of rotation in the reference configuration by the Cartesian unit vectors $\mathbf{E}_I$ (i.e. $\mathbf{V}_I = \mathbf{E}_I$). If we let the disc be body a and the arm body b the set of constraints can be written as (where $\mathbf{v}_3$ is axis of rotation)

$$\mathbf{C}_j = \left\{ \begin{array}{c} \mathbf{x}^{(a)} - \mathbf{x}^{(c)} \\ (\mathbf{v}_1^{(a)})^{\mathrm{T}}\mathbf{v}_3^{(b)} \\ (\mathbf{v}_2^{(a)})^{\mathrm{T}}\mathbf{v}_3^{(b)} \end{array} \right\} = \mathbf{0} \qquad (12.53)$$

and included in a formulation using a Lagrange multiplier form

$$\Pi_j = \boldsymbol{\lambda}_j^{\mathrm{T}}\mathbf{C}_j \qquad (12.54)$$

The modifications to the finite element equations are obtained by appending the variation and linearization of Eq. (12.54) to the usual equilibrium equations. Here five Lagrange multipliers are involved to impose the three translational constraints (spherical joint) and the angle constraints for the rotating disc. The set of constraints is known as a *revolute* joint.[2]

12.5.3 Library of joints

Translational and rotational constraints may be combined in many forms to develop different types of constraints between rigid bodies. For the development it is necessary to have only the three types of constraints described above. Namely, the spherical joint, a single translational constraint, and a single rotational constraint. Once these are available it is possible to combine them to form classical constraint joints and here the reader is referred to the literature for the many kinds commonly encountered.[2,4,7,34]

The only situation that requires special mention is the case when a series of rigid bodies is connected together to form a *closed loop*. In this case the method given above can lead to situations in which some of the joints are redundant. Using Lagrange multipliers this implies the resulting tangent matrix will be singular and, thus, one cannot obtain solutions. Here a penalty method provides a viable method to circumvent this problem. The penalty method introduces *elastic deformation* in the joints and in this way removes the singular problem. If necessary an augmented lagrangian method can be used to keep the deformation in the joint within required small tolerances. An alternative to this is to extract the closed loop rigid equations from the problem and use singular valued decomposition[35] to identify the redundant equations. These may then be removed by constructing a pseudo-inverse for the tangent matrix of the closed loop. This method has been used successfully by Chen to solve single loop problems.[34]

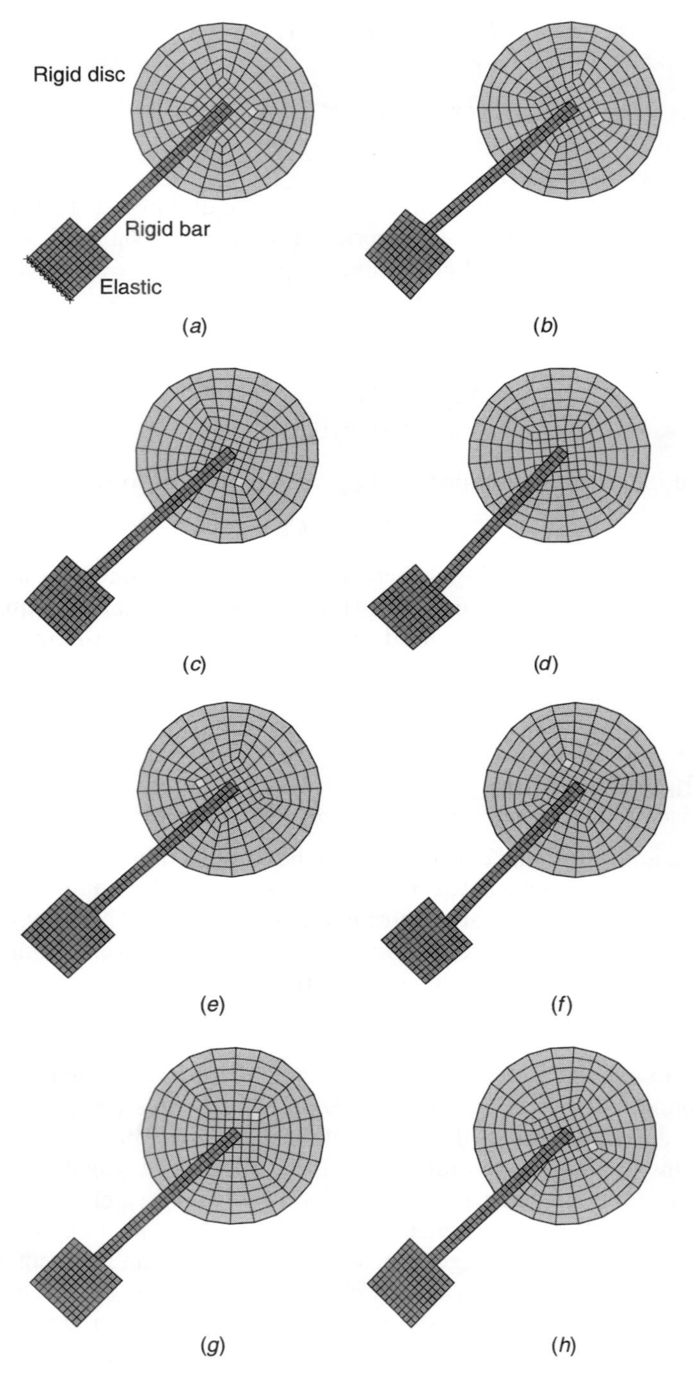

Fig. 12.4 Rigid–flexible model for spinning disc: (a) problem definition, solutions at time; (a) problem definition, solutions at time; (b) $t = 2.5$ units; (c) $t = 5.0$ units; (d) $t = 7.5$ units; (e) $t = 10.0$ units; (f) $t = 12.5$ units; (g) $t = 15.0$ units; (h) $t = 17.5$ units.

12.6 Numerical examples

12.6.1 Rotating disc

As a first example we consider a problem for the rotating disc on a rigid arm which is attached to a deformable base as shown in Fig. 12.4. The finite element model is constructed from four-noded displacement elements in which a Saint-Venant–Kirchhoff material model is used for the elastic part. The elastic properties in the model are $E = 10\,000$ and $\nu = 0.25$, with a uniform mass density $\rho_0 = 5$ throughout. The disc and arm are made rigid by using the procedures described in this chapter. The disc is attached to the arm by means of a revolute joint with the constraints imposed using the Lagrange multiplier method. The rigid arm is constrained to the elastic support by using the local Lagrange multipler method described in Sec. 12.4. The problem is excited by a constant vertical load applied at the revolute joint and a torque applied to spin the disc. Each load is applied for the first 10 units of time.

The mesh and configuration are shown in Fig. 12.4(a). Deformed positions of the model are shown at 2.5 unit intervals of time in Figs 12.4(b)–12.4(h). A marker element shows the position of the rotating disc. The displacements at the revolute joint and the radial exterior point at the marker element location are shown in Fig. 12.5.

12.6.2 Beam with attached mass

As a second example we consider an elastic cantilever beam with an attached end mass of rectangular shape. The beam is excited by a horizontal load applied at the top as a triangular pulse for two units of time. The rigid mass is attached to the top of the beam by using the Lagrange multiplier method described in Sec. 12.4 and here it is necessary to constrain both the translation and the rotation parameters of the beam. The beam is three-dimensional and has an elastic modulus of $E = 100\,000$ and a moment of inertia in both directions of $I_{11} = I_{22} = 12$. The beam mass density is low, with a value of $\rho_0 = 0.02$. The tip mass is a cube with

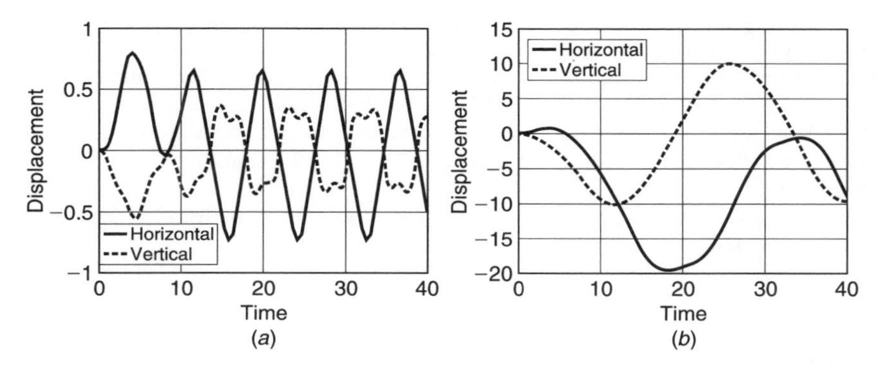

Fig. 12.5 Displacements for rigid–flexible model for spinning disc. Displacement at: (a) revolute; (b) disc rim.

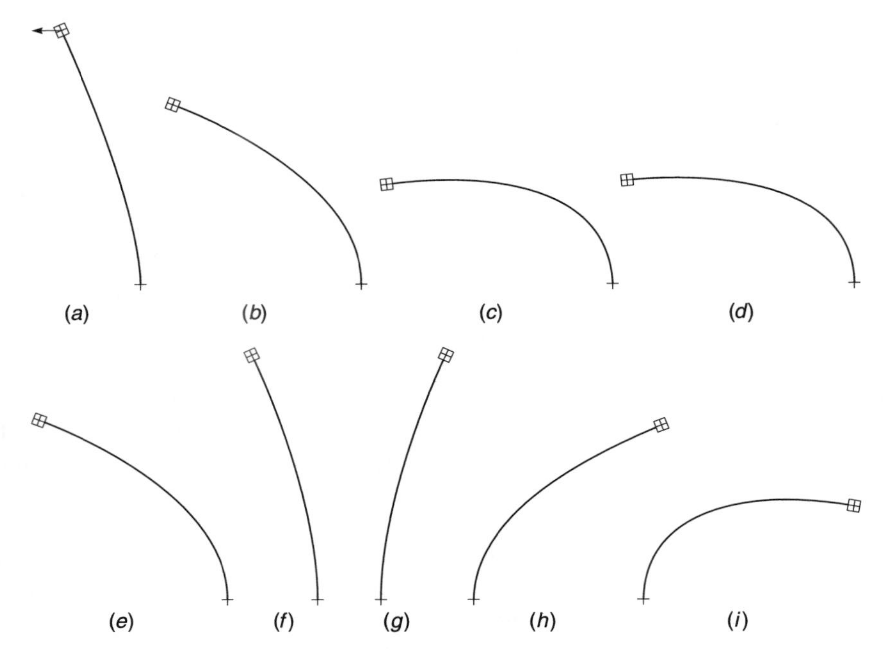

Fig. 12.6 Cantilever with tip mass: (a) $t = 2$ units; (b) $t = 4$ units; (c) $t = 6$ units; (d) $t = 10$ units; (e) $t = 12$ units; (f) $t = 14$ units; (g) $t = 16$ units; (h) $t = 18$ units; (i) $t = 20$ units.

side lengths 4 and mass density $\rho_0 = 1$. The shape of the beam at several instants of time is shown in Fig. 12.6 and it is clear that large translation and rotation is occurring and also that the rigid block is correctly following a constrained rigid body motion.

References

1. H. Cohen and R.G. Muncaster. *The Theory of Pseudo-rigid Bodies*, Springer, New York, 1988.
2. A.A. Shabana. *Dynamics of Multibody Systems*, John Wiley, New York, 1989.
3. J.M. Solberg and P. Papadopoulos. A simple finite element-based framework for the analysis of elastic pseudo-rigid bodies. *Int. J. Num. Meth. Eng.*, **45**, 1297–314, 1999.
4. D.J. Benson and J.O. Hallquist. A simple rigid body algorithm for structural dynamics programs. *Int. J. Num. Meth. Eng.*, **22**, 723–49, 1986.
5. R.A. Wehage and E.J. Haug. Generalized coordinate partitioning for dimension reduction in analysis of constrained dynamic systems. *Journal of Mechanical Design*, **104**, 247–55, 1982.
6. A. Cardona and M. Geradin. Beam finite element nonlinear theory with finite rotations. *Int. J. Num. Meth. Eng.*, **26**, 2403–38, 1988.
7. A. Cardona, M. Geradin and D.B. Doan. Rigid and flexible joint modelling in multibody dynamics using finite elements. *Comp. Meth. Appl. Mech. Eng.*, **89**, 395–418, 91.
8. J.C. Simo and K. Wong. Unconditionally stable algorithms for rigid body dynamics that exactly conserve energy and momentum. *Int. J. Num. Meth. Eng.*, **31**, 19–52, 1991; addendum, **33**, 1321–3, 1992.

9. H.T. Clark and D.S. Kang. Application of penalty constraints for multibody dynamics of large space structures. *Advances in the Astronautical Sciences*, **79**, 511–30, 1992.
10. G.M. Hulbert. Explicit momentum conserving algorithms for rigid body dynamics. *Comp. Struct.*, **44**, 1291–303, 1992.
11. M. Geradin, D.B. Doan and I. Klapka. MECANO: a finite element software for flexible multibody analysis. *Vehicle System Dynamics*, **22** (supplement issue), 87–90, 1993.
12. S.N. Atluri and A. Cazzani. Rotations in computational solid mechanics. *Archives of Computational Methods in Engineering*, **2**, 49–138, 1995.
13. O.A. Bauchau, G. Damilano and N.J. Theron. Numerical integration of nonlinear elastic multibody systems. *Int. J. Num. Meth. Eng.*, **38**, 2727–51, 1995.
14. J.A.C. Ambrosio. Dynamics of structures undergoing gross motion and nonlinear deformations: a multibody approach. *Comp. Struct.*, **59**, 1001–12, 1996.
15. R.L. Huston. Multibody dynamics since 1990. *Applied Mechanics Reviews*, **49**, S35–S40, 1996.
16. O.A. Bauchau and N.J. Theron. Energy decaying scheme for non-linear beam models. *Comp. Meth. Appl. Mech. Eng.*, **134**, 37–56, 1996.
17. O.A. Bauchau and N.J. Theron. Energy decaying scheme for non-linear elastic multi-body systems. *Comp. Struct.*, **59**, 317–31, 1996.
18. C. Bottasso and M. Borri. Energy preserving/decaying schemes for nonlinear beam dynamics causing the helicoidal approximation. *Comp. Meth. Appl. Mech. Eng.*, **143**, 393–415, 1997.
19. O.A. Bauchau. Computational schemes for flexible, nonlinear multi-body systems. *Multibody System Dynamics*, **2**, 169–222, 1998.
20. O.A. Bauchau and C.L. Bottasso. On the design of energy preserving and decaying schemes for flexible nonlinear multi-body systems. *Comp. Meth. Appl. Mech. Eng.*, **169**, 61–79, 1999.
21. O.A. Bauchau and T. Joo. Computational schemes for non-linear elasto-dynamics. *Int. J. Num. Meth. Eng.*, **45**, 693–719, 1999.
22. G.-H. Shi. *Block System Modelling by Discontinuous Deformation Analysis*, Computational Mechanics Publications, Southampton, 1993.
23. E.G. Petocz. *Formulation and Analysis of Stable Time-stepping Algorithms for Contact Problems*, PhD thesis, Department of Mechanical Engineering, Stanford University, Stanford, CA, 1998.
24. M.E. Gurtin. *An Introduction to Continuum Mechanics*, Academic Press, New York, 1981.
25. L.E. Malvern. *Introduction to the Mechanics of a Continuous Medium*, Prentice-Hall, Englewood Cliffs, NJ, 1969.
26. J. Bonet and R.D. Wood. *Nonlinear Continuum Mechanics for Finite Element Analysis*, Cambridge University Press, Cambridge, 1997.
27. J.C. García Orden and J.M. Goicolea. Dynamic analysis of rigid and deformable multibody systems with penalty methods and energy–momentum schemes. *Comp. Meth. Appl. Mech. Eng.*, 2000.
28. E.T. Wittaker. *A Treatise on Analytical Dynamics*, Dover Publications, New York, 1944.
29. J.H. Argyris and D.W. Scharpf. Finite elements in time and space. *Nuclear Engineering and Design*, **10**, 456–69, 1969.
30. A. Ibrahimbegovic and M. Al Mikdad. Finite rotations in dynamics of beams and implicit time-stepping schemes. *Int. J. Num. Meth. Eng.*, **41**, 781–814, 1998.
31. J.C. Simo. On a stress resultant geometrically exact shell model. Part VII: shell intersections with 5/6 DOF finite element formulations. *Comp. Meth. Appl. Mech. Eng.*, **108**, 319–39, 1993.
32. P. Betsch, F. Gruttmann and E. Stein. A 4-node finite shell element for the implementation of general hyperelastic 3d-elasticity at finite strains. *Comp. Meth. Appl. Mech. Eng.*, **130**, 57–79, 1996.

33. H. Goldstein. *Classical Mechanics*, 2nd edition, Addison-Wesley, Reading, MA, 1980.
34. A.J. Chen. *Energy–momentum Conserving Methods for Three Dimensional Dynamic Nonlinear Multibody Systems*, PhD thesis, Stanford University, Stanford, CA, 1998; also SUDMC Report 98-01.
35. G.H. Golub and C.F. Van Loan. *Matrix Computations*, 3rd edition, The Johns Hopkins University Press, Baltimore, MD, 1996.

13

Computer procedures for finite element analysis

13.1 Introduction

In this chapter we describe extensions to the program presented in Volume 1 to permit solution of transient non-linear problems that are modelled by a finite element process. The material included in this chapter should be considered as supplementary to information contained in Volume 1, Chapter 20.[1] Accordingly, throughout this chapter reference will be made to appropriate information in the first volume. It is suggested, however, that the reader review the material there prior to embarking on a study of this chapter.

The program described in this volume is intended for use by those who are undertaking a study of the finite element method and wish to implement and test specific elements or specific solution steps. The program also includes a library of simple elements to permit solution to many of the topics discussed in this and the first volume. The program is called FEAPpv to emphasize the fact that it may be used as a *personal version* system. With very few exceptions, the program is written using standard Fortran, hence it may be implemented on any personal computer, engineering workstation, or main frame computer which has access to a Fortran 77 or Fortran 90/95 compiler.

It still may be necessary to modify some routines to avoid system-dependent difficulties. Non-standard routines are restricted to the graphical interfaces and file handling for temporary data storage. Users should consult their compiler manuals on alternative options when such problems arise.

Users may also wish to add new features to the program. In order to accommodate a wide range of changes several options exist for users to write new modules without difficulty. There are options to add new mesh input routines through addition of routines named `UMESHn` and to include solution options through additions of routines named `UMACRn`. Finally, the addition of a user developed element module is accommodated by adding a single subprogram named `ELMTnn`. In adding new options the use of established algorithms as described in references 2–6 can be very helpful.

The current chapter is divided into several sections that describe different aspects of the program. Section 13.2 summarizes the additional program features and the command language additions that may be used to solve general linear and non-linear finite element problems. Some general solution strategies and the related command language statements for using the system to solve non-linear transient problems are presented in Sec. 13.3. The program FEAPpv includes capabilities to solve both first-order (diffusion type) and second-order (vibration/wave type) ordinary differential equations in time. In Sec. 13.3, the description of the eigensystem included in FEAPpv with the required solution statements for its use are presented. Here a simultaneous vector iteration algorithm (subspace method) is used to extract the eigenpairs nearest to a specified shift of a *symmetric tangent* matrix. Hence, the eigensystem may be used with either linear or non-linear problems. Non-linear problems are often difficult to solve and time-consuming in computer resources. In many applications the complete analysis may not be performed during one execution of the program; hence, techniques to stop the program at key points in the analysis for a later restart to continue the solution are presented in Sec. 13.4. This section completes the description of new and extended solution options that have been added to the program.

Section 13.5 describes the solution steps for some typical problems that can be solved by using FEAPpv. Finally, Sec. 13.6 includes information on how to obtain the source code as well as a user manual and support information for the program FEAPpv.

The program contained in this chapter has been developed and used in an educational and research environment over a period of nearly 25 years. The concept of the command language solution algorithm has permitted several studies that cover problems that differ widely in scope and concept, to be undertaken at the same time without need for different program systems. Unique features for each study may be provided as new solution commands. The ability to treat problems whose coefficient matrix may be either symmetric or unsymmetric often proves useful for testing the performance of algorithms that advocate substitution of a symmetrized tangent matrix in place of an unsymmetric matrix resulting from a consistent linearization process. The element interface is quite straightforward and, once understood, permits users to test rapidly new types of finite elements.

We believe that the program in this book provides a very powerful solution system to assist the interested reader in performing finite element analyses. The program FEAPpv is by no means a complete software system that can be used to solve any finite element problem, and readers are encouraged to modify the program in any way necessary to solve their particular problem. While the program has been tested on several sample problems, it is likely that errors and mistakes still exist within the program modules. The authors need to be informed about these errors so that the available system can be continuously updated. We also welcome readers' comments and suggestions concerning possible future improvements.

13.2 Description of additional program features

Description of the command language given in Chapter 20 of Volume 1 is here extended to permit solution of a broad class of non-linear applications. The principal additions relate to the solution of non-linear static and transient problems and adds the description to consider applications which have unsymmetric tangent 'stiffness' matrices. In addition, the program introduces the BFGS (Broyden–Fletcher–Goldfarb–Shanno) algorithm and a line-search algorithm which may be invoked to permit convergence

Table 13.1 List of new command language statements

Columns					Description
1–4	16–19	31–45	46–60	61–75	
BACK					Back-up to restart a time step
TRAN	SS11	V1			Set solution algorithm to SS11; V1 is value of θ
TRAN	SS22	V1	V2		Set solution algorithm to SS22; V1, V2 are values of θ_1, θ_2
TRAN	GN22	V1	V2		Set solution algorithm to GN22; V1, V2 are values of β, γ
TRAN		V1	V2		Same as TRAN GN22
BFGS		N1	V2		Performs N1 BFGS steps with line-search tolerance set to V2
EIGV		N1			Output N1 eigenpairs (after SUBS)
IDEN					Set the mass matrix to the identity
PLOT	EIGV	V1			Plot eigenvector V1
SOLV	LINE	V1			Solve for new displacements (after FORM). If LINE present, compute solution with line search; V1 controls initiation (see Sec. 13.4)
SUBS	PRIN	N1	N2		Perform eigenpair extraction for N1 values with N2 extra vectors. If PRIN print subspace arrays (after TANG and MASS or IDEN)
TANG	LINE	N1	V2	V3	Compute and factor symmetric tangent matrix (ISW = 3); see note
UTAN	LINE	N1	V2	V3	Compute and factor unsymmetric tangent matrix (ISW = 3); see note

Note: If N1 is non-zero a residual, solution, and update to the displacements are performed. If V2 is non-zero, the tangent matrix is modified by subtracting V2 multiplied by the mass matrix before computing the triangular factors. The mass matrix may be set to an identity matrix by using the command IDEN. If LINE is specified as part of the TANG or UTAN command a linear line search is performed whenever the energy ratio between two successive iterations exceeds the value of V3 (the default value is 0.8).

of Newton-type algorithms with rather large solution increments.[7] The program includes algorithms to solve transient problems by the methods discussed in Chapter 18 of Volume 1 which are applicable in a non-linear solution strategy. Finally, the program has an eigensolution algorithm based upon subspace iteration.[8–10] A description of the user information required to invoke the added commands for these features is contained in Table 13.1.

13.3 Solution of non-linear problems

The solution of non-linear problems using the program contained in this volume is designed for a Newton-type or modified Newton-type algorithm as described in Chapter 2 and reference 11. In addition, the solution for transient non-linear problems may be achieved by combining a Newton-type algorithm with the transient integration method described below.

13.3.1 Static and steady-state problems

We first consider a non-linear problem described by (see Chapter 2)

$$\mathbf{\Psi}(\mathbf{a}) = \mathbf{P}(\mathbf{a}) - \mathbf{f} \tag{13.1}$$

where $\mathbf{f}$ is a vector of applied loads and $\mathbf{P}$ is the non-linear *internal* force vector which is indicated as a function of the nodal parameters $\mathbf{a}$. The vector $\mathbf{\Psi}$ is known as the residual of the problem, and a solution is defined as any set of nodal displacements, $\mathbf{a}$, for which the residual is zero. In general, there may be more than one set of displacements which define a solution and it is the responsibility of the user to ensure that a proper solution is obtained. This may be achieved by starting from a state which satisfies physical arguments for a solution and then applying small increments to the loading vector, $\mathbf{f}$. By taking small enough steps, a solution path may usually be traced. Thus, for any step, our objective is to find a set of values for the components of $\mathbf{a}$ such that

$$\boldsymbol{\Psi}(\mathbf{a}) = \mathbf{0} \tag{13.2}$$

We assume some initial vector exists (initially in the program this vector is zero), from which we will seek a solution, and denote this as $\mathbf{a}^{(0)}$. Next we compute a set of iterates such that

$$\mathbf{a}^{(i+1)} = \mathbf{a}^{(i)} + \eta\, d\mathbf{a}^{(i)} \tag{13.3}$$

The scalar parameter, η, is introduced to control possible divergence during early stages of the iteration process and is often called *step-size control.* A common algorithm to determine η is a *line search* defined by[7]

$$g = \min_{\eta \in 0,1} |G(\eta)| \tag{13.4}$$

where

$$G(\eta) = d\mathbf{a}^{(i)} \cdot \boldsymbol{\Psi}(\mathbf{a}^{(i)} + \eta\, d\mathbf{a}^{(i)}) \tag{13.5}$$

An approximate solution to the line search is often advocated.[7]

It remains to deduce the vector $d\mathbf{a}^{(i)}$ for a given state $\mathbf{a}^{(i)}$. Newton's method is one algorithm which can be used to obtain incremental iterates. In this procedure we expand the residual $\boldsymbol{\Psi}$ about the current state $\mathbf{a}^{(i)}$ in terms of the increment $d\mathbf{a}^{(i)}$ and set the linear part equal to zero. Accordingly,

$$\boldsymbol{\Psi}(\mathbf{a}^{(i)}) + \left.\frac{\partial \mathbf{P}}{\partial \mathbf{a}}\right|_{\mathbf{a}^{(i)}} \cdot d\mathbf{a}^{(i)} = \mathbf{0} \tag{13.6}$$

We define the tangent (or Jacobian) matrix as

$$\mathbf{K}_{\mathrm{T}}^{(i)} = \left.\frac{\partial \mathbf{P}}{\partial \mathbf{a}}\right|_{\mathbf{a}^{(i)}} \tag{13.7}$$

Thus, we obtain an increment

$$d\mathbf{a}^{(i)} = -(\mathbf{K}_{\mathrm{T}}^{(i)})^{-1}\, \boldsymbol{\Psi}(\mathbf{a}^{(i)}) \tag{13.8}$$

This step requires the solution of a set of simultaneous linear algebraic equations. We note that for a linear differential equation the finite element internal force vector may be written as

$$\mathbf{P}(\mathbf{a}) = \mathbf{K}\,\mathbf{a} \tag{13.9}$$

where $\mathbf{K}$ is a *constant* matrix. Thus, Eq. (13.9) generates a constant tangent matrix and the process defined by Eqs (13.3) and (13.6) converges in one iteration provided a unit value of η is used.

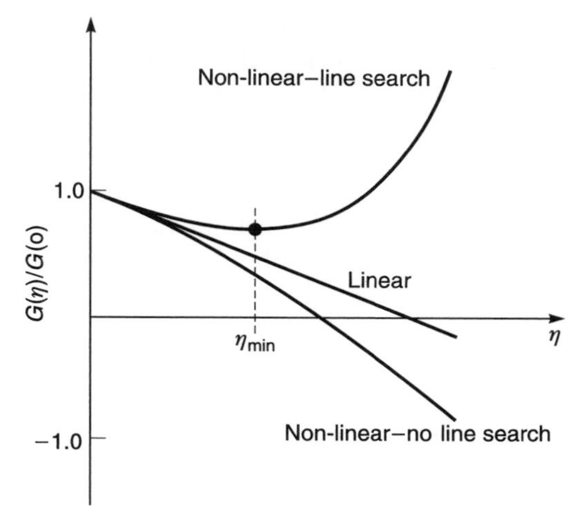

Fig. 3.1 Energy behaviour: line-search use.

For Newton's method the residual $\mathbf{\Psi}$ must have a norm which gets smaller for a sufficiently small $d\mathbf{a}^{(i)}$; accordingly, a Newton step may be projected onto G as shown in Fig. 13.1. Generally, Newton's method is convergent if

$$G^{(i)}(1) < \alpha G^{(i)}(0); \qquad 0 < \alpha < 1 \tag{13.10}$$

for all iterations; however, convergence may not occur if this condition is not obtained.*

A Newton solution algorithm may be constructed by using the command language statements included in the program described in this chapter. A solution for a single loading step with a maximum of 10 iterations is given by†

```
LOOP,newton,10
  TANG
  FORM
  SOLV
NEXT,newton
```

or

```
LOOP,newton,10
  TANG,,1
NEXT,newton
```

The second form is preferred since this will ensure that $\mathbf{K}_T^{(i)}$ and $\mathbf{\Psi}^{(i)}$ are computed simultaneously for each element, whereas the first form of the algorithm computes

* For some problems (such as those defined by non-linear theories of beams, plates, and shells) early iterations may produce shifts between one mode of behaviour (bending) and another (membrane) which can cause very large changes in $|\mathbf{\Psi}|$. Later iterations, however, generally follow Eq. (13.10).

† Recall that command information shown in upper-case letters must be given; text indicated by lower-case letters is optional. Finally, information given in italics must have numerical values assigned to define a proper command statement.

the two separately (for this case each FEAPpv element must compute both $\mathbf{K}_T^{(i)}$ and $\mathbf{\Psi}^{(i)}$ when ISW $= 3$). If convergence occurs before 10 iterations are performed, the process will transfer to the command statement following the NEXT statement. Convergence is based upon

$$G^{(i)}(0) < tol \cdot G^{(0)}(0)$$

where *tol* is specified by a TOL statement (a default value of 10^{-12} is set).

A line search may be added to the algorithm by modifying the commands to

```
LOOP,newton,10
  TANG,,1
  TANG
  FORM
  SOLV,LINE,0.6
NEXT,newton
```

or

```
LOOP,newton,10
  TANG,LINE,1,,0.6
NEXT,newton
```

where 0.6 denotes the value assigned to α in Eq. (13.10).

Line search requires repeated computations of $\mathbf{\Psi}(\mathbf{a} + \eta\, d\mathbf{a})$ which may increase solution times. Some assessment of need should be made before proceeding with large numbers of solution steps.

A modified Newton method also may be performed by removing the tangent computation from the loop. Accordingly,

```
TANG
LOOP,newton,10
  FORM
  SOLV,LINE,0.6
NEXT,newton
```

would compute only one tangent matrix $\mathbf{K}_T^{(1)}$ and its associated triangular factors. The form command computes only $\mathbf{\Psi}$ and SOLV solves the equations by using previously computed triangular factors of the tangent matrix. Algorithms between full Newton and modified Newton may also be constructed. For example

```
LOOP,,2
  TANG
  LOOP,newton,5
    FORM
    SOLV,LINE,0.6
  NEXT,newton
NEXT
```

In this algorithm it should be noticed that convergence in the first 5 iterations would transfer to the outer NEXT statement and a second TANG would be computed followed by a single iteration in the inner loop before the entire algorithm is completed.

Use of the BFGS algorithm described in Chapter 2 may lead to improved solution performance and/or reduced solution cost (see the necking example in Chapter 10). It is also particularly effective when no exact tangent matrix can be computed. In FEAPpv use of the BFGS algorithm is specified by the commands

```
LOOP,,10
  TANG,,1
  BFGS,,6,0.6
NEXT
```

where the 6 on the BFGS command indicates 6 updates using the vectors described in Sec. 2.2.4, and the 0.6 is again the line search tolerance.

The above algorithm recomputes the tangent matrix after each set of BFGS updates. This aspect usually improves the convergence rate. However, in some problems the tangent may have significant errors which lead to erroneous 'geometric' stiffness contributions. These can impede the effectiveness of the BFGS algorithm. In such situations it is possible to obtain a better estimate of the tangent by taking a very small solutions step (e.g. setting the time increment to be small) and follow this by the full step. This is easily achieved by using FEAPpv by the command sequence

```
DT,,0.001*dt
TANG,,1
DT,,dt
LOOP,,10
  TANG,,1
  BFGS,,6,0.6
NEXT
```

In the above a step of 1/1000 of the Δt is indicated to compute the trial tangent step. Thus the update correction should be very small; however, the program now has an estimate as to whether loading or unloading is occurring at each quadrature point and thus in subsequent iterations can use an appropriate tangent for the full step set by the second DT command.

The reader can use the above options to design a solution algorithm which meets the needs of most applications. However, in the realm of non-linear analysis there is no one algorithm which always 'works' efficiently and one must try various options. Indeed, in an interactive mode one can use the BACK command to restart a time step in situations where standard algorithms fail. Using various options a search for the best strategy can be found (or at least a strategy which produces a solution!).

13.3.2 Transient problems

The solution of transient problems defined by Eqs (10.1)–(10.3) may be performed by using the program described in this chapter. The program includes options to solve transient finite element problems which generate first- and second-order ordinary differential equations using the GN11 and Newmark (GN22) algorithms described in Chapter 18 of Volume 1. Options also exist to use an explicit version of the

GN22 algorithm and here the user is referred to the user instructions obtained from the publisher's website (http://www.bh.com/companions/fem).

Solution of first-order problems using GN11

Consider first a linear problem described by

$$\mathbf{C}\dot{\mathbf{a}} + \mathbf{K}\mathbf{a} + \mathbf{f} = \mathbf{0} \tag{13.11}$$

If we introduce the SS11 algorithm, we have, at each time t_{n+1}, the discrete problem given by Eq. (18.45) of Volume 1 as

$$\boldsymbol{\Psi}(\boldsymbol{\alpha}_{n+1}) = \mathbf{C}\boldsymbol{\alpha}_{n+1} + \mathbf{K}[\tilde{\mathbf{a}}_{n+1} + \theta\Delta t\boldsymbol{\alpha}_{n+1}] + \bar{\mathbf{f}}_{n+1} = \mathbf{0} \tag{13.12}$$

with

$$\tilde{\mathbf{a}}_{n+1} = \mathbf{a}_n \tag{13.13}$$

and from Eq. (18.47) of Volume 1

$$\mathbf{a}_{n+1} = \mathbf{a}_n + \Delta t\boldsymbol{\alpha}_{n+1} \tag{13.14}$$

We may also consider a non-linear, one-step extension to this problem, expressed by

$$\boldsymbol{\Psi}(\boldsymbol{\alpha}_{n+1}) = \mathbf{C}\boldsymbol{\alpha}_{n+1} + \mathbf{P}(\tilde{\mathbf{a}}_{n+1} + \theta\Delta t\boldsymbol{\alpha}_{n+1}) + \bar{\mathbf{f}}_{n+1} = \mathbf{0} \tag{13.15}$$

where $\mathbf{P}$ is again the vector of non-linear internal forces. The solution to either the linear or the non-linear problem may be expressed as

$$[\mathbf{C} + \theta\Delta t\mathbf{K}_{\mathrm{T}}^{(i)}]\Delta\boldsymbol{\alpha}_{n+1}^{(i)} = -\boldsymbol{\Psi}(\boldsymbol{\alpha}_{n+1}^{(i)}) \tag{13.16}$$

with

$$\boldsymbol{\alpha}_{n+1}^{(i+1)} = \boldsymbol{\alpha}_{n+1}^{(i)} + \eta\Delta\boldsymbol{\alpha}_{n+1}^{(i)} \tag{13.17}$$

where η is the step size as described above for non-linear problems, and for the linear problem η is always taken as unity. For linear problems $\mathbf{K}_{\mathrm{T}}^{(i)} \equiv \mathbf{K}$ whereas for non-linear problems

$$\mathbf{K}_{\mathrm{T}}^{(i)} = \left.\frac{\partial \mathbf{P}}{\partial \mathbf{a}}\right|_{\mathbf{a}^{(i)}} \tag{13.18}$$

Finally, converged values are expressed without the (i) superscript.

The solution of the transient problem is achieved by satisfying the following steps:

1. Specify θ.
2. Specify Δt.
3. Specify the time, t_{n+1}, the number of time steps, and set $i = 0$.
4. For each time, t_{n+1}:
   (a) compute $\boldsymbol{\Psi}(\boldsymbol{\alpha}_{n+1}^{(i)})$,
   (b) compute $\mathbf{C} + \theta\Delta t\mathbf{K}_{\mathrm{T}}^{(i)}$,
   (c) solve for $\Delta\boldsymbol{\alpha}_{n+1}^{(i)}$.
5. Check convergence for non-linear problems:
   (a) if satisfied terminate iteration,
   (b) if not satisfied set $i = i + 1$ and repeat Step 5.
6. Output solution information if needed.

7. Check time limit:
   (a) if $n \geqslant$ maximum number, stop, else,
   (b) if $n <$ maximum number, go to Step 4.

For a typical problem these steps may be specified by the set of statements

```
TRANs,SS11,0.5      [Step 1]
DT,,0.1             [Step 2]
LOOP,time,20        [Step 3]
  TIME
  LOOP,newton,10    [Step 4]
    TANG              [Step (4a)]
    FORM              [Step (4b)]
    SOLV              [Step (4c)]
  NEXT,newton       [Step 5]
  DISP,ALL          [Step 6]
NEXT,time           [Step 7]
```

The above algorithm works both for linear and for non-linear problems. For linear problems the residual should be a numerical zero at the second iteration (if not there is a programming error!) and for efficiency purposes the commands

```
LOOP,newton,10
```

and

```
NEXT,newton
```

may be removed. Also, for linear problems in which the time step is the same, a single tangent command may be used – in the above this can be accomplished by placing the `TANG` statement immediately after the `DT` statement.

Any of the options for solving a non-linear problem may be used (e.g. modified Newton's method or BFGS) by following the descriptions given above in the non-linear section. In particular, again for efficiency, one should use

```
LOOP,newton,10
  TANG,,1
NEXT,newton
```

for a full Newton solution step.

Specification of time-dependent loading may be given for

1. proportional loading with a fixed spatial distribution of the nodal load vector, and/or
2. general time-varying loading.

For proportional loading

$$\bar{\mathbf{f}}_{n+\theta} = p(\bar{t}_{n+\theta})\mathbf{f}_0 \tag{13.19}$$

where, in the program,

$$\bar{t}_{n+\theta} = t_n + \theta \Delta t \tag{13.20}$$

The value of $\mathbf{f}_0$ is specified either as nodal forces (during description of the mesh using the `FORCe` option) or is computed in each element as an element loading. For example, the heat source Q in Eq. (1.54) would generate element loads at node i as

$$f_i = \int_\Omega N_i Q \, d\Omega \tag{13.21}$$

The value of the proportional factor can be specified in the program as

$$p(t) = A_1 + A_2 t + A_3 \sin^L A_4 (t - t_{min}); \quad t_{min} \leqslant t < t_{max} \tag{13.22}$$

Details for input of the parameters are given in the user manual. For proportional loading the command language program given above is modified by adding a Step 0: specify proportional loading function, $p(t)$. The command for this step to specify a single proportional load function is

```
PROP,,1
```

additional data that define the A_i, t_{min}, t_{max} and L follow the `END` command in a `BATCh` solution mode. Each specification of a new time will cause the program to recompute $p(\bar{t}_{n+1})$. The value of the proportional loading is passed to each element module as a `REAL` number which is the first entry in the `ELDATA` common statement (and is named `DM`) and may be used to multiply element loads to obtain the correct loading at each time.

General loading can be achieved only by re-entering the mesh generation module and respecifying the nodal values. Accordingly, for this option a `MESH` command must be inserted in the time-step loop. For example, one can modify Step 4 above to

```
LOOP,time,20
  TIME
  MESH
```

For each time step it is then necessary to specify the new nodal values for $\bar{\mathbf{f}}_{n+1}$. If the value at a node previously set to a non-zero condition becomes zero, the value *must be specified* to reset the value. This may be achieved by specifying the node number only. For example,

```
FORCe
  12
  26,,5.0

END
```

would set all the components of the *force* at node 12 to zero and the first component of the force at node 26 to 5.0 units.

In batch execution the number of `FORC-END` paired statements must be equal to or exceed the number of time steps. In interactive execution a `MESH >` prompt will appear on the screen and the user must provide the necessary `FORC` data from the keyboard (terminating input with a blank line and an `END` statement). With a proper termination the prompt will once again indicate interactive inputs for solution command statements.

While both proportional and general loads may be combined, extreme caution must be exercised as all nodal values will be multiplied by the current value of $p(\bar{t})$.

The value of the time increment may be changed by repeating Step 2 and then performing Steps 4–8 for the new increment. If a large number of different-size time increments is involved this will be inefficient, and the program has an option to specify a new value for *dt* as data. The command language program would be modified by replacing Step 4 with

```
LOOP,time,20
  DATA,DT
  TIME
```

For each time step the DT statement is supplied as new data. For interactive computations the user inputs

```
DT,,dt
```

where dt is the size of the time step to be used. In batch executions these commands again follow the END statement which terminates the command language program. If other data are input (e.g. the FORC-END paired data), then the data statements must be in the order requested by the solution program statements.

A DATA instruction may also be used to set a solution tolerance. The form would be

```
DATA,TOL
```

and the user would input a command

```
TOL,,1.E-10
```

to set the solution convergence tolerance to 10^{-10}.

Solution of second-order systems

For simplicity the damping matrix, $\mathbf{C}$, is not included in any of the FEAPpv elements. Steps to add such effects are summarized, however, at the end of this section. Thus, we consider the differential equation

$$\mathbf{M}\ddot{\mathbf{a}} + \mathbf{K}\mathbf{a} + \mathbf{f} = \mathbf{0} \tag{13.23}$$

for linear problems, and

$$\mathbf{M}\ddot{\mathbf{a}} + \mathbf{P}(\mathbf{a}) + \mathbf{f} = \mathbf{0} \tag{13.24}$$

for non-linear problems.

Solution using the GN22 algorithm The GN22 algorithm may be selected, and from Chapter 18 of Volume 1 we have for a linear problem

$$\mathbf{\Psi}(\ddot{\mathbf{a}}_{n+1}) = \mathbf{M}\ddot{\mathbf{a}}_{n+1} + \mathbf{K}[\bar{\mathbf{a}}_{n+1} + \tfrac{1}{2}\beta_2\Delta t^2\ddot{\mathbf{a}}_{n+1}] + \mathbf{f}_{n+1} = \mathbf{0} \tag{13.25}$$

or for a non-linear problem

$$\mathbf{\Psi}(\ddot{\mathbf{a}}_{n+1}) = \mathbf{M}\ddot{\mathbf{a}}_{n+1} + \mathbf{P}(\bar{\mathbf{a}}_{n+1} + \tfrac{1}{2}\beta_2\Delta t^2\ddot{\mathbf{a}}_{n+1}) + \mathbf{f}_{n+1} = \mathbf{0} \tag{13.26}$$

where

$$\begin{aligned} \bar{\mathbf{a}}_{n+1} &= \mathbf{a}_n + \Delta t\,\dot{\mathbf{a}}_n + \tfrac{1}{2}\Delta t^2\,\ddot{\mathbf{a}}_n \\ \bar{\dot{\mathbf{a}}}_{n+1} &= \dot{\mathbf{a}}_n + \Delta t\,\ddot{\mathbf{a}}_n \end{aligned} \tag{13.27}$$

and

$$\mathbf{a}_{n+1} = \bar{\mathbf{a}}_{n+1} + \beta \Delta t^2 (\ddot{\mathbf{a}}_{n+1} - \ddot{\mathbf{a}}_n)$$
$$\dot{\mathbf{a}}_{n+1} = \dot{\bar{\mathbf{a}}}_{n+1} + \gamma \Delta t (\ddot{\mathbf{a}}_{n+1} - \ddot{\mathbf{a}}_n) \tag{13.28}$$

The above can be related to commonly used Newmark parameters by the expressions

$$\gamma = \beta_1 \qquad \text{and} \qquad \beta = 2\beta_2 \tag{13.29}$$

A solution to the GN22 problems may be expressed as

$$[\mathbf{M} + \tfrac{1}{2}\beta_2 \Delta t^2 \mathbf{K}_{\mathrm{T}}^{(i)}]\, d\ddot{\mathbf{a}}_{n+1}^{(i)} = -\boldsymbol{\Psi}(\ddot{\mathbf{a}}_{n+1}^{(i)}) \tag{13.30}$$

with

$$\ddot{\mathbf{a}}_{n+1}^{(i+1)} = \ddot{\mathbf{a}}_{n+1}^{(i)} + \eta\, d\ddot{\mathbf{a}}_{n+1}^{(i)} \tag{13.31}$$

Again, the algorithm for solution is identical to the steps for SS11 except that now it is initiated using the command statement

```
TRANs,GN22,0.5,0.5
```

where the two numerical values are for β_1 and β_2, respectively (default values are $\beta_1 = \beta_2 = 0.5$). Alternatively, the command

```
TRANs,NEWMark,beta,gamma
```

may be used for the Newmark algorithm with the default values $\beta = 0.25$ and $\gamma = 0.5$.

To start the solution process it is necessary to define an initial value for the acceleration $\ddot{\mathbf{a}}_0$. This may be performed using Eq. (13.26) and the specified initial conditions for $\mathbf{a}_0$ and $\dot{\mathbf{a}}_0$. To determine the initial solution, all the loading values must be assigned and then the command language statement

```
FORM,ACCEleration
```

may be used to compute the correct value of $\ddot{\mathbf{a}}_0$.

Adding damping effects A damping matrix may be added by including in each element routine the appropriate terms. Thus, when computing a residual (see Volume 1, Chapter 20) with `ISW` = 3 and `ISW` = 6, the term

$$\mathbf{C}\dot{\mathbf{a}}_{n+1} \tag{13.32}$$

must be added to the equilibrium equation in each element. The value of $\dot{\mathbf{a}}_{n+1}$ localized for each element and adjusted for each algorithm is passed as part of the `UL` array (see Volume 1, Chapter 20, for variable name descriptions). The `UL` array may be assumed to be dimensioned as

```
REAL*8 UL ( NDF , NEN , IT )
```

where `NDF` is the number of unknowns at each node, `NEN` is the maximum number of nodes on any element, and `IT` denotes the quantities as indicated in Table 13.2. Using these values and a definition for **C**, the appropriate terms may be computed and added to the element residual vector, **P**. Similarly, the appropriate tangent stiffness term must be added to the element array **S** for `ISW` = 3. Using, the tangent factors

Table 13.2 Values in UL array for transient algorithms; n.v.r. = no value returned

IT value	Algorithm	
	GN11	GN22
1	$\bar{\mathbf{a}} + \theta\Delta t \boldsymbol{\alpha}^{(i)}$	$\mathbf{a}^{(i)}$
2	$\theta\Delta t \boldsymbol{\alpha}^{(i)}$	$\mathbf{a}^{(i)} - \mathbf{a}^{(0)}$
3	$\theta\Delta t \Delta\boldsymbol{\alpha}^{(i)}$	$\mathbf{a}^{(i)} - \mathbf{a}^{(i-1)}$
4	n.v.r.	$\dot{\mathbf{a}}^{(i)}$
5	n.v.r.	$\ddot{\mathbf{a}}^{(i)}$

it is necessary only to compute **C** and multiply it by CTAN(2) to perform this step (see Chapter 20, Volume 1).

13.3.3 Eigensolutions

The solution of a general linear eigenproblem is a useful feature included in the program contained in this chapter. The program can compute a set of the smallest eigenvalues (in absolute value) and their associated eigenvectors for the problem

$$\mathbf{K}_T \mathbf{V} = \mathbf{M} \mathbf{V} \boldsymbol{\Lambda} \tag{13.33}$$

In the above, $\mathbf{K}_T$ is any symmetric tangent matrix which has been computed by using a TANG command statement; **M** is a mass or identity matrix computed using a MASS or IDEN command statement, respectively; the columns of **V** are the set of eigenvectors to be computed; and $\boldsymbol{\Lambda}$ is a diagonal matrix which contains the set of eigenvalues to be computed. For second-order equations the eigenvalues λ are the frequencies squared, ω^2. Accordingly, the program will also compute and report the square root of λ. Since negative values of λ can occur, the square root of the absolute values is computed. For negative λ the reported values are in fact pure imaginary numbers.

The tangent matrix often has zero eigenvalues and, for this case, the algorithm used requires the problem to be transformed to

$$(\mathbf{K}_T - \alpha\mathbf{M})\mathbf{V} = \mathbf{M}\mathbf{V}\boldsymbol{\Lambda}_\alpha \tag{13.34}$$

where α is a parameter called the *shift* (see Chapter 17, Volume 1), which must be selected to make the coefficient matrix on the left-hand side of Eq. (13.34) non-singular. $\boldsymbol{\Lambda}_\alpha$ are the eigenvalues of the shift which are related to the desired values by

$$\boldsymbol{\Lambda} = \boldsymbol{\Lambda}_\alpha + \alpha\mathbf{I} \tag{13.35}$$

FEAPpv always reports the value of the eigenvalue and not its shifted value. The shift may also be used to compute the eigenpairs nearest to some specified value. The components of $\boldsymbol{\Lambda}$ are output as part of the eigenproblem solution. In addition, the vectors may be output as numerical values or presented graphically.

The program uses a subspace algorithm[8–10] to compute a small general eigenproblem defined as

$$\mathbf{K}^*\mathbf{x} = \mathbf{M}^*\mathbf{x}\boldsymbol{\lambda} \tag{13.36}$$

where

$$\mathbf{V} = \mathbf{Q}\mathbf{x} \tag{13.37}$$

and

$$\begin{aligned} \mathbf{K}^* &= \mathbf{Q}^{\mathrm{T}}\mathbf{M}^{\mathrm{T}}(\mathbf{K}_{\mathrm{T}} - \alpha\mathbf{M})^{-1}\mathbf{M}\mathbf{Q} \\ \mathbf{M}^* &= \mathbf{Q}^{\mathrm{T}}\mathbf{M}\mathbf{Q} \end{aligned} \tag{13.38}$$

Accordingly, after the projection, the $\boldsymbol{\lambda}$ are reciprocals of $\boldsymbol{\Lambda}_\alpha$ (i.e. $\boldsymbol{\Lambda}_\alpha^{-1}$). An eigensolution of the small problem may be used to generate a sequence of iterates for $\mathbf{Q}$ which converge to the solution for the original problem (e.g. see reference 10). The solution of the projected small general problem is solved here using a transformation to a standard linear eigenproblem combined with a QL algorithm.[12]

The transformation is performed by computing the Choleski factors of $\mathbf{M}^*$ to define the standard linear eigenproblem

$$\mathbf{H}\mathbf{y} = \mathbf{y}\boldsymbol{\lambda} \tag{13.39}$$

where

$$\begin{aligned} \mathbf{M}^* &= \mathbf{L}\mathbf{L}^{\mathrm{T}} \\ \mathbf{y} &= \mathbf{L}^{\mathrm{T}}\mathbf{x} \\ \mathbf{H} &= \mathbf{L}^{-1}\mathbf{K}^*\mathbf{L}^{-\mathrm{T}} \end{aligned} \tag{13.40}$$

In the implementation described here scaling is introduced, which causes $\mathbf{M}^*$ to converge to an identity matrix; hence the above transformation is numerically stable. Furthermore, use of a standard eigenproblem solution permits calculation of positive and negative eigenvalues. The subspace algorithm implemented provides a means to compute a few eigenpairs for problems with many degrees of freedom or all of the eigenpairs of small problems. A subspace algorithm is based upon a power method to compute the dominant eigenvalues. Thus, the effectiveness of the solution strategy depends on the ratio of the absolute value of the largest eigenvalue sought in the subspace to that of the first eigenvalue not contained in the subspace. This ratio may be reduced by adding additional vectors to the subspace. That is, if p pairs are sought, the subspace is taken as q vectors so that

$$\left|\frac{\lambda_p}{\lambda_{q+1}}\right| < 1 \tag{13.41}$$

Of course, the magnitude of this ratio is unknown before the problem is solved and some analysis is necessary to estimate its value. The program tracks the magnitude of the shifted reciprocal eigenvalues $\boldsymbol{\Lambda}$ and computes the change in values between successive iterations. If the subspace is too small, convergence will be extremely slow owing to Eq. (13.41) having a ratio near unity. It may be desirable to increase the subspace size to speed the convergence. In some problems, characteristics of the eigenvalue magnitudes may be available to assist in the process. It should be especially noted that when p is specified as the total number of degrees of freedom in the problem (or q becomes this value), then λ_{q+1} is infinitely larger and the ratio given in Eq. (13.41) is zero. In this case subspace iteration converges in a single iteration, a fact which is noted by the program to limit the iterations to 1. Accordingly, it is usually more efficient to compute all the eigenpairs if q is very near the number of degree of freedoms.

Use of the subspace algorithm requires the following steps:

1. compute $\mathbf{M}$, compute the tangent matrix $\mathbf{K}_\mathrm{T}$, apply the α shift if necessary,
2. compute the eigenpairs,
3. output the results.

The commands to achieve this algorithm are:

```
MASS (or IDEN)
TANG,,,alpha
SUBS,<PRINt>,p,q
EIGV,,n
PLOT,EIGV,n
```

Note the specification of the shift value α as part of the TANG statement. The TANG command computes both the matrix and its triangular factors. The shift is performed during computation by defining in each element a tangent matrix

$$\mathbf{K}_\mathrm{T}^\mathrm{e} = c_1 \mathbf{K}_\mathrm{T} + c_2 \mathbf{C} + c_3 \mathbf{M}$$

with $c_1 = 1$, $c_2 = 0$, and $c_3 = -\alpha$. Thus, specification of the commands MASS and TANG is not order dependent. The value for q is optional and, when omitted, is computed by the program as

$$q = \min(NM, NEQ, 2 * p, p + 8)$$

where NM is the number of non-zero terms in the diagonal mass matrix and NEQ is the number of degrees of freedom in the problem (i.e. those not restrained by boundary constraints specified using BOUN).

The plot of eigenvectors may need to be increased or decreased by a factor to permit proper viewing. Each eigenvector may be viewed graphically and/or as output using the EIGV print instruction.

The eigenproblem for individual elements is a common procedure used to evaluate performance. It is necessary to describe a mesh without restraints to the degrees of freedom (i.e. BOUN should not be used in the mesh data). The element normally has zero eigenvalues, hence a shift must be used for the analysis. Failure to use a shift will result in errors in the triangular factors of $\mathbf{K}_\mathrm{T}$ (the program will detect a near singularity and output a warning) and generally all or a large number of the eigenpairs will collapse onto the singular subspace. If the shift is specified very close to an eigenvalue these types of errors may also occur. The user should monitor the outputs during the TANG and the SUBS commands to detect poor performance. If all or a large number of the eigenvalues are extremely near the shift, a second shift should be tried. In general, a shift should be picked nearly halfway between reported values. The program also counts the number of eigenvalues which are less than the shift. This may be used to ensure that the shift is not too large to determine the desired eigenvalues. Recall that only the p values nearest to the shift in absolute value are determined.

When properly used, the subspace method can produce accurate and reliable values for the eigenpairs of a finite element system problem. The method may be used to compute the vibration modes of structural systems and is implemented so that it may be used both for linear and for non-linear finite element models. For non-linear problems this permits the dependence on frequency with load to be assessed. Thus, a dynamic buckling load where a frequency goes to zero may be computed. For non-linear models the static buckling loads for a problem also may be determined by solving the eigenproblem

$$\mathbf{K}_{\mathrm{T}}\mathbf{V} = \mathbf{I}\mathbf{V}\boldsymbol{\Lambda} \tag{13.42}$$

for a set of loads and tracking the approach to zero of the smallest eigenvalue. A buckling load corresponds to a zero eigenvalue in Eq. (13.42). As the buckling load is approached, a shift may be necessary to maintain high accuracy; however, since a collapsed subspace is the desired solution, this is usually unnecessary.

13.4 Restart option

The program FEAPpv permits a user to save a solution state and subsequently use these data to continue an analysis from the point the data were saved. This is called a *restart* option. To use the restart feature, the file names given at initiation of the program must be appropriately specified.

Names for two files may be specified and are denoted on the screen as the `RESTART (READ)` and the `RESTART (WRITE)` files. The `READ` file is the name of a file which contains the data from a previous analysis. The screen will also indicate `EXIST` or `NEW` for the file. A `NEW` label indicates that no file with the specified name exists. The `WRITE` file is the name of a file to which data will be written as part of the current session. By default the same file name is given for both the `READ` and the `WRITE` files.

During solution a restart file may be saved by using the command

```
SAVE,<extender>
```

This results in a file being written to the `RESTART (WRITE)` file with an optional extender appended with the name specified in the command. For example, if the restart write file has the name ‘Rprob’, issuing the command

```
SAVE,ti0
```

saves the data on a file named `Pprob.ti0`. Alternatively, issuing the command as

```
SAVE
```

saves the data on the file named `Pprob`. For large problems the restart file can be quite large (especially if the elements use several history variables at each integration point) thus one should be cautious about use of too many files in these situations.

To restore a file the command

```
RESTart
```

is given to load the file without an extender, and the command

```
RESTart,<extender>
```

to load the file with extender. If the write and read files have the same name it is possible to restart a problem at an earlier state during a solution process. Thus, if a solution has proceeded to a state after a save where it is decided to pursue a different loading sequence it is possible to 'back up' to a state where the change is to be made and continue the analysis. Use of restart files can be particularly useful for performing complex analyses in which different design options are considered.

13.5 Solution of example problems

The first step in using FEAPpv to solve a finite element problem is to create a file with the mesh data. It is useful to specify the file with a name beginning with I (e.g. IBEAM for a beam data file). The program then automatically provides default names for the output and restart files by stripping the I and adding an O and R, respectively (e.g. OBEAM for output and RBEAM for restart). Once the data are available in the input file the program can be initiated by entering the program name. In what follows, it is assumed that the executable program is named 'feappv', thus entering this name initiates an execution of the program. During the first execution of the program in any directory (folder) the user must specify the name of the input file for the problem to be solved. The program saves this and other information in a small file named `feapname`. The program always requests names for the input/output and restart files. During the first execution, a name *must* be provided for the input data file. Default names for the output and restart are provided and accepted by striking the < ENTER > or < CR > key or may be replaced by specifying the name of a file to be used. In subsequent executions of the program, the names of the files from the last execution are read from the file `feapname` and used as default values. Once the information is specified, the user may accept the names assigned by pressing the 'Y' key, repeat the file specification by pressing the 'N' key, or stop execution by pressing the 'S' key (upper- or lower-case letters may be given for all commands). Once the 'Y' key is struck, the program proceeds to input the data contained in the input file until either an `INTEractive` execution command or a `STOP` is encountered. If the file contains `BATCh` execution commands, the solution is performed in a batch-type mode until user interactions are required. The solution sequence can contain multiple `BATCh` sets and multiple `INTEractive` sets. However, if an interactive execution mode is requested, the user must provide all the solution steps from the input keyboard. The inputs are given whenever the screen contains the line

```
Time = 09:45:33 Macro 1>x
```

or similar where the number following `Time =` is the clock or elapsed time for the computer and `x` indicates the computer cursor. Solution commands may then be entered as described in Volume 1, Table 20.16, or as in Table 13.1 in this chapter. For example entering the command

```
TANG,,1
```

performs a full solution step. Some examples of problem solutions using FEAPpv are found at the publisher's website (http://www.bh.com/companions/fem) and are presented in two files for each problem. The first file contains the input dataset and each is indicated by a file beginning with I. The second file contains the output from the problem. The first part of the output are the data defining the mesh and the second part the numerical results obtained.

13.6 Concluding remarks

In the discussion above we have presented some of the ways the program FEAPpv may be used to solve a variety of non-linear finite element problems. The classes of non-linear problems which may be solved using this system is extensive and we cannot give a comprehensive summary here. The reader is encouraged to obtain a copy of the program source statements and companion documents from the publisher's website (http://www.bh.com/companions/fem). In addition to the program discussed in the first two volumes a companion program devoted to solving fluid dynamics problems described in the third volume will also be found.

As noted in the introduction to this chapter the computer programs will undoubtedly contain some errors. We welcome being informed of these as well as comments and suggestions on how the programs may be improved. Although the programs available are written in Fortran it is quite easy to adapt these to permit program modules to be constructed in other languages. For example an interface for element routines written in C has been developed by Govindjee.[13]

The program system FEAPpv contains only basic commands to generate structured meshes as blocks of elements. For problems where graded meshes are needed (e.g. adaptive mesh refinements) more sophisticated mesh generation techniques are needed. There are many locations where generators may be obtained and two are given in references 14 and 15. The program GID offers two- and three-dimensional options for fluid and structure applications. Sub-programs to interface GiD to FEAPpv are also available at the publisher's web site (details above).

References

1. O.C. Zienkiewicz and R.L. Taylor. *The Finite Element Method: The Basis*, 5th edition, Volume 1, Butterworth-Heinemann, Oxford, 2000.
2. W.H. Press *et al.* (eds), *Numerical Recipes in Fortran 77: The Art of Scientific Computing*, 2nd edition, Cambridge University Press, Cambridge, 1992.
3. W.H. Press *et al.* (eds), *Numerical Recipes in Fortran 77 and 90: The Art of Scientific and Parallel Computing (Software)*, Cambridge University Press, Cambridge, 1997.
4. W.H. Press *et al.* (eds), *Numerical Recipes in Fortran 77: The Art of Parallel Scientific Computing*, Volume 2, Cambridge University Press, Cambridge, 1996.
5. G.H. Golub and C.F. Van Loan. *Matrix Computations*, 3rd edition, The Johns Hopkins University Press, Baltimore, MD, 1996.
6. J. Demmel. *Applied Numerical Linear Algebra*, Society for Industrial and Applied Mathematics, Philadelphia, PA, 1997.
7. H. Matthies and G. Strang. The solution of nonlinear finite element equations. *Int. J. Num. Meth. Eng.*, **14**, 1613–26, 1979.
8. J.H. Wilkinson and C. Reinsch. *Linear Algebra: Handbook for Automatic Computation*, Volume II, Springer-Verlag, Berlin, 1971.

9. K.-J. Bathe and E.L. Wilson. *Numerical Methods in Finite Element Analysis*, Prentice-Hall, Englewood Cliffs, NJ, 1976.
10. K.-J. Bathe. *Finite Element Procedures*, Prentice-Hall, Englewood Cliffs, NJ, 1996.
11. L. Collatz. *The Numerical Treatment of Differential Equations*, Springer, Berlin, 1966.
12. B.N. Parlett. *The Symmetric Eigenvalue Problem*, Prentice-Hall, Englewood Cliffs, NJ, 1980.
13. S.V. Govindjee. Interface for c-language routines for feap programs. Private communication (see also *http://www.ce.berkeley.edu/~sanjay*), 2000.
14. J. Shewchuk. Triangle, *http://www.cs.berkeley.edu/~jrs*.
15. CIMNE, GiD – The Personal Pre/Postprocesor (Version 5.0), CIMNE, Universitat Politechnica de Catalunya, Barcelona, Spain, 1999. (See also http://wwww.cimne.upc.es)

Appendix A

Invariants of second-order tensors

A.1 Principal invariants

Given any second-order Cartesian tensor **a** with components expressed as

$$\mathbf{a} = \begin{bmatrix} a_{11} & a_{12} & a_{13} \\ a_{21} & a_{22} & a_{23} \\ a_{31} & a_{32} & a_{33} \end{bmatrix} \tag{A.1}$$

the *principal values* of **a**, denoted as a_1, a_2, and a_3, may be computed from the solution of the eigenproblem

$$\mathbf{a}\mathbf{q}^{(m)} = a_m \mathbf{q}^{(m)} \tag{A.2}$$

in which the (right) eigenvectors $\mathbf{q}^{(m)}$ denote *principal directions* for the associated eigenvalue a_m. Non-trivial solutions of Eq. (A.2) require

$$\det \begin{vmatrix} (a_{11} - a) & a_{12} & a_{13} \\ a_{21} & (a_{22} - a) & a_{23} \\ a_{31} & a_{32} & (a_{33} - a) \end{vmatrix} = 0 \tag{A.3}$$

Expanding the determinant results in the cubic equation

$$a_m^3 - \mathrm{I}_a a_m^2 + \mathrm{II}_a a_m - \mathrm{III}_a = 0 \tag{A.4}$$

where:

$$\begin{aligned} \mathrm{I}_a &= a_{11} + a_{22} + a_{33} \\ \mathrm{II}_a &= a_{11}a_{22} + a_{22}a_{33} + a_{33}a_{11} - a_{12}a_{21} - a_{23}a_{32} - a_{31}a_{13} \\ \mathrm{III}_a &= a_{11}a_{22}a_{33} - a_{11}a_{23}a_{32} - a_{22}a_{31}a_{13} - a_{33}a_{12}a_{21} + a_{12}a_{23}a_{31} + a_{21}a_{32}a_{13} \\ &= \det \mathbf{a} \end{aligned} \tag{A.5}$$

The quantities I_a, II_a, and III_a are called the *principal invariants of* **a**. The roots of Eq. (A.4) give the principal values a_m.

The invariants for the deviator of **a** may be obtained by using

$$\mathbf{a}' = \mathbf{a} - \bar{a}\,\mathbf{1} \tag{A.6}$$

where $\bar{a}$ is the mean defined as

$$\bar{a} = \tfrac{1}{3}\,(a_{11} + a_{22} + a_{33}) = \tfrac{1}{3}\,\mathrm{I}_a \tag{A.7}$$

Substitution of Eq. (A.6) into Eq. (A.2) gives

$$\left[\mathbf{a}' + \bar{a}\mathbf{I}\right]\mathbf{q}^{(m)} = a_m\mathbf{q}^{(m)} \tag{A.8}$$

or

$$\mathbf{a}'\mathbf{q}^{(m)} = (a_m - \bar{a})\,\mathbf{q}^{(m)} = a'_m\,\mathbf{q}^{(m)} \tag{A.9}$$

which yields a cubic equation for principal values of the deviator given as

$$(a'_m)^3 + \mathrm{II}'_a\,a'_m - \mathrm{III}'_a = 0 \tag{A.10}$$

where invariants of $\mathbf{a}'$ are denoted as I'_a, II'_a, and III'_a.

Since the deviator $\mathbf{a}'$ differ from the total **a** by a mean term only, we observe from Eq. (A.9) that the directions of their principal values coincide, and the three principal values are related through

$$a_i = a'_i + \bar{a}; \quad i = 1, 2, 3 \tag{A.11}$$

Moreover Eq. (A.10) generally has a closed-form solution which may be constructed by using the Cardon formula.[1,2]

The definition of $\mathbf{a}'$ given by Eq. (A.6) yields

$$\mathrm{I}'_a = a'_{11} + a'_{22} + a'_{33} = 0 \tag{A.12}$$

Using this result, the second invariant of the deviator may be shown to have the indicial form[3]

$$\mathrm{II}'_a = -\tfrac{1}{2}\,a'_{ij}\,a'_{ji} \tag{A.13}$$

The third invariant is again given by

$$\mathrm{III}'_a = \det \mathbf{a}' \tag{A.14}$$

however, we show in Sec. A.2 that this invariant may be written in a form which is easier to use in many applications (e.g. yield functions for elasto-plastic materials).

A.2 Moment invariants

It is also possible to write the invariants in a form known as *moment invariants*.[4] The moment invariants are denoted as $\bar{\mathrm{I}}_a$, $\bar{\mathrm{II}}_a$, $\bar{\mathrm{III}}_a$, and are defined by the indicial forms

$$\bar{\mathrm{I}}_a = a_{ii}, \qquad \bar{\mathrm{II}}_a = \tfrac{1}{2}\,a_{ij}\,a_{ji}, \qquad \bar{\mathrm{III}}_a = \tfrac{1}{3}\,a_{ij}\,a_{jk}\,a_{ki} \tag{A.15}$$

We observe that moment invariants are directly related to the *trace* of products of **a**. The trace (tr) of a matrix is defined as the sum of its diagonal elements. Thus, the first three moment invariants may be written in matrix form (using a square matrix for **a**) as

$$\bar{\text{I}}_a = \text{tr}(\mathbf{a}), \qquad \bar{\text{II}}_a = \tfrac{1}{2}\,\text{tr}(\mathbf{aa}), \qquad \bar{\text{III}}_a = \tfrac{1}{3}\,\text{tr}(\mathbf{aaa}) \tag{A.16}$$

The moment invariants may be related to the principal invariants as[4]

$$\begin{aligned} \bar{\text{I}}_a &= \text{I}_a, & \bar{\text{II}}_a &= \tfrac{1}{2}\,\text{I}_a^2 - \text{II}_a, & \bar{\text{III}}_a &= \text{III}_a - \tfrac{1}{3}\,\text{I}_a^3 + \text{I}_a\text{II}_a \\ \text{I}_a &= \bar{\text{I}}_a, & \text{II}_a &= \tfrac{1}{2}\,\bar{\text{I}}_a^2 - \bar{\text{II}}_a, & \text{III}_a &= \bar{\text{III}}_a + \tfrac{1}{6}\,\bar{\text{I}}_a^3 - \bar{\text{I}}_a\bar{\text{II}}_a \end{aligned} \tag{A.17}$$

Using Eq. (A.12) and the identities given in Eq. (A.17) we can immediately observe that the principal invariants and the moment invariants for a deviatoric second-order tensor are related through

$$\text{II}'_a = -\bar{\text{II}}'_a \qquad \text{and} \qquad \text{III}'_a = \bar{\text{III}}'_a = \det \mathbf{a}' \tag{A.18}$$

A.3 Derivatives of invariants

We often also need to compute the derivative of the invariants with respect to their components and this is only possible when all components are treated independently – that is, we do not use any symmetry, if present. From the definitions of the principal and moment invariants given above, it is evident that derivatives of the moment invariants are the easiest to compute since they are given in concise indicial form. Derivatives of principal invariants can be computed from these by using the identities given in Eqs (A.17) and (A.18).

The first derivatives of the principal invariants for symmetric second-order tensors may be expressed in a matrix form directly, as shown by Nayak and Zienkiewicz;[5,6] however, second derivatives from these are not easy to construct and we now prefer the methods given here.

A.3.1 First derivatives of invariants

The first derivative of each moment invariant may be computed by using Eq. (A.15). For the first invariant we obtain

$$\frac{\partial \bar{\text{I}}_a}{\partial a_{ij}} = \delta_{ij} \tag{A.19}$$

Similarly, for the second moment invariant we get

$$\frac{\partial \bar{\text{II}}_a}{\partial a_{ij}} = a_{ji} \tag{A.20}$$

and for the third moment invariant

$$\frac{\partial \bar{\text{III}}_a}{\partial a_{ij}} = a_{jk}\,a_{ki} \tag{A.21}$$

Using the identities, the derivative of the principal invariants may be written in indicial form as

$$\frac{\partial \mathrm{I}_a}{\partial a_{ij}} = \delta_{ij}, \qquad \frac{\partial \mathrm{II}_a}{\partial a_{ij}} = \mathrm{I}_a \delta_{ij} - a_{ji}, \qquad \frac{\partial \mathrm{III}_a}{\partial a_{ij}} = \mathrm{II}_a \delta_{ij} - \mathrm{I}_a a_{ji} + a_{jk} a_{ki} \mathrm{III}_a a_{ji}^{-1} \quad (A.22)$$

The third invariant may also be shown to have the representation[3]

$$\frac{\partial \mathrm{III}_a}{\partial a_{ij}} = \mathrm{III}_a a_{ji}^{-1} \quad (A.23)$$

where a_{ji}^{-1} is the inverse (transposed) of the a_{ij} tensor. Thus, in matrix form we may write the derivatives as

$$\frac{\partial \mathrm{I}_a}{\partial \mathbf{a}} = \mathbf{1}, \qquad \frac{\partial \mathrm{II}_a}{\partial \mathbf{a}} = \mathrm{I}_a \mathbf{1} - \mathbf{a}^{\mathrm{T}}, \qquad \frac{\partial \mathrm{III}_a}{\partial \mathbf{a}} = \mathrm{III}_a \mathbf{a}^{-\mathrm{T}} \quad (A.24)$$

where here $\mathbf{1}$ denotes a 3×3 identity matrix.

The expression for the derivative of the determinant of a second-order tensor is of particular use as we shall encounter this in dealing with volume change in finite deformation problems and in plasticity yield functions and flow rules.

Performing the same steps for the invariants of the deviator stress yields

$$\frac{\partial \mathrm{I}'}{\partial a'_{ij}} = \frac{\partial \bar{\mathrm{I}}'}{\partial a'_{ij}} = 0, \qquad \frac{\partial \mathrm{II}'}{\partial a'_{ij}} = -\frac{\partial \bar{\mathrm{II}}'}{\partial a'_{ij}} = -a'_{ji}, \qquad \frac{\partial \mathrm{III}'}{\partial a'_{ij}} = \frac{\partial \bar{\mathrm{III}}'}{\partial a'_{ij}} = a'_{jk} a'_{ki} \quad (A.25)$$

with only a sign change occurring in the second invariant to obtain the derivative of principal invariants from derivatives of moment invariants.

Often the derivatives of the invariants of a deviator tensor are needed with respect to the tensor itself, and these may be computed as

$$\frac{\partial (\cdot)'}{\partial a_{mn}} = \frac{\partial (\cdot)'}{\partial a'_{ij}} \frac{\partial a'_{ij}}{\partial a_{mn}} \quad (A.26)$$

where

$$\frac{\partial a'_{ij}}{\partial a_{mn}} = \delta_{im} \delta_{jn} - \frac{1}{3} \delta_{ij} \delta_{mn} \quad (A.27)$$

Combining the two expressions yields

$$\frac{\partial (\cdot)'}{\partial a_{mn}} = \frac{\partial (\cdot)'}{\partial a'_{mn}} - \frac{1}{3} \delta_{mn} \left[\delta_{ij} \frac{\partial (\cdot)'}{\partial a'_{ij}} \right] \quad (A.28)$$

A.3.2 Second derivatives

In developments of tangent tensors we need second derivatives of the invariants. These may be computed directly from Eqs (A.19)–(A.21) by standard operations. The second derivatives of $\bar{\mathrm{I}}_a$, $\bar{\mathrm{II}}_a$, $\bar{\mathrm{III}}_a$ yield

$$\frac{\partial^2 \bar{\mathrm{I}}_a}{\partial a_{ij} \partial a_{kl}} = 0, \qquad \frac{\partial^2 \bar{\mathrm{II}}_a}{\partial a_{ij} \partial a_{kl}} = \delta_{jk} \delta_{il}, \qquad \frac{\partial^2 \bar{\mathrm{III}}_a}{\partial a_{ij} \partial a_{kl}} = \delta_{jk} a_{il} + a_{jk} \delta_{il} \quad (A.29)$$

The computations for principal invariants follow directly from the above using the identities given in Eqs (A.17) and (A.18). Also, all results may be transformed to the vector form used extensively in this volume for the finite element constructions. These steps are by now a standard process and are left as an excercise for the reader.

References

1. H.M. Westergaard. *Theory of Elasticity and Plasticity*, Harvard University Press, Cambridge, MA, 1952.
2. W.H. Press *et al.* (eds), *Numerical Recipes in Fortran: The Art of Scientific Computing*, 2nd edition, Cambridge University Press, Cambridge, 1992.
3. I.H. Shames and F.A. Cozzarelli. *Elastic and Inelastic Stress Analysis*, revised edition, Taylor & Francis, Washington, DC, 1997.
4. J.L. Ericksen. Tensor fields. In S. Flügge (ed.), *Encyclopedia of Physics*, Volume III/I, Springer-Verlag, Berlin, 1960.
5. G.C. Nayak and O.C. Zienkiewicz. Convenient forms of stress invariants for plasticity. *Proceedings of the American Society of Civil Engineers*, **98**(ST4), 949–53, 1972.
6. O.C. Zienkiewicz and R.L. Taylor. *The Finite Element Method*, 4th edition, Volume 2, McGraw-Hill, London, 1991.

Author index

Page numbers in **bold** refer to the list of references at the end of each chapter.

Subject index